Environmental Toxicity Testing

Environmental Toxicity Testing

Edited by

K. CLIVE THOMPSON
Chief Scientist
ALcontrol Laboratories, Rotherham, UK

KIRIT WADHIA
Principal Ecotoxicologist
ALcontrol Laboratories, Rotherham, UK

ANDREAS P. LOIBNER
Contaminated Land Management
IFA-Tulln, Environmental Biotechnology
Tulln, Austria

Blackwell
Publishing

CRC Press

© 2005 by Blackwell Publishing Ltd

Editorial offices:
Blackwell Publishing Ltd, 9600 Garsington Road, Oxford OX4 2DQ, UK
  Tel: +44 (0)1865 776868
Blackwell Publishing Asia Pty Ltd, 550 Swanston Street, Carlton, Victoria 3053, Australia
  Tel: +61 (0)3 8359 1011

ISBN-10: 1-4051-1819-9
ISBN-13: 978-1-4051-1819-4

Published in the USA and Canada (only) by
CRC Press LLC, 2000 Corporate Blvd., N.W., Boca Raton, FL 33431, USA
Orders from the USA and Canada (only) to
CRC Press LLC

USA and Canada only:
ISBN 0-8493-2399-1

The right of the Author to be identified as the Author of this Work has been asserted in accordance with the Copyright, Designs and Patents Act 1988.

All rights reserved. No part of this publication may be reproduced, stored in a retrieval system, or transmitted, in any form or by any means, electronic, mechanical, photocopying, recording or otherwise, except as permitted by the UK Copyright, Designs and Patents Act 1988, without the prior permission of the publisher.

This book contains information obtained from authentic and highly regarded sources. Reprinted material is quoted with permission, and sources are indicated. Reasonable efforts have been made to publish reliable data and information, but the author and the publisher cannot assume responsibility for the validity of all materials or for the consequences of their use.

Trademark notice: Product or corporate names may be trademarks or registered trademarks, and are used only for identification and explanation, without intent to infringe.

First published 2005

Library of Congress Cataloging-in-Publication Data:
A catalog record for this title is available from the Library of Congress

British Library Cataloguing-in-Publication Data:
A catalogue record for this title is available from the British Library

Set in 10/12 pt Times
by Kolam Information Services Pvt. Ltd, Pondicherry, India
Printed and bound in Great Britain
by MPG Books Ltd, Bodmin, Cornwall

The publisher's policy is to use permanent paper from mills that operate a sustainable forestry policy, and which has been manufactured from pulp processed using acid-free and elementary chlorine-free practices. Furthermore, the publisher ensures that the text paper and cover board used have met acceptable environmental accreditation standards.

For further information on Blackwell Publishing, visit our website:
www.blackwellpublishing.com

Contents

Preface	xv
Contributors	xvii

1 Historical perspective and overview — 1
JIM WHARFE

1.1	Introduction		1
1.2	Man and his environment – a growing dependency on chemicals		2
	1.2.1	Early times	2
	1.2.2	Chemicals development and environmental impact	3
	1.2.3	The chemical industry today	5
1.3	Ecotoxicity testing and its role in decision-making		9
	1.3.1	The development of test methods	9
	1.3.2	The use of bioassays in the management and control of hazardous substances	13
1.4	Chemical legislation and drivers for change		15
1.5	Change and challenges ahead		20
	1.5.1	Developments in the legislation concerning the pre-release of chemicals	20
	1.5.2	Developments in the legislation concerning the controlled release of chemicals	22
	1.5.3	Some of the challenges ahead	22
	References		24

2 Effective monitoring of the environment for toxicity — 33
IAN JOHNSON, PAUL WHITEHOUSE and MARK CRANE

2.1	Introduction			33
2.2	Design of monitoring programmes			34
	2.2.1	Introduction		34
	2.2.2	Setting of information goals		35
	2.2.3	Selection of indicators of environmental quality		36
	2.2.4	Location and frequency of samples, and data analysis		38
		2.2.4.1	Comparison of ambient samples	39

		2.2.4.2	Trend analysis	40
		2.2.4.3	Breach of regulatory limits/compliance monitoring	41
		2.2.4.4	Assessment of environmental impact	41
	2.3	Quality issues in the use of bioassays		42
		2.3.1	Sample collection, handling and pretreatment	42
			2.3.1.1 Sample collection and handling	42
			2.3.1.2 Sample pretreatment	43
		2.3.2	Test standardisation	44
		2.3.3	Variability in bioassay data	44
			2.3.3.1 How does variability arise?	46
			2.3.3.2 Why does variability matter?	47
			2.3.3.3 How much variability is there?	48
			2.3.3.4 Sources of variability	50
			2.3.3.5 How much variability is acceptable?	50
			2.3.3.6 How can variability be controlled?	52
			2.3.3.7 Defining limits for accuracy	52
			2.3.3.8 Defining limits for precision	53
			2.3.3.9 Test method development and the derivation of quality control criteria	53
	2.4	Summary		55
	References			55

3 The nature and chemistry of toxicants 61
ULF LIDMAN

	3.1	Introduction	61
		3.1.1 History	61
		3.1.2 Properties	62
		3.1.3 Exposure	62
		3.1.4 Bioavailability	62
		3.1.5 Bioaccumulation	63
		3.1.6 Biomagnification	63
		3.1.7 Metabolism	63
		3.1.8 Effects of environmental toxicants	66
		3.1.9 Interactions between envirotoxicants	66
	3.2	Toxic metals	67
		3.2.1 Introduction	67
		3.2.2 Cadmium	68
		3.2.3 Mercury	69
		3.2.4 Lead	70
		3.2.5 Copper	71
		3.2.6 Tin	71

3.3	Halogenated hydrocarbons		72
	3.3.1	Introduction	72
	3.3.2	Polychlorinated biphenyls (PCBs)	73
	3.3.3	Polychlorinated dibenzodioxins (PCDDs) and dibenzofurans (PCDFs)	75
	3.3.4	Polybrominated flame retardants (PBFRs)	76
	3.3.5	Chlorinated pesticides/insecticides	78
	3.3.6	Other halogenated organic compounds of environmental concern	82
		3.3.6.1 Chlorophenols	82
		3.3.6.2 Chlorinated paraffins	83
3.4	Polycyclic aromatic hydrocarbons (PAHs)		83
3.5	Medical and veterinary drugs		85
3.6	Acid rain and acidification of the environment		89
References			90

4 Frameworks for the application of toxicity data — 94
MARIA CONSUELO DIAZ-BAEZ and BERNARD J. DUTKA

4.1	Introduction		94
	4.1.1	Background and objectives	94
4.2	The purpose of bioassays		97
	4.2.1	Toxicity tests within a triad of techniques	97
	4.2.2	Advantages and disadvantages of toxicity testing	100
4.3	Interpretation of toxicological data		100
	4.3.1	Field validation	100
	4.3.2	Application factors	102
	4.3.3	Acute to chronic ratio (ACR)	103
	4.3.4	Toxic units	104
	4.3.5	Toxicity indices	105
		4.3.5.1 Potential ecotoxic effects probe	105
		4.3.5.2 Ranking scheme and 'battery of tests' approach	107
		4.3.5.3 Toxicity classification system	110
		4.3.5.4 The Chimiotox system	113
	4.3.6	Structure–activity relationships (SARs)	114
	4.3.7	Toxic emissions	116
4.4	Ecological risk assessment		117
	4.4.1	Risk characterization	118
		4.4.1.1 Chemical toxicity line	120
		4.4.1.2 Toxicity tests line	121
		4.4.1.3 Ecoassessment line	121
		4.4.1.4 Biomarkers line	122
References			125

5 The aquatic environment
WILLIAM L. GOODFELLOW Jr. — 131

- 5.1 Introduction — 131
- 5.2 Aquatic toxicity — 133
    - 5.2.1 Freshwater species — 134
    - 5.2.2 Estuarine and marine species — 134
    - 5.2.3 Other organisms used for toxicity testing — 136
    - 5.2.4 Multiple species assessments — 137
    - 5.2.5 Dilution water — 138
        - 5.2.5.1 Freshwater — 138
        - 5.2.5.2 Marine and estuarine waters — 139
    - 5.2.6 Selection of end-points — 139
- 5.3 Wastewater toxicity identification evaluations — 140
- 5.4 Sediment toxicity — 143
    - 5.4.1 Water column or elutriate testing — 146
    - 5.4.2 Pore water — 146
    - 5.4.3 Bulk sediment — 147
    - 5.4.4 *In situ* testing or ambient testing — 147
- 5.5 Assessment of bioavailability — 147
    - 5.5.1 Alternative extraction methods — 148
    - 5.5.2 Membrane analog methods — 149
    - 5.5.3 Pore water assessments — 149
- 5.6 Factors controlling bioavailability — 150
    - 5.6.1 Acid-volatile sulfide — 150
    - 5.6.2 Organic carbon — 151
- 5.7 Processes affecting bioavailability in sediments — 152
    - 5.7.1 Ageing or weathering — 152
    - 5.7.2 Sorption — 153
    - 5.7.3 Seasonality — 153
    - 5.7.4 Bioturbation and sediment resuspension — 153
- 5.8 Estimating bioavailability — 154
- 5.9 Summary and conclusions — 156
- References — 156

6 Biological methods for assessing potentially contaminated soils — 163
DAVID J. SPURGEON, CLAUS SVENDSEN and PETER K. HANKARD

- 6.1 Why biological testing — 163
- 6.2 Standardised procedures — 164
    - 6.2.1 Ecological indicators — 164
        - 6.2.1.1 Soil functional assessments — 164

		6.2.2	Bioassays	165
			6.2.2.1 Aquatic plant tests	165
			6.2.2.2 Terrestrial plant tests	166
			6.2.2.3 Aquatic invertebrate tests	166
			6.2.2.4 Terrestrial invertebrate tests	166
		6.2.3	Biosensors	168
	6.3	Academically established methods		169
		6.3.1	Ecological indicators	169
			6.3.1.1 Invertebrate feeding activity using bait lamina strip	169
			6.3.1.2 Community-level physiological profiling (CLPP) using BIOLOG plates	170
			6.3.1.3 Litterbags	170
			6.3.1.4 Minicontainers	171
		6.3.2	Microbial assemblage functions	171
			6.3.2.1 Nitrification	171
			6.3.2.2 Nitrogen fixation	172
			6.3.2.3 Soil enzyme activity	172
		6.3.3	Bioassays in non-standardised species	173
		6.3.4	Biomarkers	173
			6.3.4.1 Tissue/cellular histopathological changes	173
			6.3.4.2 Lysosomal membrane stability	174
			6.3.4.3 Immune system activity	174
			6.3.4.4 DNA alterations	175
			6.3.4.5 Enzyme activity/induction	176
			6.3.4.6 Enzyme inhibition	178
			6.3.4.7 Protein-based biomarkers	178
	6.4	'Emerging' techniques with future potential		181
		6.4.1	Bioindicators	181
			6.4.1.1 Microbial community profiling	181
		6.4.2	Bioassays	182
		6.4.3	Biomarkers	183
			6.4.3.1 Molecular genetic assays	183
			6.4.3.2 The 'omic' technologies	185
		6.4.4	Biosensors	187
		6.4.5	Further methods	189
	6.5	Community census analysis using macrofauna/flora		189
		6.5.1	Selection of groups for census studies	189
		6.5.2	Community census analysis with macroinvertebrates	190
		6.5.3	Community census analysis with plants	191
	6.6	Summary and selection of suitable assays		191
Acknowledgement				194
References				194

7	**Review of biomarkers and new techniques for *in situ* aquatic studies with bivalves** FRANCOIS GAGNÉ and CHRISTIAN BLAISE	**206**
	7.1 Introduction	206
	7.2 Biomarkers of exposure	209
	7.3 Biomarkers of early biological effects and defence	209
	7.3.1 Defence mechanisms	210
	7.4 Biomarkers of damage	213
	7.5 Biomarkers of reproduction	214
	7.6 Integrating effects	216
	7.7 Linking biomarkers at higher levels of biological organization	220
	7.8 Emerging issues	222
	References	224
8	**Environmental monitoring for genotoxic compounds** JOHAN BIERKENS, ETHEL BRITS and LUC VERSCHAEVE	**229**
	8.1 Introduction	229
	8.2 Types of genotoxic effect	229
	8.2.1 Direct genotoxic effects	230
	8.2.2 DNA repair	231
	8.2.3 Indirect genotoxic effects	232
	8.3 Genotoxicity testing methods	233
	8.3.1 Test battery approach	233
	8.3.2 Selection criteria for genotoxicity assays	234
	8.3.3 Individual fast-screening test systems	234
	8.3.3.1 Genotoxicity tests for monitoring DNA damage and repair (exposure assessment)	235
	8.3.3.2 Genotoxicity tests for monitoring cytogenetic effects (effect assessment)	243
	8.4 Exposure assessment of genotoxic compounds	246
	8.5 Ecological implications of genotoxic effects	249
	8.6 Conclusions	250
	References	251
9	**Approach to legislation in a global context**	**257**
	A UK perspective JIM WHARFE	**257**
	9A.1 Introduction	257
	9A.2 History and tradition in the UK	258

	9A.3	Development of chemical regulations in Europe and the UK	261
	9A.4	The role of a National Regulatory Agency	261
	9A.5	Future developments	262
	Notes		268
	References		269

B The Netherlands perspective – soils and sediments — 269
MICHIEL RUTGERS and PIET DEN BESTEN

	9B.1	Developments in soil contamination policy in The Netherlands	269
	9B.2	Environmental quality criteria	270
	9B.3	Towards site-specific approaches	271
	9B.4	Risk perceptions and negotiation formats	271
	9B.5	Many ways to improve site specificity in ERA	272
	9B.6	Weight-of-evidence (WOE) approaches	273
	9B.7	Aquatic ecosystems	273
	9B.8	Sediments	274
	9B.9	Triad in terrestrial ecosystems and selection of biological tests	276
	9B.10	Reference data from reference sites, reference samples and literature	277
	9B.11	Quantification of results from terrestrial sites	278
	9B.12	Site, sampling, soil characteristics and biological assays	279
	9B.13	Calculation of toxic pressure from the contamination levels	279
	9B.14	Determination of toxicity in samples using bioassays	281
	9B.15	Ecological field observations	281
	9B.16	Integration of Triad results and calculation of ecological effects	282
	9B.17	Results from the pilot at Tilburg	282
	9B.18	Issues and recommendations	285
	9B.19	Future prospects	287
	Acknowledgements		287
	References		287

C German perspective — 290
HANS-JÜRGEN PLUTA and MONIKA ROSENBERG

	9C.1	Wastewater	290
	9C.2	Waste and soil	297
		9C.2.1 Waste	297
		9C.2.2 Soil	299

	9C.3	Biological tests	300
	Notes		300

D USA perspective — 301
BARBARA BROWN and MARGARETE HEBER

	9D.1	History of environmental legislation regulating toxics in water in the USA	301
	9D.2	Current legislative and regulatory framework	302
		9D.2.1 The Clean Water Act	302
		9D.2.2 The Safe Drinking Water Act	304
	9D.3	Current implementation: institutional responsibilities (national and state)	305
	9D.4	Future	306
		9D.4.1 Major issues	306
		9D.4.2 New developments: coordinated management of toxics between legislative mandates	307
	9D.5	Conclusion	308
	References		308

10 Case study: Whole-effluent assessment using a combined biodegradation and toxicity approach — 310
GRAHAM F. WHALE and NIGEL S. BATTERSBY

10.1	Introduction		310
10.2	Considerations prior to initiating the study		312
10.3	Materials and methods		314
	10.3.1	Effluents tested	314
		10.3.1.1 Effluent A – refinery process effluent	314
		10.3.1.2 Effluent B – petrochemical effluent	315
		10.3.1.3 Effluent C – untreated refinery wastewater	315
	10.3.2	Sample collection	315
	10.3.3	Assessment of biodegradable and persistent ecotoxicity	316
		10.3.3.1 Biodegradability studies	316
		10.3.3.2 Chemical analyses	317
		10.3.3.3 Aquatic ecotoxicity tests	318
10.4	Results		321
	10.4.1	Effluent A – refinery process effluent	321
	10.4.2	Effluent B – petrochemical effluent	323
	10.4.3	Effluent C – untreated refinery wastewater	326
10.5	Discussion		330

		10.6	Conclusions	334
		Acknowledgements		335
		Disclaimer		335
		References		335

11 Potential future developments in ecotoxicology — 337
WIM DE COEN, GEERT HUYSKENS, ROEL SMOLDERS,
FREDDY DARDENNE, JOHAN ROBBENS,
MARLEEN MARAS and RONNY BLUST

	11.1	Introduction		337
	11.2	Future research needs in ecotoxicology		340
		11.2.1 Biomarkers in ecotoxicology: where do we go from here?		340
		11.2.2 Transgenic systems in ecotoxicology		341
			11.2.2.1 Reporter systems	342
			11.2.2.2 Promoters in multi(cellular) systems	343
		11.2.3 Novel markers at the proteome level: glycosylation of proteins		346
		11.2.4 Organism-level effects: mechanisms of reproductive toxicology		350
		11.2.5 Realistic effect assessments: predicting effects of mixtures rather than single chemicals		355
		11.2.6 Ecological complexity in toxicity testing: interactions between pollutant stress and food availability		358
	11.3	Conclusions		361
	References			363

Glossary — 372
Index — 376

Preface

The effect of anthropogenic activity relating to industrial and economic development has had a significant impact on the environment. In recent decades, public awareness of environmental pollution has increased markedly, and the introduction of guidelines and legislation for the protection of water, air and soil quality – of major importance in the political arena – is imminent. As an integral component of environmental policy, it has become essential to regulate and monitor toxic substances.

Past emphasis has been primarily on analytical approaches to the detection of specific, targeted contaminants, thus allowing chemical characterisation. However, toxicity testing or biological assessment is necessary for ecotoxicological evaluation, and this offers marked benefits and advantages that complement chemical analysis.

The extent of routine toxicity testing on routine environmental samples has until now been limited. It is generally agreed that there is an incontrovertible need for fit-for-purpose environmental toxicity testing. To attain this requirement, key issues to be addressed include:

- identification of pertinent tests
- reproducibility and robustness of these tests
- cost considerations

This book examines these issues and describes and explains the approaches that have been developed for environmental toxicity evaluations. Advantages, benefits and drawbacks of the strategies and methods are highlighted.

A historical perspective on effective management of the environment is presented in Chapter 1, which provides a comprehensive overview of the subject. This is followed by a chapter on effective monitoring of environmental toxicity, including aspects of quality control. Quality control is of fundamental importance in environmental toxicity testing, but it fails to achieve a prominence comparable with routine chemical analysis parameters. In Chapter 3, the fundamental concepts of ecotoxicological testing and evaluation are described, with explanations of the relevant methodology and systems. The extent of variability and standardisation of testing are clarified.

The rationale for the utilisation of toxicity tests and the inference of data employing different techniques is discussed in Chapter 4. Monitoring of the quality of water and soil ecotoxicological techniques are likely to assume greater importance with the implementation of the EU Water Framework Directive and the EU soil assessment strategy. Aspects relevant to the aquatic environment are conveyed in Chapter 5, and biological methods available for the assessment of the terrestrial environment are described in Chapter 6.

Chapters 7 and 8 on biomarkers and genotoxic substances clarify these two controversial areas of increasing importance.

Chapter 9 examines legislation in a global context, with examples from the UK, the Netherlands, Germany and the USA. It is evident that the strategies adopted are country-dependent. The penultimate chapter is an illustrative case study from the petroleum industry, which illustrates the use of a robust, pragmatic approach to a complex problem.

The final chapter provides an insight into the future, highlighting likely new developments that should improve environmental toxicity testing in respect of relevance of tests, improvements in efficiency and, ultimately, reductions in costs.

K. Clive Thompson
Kirit Wadhia
Andreas P. Loibner

Contributors

Dr Nigel S. Battersby Shell Global Solutions (UK), PO Box 1, Chester CH1 3SH, UK

Dr Johan Bierkens VITO Flemish Institute for Technological Research, Boeretang 200, B-2400 Mol, Belgium

Professor Christian Blaise Environment Canada, St Lawrence Centre, 105 McGill, 7th Floor, Montreal, Quebec H2Y 2E, Canada

Professor Ronny Blust Laboratory for Ecophysiology, Biochemistry and Toxicology, University of Antwerp, Groenenborgerlaan 171, 2020 Antwerp, Belgium

Dr Ethel Brits VITO Flemish Institute for Technological Research, Boeretang 200, B-2400 Mol, Belgium

Ms Barbara Brown Environmental Monitoring Consultant, Attleboro, MA, USA

Dr Mark Crane WRc–NSF Ltd, Henley Road, Medmenham, Marlow, Buckinghamshire SL7 2HD, UK

Dr Freddy Dardenne Laboratory for Ecophysiology, Biochemistry and Toxicology, University of Antwerp, Groenenborgerlaan 171, 2020 Antwerp, Belgium

Professor Wim De Coen	Laboratory for Ecophysiology, Biochemistry and Toxicology, University of Antwerp, Groenenborgerlaan 171, 2020 Antwerp, Belgium
Dr Piet den Besten	Institute for Inland Water Management and Waste Water Treatment (RIZA), Ministry of Transport, Public Works and Water Management, PO Box 17, 8200 AA Lelystad, The Netherlands
Professor Maria Consuelo Diaz-Baez	Universidad Nacional de Colombia, Facultad de Ingenieria Ambiental, AA 14490 Bogota, Colombia, South America
Dr Bernard J. Dutka	Research scientist Emeritus, National Water Research Institute (NWRI), Canada Centre for Inland Waters, 867 Lakeshore Road, PO Box 5050, Burlington, Ontario L7R 4A6, Canada
Dr Francois Gagné	Environment Canada, St Lawrence Centre, 105 McGill, 7th Floor, Montreal, Quebec H2Y 2E, Canada
Mr William L. Goodfellow Jr	EA Engineering Science and Technology, 15 Loveton Circle, Sparks, MD 21152, USA
Dr Peter K. Hankard	Centre for Ecology & Hydrology, Monks Wood, Abbots Ripton, Huntingdon, Cambridgeshire PE28 2LS, UK
Ms Margarete Heber	United States Environmental Protection Agency, Office of Water, 1200 Pennsylvania Avenue NW, Washington, DC 20460, USA
Dr Geert Huyskens	Laboratory for Ecophysiology, Biochemistry and Toxicology,

	University of Antwerp, Groenenborgerlaan 171, 2020 Antwerp, Belgium
Dr Ian Johnson	WRc–NSF, Henley Road, Medmenham, Marlow, Buckinghamshire SL7 2HD, UK
Dr Ulf Lidman	Department of Biology and Environmental Science, Kalmarsundslaboratoriet, University of Kalmar, SE-391 82 Kalmar, Sweden
Dr Andreas P. Loibner	Contaminated Land Management, IFA-Tulln, Environmental Biotechnology, Konrad Lorenz Strasse 20, A-3430 Tulln, Austria
Dr Marleen Maras	Laboratory for Ecophysiology, Biochemistry and Toxicology, University of Antwerp, Groenenborgerlaan 171, 2020 Antwerp, Belgium
Dr Hans-Jürgen Pluta	Umweltbundesamt, FG III 3.4, Schichauweg 58, D-12307 Berlin, Marienfelde, Germany
Dr Johan Robbens	Laboratory for Ecophysiology, Biochemistry and Toxicology, University of Antwerp, Groenenborgerlaan 171, 2020 Antwerp, Belgium
Monika Rosenberg	Umweltbundesamt, FG III 3.4, Schichauweg 58, D-12307 Berlin, Marienfelde, Germany
Dr Michiel Rutgers	Laboratory for Ecological Risk Assessment, National Institute for Public Health and the Environment, PO Box 1, 3720 BA Bilthoven, The Netherlands

Dr Roel Smolders	Laboratory for Ecophysiology, Biochemistry and Toxicology, University of Antwerp, Groenenborgerlaan 171, 2020 Antwerp, Belgium
Dr David Spurgeon	Centre for Ecology & Hydrology, Monks Wood, Abbots Ripton, Huntingdon, Cambridgeshire PE28 2LS, UK
Dr Claus Svendsen	Centre for Ecology & Hydrology, Monks Wood, Abbots Ripton, Huntingdon, Cambridgeshire PE28 2LS, UK
Professor K. Clive Thompson	ALcontrol Laboratories, Templeborough House, Mill Close, Rotherham S60 1BZ, UK
Dr Luc Verschaeve	VITO Flemish Institute for Technological Research, Boeretang 200, B-2400 Mol, Belgium
Dr Kirit Wadhia	ALcontrol Laboratories, Templeborough House, Mill Close, Rotherham S60 1BZ, UK
Mr Graham Whale	Shell Global Solutions (UK), PO Box 1, Chester CH1 3SH, UK
Dr Jim Wharfe	Science Group, Environment Agency, Evenlode House, Howbery Park, Wallingford, Oxfordshire OX10 8BD, UK
Dr Paul Whitehouse	WRc–NSF, Henley Road, Medmenham, Marlow, Buckinghamshire SL7 2HD, UK

1 Historical perspective and overview
Jim Wharfe

1.1 Introduction

People from all walks of life are dependent on chemicals for a wide range of services and products, and the chemicals industry has grown to become one of the largest manufacturing industries in the world. Concerns regarding the effects of hazardous substances released into our environment have been raised increasingly since the introduction of synthetic pesticides in the 1940s. Since then, ecotoxicology has emerged from the fields of classical toxicology and environmental chemistry and has greatly influenced the recent developments in risk assessment procedures. Expansion in this field of science undoubtedly has benefited from the rapid exchange of information made possible by electronic communication, and from the increased opportunities for the global science community to share and exchange views at international meetings. The continued development of risk-based procedures relies on a multidisciplinary approach and now embraces other important areas of expertise, including the socioeconomic scientists, the environmental legislators and the policy-makers. Ecotoxicity testing, the subject of this timely publication, is a central element to any chemical risk assessment and has grown from the early work on fish testing that began in earnest little more than half a century ago.

The speed of development and expansion of many new areas of science associated with environmental toxicology makes comprehensive coverage difficult. This chapter provides a perspective and historical overview and sets the scene for the following chapters by considering:

- Man's growing dependency on chemicals as the world population and demand both increase.
- The development of ecotoxicity testing and its role in regulation.
- Internationalisation of chemicals regulation and the associated legislative drivers behind risk reduction strategies.
- Future challenges that will require a more integrated approach between the many scientific, social and political interests together with improvements needed to focus the global research effort in ecotoxicology and to allow the necessary transfer of technologies into operational practice.

1.2 Man and his environment – a growing dependency on chemicals

1.2.1 Early times

Since early times, man has used chemicals and exploited their many properties to help provide key services such as food production, the supply of clean water, medical care and transport. As population levels have increased, prosperity and life expectancy in many parts of the world have also risen. Modern-day society is dependent on the products and services that chemicals help to provide but, without due care and attention, they pose a threat to human health and to our environment.

Early man inhabited the earth for many hundreds of thousands of years before becoming less reliant on hunting and fishing and more reliant on the land to provide food. By 7000 BC survival depended largely on domesticated livestock and cultivated crops in many parts of the world. The number of people inhabiting the Earth grew from an estimated 5 to 250 million people during the 10 000 years before AD and the majority of all food plants found today were cultivated somewhere on the globe. By this time many plant diseases had been recorded, the Chinese had classified plants with medicinal value and early pest control included the application of salt and ashes to rid the land of weeds and the use of chrysanthemum dust to repel insects.

Throughout history, various challenges associated with population growth, including drought, plague and famine, made man more aware of his environment. There is some evidence of the use of chemicals since early times, including sulphur as a fumigant and arsenic as an insecticide. Improved agricultural technology and crop rotation steadily increased food production and, long before the population explosion that was to accompany the industrial revolution of the 18th and 19th centuries, industries were springing up as discoveries were made and new technologies were introduced. The 17th century saw rapid developments in science, including medicine, chemistry and the beginning of genetics some 200 years before the natural laws of heredity proposed by Mendel. The application of fertilisers and pest control chemicals, and the cross-fertilisation of plants to introduce resistant varieties, became increasingly necessary to improve food production and to sustain the population of the Western world.

Inventions such as the steam engine in 1765 and the spinning Jenny a few years later revolutionised industry, as did improvements in agriculture during the 1770s. The agrarian revolution saw scientific crop rotation (a practice followed since AD 600), improved livestock and efficient tools and the introduction of more productive seeds; also, with the continued expansion of industry, the world population reached three-quarters of a billion by the mid-18th century. Almost three centuries after such famous explorers as Columbus, Cabot and Magellan had discovered the Americas, the population of Britain, at around 13 million, was equal to that of North America. Industry in the civilised Western world advanced rapidly, food production increased with demand and the world population continued to escalate (Figures 1.1 and 1.2).

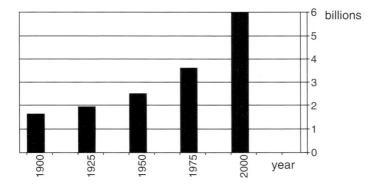

Figure 1.1 World population (billions) increase in the 20th century.

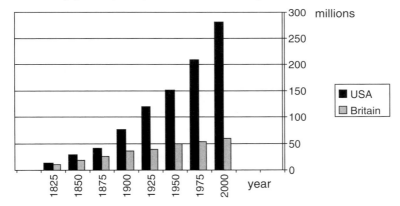

Figure 1.2 Population increase (millions) in the USA and Britain.

1.2.2 Chemicals development and environmental impact

Long before the synthesis of the first organic compounds in the 1820s and some way into the early 20th century a number of naturally occurring substances, including nicotine from extract of tobacco leaves, rotenone from the roots of the derris plant and pyrethrum from the flowers of chrysanthemums, were used as insecticides. Inorganic substances such as lime sulphur, arsenic, mercury and copper compounds also were used extensively to combat insect pests. As time passed, chemicals in the form of fumigants, fungicides and seed dressings were used more and more to benefit society. Selective herbicides were introduced to enhance cereal production, and alternative pesticides were sought in response to increasingly resistant strains of insects.

Public concern in the 1920s relating to chemical residues in vegetables and fruit saw the introduction of oils and tars and eventually led to a search for synthetic organic compounds. Early examples include alkyl phthalate insecticides, anabasine fungicides and methyl bromide as a fumigant. Tetraethyl pyrophosphate (TEPP), the first organophosphate insecticide, was discovered in 1938 and in the following year the insecticidal properties of 1,1,1-trichloro-2,2-bis(*p*-chlorophenyl)ethane (DDT), a chemical discovered in Germany in 1873, were realised by the Swiss entomologist Paul Müller. This relatively low cost production chemical was manufactured in 1943 and introduced to the USA and rapidly became the most widely used insecticide in the world.

Following the success of DDT, many other chemicals emerged, including the plant growth regulator 2,4-dichlorophenoxyacetic acid (2,4-D) used as a selective broad-leafed weed killer. The introduction of other well known pesticides included chlordane and parathion, used largely to kill aphids and mites, and toxaphene, used widely in the USA to help control grasshoppers and boll weevils. By the late 1940s a number of other organochlorine compounds were developed as contact pesticides, including the persistent soil pesticides aldrin and dieldrin. The organophosphorus compounds developed as wartime nerve gases were also introduced as potent systemic and contact insecticides. The first organophosphate insecticides were highly toxic to mammalian species but by 1950 malathion, the first wide-spectrum phosphate-based insecticide with low mammalian toxicity, appeared. A wide range of biocides and medicines were developed during the second half of the 20th century as the global chemicals market rapidly expanded. These included the carbamate and synthetic pyrethroid insecticides, the triazine and bipyridylium herbicides, a range of systemic fungicides and other chemicals such as veterinary medicines, anti-fouling paints and timber preservatives widely used in the environment.

Awareness and concern about the risks of exposure to chemicals released into the environment were heightened during the latter half of the 1940s with the discovery of insect resistance to DDT, long before the publication of Rachel Carson's *Silent Spring* in 1962. There followed a period during which excessive reports appeared on chemical residues in food crops and the effects on wildlife, including eggshell thinning and the associated widespread decline in the numbers of birds of prey (Hickey and Anderson, 1968; Hickey, 1969). Ratcliffe (1967) was able to establish that eggshell thinning started in the mid-1940s at the time when DDT was first introduced.

Since *Silent Spring*, a number of discourses have focused on these issues and legislative action has been taken by regulatory authorities. These include an account by Mellanby (1992) and an alternative version of the DDT story by Kemm (1999). Kemm noted the global implication of restricting the use of or removing such chemicals from the market, promoted the need to take account of socioeconomic consequences in risk-based decision-making and appraised a range of options concerning the suitability of subsequent control measures worldwide.

Since the 1950s there have been many well-documented incidents of damage to human health and wildlife associated with the release of chemicals into our environment. A few examples, both old and new, with selective references, serve to illustrate these environmental concerns:

- The bioaccumulative effects of persistent organochlorine pesticides in raptors and other predatory wildlife (Moore, 1967; Hickey, 1969).
- Mercury poisoning in humans and seed-eating birds (Kurland *et al.*, 1960; Harada, 1982; Scheuhammer, 1987).
- Oil spills and detergent use from many famous shipping accidents such as the *Torrey Canyon* in 1967, the *Amoco Cadiz* in 1978 and the *Sea Empress* in 1996 (Bellamy *et al.*, 1967; White and Baker, 1998; Edwards and White, 1999).
- Imposex in dogwhelks caused by tributyl tin (Langston, 1995).
- Endocrine disruption in a wide range of species including alligators, fish and invertebrates linked to a range of chemicals, most notably the natural and synthetic oestrogenic hormones (Moore and Stevenson, 1991; Guillette *et al.*, 1994, 1995; Jobling *et al.*, 1998).
- Amphibian deformities related to chemicals, disease and radiation (Harris *et al.*, 1998a, b; Dalton, 2002).

A brief chronology of the more notable historic events is shown in Table 1.1, and includes some of the more notable pieces of legislation covered later in this chapter.

To gauge both the historical and the emerging concerns regarding risks to human health and wildlife from exposure to chemicals through environmental pathways and to help regulatory decision-making in the future, it is necessary to place in context the global importance of the multinational chemical industry.

1.2.3 The chemical industry today

The chemical industry has grown to become one of the largest manufacturing industries in the world, with sales estimated in the year 2000 to be well in excess of 1500 billion dollars. Some projections indicate an increase in the number of commercially available chemicals in the immediate future, with the possibility that they will be administered in smaller but more potent doses.

The output from the industry covers the very large pharmaceutical, petrochemical and agrochemical businesses, as well as many others that provide speciality chemicals and consumer products such as paints, dyes, plastics and textiles. The industry is a large employer with an estimated 1.7 million people and many more that are reliant on its services. In Europe, sales in the year 2000 exceeded 480 billion euros in an industry that contributed almost one-third of the world's output.

Table 1.1 Key events in the history of chemical usage.

Period	Event
10000 BC–AD 0	World population grew from 5 to 250 million Cultivated crops and domesticated livestock
1000 BC–AD 1000	• Sulphur used as a fumigant • Arsenic used for pest control
1649	Rotenone used in South Africa to paralyse fish
1690	Nicotine, from tobacco leaf extract, used as a contact insecticide
1750	World population reached 750 million
1750–early 1800s	• Landmark inventions of the industrial revolution include the steam engine, the spinning Jenny and the spinning mule • Agrarian revolution includes improved livestock, more productive seed, scientific crop rotation and more efficient tools
1800–1850	• Use of louse powder (pyrethrum) • Arsenic dip for sheep scab control • Gypsum applied as a fertiliser • Soap and fish oil used to repel insects • Iodine used as an antiseptic • Mercuric chloride and alcohol used as a bedbug control • Phosphorus pastes used as a rodenticide • Derris (rotenone) for insect control in Asia
1850–1900	• DDT 1,1,1-trichloro-2,2-bis(p-chlorophenyl)ethane prepared by Zeidler, insecticide properties not known • Pyrethrum first used in the USA • Paris Green (copper arsenite), and other inorganic substances, used as insecticides and herbicides include lime sulphur, lead arsenate, hydrogen cyanide, copper sulphate and kerosene • 1864 Mendel's laws of heredity, 1875 new food plants from selection and cross-fertilisation
1900	World population reached 1.65 billion
1900–1925	• Nicotine sulphate used dry for dusting crops • Mercury seed dressing developed • Selenium compounds tested for insecticidal properties
1920	World population reached 1.86 billion
1926–1950	• Alkyl phthalates patented as insect repellents • Methyl bromide used as a fumigant • Nicotine bentonite used as a dry dust • 1938 Tetraethyl pyrophosphate (TEPP) discovered – first organophosphate pesticide • 1939 Insecticidal properties of DDT discovered by Müller • 1940s Early fish testing • 1941 Hexachlorocyclohexane (HCH) insecticide

Table 1.1 *Continued*

Period	Event
	• 1943 2,4-D (2,4-dichlorophenoxyacetic acid) patented as a plant growth regulator and later as a general herbicide; dithiocarbamate fungicide marketed • 1946 Chlordane introduced in the USA • 1947 Parathion and toxaphene introduced • 1947 The anti-coagulant warfarin discovered • 1948 Aldrin and dieldrin appeared • 1950 Malathion organophosphate pesticide introduced
1950	World population reached 2.52 billion
1951–1975	• 1950s Increased scientific publication on toxicity testing, the establishment of dedicated laboratories and standard test protocols • Early 1950s First carbamate insecticides appeared, including isolan and pyrolan, followed later by carbaryl • Insecticidal properties of diazinon described • 1957 European Economic Community established under the Treaty of Rome • 1958 Atrazine, first of the triazine herbicides, introduced together with paraquat, the first bipyridylium herbicide • 1962 Rachel Carson's *Silent Spring* warns of the dangers of chemicals in the environment • Systemic fungicides and the synthetic pyrethroid insecticides appear in the 1960s • 1967 Torrey Canyon oil disaster off Cornish Coast – 119 000 tons of crude oil and excessive use of detergent kill many marine animals • 1969 Massive fish kill on river Rhine following dumped cans of the insecticide thiodan • 1970 Elevated mercury levels in livers of Alaskan fur seal and in tuna fish • 1970s An increase in legislative control on chemicals is seen
1970	World population reached 3.63 billion
1981	European Inventory of Existing Commercial Chemical Substances
1987	World population reached 5 billion
	Montreal protocol on ozone depleting substances
1988	Rotterdam convention on prior informed consents
1992	Earth Summit in Rio establishes the Intergovernmental Forum on Chemical Safety
	UN Economic Commission protocol to control releases of persistent organic pollutants
1998	OSPAR strategy to reduce discharges, emissions and losses of hazardous substances to the marine environment

Cont.

Table 1.1 *Continued*

Period	Event
1999	World population reached 6 billion
2000	EU Water Framework Directive (2000/60/EC)
2001	European Commission White Paper proposes a new regulatory system for chemicals comprising three components; registration, evaluation and authorisation of chemicals (REACH)
2002	Johannesburg World Summit on Sustainable Development. Countries committed to "achieve by 2020, that chemicals are used and produced in ways that lead to the minimisation of significant adverse effects on human health and the environment"
2005?	REACH system introduced

Not surprisingly, there are many synthetic chemicals and formulations commercially available. In Europe alone more than 100 000 chemicals (not formulations) are on the European Inventory of Existing Commercial Chemical Substances (EINECS), which lists chemicals on the market before September 1981. This list differentiates between existing and all new chemicals produced since 1981 that have to be notified. During the period between 1981 and 2000 more than 2700 new substances were notified in Europe (Figure 1.3). For new substances, the obligatory notification system requires the manufacturer or importer to provide information suitable for risk assessment to be submitted to the competent authorities. The details required are dependent on the production volume or import quantities of the chemical.

More than 30 000 of the commercially available chemicals have recorded production volumes of greater than 1 tonne and, of these, 5200 are known as high volume production chemicals, produced in quantities of more than 1000 tonnes. It is difficult to know exactly how many chemicals are available in the marketplace at any one time but estimates are in tens of thousands and for the high volume production chemicals the manufacturers or importers are required to submit information suitable for risk assessment to the European Commission. The details include information available on the uses of the chemical and on the physicochemical and toxicological properties. From these details the Commission has prepared priority lists of potentially hazardous substances that require more detailed testing and assessment.

A White Paper published by the European Commission in February 2001 identified inadequacies in the current arrangements and proposed a new system for the registration, evaluation and authorisation of chemicals (REACH). The proposal will change the current procedures for submissions of risk assessments and is discussed later in this chapter. Risk-based decision-making thus has become

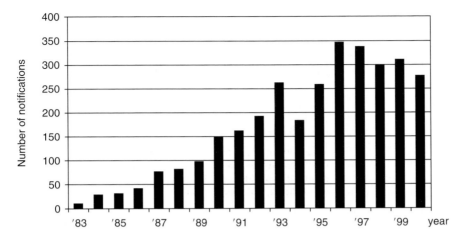

Figure 1.3 New chemicals notified in Europe each year since the introduction of NONS (notification of new substances).

an important tool for the regulators, with ecotoxicity testing having a central role in the risk assessment and evaluation of chemicals.

1.3 Ecotoxicity testing and its role in decision-making

1.3.1 The development of test methods

Between the 12th and 15th centuries the establishment of hospitals and medical schools in Europe helped to advance medical science, and the origins of modern toxicology date back to the early 19th century. Gallo and Doull (1991) provide an account of the historical aspects of the science. Basic research during the 1800s, including testing on a range of animals, was reported in a number of countries but it was not until the mid-1900s that environmental effects of chemicals became a concern. Some basic fish toxicity tests were developed in the 1940s (Anderson, 1944; Hunn, 1989) and the post World War II years established dedicated laboratory facilities in America and Europe. At this time, laboratory testing was conducted largely in static systems employing a range of test organisms and different control procedures.

A number of scientific publications have emerged since the 1950s, including the work of Doudoroff (1976) and Doudoroff and Katz (1950, 1953) on the use of so-called 'pickle-jar tests' and the effects of industrial wastes. Doudoroff *et al.* (1951) developed a single-species fish bioassay, Cairns (1956, 1957) reported the physiological effects of temperature changes on fish and Henderson (1957) considered the application of bioassays for the safe disposal of toxic wastes. This early work on fish helped to establish aquatic toxicity testing and was followed quickly by the concept of trophic-level testing, whereby a range or organisms was tested to represent different levels of the food chain.

During the 1960s the scientific effort increased as public awareness grew and the environmental lobby raised issues of concern about releases of synthetic chemicals. Many classic papers were published on method developments and effects testing in the ensuing years, including those by Sprague and Ramsay (1965), Mount and Brungs (1967), Mount and Stephen (1967), Mount and Boyle (1969), Sprague (1969, 1970, 1971, 1973), Alabaster and Lloyd (1980), Cairns (1985, 1986, 1992) and Cairns and Mount (1990). During this period of heightened activity, more expert facilities were established and fish acute toxicity measures were introduced to meet legislative and investigative needs. Improvements were seen in flow-through systems, multispecies tests, sublethal end-points and interlaboratory calibration exercises and the interpretation of data.

There followed developments in many other areas of environmental science as pollution-related issues and the chemicals effects agenda secured more research funding. In academia, modular degrees were introduced that allowed better integration of the various science skills and disciplines. In the field, extensive bioaccumulation surveys were undertaken to obtain information on tissue residues in wildlife and to study the effects of biomagnification up the food chain (Robinson *et al.*, 1967; Reish, 1972; Preston, 1973; Stephenson *et al.*, 1986; Phillips, 1990, 1993). National biological monitoring programmes helped to assess and classify water quality (Armitage *et al.*, 1983, 1987; Furse *et al.*, 1984; Wright *et al.*, 1984, 1988, 1993a, b; Moss *et al.*, 1987; Armitage, 1989; Wright, 1995) and subsequently allowed associations with ecotoxicological data to evaluate community-level responses (Maltby *et al.*, 2000). Microcosm and mesocosm tests were undertaken to mimic field conditions (Cairns and Pratt, 1989; Pratt and Bowers, 1990; Graney *et al.*, 1993) and a range of biological effect measures were introduced to help establish cause-and-effect relationships.

Biomarker research emerged during the 1980s and 1990s to help understand the mechanistic links between exposure and effect (McCarthy and Shuggart, 1990; Peakall, 1992). Many papers and reviews have been published in the last decade, some of which provide definitions and overviews (Depledge, 1993; Institute for Environmental Health, 1996). Biomarkers are a recent addition to the armoury of available techniques that provide information on individual and population responses either in the field or under controlled laboratory conditions. Opinion differs on their suitability for application in ecological risk assessment, and McCarty *et al.* (2002), introduced a definition for a bioindicator that attempts to overcome the lack of any requirement in the normative definitions of biomarkers to establish cause–effect linkages.

The development of risk assessment procedures identified a need for expert contribution from a range of disciplines to help determine the mechanistic links between exposure and effect, opening up multiple channels of investigation. The global nature of many of the issues has seen expert groups established, the membership of international learned societies expand and the appearance of many reference publications (Rand and Petrocelli, 1985; Munawar *et al.*, 1989; Calow, 1993; Forbes and Forbes, 1994).

Some of the limitations of ecotoxicity data sets have stimulated alternative approaches, including the development of structure–activity relationships (Niemi, 1990; USEPA, 1994a; Hansch and Leo, 1995) and assumptions in exposure models to help fill data gaps and interpret acute toxicity data (Kenaga, 1982). Improvements in the quality of ecotoxicity test data since the mid-1950s has continued with the recognition of internationally accepted standard protocols and procedures adopted by societies such as the American Society for Testing and Materials (ASTM), the International Organization for Standardization (ISO) and the Organization for Economic Cooperation and Development (OECD). The fish toxicity test was the basis of the first ASTM protocol (1954) for aquatic toxicity testing. Since then, many others have been adopted and some of the more up to date and relevant references are shown in Table 1.2.

An extensive and fully comprehensive review of all new developments in ecotoxicology is not possible in the scope of this chapter but continued effort in some areas that will help to integrate the various science strands are mentioned. These include continued growth in the range of available techniques both for investigation and routine monitoring. Examples include *in situ* deployments (Crane *et al.*, 1996; Roddie *et al.*, 1996; Olsen *et al.*, 2001; Maltby *et al.*, 2002,) and tests for compartments other than water, such as soils (Lokke and Van Gestel, 1998; Van Gestel *et al.*, 2001) and sediments (ASTM, 1995b; Crane *et al.*, 2000; ISO, 2000e; OECD, 2000a; Environment Canada, 2002). Other developments include biosensor technology (Polak *et al.*, 1996; Rogerson *et al.*, 1996), portable field equipment and rapid-throughput toxicity tests that offer a quick turnaround of information and larger sample throughput, although few of these are currently used in routine monitoring programmes. Emerging technologies, including gene arrays and protein expression, are helping to unravel molecular-level activity and establish mechanistic links from exposure, availability and uptake to responses in the individual, including irreversible effects and population-level impacts. Information technology and modelling techniques are advancing our ability to model complex systems and to use extensive data sets to help understand community and ecosystem effects and to improve our diagnostic capability (Walley and Fontama, 1998; Walley and O'Connor, 2000). As a result of these and other developments, risk frameworks are continually under development to improve the level of certainty in risk-based decision-making.

There remain many challenges for these multiple lines of investigation and, despite the wealth of literature that continues to grow and the rapid advances in our understanding, there is a need to improve risk-based decision-making associated with the legislative framework for controlling chemicals. Hopefully, targeted research effort in the future will introduce new scientific information to help meet

Table 1.2 Key references.

Test group	ASTM	ISO	OECD	Other
Bacteria	1990, 1996a	1989a, 1995a, 1998, 2000a,b	1984a	Environment Canada 1992c
Algae and plants	1980, 1991a, 1992d, 1996b, 1997c,d	1989b, 1995b, 1999a, 2000c	1984b, 1999a	Environment Canada 1992d, 1999
Invertebrates	1989, 1991b, 1992b,c, 1994a,b,c,d, 1995a,b, 1997a,b	1996a, 1999b, 2000d,e,f,g,h	1984c, 1998a, 1999b, 2000a	Environment Canada 1990c, 1992a,e
Fish	1992a, 1994b, 1995c,d	1994, 1996b,c,d	1984d, 1992a,b, 1998b, 2000b	Environment Canada 1990a,b, 1992b,e, 1998a,b
Multigroup complex effluents and sediments				United States Environmental Protection Agency 1985, 1989a,b, 1991, 1992, 1993a,b,c, 1994b,c, 1995 Environment Canada 1998b, 2000a,b, 2001, 2002

the needs of the end-user, and effective dialogue between scientists, policy-makers and operational staff will enhance our understanding and underpin decision-making with sound science.

The following brief review of current practice illustrates the increasing application of bioassays around the world.

1.3.2 The use of bioassays in the management and control of hazardous substances

The conventional approach to controlling releases of hazardous substances involves establishing safe levels for specific substances that allow environmental quality objectives and associated standards to be set and licence conditions or fixed emission limits to be determined. Environmental safe levels take account of the toxicity of the substance, its persistence and ability to bioaccumulate, and in some cases can include mutagenicity, carcinogenicity and reproductive impairment. Whitehouse and Cartwright (1998) discuss the need for environmental standards and identify the following purposes:

- Environmental benchmarks against which environmental monitoring data can be assessed.
- Setting goals for pollution control activities.
- Acting as triggers for remedial action.
- Environmental management tools that can be applied across different locations and times.

The effects of individual substances on a wide range of species in the environment are rarely understood and, depending on the adequacy and the quality of the information, an uncertainty factor usually is applied. There are a number of limitations to this approach. It is substance specific and of the total number of listed chemicals less than 1% have sufficient information to derive a safe level. Uncertainty factors can move towards overly stringent control, the approach is unable to consider toxic effects of complex mixtures (which is the usual route of entry) and it takes no account of additive or synergistic effects.

Bioassays applied to whole samples help to overcome many of the problems relating to the release of complex mixtures. Whole-sample toxicity test programmes, largely from point source effluents, were promoted in the USA during the late 1970s and the 1980s and later in Europe. Wharfe (1996), Tinsley *et al.* (1996) and Wharfe and Heber (1998) provide an overview of the development of the procedures used in the USA and the UK and review the role of whole-sample toxicity assessment in the regulatory control of complex effluents. The approach has since found increasing recognition around the world and Power and Boumphrey (2004) reviewed the global application of bioassays for both regulatory and non-regulatory effluent testing. Table 1.3 is a synthesis from their work, reproduced with their kind permission.

Table 1.3 The use of bioassays around the world.

North America	
USA	Whole Effluent (Toxicity) Testing (WET) is well established and is used to help fulfil the requirements of the National Pollutant Discharge Elimination System (NPDES); it is a regulatory requirement under the Clean Water Act. Primarily used for source control, trophic level testing employs algae, invertebrate and fish species with both acute and chronic end-points. Application of WET criteria varies from region to region. Since the early 1980s more than 6500 permits containing WET criteria have been issued.
Canada	Toxicity testing is conducted on an industry sector basis to meet regulatory requirements. It is a general provision of the Fisheries Act and an industry-specific requirement under provisions of the Pulp and Paper Regulations and the Mining Effluent Regulations. Trophic level testing is used with both acute and chronic measures. Primarily used for source control but provisions can include environmental monitoring.
Europe	
Belgium	The use of bioassays for effluent testing is not well established although currently being considered. There is no regulatory requirement for such testing.
Denmark	Toxicity testing is used to help characterise effluent discharges and to monitor marine and fresh-waters. Bacteria, algae, macrophytes, invertebrates and fish have been used for investigation purposes but there is no regulatory requirement.
France	Bioassays are often used to monitor industrial effluents and less frequently for regulatory purposes, although conditions are imposed in some permits. Bacteria, algae, invertebrates and fish are used and mutagenicity is measured in addition to the more traditional acute and chronic end-points.
Germany	Well established use of toxicity testing to meet regulatory requirements under the Wastewater Ordinance and Wastewater Charges Act. Bacteria, algae, macrophytes, invertebrates and fish are used to provide acute and chronic end-point measures and information on mutagenicity and genotoxicity. Used mainly for effluent testing; some monitoring of the receiving waters is also undertaken.
Netherlands	Toxicity testing is used primarily for investigation and there is currently no regulatory requirement of such tests. Bacteria, algae, macrophytes, invertebrates and fish are used and, in addition to the acute and chronic end-points, mutagenicity, endocrine and enzymatic responses are measured.
Northern Ireland	Toxicity testing is conducted on both effluents and receiving waters and can be applied as a regulatory condition. Bacteria, algae, macrophyte, invertebrate and fish test are used to generate both acute and chronic data.
Norway	Limited use of toxicity testing is made and can be applied as a statutory enforceable requirement. Trophic-level testing is conducted.
Spain	Effluent toxicity testing is undertaken on a regional basis. There is no national regulatory requirement but legally enforceable conditions can be set locally.

Table 1.3 *Continued*

	Bacteria, algae and fish have been used to generate the acute data that are used primarily for source application, including discharges to municipal sewer systems.
Sweden	Bioassays are used to help fulfil requirements for the characterisation of industrial discharges, mainly from large industries. Statutory application is limited, although treatment plants can refuse to accept or restrict toxic industrial discharges. Bacteria, algae, macrophytes, invertebrates and fish are used to generate information on acute and chronic end-points and on mutagenicity and enzymatic responses.
UK	Effluent and receiving water toxicity is conducted, primarily for investigation and site characterisation. Regulatory requirement is limited to approximately 20 consent conditions, including toxicity measures, although further application is now part of an implementation process following a demonstration programme that involved both industry and regulators. Trophic-level testing is employed to generate both acute and chronic data.
Australasia	
Australia	Bioassays are conducted on effluents on a site-specific basis. Permit conditions can be enforced by states or territories based on national guidance. Trophic-level testing generates both acute and chronic effects data.
New Zealand	Effluent toxicity testing is conducted widely and can be applied as part of a permit condition. The system is flexible and leaves decisions at the regional level. Trophic-level testing generates acute and chronic data.

1.4 Chemical legislation and drivers for change

There is little doubt that Rachel Carson's publication *Silent Spring* raised awareness of the impact of chemicals in the environment to an international level. Since then, environment groups worldwide have picked up the baton. Arguably, some activities and government-led responses were misjudged but their influence on the need for, and development of, environmental legislation concerning chemicals is unquestionable.

There is now a large and diverse number of regulations concerning chemicals nested in a hierarchy of international agreements, continent-wide legislation and directives and national and regional instruments (Figure 1.4). Many associated priority lists of substances of concern have arisen from these pieces of legislation and more recently longer term strategies have emerged.

An outline of this hierarchical approach, using Europe as an example, is described briefly and refers to some of the more important pieces of legislation. Table 1.4 provides more detail on the key pieces of European legislation concerning the control of chemicals.

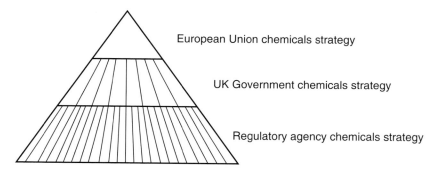

Figure 1.4 The chemicals hierarchy.

The 1992 Earth Summit in Rio de Janeiro laid the foundation for an Intergovernmental Forum on Chemical Safety (IFCS) and Chapter 19 of Agenda 21 dealt with the environmentally sound management of toxic chemicals. The IFCS oversees the implementation of six programme areas:

- The expansion and acceleration of international chemical risk assessments.
- The harmonisation of chemical classification and labelling.
- Information exchange on toxic chemicals and chemical risks.
- The establishment of risk reduction programmes.
- The strengthening of national capabilities and capacities for the management of chemicals.
- Stopping illegal international traffic in toxic and dangerous products.

The IFCS coordinates an international structure that includes the United Nations Environment Programme (UNEP), the OECD, the UN Food and Agriculture Organization (FAO), the World Health Organization (WHO), the UN Industrial Development Organization (UNIDO) and the UN Institute for Training and Research (UNITAR).

The chemical industry spans many countries of the world and chemicals are used worldwide. Despite the plethora of regulations on chemicals, many countries do not have adequate legislation or the necessary infrastructure to ensure their safe use. To help overcome this situation there is an international initiative for wide sharing of information and to provide support through training programmes and technical expertise. The Rotterdam Convention on the prior informed consent procedure for certain hazardous chemicals and pesticides in international trade, adopted in 1998, is an example. Other international agreements worthy of mention

Table 1.4 European Directives concerning chemicals.

Directive/Regulation	Commentary
Key European Directive Pre-release Controls	
European Commission White Paper published in 2001, setting out change to existing Chemicals Policy	In February 2001, the European Commission published a White Paper entitled *Strategy for a Future Chemicals Policy*. The White Paper presented a new regulatory system, called REACH, for the "Registration, Evaluation and Authorisation of Chemicals".
Notification of New Substances Regulations 1993 (NONS 93) implementing part of EC Directive 92/32/EEC, which is the Seventh Amendment of Directive 67/548/EEC	All new chemicals placed on the market in Europe require notification to the competent authority of a Member State. The notification sets out the information requirements, including toxicity testing, and this information is used to conduct risk assessments. Will be replaced by REACH.
Existing Substance Regulation 793/93/EEC came into effect in 1993	Existing chemicals are those that were on the EU market before 18 September 1981 and listed in EINECS (the European Inventory of Existing Commercial Substances). The Regulation requires industry to report data on existing substances, the Commission to prioritise substances for risk assessment and Member States to conduct risk assessments of those priority substances. Will be replaced by REACH.
Chemicals (Hazardous Information and Packaging for Supply – CHIP) Regulations 2002, which implements Directive 67/548/EEC and its seventh amendment 92/32/EEC, the Dangerous Preparations Directive (99/45/EC) and its amendments, and the Safety Data Sheet Directive (91/155/EEC)	The classification and labelling requirements under the Dangerous Preparations Directive 67/548/EEC are implemented by the CHIP Regulations. The Regulations require chemical suppliers to identify the hazards of substances and preparations. Dangerous chemicals then must be packaged suitably, provided with labels and accompanied by additional information for safe use (such as safety data sheets).
The Plant Protection Products (PPP) Directive 91/414/EEC, Directive 79/117/EEC on the marketing and use of pesticides, Directive 67/548/EEC on classification, packaging and labelling requirements, and Directives 76/895, 86/362/EEC, 86/363/EEC and 90/642/EEC on pesticide residues in food	Council Directive 91/414/EEC aims to harmonise the authorisation of plant protection products (PPPs). This also provides for the review of active substances currently on the market. The Control of Pesticides Regulations 1986 requires that only approved pesticides are sold, supplied, used, stored or advertised. Directive 79/117/EEC prohibits the placing on the market and the use of PPPs containing certain active substances. Directive 67/548/EEC is implemented via the CHIP Regulations (as above) and the Pesticide Residues Directives that set Maximum Residue Levels (MRLs) in food regulations.
The Biocidal Products Directive (98/08/EC)	The Biocidal Products Directive aims to harmonise the authorisation and marketing of biocidal products and provide a high level of protection for humans and the environment across Europe.

Cont.

Table 1.4 *Continued*

Directive/Regulation	Commentary
Key European Directive Release Controls	
Dangerous Substances Directive (76/464/EEC) and amendments (1997)	The Directive establishes a system for controlling releases of dangerous substances to water through the setting of environmental quality standards for certain substances included in a list of daughter directives.
Water Framework Directive (2000/60/EC)	One of the largest and most significant pieces of EU water policy to be developed for at least 20 years. It uses a new and integrated approach to the protection, improvement and sustainable use of river basins for the benefit of people and biodiversity. The provisions of the Directive take over the framework for the control of pollution by dangerous substances under Directive 76/464/EEC and will be supplemented by a new groundwater daughter directive.
Groundwater Directive (80/68/EEC)	Relates to the protection of groundwater quality from certain substances. It prohibits the discharge of substances/groups of substances in List 1 and requires pollution to be minimised from other substances in List 2. New arrangements will include requirements on monitoring and reporting.
Directive (96/61/EC) concerning integrated pollution prevention and control	The IPPC Directive is designed to prevent, reduce and eliminate pollution at source through the efficient use of natural resources. It is intended to help industrial operators move towards greater environmental sustainability.
Directive on Hazardous Wastes (91/689/EEC)	Provides a standard definition of, and laid down additional controls over, hazardous waste. Establishes a system for identifying, recording and controlling the collection, transport, storage and disposal of hazardous waste. Waste disposal sites must have a licence specifying the type and quantity of waste that they can accept.
Waste Incineration Directive (2000/76/EC)	Introduced new controls over waste incinerators and set new objectives for the reduction of emissions of dioxins, cadmium, mercury and lead.
Landfill Directive (99/31/EC)	Introduced new technical and operational requirements for landfills across Europe and targets for the reduction in landfill of biodegradable wastes. The Decision (2003/33/EC), which comes into force in July 2004, is crucial for the implementation of the 1999 Landfill Directive. It outlines criteria for the waste that can be accepted at each of the defined types of site and for underground storage, sets out strict EU-wide leaching limit values and defines testing methods. The criteria are to be applied by Member States by July 2005.

Table 1.4 *Continued*

Directive/Regulation	Commentary
Other Important Directives Urban Waste Water Treatment Directive (91/271/EEC) Nitrate Directive (91/676/EEC) Bathing Water Directive 76/160/EEC Detergents Directive 73/404/EEC Shellfish water Directive 79/923/EEC Freshwater Fish Directive 78/659/EEC Surface Water Abstraction Directive (75/440/EEC)	This is not an exhaustive list but includes a number of directives with specific requirements to monitor and report on levels of particular contaminants in the aquatic environment and to take appropriate action to control releases.
Habitats Directive (92/43/EEC)	Requires measures to maintain or restore, at favourable conservation status, natural habitats and species of fauna/flora of community interest.

include the Montreal Protocol on substances that deplete the ozone layer, agreed in 1987, and, more recently, the Stockholm Convention on Persistent Organic Pollutants 2001. The UN Economic Commission for Europe convention on long-range transboundary air pollution and a protocol to the convention, which also covers North America, introduced controls on releases of persistent organic pollutants (POPs).

International agreements have set the agenda for more detailed continent-wide regulation. In Europe, the legislation on chemicals covers many aspects of human health and the environment; it includes worker protection, major accidents and the transport of dangerous goods. Legislation concerned with risks posed through environmental exposure includes: marketing and use of legislation on the formulation, production, use and disposal of industrial chemicals; the export and import of chemicals; and additional legislation that covers associated practices, such as packaging, labelling, transport and storage.

Strategic approaches cover wider issues concerned with the release of substances to the environment. The OSPAR Convention (formerly the Oslo and Paris Commissions) agreed a strategy in 1998 for the protection of the marine environment of the North East Atlantic. The objective is to prevent pollution of the maritime area by continuously reducing discharges, emissions and losses of hazardous substances, with the ultimate aim of achieving concentrations near the background values for naturally occurring substances and close to zero for synthetic substances. The Convention for the Protection of the Marine

Environment of the Baltic Sea (Baltic Marine Environment Protection Commission) adopts appropriate legislative, administrative or other relevant measures to prevent and eliminate pollution in order to promote ecological restoration of the Baltic Sea area and preservation of its ecological balance.

European directives (Table 1.4) are usually implemented through binding regulation on all Member States or through national legislation. This national legislation varies from one Member State to another, and it can involve the setting of environmental objectives and targets associated with substance-specific standards. Some standards are adopted by all Member States as statutory limits, whereas others cover specific regional requirements. The legislation can make provision for a system permitting fixed-point discharges and associated compliance measures. In recent times, a greater determination to make such information available has seen the wider dissemination of available information with the publication of registers and the development of pollution release inventories.

International agreements and legislation at all levels provide an important basis and direction for the work of the ecotoxicologist and can involve the prioritisation of substances of concern, hazard assessment, risk evaluation and management and expert opinion. To be effective, the toxicologist (both human toxicologists and ecotoxicologists) must work with many other scientists to help assess the impact of chemicals on both human health and the environment. These include professional risk assessors (under the various guises of chemical, ecological and environmental), environmental scientists working in and across an array of disciplines, from the molecular to the ecological sciences, and socioeconomists, reminding us of the need for cost-effective appraisal of remedial options. The continued development of an integrated approach in risk assessment will be needed to meet the many challenges ahead.

1.5 Change and challenges ahead

Ecotoxicity testing has become internationalised in recent times and, with it, the recognition of its major contribution to risk-based procedures for the assessment and management of chemicals. National and international legislation has influenced the rapid advances we have seen and, with many unresolved challenges, it seems likely that this pace of progress will be maintained.

1.5.1 Developments in the legislation concerning the pre-release of chemicals

In Europe the volume of international legislation increased during the last half of the twentieth century; Table 1.4 shows the adopted pieces of environmental regulation that emerged during this period. Current European Union (EU) policy on chemicals separates the existing chemicals – those that were on the market before September 1981 and listed in EINECS – and new chemicals – those subject to notification requirements since 1981. It is broadly understood that the notification and evaluation of post-1981 chemicals has been more or less an effective regime.

Of the 106 000 existing substances (pre-1981) that account for over 95% of the total volume of substances on the market, only 140 had been identified for risk assessment on the four existing priority lists by 2002. During the period awaiting further assessment there is no marketing restriction on the sales of these substances and there is no deadline under the regulation for the risk assessment or for possible trade sanctions where the producer fails to provide the necessary information.

Under the current regulations the Member State authorities are responsible for the assessment; the companies producing, importing or using the substances bear no responsibility and it has been difficult to obtain precise information on the use of chemicals and the exposure arising from downstream uses. Assessment of the substances on the priority lists is regarded by many to be unacceptably slow and costly and the allocation of responsibilities to be inappropriate. The existing chemicals policy on the whole has proved unsatisfactory and the limitations have led to a White Paper published by the European Commission in February 2001 entitled *Strategy for a Future Chemicals Policy*. The paper proposes a new single regulatory system for all chemicals, comprising registration, evaluation and authorisation of chemicals (REACH):

- *Registration.* Producers will be obliged to provide basic safety data, by fixed deadlines to authorities, on all chemicals produced in quantities above 1 tonne per year.
- *Evaluation.* For higher production volume chemicals above 100 tonnes, and for chemicals of concern, experts from Member States in association with a central co-ordinating body will evaluate the registration data. The evaluation may lead to authorisation (in the case of chemicals of 'very high concern'), risk reduction (where dangerous uses are restricted) or no further regulatory action.
- *Authorisation.* The use of chemicals considered to be of 'very high concern' would be subject to authorisation. The aim is for such chemicals to be phased out and substituted, unless industry can show that the use presents negligible risk or that it is acceptable, taking into account its socioeconomic benefits, the lack of safer chemicals and measures to minimise exposure. Chemicals of 'very high concern' are likely to include carcinogens, mutagens or reprotoxic substances (CMRs), particularly persistent, bioaccumulative and toxic substances.

There will be many debates on the implications of the new proposals and on the increased burden to industry. The current overall timetable for chemicals legislation is to be published in the official journal by the end of 2005, although protracted discussion is already threatening this deadline.

1.5.2 Developments in the legislation concerning the controlled release of chemicals

Much of the existing environmental legislation concerning chemicals once they are released into the environment is targeted at point source inputs and requires the identification of priority substances of concern and the use of risk-based approaches to determine management strategies. The regulations are substance specific and often based on the achievement of a target load or concentration. Some of the newer European Directives to emerge move to the achievement of defined environmental outcomes in the form of ecological status (Water Framework Directive 2000/60/EC) and habitat protection (Habitats Directive 92/43/EEC).

In December 2000, the EU Water Framework Directive entered into force. It is the largest and most significant piece of EU water policy to be developed in 20 years. Reporting the requirements of the Directive are challenging and risk-based decision-making will require the further development of practical measures of ecological outcome and associated diagnostic procedures to help appraise the options for improvement. Uncertainty is a feature of all ecological risk assessments, requiring the evaluation of multiple lines of evidence to improve the decision-making process. Often there is no definitive answer to a particular problem and questions of ecological relevance remain elusive.

More recently, issues of low-dose effects on hormonal systems, particularly reproductive and genetic end-points, have brought into question the adequacy of current toxicity test procedures. These issues have significant implications for chemicals legislation that will have an impact on the chemicals industry and on society.

1.5.3 Some of the challenges ahead

Many national and international strategies concerned with sustainable development, particularly the assessment and control of chemicals, identify the need to use the resources offered by our environment while protecting ecological processes and ecosystems. It seems certain that the development of international legislation concerning the assessment and management of chemicals will continue to raise questions needing new knowledge from many branches of environmental and social sciences to help improve risk-based procedures and to reduce uncertainty in decision-making. Ecotoxicity testing has advanced rapidly since its early beginnings and has a central role in biological effects programmes. The continued development and application of methods and their acceptability will be directed towards the many challenges of emerging legislation. Some of the key questions include:

- How well are we able to determine effects on receptors resulting from continued low-dose and long-term exposure to chemicals and mixtures?
- What improvements are needed to current test regimes and biological endpoints to provide adequate information and protection through current and emerging environmental regulation?
- Should the prioritisation of chemicals of concern and their subsequent assessment and control take a more precautionary hazard-based approach or should a risk assessment always be undertaken?
- Can new genomic, proteomic and information based technologies help establish mechanistic links between exposure and effect? Can they provide rapid information for monitoring and regulatory control?
- Do current biological measures provide robust environmental indicators, show clear environmental outcome from legislative action, and are they sufficiently well built into ecological risk frameworks?
- Can we adequately regulate complex mixtures and diffuse inputs across all environmental compartments?
- Will emerging nanotechnologies present a threat to our environment? Are we able to assess the toxic effect of very small particles?
- Do current approaches to assess and control chemicals consider cradle to grave options?
- Do research outputs adequately consider implementation needs and operational practice?

Some solutions are already available but uptake and practical application are often difficult and other questions will require improved and effective dialogue between the scientists and the policy-makers. For example, genomic and proteomic technologies will help to improve our understanding of the effects of chemicals and will challenge traditional toxicological end-points. New lines of exploration in fields that include transcript profiling, proteomics and metabolomics seem likely to alter the traditional use of ecotoxicological test data in risk evaluation (Butler, 2001; Curtis et al., 2003). This has implications for current regulatory policies on chemicals. Our diagnostic and monitoring capability will be enhanced by new techniques as multispecies genomic information becomes available to help unravel community interactions between species at the genetic level. Interactions with environmental stressors at the individual and population level also will be better understood.

At present, environmental regulators are often faced with scientific information that is inconclusive and incomplete and they are required to make judgement on whether or not the case has been made for decisive action. Such decisions help to direct billion-dollar investment programmes and influence the continued competitive ability of the manufacturing industry and agriculture in a world market and in society at large. The end-user role that provides scientific direction and decision-making is important to successful implementation and needs to be understood and established to help frame research proposals at an early stage. The drafting of an applied research proposal and the associated project initiation documentation should identify the responsibilities of both the investigator and the end-user. All too often science deliverables showing promise are left without any mechanism to transfer the output into a meaningful outcome through operational practice. The continued need for scientists to publish novel research can constrain development, and further demonstration often is required to consider issues of transferability that include equipment, facilities, training, skill acquisition, quality assurance and reporting requirements. Insufficient dialogue at an early stage will fail to identify the resources and funding needed to implement and sustain the science in operational mode. Ecotoxicity testing is at the heart of risk-based assessment and management of chemicals. Together with other intrinsically linked scientific areas, this will help provide solutions to the environmental challenges that have been posed in this chapter and considered in greater detail in subsequent chapters.

References

Alabaster, J.S. and Lloyd, R. (1980) *Water Quality Criteria for Freshwater Fish*. Butterworths for Food and Agriculture Organisation, London.

Anderson, B.C. (1944) Toxicity thresholds of various substances found in industrial wastes as determined by the use of *Daphnia magna. Sewage Works J.*, **16**, 1156–1165.

American Society for Testing and Materials (ASTM) (1954) *Standards Test Methods for Evaluating Acute Toxicity of Water to Fresh-water Fishes. Annual Book of Standards – Water D1345–45T*. ASTM, Philadelphia.

American Society for Testing and Materials (ASTM) (1980) *Standard Practice for Algal Growth Potential Testing with Selenastrum capricornutum*, D3978–3980. ASTM, Philadelphia.

American Society for Testing and Materials (ASTM) (1989) *Standard Guide for Conducting Three-brood, Renewal Toxicity Tests with Ceriodaphnia dubia*, E1415–1491. ASTM, Philadelphia.

American Society for Testing and Materials (ASTM) (1990) *Standard Test Method for Inhibition of Respiration in Microbial Cultures in Activated Sludge Process*, D5120–5190. ASTM, Philadelphia.

American Society for Testing and Materials (ASTM) (1991a) *Standard Guide for Conducting Static Toxicity Tests with Lemna gibba*, G3 E1415–1491. ASTM, Philadelphia.

American Society for Testing and Materials (ASTM) (1991b) *Standard Guide for Acute Toxicity Tests with the Rotifer Brachionus*, E1440–1491. ASTM, Philadelphia.

American Society for Testing and Materials (ASTM) (1992a) *Standard Guide for Conducting Early Life-stage Toxicity Tests with Fishes*, E1241–1292. ASTM, Philadelphia.

American Society for Testing and Materials (ASTM) (1992b) *Standard Guide for Conducting 10-Day Static Sediment Toxicity Tests with Marine and Freshwater Amphipods*, E1367–1392. ASTM, Philadelphia.

American Society for Testing and Materials (ASTM) (1992c) *Standard Guide for Conducting Static and Flow-through Acute Toxicity Tests with Mysids from the West Coast of the United States*, E1463–1492. ASTM, Philadelphia.

American Society for Testing and Materials (ASTM) (1992d) *Standard Guide for Conducting Sexual Reproduction Tests with Seaweeds*, E1498–1492. ASTM, Philadelphia.

American Society for Testing and Materials (ASTM) (1994a) *Standard Guide for Conducting Static Acute Toxicity Tests Starting with Embryos of Four Species of Saltwater Bivalve Molluscs*, E724–794. ASTM, Philadelphia.

American Society for Testing and Materials (ASTM) (1994b) *Standard Guide for Conducting Bioconcentration Tests with Fishes and Saltwater Bivalve Molluscs*, E1022–1094. ASTM, Philadelphia.
American Society for Testing and Materials (ASTM) (1994c) *Standard Guide for Conducting Acute, Chronic, and Life-cycle Aquatic Toxicity Tests with Polychaetous Annelids*, E1562–1594. ASTM, Philadelphia.
American Society for Testing and Materials (ASTM) (1994d) *Standard Guide for Conducting Sediment Toxicity Tests with Marine and Estuarine Polychaetous Annelids*, E1611–1694. ASTM, Philadelphia.
American Society for Testing and Materials (ASTM) (1995a) *Standard Guide for Conducting Static Acute Toxicity Tests with Echinoid Embryos*, E1563–1595. ASTM, Philadelphia.
American Society for Testing and Materials (ASTM) (1995b) *Standard Guide for Measuring the Toxicity of Sediment-associated Contaminants with Freshwater Invertebrates*, E1706–1795. ASTM, Philadelphia.
American Society for Testing and Materials (ASTM) (1995c) *Standard Guide for Measurement of Behaviour during Fish Toxicity Tests*, E1711–1795. ASTM, Philadelphia.
American Society for Testing and Materials (ASTM) (1995d) *Standard Guide for Ventilatory Behavioural Toxicology Testing of Freshwater Fish*, E1768–1795. ASTM, Philadelphia.
American Society for Testing and Materials (ASTM) (1996a) *Standard Test Method for Assessing the Microbial Detoxification of Chemically Contaminated Water and Soil using a Toxicity Test with Luminescent Marine Bacterium*, D5660–5696. ASTM, Philadelphia.
American Society for Testing and Materials (ASTM) (1996b) *Standard Guide for Conducting Renewal Phytotoxicity Tests with Freshwater Emergent Macrophytes*, E1841–1896. ASTM, Philadelphia.
American Society for Testing and Materials (ASTM) (1997a) *Standard Guide for Conducting Life-cycle Toxicity Tests with Saltwater Mysids*, E1191–1197. ASTM, Philadelphia.
American Society for Testing and Materials (ASTM) (1997b) *Standard Guide for Conducting Daphnia magna Life-cycle Toxicity Tests*, E1193–1197. ASTM, Philadelphia.
American Society for Testing and Materials (ASTM) (1997c) *Standard Guide for Conducting Static 96-h Toxicity Tests with Microalgae*, E1218–1297. ASTM, Philadelphia.
American Society for Testing and Materials (ASTM) (1997d) *Standard Guide for Conducting Static, Axenic, 14-day Phytotoxicity Tests in Test Tubes with the Submersed Aquatic Macrophyte, Myriophyllum sibiricum Komarov*, E1913–1997. ASTM, Philadelphia.
Armitage, P.D. (1989) The application of a classification and prediction technique based on macroinvertebrates to assess the effects of river regulation. In *Alternatives in Regulated River Management*, Gore, J.A. and Petts, G.E. (eds), p. 267. CRC Press, Boca Raton, FL.
Armitage, P.D., Moss, D., Wright, J.F. and Furse, M.T. (1983) The performance of a new biological water quality score system based on macroinvertebrates over a wide range of unpolluted running-water sites. *Water Res.*, **17**(3), 333–347.
Armitage, P.D., Gunn, R.J.M., Furse, M.T., Wright, J.F. and Moss, D. (1987) The use of prediction to assess macroinvertebrate response to river regulation. *Hydrobiologia*, **144**, 25–32.
Bellamy, D.J., Clarke, P.H., John, D.M., Jones, D.,Whittick, A. and Darke, T. (1967) Effects of pollution from the *Torrey Canyon* on littoral and sub-littoral ecosystems. *Nature*, **16**, 1170–1173.
Butler, D. (2001) Are you ready for the revolution? *Nature*, **409**, 758–760.
Cairns Jr, J. (1956) Effect of heat on fish. *Ind. Wastes*, **1**, 180–183.
Cairns Jr, J. (1957) Environment and time in fish toxicity. *Ind. Wastes*, **2**, 1–5.
Cairns Jr, J. (1985) *Biological Monitoring in Water Pollution*. Pergamon, New York.
Cairns Jr, J. (1986) *Overview: Community Toxicity Testing*, ASTM STP 920. American Society for Testing and Materials, Philadelphia, PA.
Cairns Jr, J. (1992) Paradigms flossed: the coming of age of environmental toxicology. *Environ. Toxicol. Chem.*, **11**, 285–287.
Cairns Jr, J. and Mount, D.I. (1990) Aquatic toxicology, Part 2. *Environ. Sci. Technol.*, **24**(2), 154–161.

Cairns Jr, J. and Pratt, J.R. (1989) The scientific basis of bioassays. *Hydrobiologia*, **188/189**, 5–20.

Calow, P. (1993) *Handbook of Ecotoxicology*, Vol. 1, Blackwell Science, Oxford.

Carson, R. (1962) *Silent Spring*. Houghton Miffin, Boston.

Crane, M., Johnson, I., and Maltby, L. (1996) *In-situ* assays for monitoring the toxic impacts of waste in rivers. In *Toxic Impacts of Wastes on the Aquatic Environment*, Tapp, J.F., Hunt, S.M. and Wharfe, J.R. (eds), pp. 116–124. Royal Society of Chemistry, London.

Crane, M., Higman, M., Olsen, K., Simpson, P., Callaghan, A., Fisher, T. and Kheir, R. (2000) An *in situ* system for exposing aquatic invertebrates to contaminated sediments. *Environ. Toxicol. Chem.*, **19**, 2715–2719.

Curtis, C.T., Bishop, W.E. and Clarke, D.P. (2003) The genomics revolution: what does it mean for human and ecological risk assessment? *Ecotoxicology*, **12**(6), 489–495.

Dalton, R (2002) Frogs put in the gender bender by America's favourite herbicide. *Nature*, **416**, 665–666.

Depledge, M.H. (1993) The rational basis for the use of biomarkers as ecotoxicological tools. In *Nondestructive Biomarkers in Vertebrates*, Fossi, M.C. and Leonzio, C. (eds), pp. 271–296. Lewis Publishers, Boca Raton, FL.

Doudoroff, P. (1976) Keynote address: reflections on 'pickle-jar' ecology. In *Biological Monitoring of Water and Effluent Quality*, Cairns Jr, J., Dickson, P.L. and Westlake, G.F. (eds), ASTM STP 607. American Society for Testing and Materials, Philadelphia, PA.

Doudoroff, P. and Katz, M. (1950) Critical review of literature on the toxicity of industrial wastes and their components to fish. I. Alkalies, acids, and inorganic gases. *Sewage. Ind. Waste*, **22**, 1432–1458.

Doudoroff, P. and Katz, M. (1953) Critical review of literature on the toxicity of industrial wastes and their components to fish. II. The metals, as salts. *Sewage. Ind. Waste*, **25**, 802–839.

Doudoroff, P., Anderson, B.G., Burdick, G.E., Galstoff, P.S., Hart, W.B., Pattrick, R., Stronge, E.R., Surber, E.W. and Van Horn, W.M. (1951) Bioassay for the evaluation of acute toxicity of industrial wastes to fish. *Sewage. Ind. Waste*, **23**, 1380–1397.

Edwards, R. and White, I. (1999) *The Sea Empress Oil Spill: Environmental Impact and Recovery. International Oil Spill Conference*, March 1999, Seattle, USA.

Environment Canada (1990a) *Acute Lethality Test using Rainbow Trout*, EPS 1/RM/9. Environment Canada, Ottawa, Ontario.

Environment Canada (1990b) *Acute Lethality Test using Threespine Stickleback*, EPS 1/RM/10. Environment Canada, Ottawa, Ontario.

Environment Canada (1990c) *Acute Lethality Test using Daphnia spp.*, EPS 1/RM/11. Environment Canada, Ottawa, Ontario.

Environment Canada (1992a) *Test of Reproduction and Survival using the Cladoceran Ceriodaphnia dubia*, EPS 1/RM/21. Environment Canada, Ottawa, Ontario.

Environment Canada (1992b) *Test of Larval Growth and Survival using Fathead Minnows*, EPS 1/RM/22. Environment Canada, Ottawa, Ontario.

Environment Canada (1992c) *Toxicity Test using Luminescent Bacteria (Photobacterium phosphoreum)*, EPS 1/RM/24. Environment Canada, Ottawa, Ontario.

Environment Canada (1992d) *Growth Inhibition Test using the Freshwater Alga Selenastrum capricornutum*, EPS 1/RM/25. Environment Canada, Ottawa, Ontario.

Environment Canada (1992e) *Fertilisation Assay using Echnoids (Sea Urchins and Sand Dollars)*, EPS 1/RM/27. Environment Canada, Ottawa, Ontario.

Environment Canada (1998a) *Toxicity Tests using Early Life Stages of Salmonid Fish (Rainbow Trout)*, EPS 1/RM/28, 2nd edn. Environment Canada, Ottawa, Ontario.

Environment Canada (1998b) *Reference Method for Determining Acute Lethality of Sediment to Marine or Estuarine Amphipods*, 1/RM/35. Environment Canada, Ottawa, Ontario.

Environment Canada (1999) *Test for Measuring the Inhibition of Growth using the Freshwater Macrophyte Lemna minor*, EPS 1/RM/37. Environment Canada, Ottawa, Ontario.

Environment Canada (2000a) *Reference Method for Determining Acute Lethality of Effluents to Rainbow Trout*, EPS 1/RM/13, 2nd edn. Environment Canada, Ottawa, Ontario.

Environment Canada (2000b) *Reference Method for Determining Acute Lethality of Effluents to Daphnia magna*, EPS 1/RM/14, 2nd edn. Environment Canada, Ottawa, Ontario.

Environment Canada (2001) *Test for Survival and Growth in Sediment using Spionid Polychaete Worms (Polydora cornuta)*, 1/RM/41. Environment Canada, Ottawa, Ontario.

Environment Canada (2002) *Reference Method for Determining the Toxicity of Sediment using Luminescent Bacteria in a Solid-phase Test*, 1/RM/42. Environment Canada, Ottawa, Ontario.

European Commission (2001) *White Paper on a Strategy for a Future Chemicals Policy*, COM(2001) 88 final. Commission of European Communities, Brussels.

Forbes, V.E. and Forbes, T.L. (1994) *Ecotoxicology in Theory and Practice*. Chapman and Hall, London.

Furse, M.T., Moss D., Wright, J.F. and Armitage, P.D. (1984) The influence of seasonal taxonomic factors on the ordination and classification of running-water sites in Great Britain and on the prediction of their macro-invertebrate communities. *Freshwater Biol.*, **14**, 257–280.

Gallo, M.A. and Doull, J. (1991) History and scope of toxicology. In *Toxicology. The Basic Science of Poisons*, Amdur, M.O., Doull, J. and Klassen, C.D. (eds), pp. 3–11. Pergamon, New York.

Graney, R.L., Kennedy, J.H. and Rodgers, J.H. (1993) *Aquatic Mesocosm Studies on Ecological Risk Assessment*. Lewis Publishers, Boca Raton, FL.

Guillette Jr, L.J., Gross, T.S., Masson, G.R., Matter, J.M., Percival, H.F. and Woodward, A.R. (1994) Developmental abnormalities of the gonad and abnormal sex hormone concentrations in juvenile alligators from contaminated and control lakes in Florida. *Environ. Health Perspect*, **102**, 680–688.

Guilette Jr, L.J., Gross, T.S., Gross, D.A., Rooney A.A. and Percival, H.F (1995) Gonadal steroidogenesis *in vitro* from juvenile alligators obtained from contaminated or control lakes. *Environ. Health Perspect.*, **103** (suppl. 4), 31–36.

Hansch, C. and Leo, A. (1995). Exploring QSAR: *Fundamentals and Applications in Chemistry and Biology*. American Chemical Society, Washington, DC.

Harada, M. (1982). Minamata disease: organic mercury poisoning caused by ingestion of contaminated fish. In *Adverse Effects of Food*, Jellife, E.F. and Jellife, D.B. (eds), pp. 135–148. Plenum Press, New York.

Harris, M.L., Bishop, C.A., Struger. J., van den Heuvel, M.R., van der Kraak, G.J., Dixon, D.G., Ripley, B. and Bogart, J. (1998a) The functional integrity of northern leopard frog (*Rana pipiens*) and green frog (*Rana clamitans*) populations in orchard wetlands. I Genetics physiology and biochemistry of breeding adults and young-of-the-year. *Environ. Toxicol. Chem.*, **17**, 1338–1350.

Harris, M.L., Bishop, C.A., Struger. J., Ripley, B. and Bogart, J. (1998b) The functional integrity of northern leopard frog (*Rana pipiens*) and green frog (*Rana clamitans*) populations in orchard wetlands. II Effects of pesticides and eutrophic conditions on early life stage development. *Environ. Toxicol. Chem.*, **17**, 1351–1363.

Henderson, C. (1957) Application factors to be applied to bioassays for the safe disposal of toxic wastes. In *Biological Problems in Water Pollution*, pp. 31–37. US Department of Health, Education and Welfare, Public Health Service, Washington, DC.

Hickey, J.J. (1969) *The Peregrine Falcon Populations: Their Biology and Decline*. University of Wisconsin Press, Madison, WI.

Hickey, J.J. and Anderson, D.W. (1968) Chlorinated hydrocarbons and eggshell changes in raptorial and fish-eating birds. *Science*, **162**, 271–273.

Hunn, J.B. (1989) History of acute toxicity tests with fish 1863–1987. Investigations in fish control (Internal Report), US Department of the Interior, Fish and Wildlife Service.

Institute for Environmental Health (1996) *The Use of Biomarkers in Environmental Exposure Assessment*, Medical Research Council Report R5, ISBN 1-89911007-0.

International Organization for Standardization (ISO) (1989a) *Water Quality – Method for Assessing the Inhibition of Nitrification of Activated Sludge Micro-organisms by Chemicals and Wastewaters*, ISO 9509. ISO, Paris.

International Organization for Standardization (ISO) (1989b) *Water Quality – Freshwater Algal Inhibition Test with Scenedesmus subspicatus* and *Selenastrum capricornutum*, ISO 8692. ISO, Paris.

International Organization for Standardization (ISO) (1994) *Water Quality – Determination of the Prolonged Toxicity of Substances to Freshwater Fish – Method for Evaluating the Effects of Substances on the Growth Rate of Rainbow Trout (Oncorhynchus mykiss Walbaum (Teleostei, Salmonidae)*, ISO 10229. ISO, Paris.

International Organization for Standardization (ISO) (1995a) *Water Quality – Pseudomonas putida Growth Inhibition Test (Pseudomonas cell multiplication inhibition test)*, ISO 8689–1. ISO, Paris.

International Organization for Standardization (ISO) (1995b) *Water Quality – Marine Algal Growth Inhibition Test with Skeletonema costatum and Phaeodactylum tricornutum*, ISO 10253. ISO, Paris.

International Organization for Standardization (ISO) (1996a) *Water Quality – Determination of the Inhibition of the Mobility of Daphnia magna Straus (Cladocera, Crustacea) – Acute Toxicity Test*, ISO 6341. ISO, Paris.

International Organization for Standardization (ISO) (1996b) *Water Quality – Determination of the Acute Lethal Toxicity of Substances to a Freshwater Fish (Brachydanio rerio Hamilton-Buchanan (Teleostei, Cyprinidae) – Part 1: Static Method*, ISO 7346–1. ISO, Paris.

International Organization for Standardization (ISO) (1996c) *Water Quality – Determination of the Acute Lethal Toxicity of Substances to a Freshwater Fish (Brachydanio rerio Hamilton-Buchanan (Teleostei, Cyprinidae) – Part 2: Semi-static Method*, ISO 7346–2. ISO, Paris.

International Organization for Standardization (ISO) (1996d) *Water Quality – Determination of the Acute Lethal Toxicity of Substances to a Freshwater Fish (Brachydanio rerio Hamilton-Buchanan (Teleostei, Cyprinidae) – Part 3: Flow-through Method*, ISO 7346–3. ISO, Paris.

International Organization for Standardization (ISO) (1998) *Water Quality – Determination of the Inhibitory Effect of Water Samples on the Light Emission of Vibrio fischeri (Luminescent Bacteria Test) – Parts 1–3*, ISO 11348–1/3. ISO, Paris.

International Organization for Standardization (ISO) (1999a) *Water Quality – Guidelines for Algal Growth Inhibition Tests with Poorly Soluble Materials, Volatile Compounds, Metals and Waste Water*, ISO 14442. ISO, Paris.

International Organization for Standardization (ISO) (1999b) *Water Quality – Determination of the Acute Lethal Toxicity to Marine Copepods (Copepoda, Crustacea)*, ISO 14669. ISO, Paris.

International Organization for Standardization (ISO) (2000a) *Water Quality – Determination of the Genotoxicity of Water and Waste Water using the UMU-test*, ISO 13829. ISO, Paris.

International Organization for Standardization (ISO) (2000b) *Water Quality – Determination of the Genotoxicity of Water and Waste Water – Salmonella/Microsome Test*, ISO 16240. ISO, Paris.

International Organization for Standardization (ISO) (2000c) *Water Quality – Determination of the Non-poisonous Effect of Water Constituents and Waste Water to Duckweed (Lemna minor, Lemna gibba)*, ISO 20079. ISO, Paris.

International Organization for Standardization (ISO) (2000d) *Water Quality – Determination of Long-term Toxicity of Substances to Daphnia magna Straus (Cladocera, Crustacea)*, ISO 10706. ISO, Paris.

International Organization for Standardization (ISO) (2000e) *Water Quality – Determination of Acute Toxicity of Marine Sediment to Amphipods*, ISO 16712. ISO, Paris.

International Organization for Standardization (ISO) (2000f) *Water Quality – Determination of Chronic Toxicity to Daphnia magna (Straus) in 7 days – Simplified Growth Inhibition Test*, ISO 20664. ISO, Paris.

International Organization for Standardization (ISO) (2000g) *Water Quality – Determination of Chronic Toxicity to Ceriodaphnia dubia in 7 days – Population Growth Inhibition Test*, ISO 20665. ISO, Paris.

International Organization for Standardization (ISO) (2000h) *Water Quality – Determination of Chronic Toxicity to Brachionus calyciflorus in 48 h – Population Growth Inhibition Test*, ISO 20666. ISO, Paris.

Jobling, S., Nolan, M., Tyler, C., Brighty, G. and Sumpter, J.P. (1998) Widespread sexual disruption in wild fish. *Environ. Sci. Technol.*, **32**, 2498–506.

Kemm, K. (1999) Malaria and the DDT story. In *Environmental Health – Third World Problems – First World Preoccupations*, Mooney, L. and Bate, R. (eds), pp. 1–16. Butterworth – Heinemann, Oxford.

Kenaga, E.E. (1982) Predictability of chronic toxicity from acute toxicity of chemicals in fish and aquatic invertebrates. *Environ. Toxicol. Chem.*, **1**, 347–358.

Kurland, L.T., Farro, S.N. and Siedler, H. (1960) Minamata disease. The outbreak of a neurologic order in Minamata, Japan and its relationship to the ingestion of seafood contaminated by mercuric compounds. *World Neurol.*, **1**, 370–395.
Langston, W.J. (1995) Tributyl tin in the marine environment: a review of past and present risks. *Pestic. Outlook*, **6**, 18–24.
Lokke, H. and Van Gestel, C.A.M. (1998) *Handbook of Soil Invertebrate Toxicity Tests*. John Wiley & Sons, Chichester.
Maltby, L., Clayton, S.A., Yu, H., McLoughlin, N., Wood, R.M. and Yin, D. (2000). Using single-species toxicity tests, community-level responses and toxicity identification evaluations to investigate effluent impacts. *Environ. Toxicol. Chem.*, **19**, 151–157.
Maltby, L., Clayton, S.A., Wood, R.M. and McLoughlin, N. (2002). Evaluation of the *Gammarus pulex in situ* feeding assay as a biomonitor of water quality: robustness, responsiveness, and reliance. *Environ. Toxicol. Chem.*, **21**, 361–368.
Mellanby, K. (1992) *The DDT Story*. British Crop Protection Council, Farnham, UK.
McCarty, L.S., Power, M. and Munkittrick, K.R. (2002) Bioindicators versus biomarkers in ecological risk assessment. *Hum. Ecol. Risk Assess.*, **8**(1), 159–164.
McCarthy, J.F. and Shuggart, L.R. (1990) *Biomarkers of Environmental Contamination*. CRC Press, Boca Raton, FL.
Moore C.G. and Stevenson J.N. (1991). The occurrence of intersexuality in harpacticoid copepods and its relationship with pollution. *Mar. Pollut. Bull.*, **22**, 72–74.
Moore, N.W. (1967) A synopsis of the pesticide problem. *Adv. Ecol. Res.*, **4**, 75–129.
Moss, D., Furse, M.T., Wright, J.F. and Armitage, P.D. (1987) The prediction of the macro-invertebrate fauna of unpolluted running-water sites in Great Britain using environmental data. *Freshwater Biol.*, **17**, 41–52.
Mount, D.I. and Boyle, H.W. (1969) Parathion – use of blood concentration to diagnose mortality of fish. *Environ. Sci. Technol.*, **3**, 1183–1185.
Mount, D.I. and Brungs, W.A. (1967) A simplified dosing apparatus for fish toxicology studies. *Water Res.*, **1**, 21–29.
Mount, D.I. and Stephen, C.E. (1967) A method for establishing acceptable limits for fish – malathion and the butoxy-ethanol ester of 2,4-D. *Trans. Am. Fish Soc.*, **96**, 185–193.
Munawar, M., Dixon, D.G., Mayfield, C.I., Reynoldson, T. and Sadar, M.H. (1989) Environmental bioassay techniques and their application. Proceedings of the 1st International Conference, Lancaster, England. *Hydrobiologia*, **188/189**, 93–116.
Niemi, G.J. (1990) Multivariate analysis and QSAR: applications of principal components analysis. In *Practical Applications of Quantitative Structure–Activity Relationships (QSAR) in Environmental Chemistry and Toxicology*, pp. 153–170. Kluwer Publishing, Dordrecht.
Organization for Economic Cooperation and Development (OECD) (1984a) *Guideline for the Testing of Chemicals. Section 2 – Effects on Biotic Systems. 209 – Activated Sludge, Respiration Inhibition Test*. OECD, Paris.
Organization for Economic Cooperation and Development (OECD) (1984b) *Guideline for the Testing of Chemicals. Section 2 – Effects on Biotic Systems. 201 – Alga, Growth Inhibition Test*. OECD, Paris.
Organization for Economic Cooperation and Development (OECD) (1984c) *Guideline for the Testing of Chemicals. Section 2 – Effects on Biotic Systems. 202 – Daphnia sp. Acute Immobilisation Test and Reproduction Test*. OECD, Paris.
Organization for Economic Cooperation and Development (OECD) (1984d) *Guideline for the Testing of Chemicals. Section 2 – Effects on Biotic Systems. 204 – Fish. Prolonged Toxicity Test: 14 Day Study*. OECD, Paris.
Organization for Economic Cooperation and Development (OECD) (1992a) *Guideline for the Testing of Chemicals. Section 2 – Effects on Biotic Systems. 203 – Fish, Acute Toxicity Test*. OECD, Paris.
Organization for Economic Cooperation and Development (OECD) (1992b) *Guideline for the Testing of Chemicals. Section 2 – Effects on Biotic Systems. 210 – Fish, Early Life-stage Toxicity Test*. OECD, Paris.
Organization for Economic Cooperation and Development (OECD) (1998a) *Guideline for the Testing of Chemicals. Section 2 – Effects on Biotic Systems. 211 – Daphnia magna Reproduction Test*. OECD, Paris.

Organization for Economic Cooperation and Development (OECD) (1998b) *Guideline for the Testing of Chemicals. Section 2 – Effects on Biotic Systems. 212 – Fish, Short-term Toxicity Test on Embryo and Sac-fry Stages.* OECD, Paris.

Organization for Economic Cooperation and Development (OECD) (1999a) *Draft Guideline for the Testing of Chemicals. Section 2 – Effects on Biotic Systems. Lemna spp. Growth Inhibition Test.* OECD, Paris.

Organization for Economic Cooperation and Development (OECD) (1999b) *Draft Guideline for the Testing of Chemicals. Section 2 – Effects on Biotic Systems. 202 – Daphnia spp. Acute Immobilisation Test.* OECD, Paris.

Organization for Economic Cooperation and Development (OECD) (2000a) *Draft Guideline for the Testing of Chemicals. Section 2 – Effects on Biotic Systems. 218/219 – Sediment-water Chironomid Test using Spiked Sediment.* OECD, Paris.

Organization for Economic Cooperation and Development (OECD) (2000b) *Guideline for the Testing of Chemicals. Section 2 – Effects on Biotic Systems. 215 – Fish Juvenile Growth Test.* OECD, Paris.

Olsen, A., Ellerbeck, L., Fisher, T., Callaghan, A. and Crane, M. (2001). Variability in acetylcholinesterase and glutathione-s-transferase activities in *Chironomus riparius* Meigen deployed *in situ* at uncontaminated field sites. *Environ. Toxicol. Chem.* **20**, 1725–1732.

Peakall, D.B. (1992) *Animal Biomarkers as Pollution Indicators.* Chapman and Hall, New York.

Phillips, D.J.H. (1990) Use of macroalgae and invertebrates as monitors of metal levels in estuaries and coastal waters. In *Heavy Metals in the Marine Environment*, Furness, R.W. and Rainbow, P.S. (eds), pp. 81–99. CRC Press, Boca Raton, FL.

Phillips, D.J.H (1993) Bioaccumulation. In *Handbook of Ecotoxicology*, Calow, P. (ed.), pp. 378–396. Blackwell Science, Oxford.

Polak, M.E., Rawson, D.M. and Haggett, B.G.D. (1996) Redox mediated biosensors incorporating cultured fish cells for toxicity assessment. *Biosens. Bioelectron.* **11**(12), 1253–1257.

Power, E.A. and Boumphrey, R.S. (2004) International trends in bioassay use for effluent management. *Ecotoxicology*, **13**(5), 377–398.

Pratt, J.R. and Bowers, N.J. (1990) A microcosm test for estimating ecological effects of chemicals and mixtures. *Toxicol. Assess.*, **5**, 189–205.

Preston, A. (1973) Heavy metals in British waters. *Nature*, **242**, 95–97.

Rand, G.M. and Petrocelli, S.R. (1985) *Fundamental Aquatic Toxicology – Methods and Applications.* Hemisphere, Washington, DC.

Ratcliffe, D.A. (1967) Decrease in eggshell weight in certain birds of prey. *Nature*, **215**, 208–210.

Reish, D.J. (1972) The use of marine invertebrates as indicators of varying degrees of marine pollution. *J. Mar. Biol. Assoc. UK*, **24**, 185–189.

Robinson, J., Richardson, A., Crabtree, A.N., Coulson, J.C. and Potts, G.R. (1967) Organochlorine residues in marine organisms. *Nature*, **214**, 1307.

Roddie, B.D., Redshaw, C.J. and Nixon, S. (1996) Sublethal biological effects monitoring using the common mussel (*Mytilus edulis*): comparison of laboratory and *in situ* effects of an industrial effluent discharge. In *Toxic Impacts of Wastes on the Aquatic Environment*, Tapp, J.F., Hunt, S.M. and Wharfe, J.R. (eds), pp. 125–137. Royal Society of Chemistry, London.

Rogerson, J.G., Atkinson, A., Evans, M.R. and Rawson, D.M. (1996) The potential for biosensors to assess the toxicity of industrial effluents. In *Toxic Impacts of Wastes on the Aquatic Environment*, Tapp, J.F., Hunt, S.M. and Wharfe, J.R. (eds), pp. 125–137. Royal Society of Chemistry, London.

Scheuhammer, A.M. (1987) The chronic toxicity of aluminium, cadmium, mercury, and lead in birds: a review. *Environ. Pollut.*, **46**, 263–295.

Sprague, J.B. (1969) Review paper. Measurement of pollutant toxicity to fish. I. Bioassay methods for acute toxicity. *Water Res.*, **3**, 793–821.

Sprague, J.B. (1970) Review paper. Measurement of pollutant toxicity to fish. II. Utilizing and applying bioassay results. *Water Res.*, **4**, 3–32.

Sprague, J.B. (1971) Review paper. Measurement of pollutant toxicity to fish. III. Sublethal effects and 'safe' concentrations. *Water Res.*, **5**, 245–266.

Sprague, J.B. (1973) The ABCs of pollutant bioassay with fish. In *Biological Methods for the Assessment of Water Quality*, Cairns Jr, J. and Dickson, K.L. (eds), pp. 6–30, STP 528. American Society for Testing and Materials, Philadelphia, PA.

Sprague, J.B. and Ramsay, B.A. (1965) Lethal effects of mixed copper and zinc solutions for juvenile salmon. *J. Fish Res. Bd. Can.*, **22**, 425–432.

Stephenson, M., Smith, D., Ichikawa, G., Goetzl, J. and Martin, M. (1986) State mussel watch program: preliminary data report 1985–1986 (Internal Report), California Department of Fish and Game.

Tinsley, D., Johnson, I., Boumphrey, B., Forow, D. and Wharfe, J. (1996) The use of direct toxicity assessment to control discharges to the aquatic environment in the United Kingdom. In *Toxic Impacts of Wastes on the Aquatic Environment*, Tapp, J.F., Hunt, S.M. and Wharfe, J.R. (eds), pp. 36–43. Royal Society of Chemistry, London.

United States Environmental Protection Agency (USEPA) (1985) *Methods for Measuring the Acute Toxicity of Effluents and Receiving Waters to Freshwater and Marine Organisms*, 3rd edn, USEPA-600-4-85-013. USEPA, Washington, DC.

United States Environmental Protection Agency (USEPA) (1989a) *Toxicity Reduction Evaluation Protocol for Municipal Wastewater Treatment Plants*, Botts, J.A., Braswell, J.W., Zyman, J., Goodfellow, W.L. and Moore, S.B. (eds), USEPA 600/2-88/062. Reduction Engineering Laboratory, Cincinnati, OH.

United States Environmental Protection Agency (USEPA) (1989b) *Generalized Methodology for Conducting Industrial Toxicity Reduction Evaluations* (TREs), Fava, J.A., Lindsay, D., Clement, W.H., Clark, R., DeGraeve, G.M., Cooney, J.D., Hansen, S., Rue, W., Moore, S. and Lankford, P. (eds), USEPA EPA/600/2-88/070. Risk Reduction Engineering Laboratory, Cincinnati, OH.

United States Environmental Protection Agency (USEPA) (1991) *Methods for Aquatic Toxicity Identification Evaluations: Phase I Toxicity Characterization Procedures*, Norberg-King, T.J., Mount, D.I., Durhan, E.J., Ankley, G.T., Burkhard, L.P., Amato, J., Lukasewycz, M., Schubauer-Berigan, M. and Anderson-Carnahan, L. (eds), USEPA 600/6-91/003. USEPA, Duluth, MN.

United States Environmental Protection Agency (USEPA) (1992) *Toxicity Identification Evaluations: Characterization of Chronically Toxic Effluents, Phase I*, Norberg-King, T.J., Mount, D.I., Amato, J.R., Jensen, D.A. and Thompson, J. (eds), USEPA 600/6-91/005F. USEPA, Duluth, MN.

United States Environmental Protection Agency (USEPA) (1993a) *Methods for Aquatic Toxicity Identification Evaluations: Phase II Toxicity Identification Procedures for Acutely and Chronically Toxic Samples*, Durhan, E.J., Norberg-King, T.J. and Burkhard, L.P (eds), USEPA-600/R-92/080. USEPA, Duluth, MN.

United States Environmental Protection Agency (USEPA) (1993b) *Methods for Aquatic Toxicity Identification Evaluations: Phase III Toxicity Identification Procedures for Acutely and Chronically Toxic Samples*, Mount, D.I. and Norberg-King, T.J. (eds), EPA-600/R-92/081. USEPA, Duluth, MN.

United States Environmental Protection Agency (USEPA) (1993c) *Methods for Measuring the Acute Toxicity of Effluents and Receiving Waters to Freshwater and Marine Organisms*, 4th edn, USEPA 600-4-90-027F. USEPA, Washington, DC.

United States Environmental Protection Agency (USEPA) (1994a) *USEPA/EC Joint Project on the Evaluation of (Quantitative) Structure Activity Relationships*, USEPA 743-R-94-001. Office of Pollution Prevention and Toxics, Washington, DC.

United States Environmental Protection Agency (USEPA) (1994b) *Short-term Methods for Estimating the Chronic Toxicity of Effluents and Receiving Waters to Freshwater Organisms*, 3rd edn, USEPA 600/4-91/002. USEPA, Cincinnati, OH.

United States Environmental Protection Agency (USEPA) (1994c) *Short-term Methods for Estimating the Chronic Toxicity of Effluents and Receiving Waters to Marine and Estuarine Organisms*, 2nd edn, USEPA 600/4-91/003. USEPA, Cincinnati, OH.

United States Environmental Protection Agency (USEPA) (1995) *West Coast Marine Species WET Test Methods: Short-Term Methods for Estimating the Chronic Toxicity of Effluents and Receiving Waters to West Coast Marine and Estuarine Organisms*, USEPA/600/R-95-136. USEPA, Cincinnati, OH.

Van Gestel, C.A.M., van der Waarde, J.J., Derksen, J.G.M., van der Hoek, E.E., Veul, M.F.X.W., Bouwens, S., Rusch, B., Kronenburg, R. and Stokman, G.N.M. (2001) The use of acute and chronic bioassays to determine the ecological risk and bioremediation efficiency of oil-polluted soils. *Environ. Toxicol. Chem.*, **20**(7), 1438–1449.

Walley, W.J. and V.N. Fontama (1998) Neural network predictors of average score per taxon and number of families at unpolluted river sites in Great Britain. *Water Res.*, **32**(3), 613–622.

Walley W. J. and O'Connor M. A. (2000) Unsupervised pattern recognition for the interpretation of ecological data. 2nd Int. Conf. on Applications of Machine Learning to Ecological Modelling, Adelaide, November 2000. *J. Ecol. Model.*, **146**(1–3), 219–230.

Wharfe, J.R. (1996). Toxicity-based criteria for the regulatory control of waste discharges and for environmental monitoring and assessment in the United Kingdom. In *Toxic Impacts of Wastes on the Aquatic Environment*, Tapp, J.F., Hunt, S.M. and Wharfe, J.R. (eds), pp. 26–35. Royal Society of Chemistry, London.

Wharfe, J.R. and Heber, M. (1998) Toxicity assessment: its role in regulation. In *Pollution Risk Assessment and Management*, Douben, P.E.T. (ed.), pp. 311–330. John Wiley and Sons, Chichester.

White, I. and Baker, J. (1998) The *Sea Empress* oil spill in context. *International Conference on the Sea Empress Oil Spill*, February 1998, Cardiff, Wales.

Whitehouse, P. and Cartwright, N. (1998) Standards for environmental protection. In *Pollution Risk Assessment and Management*, Douben, P.E.T. (ed.), pp. 235–272. John Wiley and Sons, Chichester.

Wright, J.F. (1995) Development and use of a system for predicting the macroinvertebrate fauna in flowing waters. *Aust. J. Ecol.*, **20**, 181–197.

Wright, J.F., Moss, D., Armitage, P.D. and Furse, M.T. (1984) A preliminary classification of running-water sites in Great Britain based on macro-invertebrate species and the prediction of community type using environmental data. *Freshw. Biol.*, **14**, 221–356.

Wright, J.F., Armitage, P.D., Furse, M.T. and Moss, D. (1988) A new approach to the biological surveillance of river quality using macroinvertebrates. *Verh. Int. Verein. Limnol.*, **23**, 1548–1552.

Wright, J.F., Furse, M.T. and Armitage, P.D. (1993a) RIVPACS – a technique for evaluating the biological quality of rivers in the U.K. *Eur. Water Pollut. Control*, **3** (4), 15–25.

Wright, J.F., Furse, M.T., Armitage, P.D. and Moss, D. (1993b) New procedures for identifying running-water sites subject to environmental stress for evaluating sites for conservation, based on the macroinvertebrate fauna. *Arch. Hydrobiol.*, **127**, 319–326.

2 Effective monitoring of the environment for toxicity

Ian Johnson, Paul Whitehouse and Mark Crane

2.1 Introduction

In principle, ecotoxicological testing of environmental samples (whether from the aquatic, terrestrial or aerial compartments) can be carried out at any biological level of organisation – from the molecular level, through whole organisms, to populations and, ultimately, communities or assemblages of organisms (Calow, 1989). The majority of pollutants act initially at the molecular level following accumulation into the exposed organism, with any effects then becoming apparent as physiological changes and effects on key individual parameters such as growth, reproduction or survival. Table 2.1 summarises the types of responses measurable at different levels of biological organisation.

All of these responses are quantifiable and this chapter discusses some of the issues associated with ensuring that such toxicity data are of the highest possible quality. In the spirit of the saying 'garbage in, garbage out' we first discuss

Table 2.1 Types of responses measurable at different levels of biological organisation.

Level of biological organisation	Types of measurable responses
Sub-organism	
Molecular	Structural and/or functional changes (e.g. amplification, translocation or mutation of genes, enzyme function, protein production)
Cellular	Structural and/or functional changes (e.g. carcinogenic, immunological and neurological responses)
Tissue	Structural and/or functional changes (e.g. gonadal intersex in fish)
Whole organism	Behavioural Developmental (growth and reproduction) Physiological Lethality
Population	Species densities and profiles (numbers of different sex/size classes)
Community	Composition of communities and densities of particular species

how samples should be taken from the environment so that subsequent toxicity measurements are representative and interpretable. We then illustrate the necessity of good quality assurance and quality control systems if one is to have faith in the quantitative results of these biological tests. Such quality systems are vital adjuncts to some of the more interesting and exciting aspects of environmental toxicology, and we explore the implications for decision-making if they are ignored.

2.2 Design of monitoring programmes

2.2.1 Introduction

All environmental monitoring tools, including bioassays, must be embedded within a coherent monitoring framework if results are to be related meaningfully to monitoring objectives. Biological, chemical and statistical criteria must be considered before designing such programmes.

Several comprehensive guides have been produced on the design of environmental quality monitoring programmes (e.g. Crawford *et al.*, 1983; Horner *et al.*, 1986, Mar *et al.*, 1987; Ellis, 1989; Loftis *et al.*, 1989; Ward *et al.*, 1990; Hipel and McLeod, 1994). Dixon and Chiswell (1996) reviewed over 150 papers on the subject of aquatic monitoring programme design and provide a useful framework for evaluating the rigour of a design. They divide programme design into three distinct stages: setting of information goals; selection of indicators of environmental quality; and data analysis. They argue strongly that it is only after these matters have been considered that the location and timing of sampling can be decided. Whitfield (1988) also emphasises the importance of setting clear information goals, but believes that often there is too little concern over why a particular site is chosen. In his view, reasons are too frequently given that data should be available 'just in case', or to feed data banks, or to justify the existence of a particular regulatory organisation. This sort of approach to the design of monitoring programmes can lead to results that are data rich but information poor. In an earlier paper Whitfield (1988) recommended several other stages that may be used to extend the model proposed by Dixon and Chiswell (1996).

1. Establishment of a monitoring goal.
2. Selection of a sampling strategy to meet the goal.
3. Periodic review of adequacy of sampling, including quality control studies.
4. Optimisation of sampling related to the goal over time.
5. Review of adequacy of monitoring goal.

Harmancioglu and Alpaslan (1992) agree with these sentiments and extend them by recommending a quantitative approach to measuring the information that is conveyed during environmental monitoring. Their approach, based upon information theory, can be used to develop new monitoring networks and also to pare down existing networks in the most efficient and cost-effective way.

2.2.2 Setting of information goals

There is a common complaint that environmental monitoring systems are often set up with too little thought about their purpose (Timmerman *et al.*, 2000). Even when thought is applied, it is often difficult to reconcile the wishes of different stakeholders, or to establish effective communication between information producers such as scientists and information users such as regulators, politicians and the public. This is why Timmerman *et al.* (2000) recommend the use of an iterative 'information cycle' in which both information producers and users are involved. Modern techniques for visualising data also will help scientists to communicate with non-scientists. The sophistication of these techniques has advanced beyond simple graphs, to complex three-dimensional animations of changing water quality through time and space (e.g. Boyer *et al.*, 2000), although simple displays will often be the most cost-effective means of communication.

Different goals of an environmental monitoring programme might include information on ambient conditions and trends or whether regulatory limits have been breached (El-Shaarawi, 1993; Dixon and Chiswell, 1996; Brydges and Lumb, 1998). Assessment of particular environmental impacts caused by human activities and estimation of mass transport may also be valid programme goals (Whitfield, 1988). Human society, or at least a section of it, will decide upon the goals of environmental management (Pegram *et al.*, 1997) because other organisms have no voice. How wider society should be consulted, so that protection targets and information goals can be decided, is an active area of sociological research.

Against this background of debate, it is clear that bioassays are already used – or poised to play a role – in a number of statutory roles in the UK. These include:

- Effluent control as part of a Direct Toxicity Assessment (DTA) approach (UKWIR, 2001), based on the recommendations of a DTA Demonstration Programme completed in 2000. Toxicity-based conditions may be implemented as part of Integrated Pollution Prevention and Control (IPPC) authorisations or to control emissions under the Water Resources Act.
- Identification of 'Special Wastes' under the Special Waste Regulations (SWR) under Section 2 of the Environmental Protection Act 1990. Hazardous categories include the category 'Ecotoxicity', defined as substances and preparations that present or may present immediate or delayed risks for one or more sectors of the environment. Most waste is likely to be classified according to its individual chemical components and information abstracted from toxicity databases. However, there is provision in the SWR for bioassays to be performed, specifically those in Annex V of EC Directive 67/548/EEC.

- The UK National Marine Monitoring Programme (NMMP), in which samples of seawater, sediment and biota are collected for chemical analysis and application of a number of biological effects techniques, including water column and sediment bioassays and the measurement of biomarkers in fish.
- The Framework Directive for Water (2000/60/EC) makes it possible that the environment agencies across Europe will come under increasing pressure to use bioassays. Indeed, after considering the surface water monitoring requirements of the Directive, the Scientific Committee for Toxicity, Ecotoxicity and the Environment (CSTEE, 1998) recommended that bioassays might be helpful in classifying the ecological status of waters, separating physical and chemical impacts on water quality and linking biological and chemical effects.

Further information on the regulatory roles in which bioassays may be applied is given in Chapter 9.

2.2.3 Selection of indicators of environmental quality

The selection of appropriate environmental quality indicators depends upon the overall goal of a monitoring programme, as discussed above (Harmancioglu and Alpaslan, 1992). These goals themselves will depend upon decisions that are made about what it is that society wishes to protect.

Once the overall goal has been selected, a feature must be selected that is both related to this goal and amenable to measurement (i.e. a 'measurement' end-point). Natural systems are often variable, so measurement end-points need to be selected that are as low in variability as possible. In practice, this may mean aggregating some data, such as information on individual species, so that a trade-off between useful biological information and low variability is obtained (Stow et al., 1998).

In the selection of bioassays, much is often made of their 'ecological relevance', based on the understandable view that assays should help to inform us about the responses of biota at the site of concern. However, if the purpose of a test is to detect whether a sample is toxic or not, then any test system will be appropriate provided that it is sufficiently sensitive to detect toxicity. Some researchers claim that it is also important to use a test organism in a bioassay that has some kind of 'ecological relevance'. This relevance may be defined as a test species that is also present in the habitat being investigated, one that bears physical or physiological similarities to those present, one that is functionally important, e.g. in nutrient cycling (perhaps a 'keystone' species), or a test that yields demographically important end-points that directly influence population size, e.g. survival and fecundity.

Because test organisms are usually bred in the laboratory and exposed under rather artificial conditions, it is more appropriate to regard them as surrogates for the many species that will remain untested. Under these circumstances it is hard to see how any researcher can claim superior 'ecological relevance' for their favoured test species and end-points, with the possible exception of demographic end-points (survival, fecundity and migration) used as input parameters for ecological models, which estimate and predict population size. In contrast to 'ecological relevance', sensitivity to toxic chemicals normally should be the most important criterion in selecting test systems, especially when bioassaying ambient samples.

We can extend this further by considering whether decision-making should be based on the responses of a single species or on a battery of test species. It is generally not appropriate to rely on the results of a single toxicity test or bioassay when making regulatory decisions (USEPA, 1991). The greatest risk is that of false negatives (Hill *et al*., 1993; Ankley, 1995), where results indicate that samples are non-toxic when in fact they are polluted. As a result, toxic samples could be classified as 'good quality' and probably would not be considered further in regulatory assessments.

The most common cause of false-negative results is the insensitivity of the chosen species to contaminants in the test matrix. This may be because the test species is not susceptible to the primary mode of toxic action of the chemical(s) present in the sample or because it can avoid maximum exposure to the chemical. For example, organic compounds with a high log K_{oc} will bind preferentially to sediment particles. In a sediment–water exposure system, toxic effects may be seen in the sediment ingestors but not in the benthic filter feeders or those that live in the overlying water. Therefore, the latter test organisms are more likely to give false-negative results when these high log K_{oc} compounds are present. Therefore, a battery of complementary tests (which together are sensitive to a wide range of chemical classes) is needed.

However, the key question facing those responsible with carrying out testing for a particular role is 'which battery of methods should be used?'. In short, the answer has to be 'those that are most appropriate for the operational role'. Because the number of available test methods is large and increases each year, it can be difficult to identify which methods from the range of potential sub-organism, whole organism and population-based tests are most appropriate. However, conventional wisdom indicates that the most objective approach is to evaluate different potential methods against a series of performance criteria (such as relevance, reproducibility, reliability, robustness, repeatability and sensitivity) and identify those methods that are most effective (Calow, 1993). We explore these issues in more detail below.

Cairns *et al*. (1993) list several criteria that they believe should be present in a biological indicator:

- *Biological relevance* – it should be easily related to what society wishes to protect in nature.
- *Social relevance* – the public and decision-makers should value it.
- *Broad application* – to many potential stressors at many different sites.
- *Sensitivity to stressors* – but without large variability, or a simple 'all or nothing' response.
- *Measurability* – it should be definable and quantifiable with known precision and accuracy, using an accepted procedure.

- *Interpretability* – acceptable and unacceptable impacts on the indicator should be identifiable in a way that will stand up in a court of law and be supported by the scientific community.
- *Continuity of measurement* – should be possible both spatially and temporally, and at appropriate scales of space and time.
- *Low redundancy* – it should provide information that is additional to that provided from other sources.
- *Integrative* – it should summarise information from other important natural phenomena that cannot be measured currently.
- *Anticipatory* – it should provide information before serious harm has occurred to the natural system that it was chosen to protect.
- *Timely* – information should be provided within the time-scale required for a management decision and appropriate action.
- *Diagnostic* – of the particular type of stressor that is causing the response.
- *Cost-effective* – in producing the maximum amount of information for the resources invested.
- *Historically based* – there should be sufficient historical information to enable determination of the normal range of responses.
- *Non-destructive* – to the natural system.

Both ecological survey and bioassay approaches fulfil most of these criteria, albeit to different degrees.

2.2.4 Location and frequency of samples, and data analysis

It is only after information goals and appropriate indicators have been defined accurately and precisely that sample location and frequency can be decided. Continuous monitoring of some water quality parameters is now commonplace in many parts of the world. However, the cost of these systems and limits to the parameters that can be measured with them means that spot or grab samples for testing probably always will form part of environmental quality monitoring strategies (Hazleton, 1998), and this remains true for bioassays.

It is a cliché that correlation or association does not imply causality. However, environmental quality managers need to discover the causes of any positive or negative changes in environmental quality. Are they doomed, therefore, never to achieve this goal? Not necessarily. If correlations between two parameters are consistent, responsive and have a mechanistic basis, then it is highly likely that causality can be inferred (Spooner and Line, 1993). A consistent response means that the association is similar in each data set. Responsive associations occur when experimental manipulation of one parameter leads to the expected change in the other parameter. Mechanistic bases to associations are known when it is possible to describe a physical link between one parameter and another, such as knowing that a particular concentration of a pollutant will reduce energy allocation within an individual organism, leading to a decline in reproduction (Maltby, 1999). Bioassays are well placed to meet these criteria when compared with ecological surveys.

Of course, even if associations are consistent, responsive and mechanistic, there may not necessarily be a cause-and-effect relationship. However, it could be argued that this is also true of many laboratory experiments on biological responses to environmental parameters. It therefore seems sensible to accept that consistent, responsive and mechanistic associations detected in the field are strongly indicative of a causal relationship. For example, Crane *et al.* (1995) found that cholinesterase activity in caged freshwater amphipods (*Gammarus pulex*) consistently was inhibited immediately below different watercress beds sprayed with organophosphate insecticide. This effect has a well known mechanistic basis and cholinesterase activity can be reduced in the laboratory by exposure to similar compounds. The result of this field study with *in situ* bioassays was that management practices (the installation of settlement ponds) were adopted by watercress growers to prevent pesticide-contaminated water from flowing into nearby chalk streams.

When answering questions about the number, location and frequency of samples, data from earlier studies are usually essential to maximise the cost-effectiveness of sampling. Dixon and Chiswell (1996), in a comprehensive survey of the literature, found only one method in which prior information on water quality was not needed; in this case, sampling locations were determined by an algorithm based upon stream order number (Sharp, 1971). Harmancioglu and Alpaslan (1992) describe how this basic approach can be enhanced by only rather limited information on the number of pollutant discharges and their likely contribution to pollution (e.g. biochemical oxygen demand). It may be possible also to use topographic maps and aerial photography to identify sites most at risk from human impact (Bryce *et al.*, 1999).

In most cases previous information is required, and researchers often need to visit a large number of proposed monitoring sites before using judgement, or a statistically based design, to settle on the best sites for future sampling (Dupont, 1992). This may not be too onerous a task in a small area, but could prove prohibitively expensive if a large area or a large number of smaller areas need to be assessed. However, researchers do not necessarily have to travel over an entire catchment to determine the best sampling sites. Hydrological models coupled with Geographical Information Systems have proved useful in identifying optimal water quality monitoring sites (Rosenthal and Hoffman, 1999).

2.2.4.1 Comparison of ambient samples
Ward and Loftis (1986) recommend the calculation of measures of central tendency, such as arithmetic means, or geometric means if data are highly skewed, and the plotting of these on charts or graphs as the easiest method for communicating differences in environmental quality to the public and environmental managers. Confidence limits should be reported always with measures of central tendency, such as means and lethal concentration (LC)/effective concentration

(EC) toxicity values, so that the reader has some indication of uncertainty in the estimate.

These measures of central tendency can be compared for two samples using a parametric t-test or non-parametric Mann-Whitney test. When there are more than two samples, tests such as the parametric analysis of variance (ANOVA) or non-parametric Kruskal–Wallis test can be used, followed by a multiple comparison test. The power of these tests can be calculated readily and used to determine optimal sample sizes for the future. However, care should be exercised when using ANOVA on survey data, because incorrect models are frequently used (Evans and Coote, 1993). Data from the field often consist of fixed factors chosen by an investigator (e.g. particular sampling points) and random factors that are sampled from a larger population (e.g. sampling date, if samples have to be taken over several days, but the effect of date on the dependent variable is of no particular interest to the investigator). Because of this, mixed-model ANOVAs are often most appropriate for analysing environmental monitoring data (Evans and Coote, 1993).

Because analyses of water quality usually involve the collection of data on several variables, multivariate statistical analyses often are relevant. For example, multivariate analysis of variance (MANOVA) and discriminant analysis were used by Alden (1997) to investigate water quality trends in Chesapeake Bay.

2.2.4.2 Trend analysis

Detection of trends is an objective of many environmental monitoring schemes (e.g. Hurley *et al.*, 1996) and normally will require regular, long-term monitoring of environmental quality at fixed sites (van Belle and Hughes, 1983; Whitfield, 1988). Even when quality varies diurnally at a particular station (e.g. for water or air), it will be best usually to sample at that station at the same time of day if trend detection is the goal (van Belle and Hughes, 1983). Techniques are available to compensate for unequal sampling dates or missing values (Champely and Doledec, 1997). However, Comber and Gardner (1999), in a study of trends in environmental chemical concentrations in water and biota across Europe, found that it was difficult to assess data quality or even to access data that had been collected in different countries. The duration, location and information goals of monitoring programmes also varied across different European countries. The unhelpful diversity of environmental monitoring approaches and the variable quality of results have been widely recognised (Clark *et al.*, 1996), and international environmental quality programmes now often begin with a phase designed to harmonise information gathering activities (e.g. Botterweg and Rodda, 1999).

Initially, spatially and temporally intensive surveys may be needed to establish appropriate sampling locations and frequencies (e.g. Caruso and Ward, 1998). These then can be reduced in number when data redundancy becomes apparent. However, if the goal is to assess long-term trends over large areas involving many sites, intensive surveys are unlikely to be of much practical use in detecting trends and would be a very expensive method for determining sampling sites (van Belle and Hughes, 1983).

Simple time-series plots of water or air quality can reveal a great deal of information (Ward and Loftis, 1986), such as:

- Detection of extreme values
- Trends
- Known and unknown impacts
- Dependencies between observations
- Seasonality
- A need for data transformation
- Heterogeneous trends across stations
- Long-term cycles

Various parametric and non-parametric statistical methods have been recommended for detecting changes over time in environmental time-series data (e.g. Ward and Loftis, 1986; Kwiatkowski, 1991; El-Shaarawi, 1993; Esterby, 1993; Dixon and Chiswell, 1996; Thas et al., 1998; Comber and Gardner, 1999), including multivariate approaches (e.g. Loftis et al., 1991). Many of these, such as the Seasonal Kendall test, are available in easy-to-use general statistical software packages, and their statistical power can be estimated so that optimal sampling can be agreed. The Seasonal Kendall test is used frequently because it is robust when there are non-normal, censored, missing or seasonal data, such as those commonly found in environmental quality monitoring programmes, although only monotonic trends (i.e. trends either upwards or downwards) can be detected with this test (El-Shaarawi, 1993; Alden, 1997; Antelo et al., 1998; Stoddard et al., 1998). Newell et al. (1993) show how this test can be modified for use when changes in sampling or analytical methods appear to prevent analysis over an entire times series, and modifications can be made to account for serially correlated data (Thas, 1999). Thas et al. (1998) provide information on power calculations for several non-parametric time-series techniques.

2.2.4.3 Breach of regulatory limits/compliance monitoring
Although fixed-frequency sampling is likely to be best for detecting longer term trends in environmental quality, it will rarely be the optimal strategy for detecting breaches in regulatory limits, whether chemically or biologically based. Other approaches, such as sequential, Markovian and exceedance-driven sampling, are likely to be more effective (Whitfield, 1988). It may be useful also for an environmental manager to estimate the probability of a breach at a particular location. This will allow evaluation of the level of success in preventing breaches and help to focus resources on priority areas (Ward and Loftis, 1986).

2.2.4.4 Assessment of environmental impact
Optimal strategies for detecting the impacts of particular activities, such as damming, dredging or site development, often need to be decided on a case-by-case basis. However, despite some criticisms (Hurlbert, 1984; Stewart-Oaten et al.,

1992; Smith *et al.*, 1993), most sampling designs involve a Before and After, Control and Impact (BACI) type of design (Green, 1979; Stewart-Oaten *et al.*, 1992). The BACI design requires the selection of similar sites: at least one control and one site that will be impacted. Samples then are taken at both sites before and after the impact (Whitfield, 1988). Underwood (1991) provides examples of appropriate experimental designs for BACI and similar studies. El-Shaarawi (1993) shows how time-series analysis can be used in this context, and Reckhow (1990) provides a Bayesian framework for analysing this type of problem.

If the aim is to detect time-series trends associated with environmentally impacting activities, then samples should be taken at regular intervals before and after the impact, with more samples taken after in case there is a lag in any effect. Of course, in many cases the impact may have occurred already if it was accidental, in which case the best design is that for comparison of ambient samples described above. It may be difficult to find truly independent sites that are sufficiently similar. For example, rivers are often monitored by comparing sites upstream of an impact with sites downstream of the impact. It is arguable whether such sites are truly independent.

In summary, when designing any environmental monitoring strategy, including one with bioassays, considerable advice is available on optimal location and timing of samples. It is no longer necessary simply to rely on tradition or guesswork.

2.3 Quality issues in the use of bioassays

We have addressed issues of ecological relevance and now turn our attention towards measures for ensuring the reliability of test data that are generated from bioassays. This can be thought of in three distinct areas that, together, can make a significant contribution to the reliability of toxicity estimates that emerge from testing, and therefore the robustness of regulatory decisions made as a result:

- Collection, handling and pretreatment of samples prior to testing
- Standardisation of test protocols
- Performance characteristics (particularly variability) of bioassays

2.3.1 *Sample collection, handling and pretreatment*

2.3.1.1 *Sample collection and handling*
It is vital that environmental samples (such as effluents, leachates, receiving waters, sediments and soils) taken for testing with bioassays are considered representative and that the procedures adopted for the collection, storage and preparation of samples ensure that the toxicity of the sample obtained at source does not change markedly before a test is conducted. It is also vital that supporting documentation, in the form of a chain of custody record, accompanies the sample.

Detailed information on procedures for the collection, handling and preservation of aqueous and sediment samples is given in a series of ASTM (1994a,b) and ISO (1990, 1991, 1992a,b, 1994, 1995, 1998a,b, 1999, 2000) documents. General issues for the collection, storage and preparation of environmental samples for bioassays that need to be considered are summarised below:

- Environmental samples for bioassays should be collected in accordance with any existing regulatory procedures.
- Environmental samples should be collected in containers made of materials certified by manufacturers as being inert.
- Measurements of appropriate basic physicochemical properties should be made both in the field at the time when samples are collected and on receipt at the laboratory.
- Samples must be labelled with the name and location (e.g. grid reference), the date and time the sample was taken, the duration of sampling, the initials of the sampler and the number of the chain of custody record forms.
- Samples should be transported to the testing facility within 24 h and testing should commence as soon as possible after collection. Samples must be kept in the dark during transport and the sample temperature should not deviate markedly ($\pm$ 2°C) from that at the time of collection. The temperature history of the samples from collection to arrival at the testing facility should be recorded, ideally using disposable temperature recorders, where appropriate.
- Aqueous samples should be accompanied by triplicate trip blanks (e.g. reference water samples) to allow cross-contamination in transit to be identified.
- Samples of effluents or receiving waters requiring immediate testing should be adjusted to the required temperature for the relevant toxicity test(s) immediately on receipt at the test facility. If the sample is not to be tested immediately, it should be stored in darkness at 2–8°C.
- Initial characterisation studies on aqueous samples (effluents, leachates or receiving waters) should address the issue of sample stability and temporal changes in the toxicity of collected samples.

2.3.1.2 Sample pretreatment
The extent to which environmental samples are treated prior to testing depends on the objectives of the study and should be the subject of discussions between interested parties (e.g. dischargers and regulators). There are two possible approaches:

1. Testing of samples unadjusted in order to gain information on the total biological effects, including the influence of potentially confounding physicochemical parameters such as pH, dissolved oxygen, suspended solids and turbidity, hardness or salinity and colour. This approach could mean that in certain instances it will not be possible to carry out certain methods because

physicochemical parameters will fall outside the limits specified for the procedures (see Table 2.2).
2. Adjusting either the sample or specific test solutions so that all potentially confounding physicochemical parameters specified for a particular method are met (see Table 2.2). Modification of the sample or test solutions will remove the influence of these parameters and reflect residual chemical toxicity.

In selecting an approach there is an issue of how representative a test methodology might be of conditions in the environment and whether the sample was being modified to meet test requirements with a subsequent loss of environmental realism (for example, due to changes in effluent character on adjustment). The influence of physicochemical parameters on aqueous samples typically will be more pronounced for effluents or leachates than receiving waters, where dilution may have occurred. Furthermore, problems of sample or test solution treatment for confounding physicochemical parameters generally will become important only if toxicity occurs at higher effluent concentrations. For samples where toxicity is evident at lower sample concentrations, dilution with reference water often will mean that physicochemical parameters in the test solutions meet the test method criteria.

2.3.2 Test standardisation

Consistency in the way testing is performed is vital if excessive variation in methodology is not to undermine the reliability of the test data that emerge. The next section explains how bias and variability can result, but this can be constrained by identifying critical aspects of a test method and defining how these should be performed. This is clearly of greatest value if it can be done at the stage of test method development and, indeed, often features in the drafting of new test guidelines. As well as describing how a test is to be performed, these test guidelines typically advise on sources of test organisms and minimum performance criteria for control responses.

Such test guidelines often form a key part of a laboratory's formal quality assurance procedure (e.g. Good Laboratory Practice or other formal accreditation schemes) and enable test results to be audited. However, to be fully effective these quality assurance procedures need to take account of the performance characteristics of the test (quality control). Such considerations form the basis of the following section.

2.3.3 Variability in bioassay data

If the same bioassay is performed on a sample of a substance, waste or environmental medium on a number of occasions, a range of toxicity estimates would be generated. The same would be true if different laboratories were asked to perform

Table 2.2 Threshold criteria for different physicochemical parameters in test solutions in various test procedures.

Test procedure	Threshold criteria for different physicochemical parameters in test solutions					
	pH	Dissolved oxygen	Hardness (mg CaCO$_3$/l)	Salinity (‰)	Suspended solids (mg/l)	Colour
Algal growth inhibition	7.4–8.5	NA[a]	NA	>27–36	<2[b]	No value; can affect growth
D. magna immobilisation	7.4–8.5	≥60% ASV[c] in lowest concentration causing 100% immobilisation	140–250	—	<20	No value; can affect ability to observe organisms
Oyster embryo–larval development test	7.8–8.5	≥60% ASV in lowest concentration causing 100% immobilisation	—	≥22–36	<20	No value; can affect ability to count organisms
Marine copepod lethality	7.8–8.5	≥50% ASV in lowest concentration causing 100% lethality	—	≥20–36	<20	No value; can affect ability to observe organisms
Fish lethality	6.0–8.5	≥60% ASV in lowest concentration causing 100% lethality	10–250	≥27–36	<25	NA

[a] NA, not applicable.
[b] USEPA recommend filtering samples for algal growth inhibition tests through 0.45 μm membrane filters.
[c] ASV, air-saturated value.

a test, some variability between replicates is to be expected as a result of individuals or groups of individuals varying in their sensitivity to the material being tested.

This variability is not unique to biological testing and is well understood in many industrial activities and chemical analyses. Indeed, substantial resources are devoted to measuring and controlling variability in outputs in these sectors. However, the same cannot be said of toxicity testing. In the following sections we consider why variability in aquatic bioassays is important and what can be done to constrain it within reasonable bounds.

2.3.3.1 How does variability arise?
In chemical analysis we can assess the accuracy of a method by comparing the measured result against known standards of known concentration, i.e. if we dissolve 1 mg of a substance in a litre of water we can calibrate our analysis to estimate a concentration of 1 mg/l. Thus the *accuracy* of the analysis can be determined by the degree of agreement between the measured value and the 'true' value. An accurate measurement is one in which there is a lack of bias or systematic error. Accuracy in bioassays, on the other hand, is impossible to assess because the measured toxicity is determined empirically. There is no 'true' underlying toxicity that can be used as a reference point. In this case, the only realistic way of understanding the underlying toxicity of a substance in a particular test (and thereby understanding how accurate a particular toxicity measurement is) is to make many estimations, ideally in different laboratories. The overall mean then may be taken as an estimate of the sample's underlying toxicity in this particular test. The more tests that are performed, the better this estimate should be.

Another important characteristic is that of *precision*. This becomes evident only when repeat measurements are made, because precision refers to the amount of agreement between repeated measurements (the standard deviation around the mean estimate). Precision is subject to both random and systematic errors. In industrial quality control and chemical analysis, Shewhart Control Charts provide a means of assessing the precision of repeat measurements but these approaches are rarely used in ecotoxicity testing. The effect is that we generally understand little about either the accuracy or the precision of most bioassays.

Fulk (1995) usefully identifies two types of variability (random and systematic error) and three ways in which this may become manifest (within-test, within-laboratory and between laboratories). In a conventional concentration–response experiment, the random occurrence of variability *within* an experiment will give rise to error around the fitted regression line, i.e. it will make the estimate of toxicity less precise. By contrast, non-random occurrence of these factors *between* experiments or laboratories can result in different estimates of toxicity, leading to *bias*. However, testing on different occasions or in different laboratories is also subject to random errors. As a result, the variability that we see is usually a combination of random and systematic errors. Variability resulting from random errors is, by definition, difficult to address but systematic errors can be investigated by reviewing current practices and comparing them with test protocols or practices adopted in other laboratories.

2.3.3.2 Why does variability matter?

Variability in ecotoxicity testing can have important implications, especially if we are using the data to make a decision about the acceptability (or otherwise) of a chemical, site, waste stream or process on the basis of the result. Thus, whenever we compare the outcome of a test with a threshold, there is a chance that a sample will appear to have exceeded the threshold simply because, on a particular occasion, the test yielded higher toxicity than the norm. Conversely, a bioassay may produce a result that indicates lower toxicity than the threshold, but only because the test on this occasion yielded an EC_{50} that was particularly high. In both cases, the same test on another occasion, or the same test performed by another laboratory, could result in a different outcome.

The results from toxicity tests and bioassays are usually compared with a predefined threshold, such as a compliance limit or classification threshold. When toxicity lies comfortably away from the threshold there is unlikely to be much doubt about which side of the threshold it lies. Put another way, the level of variability needs to be much higher to result in misclassification. However, the risk of misclassification increases as the underlying ('true') toxicity of a material approaches the threshold or where the test method gives rise to a particularly wide range of possible values. Although test regimes tend to involve the use of a battery of test species, often not all these will be used in the classification of a sample. Unless a weight-of-evidence approach is used, the determination of risk or hazard classification frequently is based on the most sensitive test datum and a simple face-value comparison with the threshold defining an unacceptable risk or separating categories of environmental quality.

We are not arguing that bioassay data must be completely free of systematic and random error. Indeed, if the degree of variability is known, this information may be incorporated into the decision-making process. It can be used to minimise the incidence of false negatives, by narrowing the margin between the measured toxicity and the limit value to a point where the chances of a false negative are reduced to some predefined level. The effect of such a strategy is to accord the 'benefit of the doubt' to the environment. Alternatively, in the case of an emission to which a toxicity limit has been assigned, the risk assessor may wish to accord the benefit of the doubt to a discharger by minimising the incidence of false positives. In this case, the margin between the measured toxicity and the limit value is effectively widened. For a fuller discussion, the reader is referred to Warren-Hicks and Parkhurst (1995).

However, this strategy does little more than 'make the best of a bad job' and, in any case, requires an investment in understanding the degree of variability present. We argue that, if such information is to be gathered, it may be used more effectively to achieve some *control* over variability rather than merely making *allowances* for it.

To summarise, it is clear that:

- Currently, the degree of accuracy and precision actually achieved in bioassays is largely unknown, and these performance characteristics matter.
- The accuracy and precision of test data *can* be constrained and this will improve regulatory decision-making.

2.3.3.3 How much variability is there?

We have seen that variability in toxicity testing can arise from repeat measurements made within a laboratory and also between laboratories. In reality, the variability seen between laboratories is a consequence of both within- and between-laboratory sources of variability, and both are also subject to the within-test variability referred to earlier, as evident from differences between test replicates. Research based on a series of acute aquatic toxicity tests (Whitehouse *et al.*, 1996) shows that variation between laboratories is higher than that between repeat tests in the same laboratory. This, in turn, accounts for more variability than that seen between replicates within a test. Similar findings are evident from the work of others in connection with the introduction of whole-effluent toxicity tests in the USA (e.g. Warren-Hicks and Parkhurst, 1992; Fulk, 1995). Over the years, a number of authors have examined variability in aquatic toxicity testing. Typically these describe variability in terms of the coefficient of variation (standard deviation divided by the mean) in EC_{50} or LC_{50} values that is achieved when the same toxicant is tested several times (or by several laboratories) using the same method. Table 2.3 summarises the results of a review of published data.

In 1995, a more detailed study was carried out, based on the widely used *Daphnia magna* 48-h immobilisation test. When a series of laboratories across Europe were asked to perform this test according to OECD Test Guideline 202 using a subsample of the same batch of 3,4-dichloroaniline as a reference toxicant, a range of 48-h EC_{50} values were reported (Figure 2.1). An overall mean EC_{50} of 0.83 mg/l was found but, even with this well-established test method, the range of 48-h EC_{50} values for 3,4-dichloroaniline ranged between 0.14 and 2.51 mg/l. It is clear also that variability is relatively low in highly toxic or non-toxic media and is highest at concentrations near the EC_{50} or LC_{50} (Parkhurst *et al.*, 1992; Warren-Hicks and Parkhurst, 1992). The examples given in Figure 2.1 may therefore represent 'worst case' scenarios.

Other test methods, in which laboratories had less experience, tended to generate higher levels of variability. Interestingly, variability both within and between laboratories was smallest for a 15-min IC_{50} derived from the Microtox® test – a rapid bioluminescence assay based on the marine bacterium *Vibrio fischeri*. For comparative purposes, 15-min IC_{50} data for 3,4-dichloroaniline are also shown in Figure 2.1. Unlike most of the other tests evaluated, the Microtox® methodology

Table 2.3 Summary of variability in toxicity resulting from testing single chemicals in aquatic toxicity tests (at least six data are required)[a].

Test organism	Source of variability	No. of studies	Coefficients of variation		
			25%	50%	75%
Microtox® (15-min EC_{50})	Within test	29	2	4	18
	Between test	3		Insufficient data	
	Between laboratory	7	19	22	43
Pacific oyster embryo larvae (24-h EC_{50})	Within test	0		No data	
	Between test	3		Insufficient data	
	Between laboratory	4		Insufficient data	
Acartia tonsa (48-h LC_{50})	Within test	0		No data	
	Between test	16[b]	4	17	33
	Between laboratory	4		Insufficient data	
Tisbe battaglia (48-h LC_{50})	Within test	0		No data	
	Between test	6	5	13	21
	Between laboratory	3		Insufficient data	
Daphnia magna (48-h EC_{50})	Within test	45	1	4	7
	Between test	47	10	18	33
	Between laboratory	18	15	26	46
Selenastrum capricornutum (72-h EC_{50})	Within test	0		No data	
	Between test	4		Insufficient data	
	Between laboratory	1		Insufficient data	
Skeletonema costatum (72-h EC_{50})	Within test	0		No data	
	Between test	6	2	14	32
	Between laboratory	3		Insufficient data	
Phaeodactylum tricornutum (72-h EC_{50})	Within test	0		No data	
	Between test	5		Insufficient data	
	Between laboratory	3		Insufficient data	
Rainbow trout (96-h LC_{50})	Within test	0		No data	
	Between test	10	10	17	29
	Between laboratory	11	39	58	98
Turbot (96-h LC_{50})	Within test	0		No data	
	Between test	5		Insufficient data	
	Between laboratory	3		Insufficient data	

[a] This table is based on studies by the following:
Anderson and Norberg-King (1991)
Baird et al. (1991)
Bjornestad and Petersen (1992)
Buikema (1983)
Butler et al. (1992)
Canton and Adema (1978)
Casseri et al. (1983)
Curtis et al. (1982)
Davis and Hoos (1975)
De Zwart and Sloof (1983)
Dixon and Sprague (1981)
Environment Canada (1990)
Gersich et al. (1986)
Hansen et al. (1979)
Horning and Weber (1985)
ISO (1989)
Lemke (1981)
Lewis and Horning (1991)
Lewis and Weber (1985)
Qureshi et al. (1982)
Walker (1988)

[b] It is not clear whether these studies generate information on within-test or within-laboratory variability.

is described in detail in a manufacturer's manual and gives relatively little room for flexibility in how the test is performed or how the data are manipulated to estimate toxicity.

2.3.3.4 Sources of variability
Variability in toxicity measurements is recognised at every level of biological organisation, from the subcellular (Ratner and Fairbrother, 1991) to the community level (Taub *et al.*, 1989). Differences in the response of individuals to pollutants may be due to environmental or genetic factors or a combination of the two (Hoffmann and Parsons, 1991). The total variability in the response of an organism to a chemical (V_p) can be represented by the equation:

$$V_p = V_g + V_e + V_{ge}$$

(Baird *et al.*, 1989)

where: V_g = variability due to genetic heterogeneity
V_e = variability due to environmental heterogeneity
V_{ge} = variability due to interaction between genetic and environmental factors

Taking the 48-h *Daphnia magna* immobilisation test as an example, V_g should be small because this species reproduces parthenogenetically. Although clonal differences can lead to significant differences in sensitivity (Baird *et al.*, 1990, 1991), in practice, test laboratories maintain a relatively limited range of clones (Forbes and DePledge, 1992). Various environmental factors have been suggested or shown to influence the sensitivity of *Daphnia* to toxicants (V_e), including culture medium, diet type and ration, temperature, water hardness, light intensity and photoperiod. Many of these were investigated as part of a programme to update the test guideline for the OECD 21-day *Daphnia magna* reproduction test (Sims *et al.*, 1993), but the contribution made by many of these factors remains unclear. It is thus perhaps not surprising that interactions between these factors (V_{ge}) remain almost completely unknown.

Forbes and DePledge (1992) add a third consideration, that of experimental error. Our experience of ring-testing would tend to support this as an important consideration: there is evidence to indicate that greater precision is achieved by the more experienced laboratories and that closely defined protocols, such as the Microtox® example mentioned earlier, are associated with greater precision than those where there is a high degree of latitude in the test protocol (Whitehouse *et al.*, 1996).

2.3.3.5 How much variability is acceptable?
Clearly there can come a point when estimates of toxicity are so imprecise or subject to bias that any conclusions drawn from them are meaningless. At the same time, it would be inappropriate to devote resources to constraining variability unless it was going to benefit decision-making in risk assessment.

(a) *Daphnia magna* immobilisation

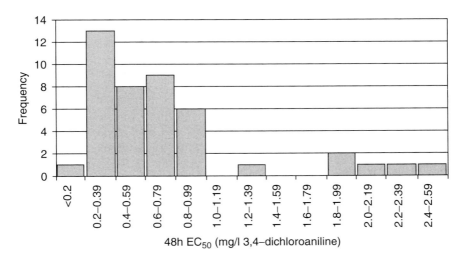

(b) Microtox® (*vibrio fischeri*)

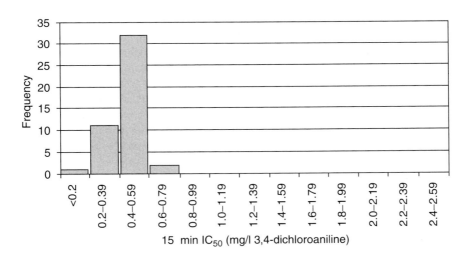

Figure 2.1 Probability density functions for EC_{50}/IC_{50} values derived from an interlaboratory ring-test for *Daphnia* and Microtox® tests using 3,4-dichloroaniline as a reference toxicant.

A complete absence of any constraint on variability would seem to be irresponsible. The development of test protocols is an important contribution to constraining variability because it limits the latitude for systematic variation between operators. However, the resulting variability within or between laboratories remains unknown. As we explain below, variability can be monitored and, if necessary, controlled by defining acceptable levels of variability based on those achieved by experienced laboratories.

2.3.3.6 How can variability be controlled?

A conventional response to issues of variability in bioassays is to construct Shewhart Control Charts based on the results achieved in repeat tests within a laboratory using a reference toxicant. This effectively describes the range of results typically found within the laboratory and hence can be used to define limits within which the laboratory normally expects to operate. However, there is a flaw in such 'internal' quality control because the more variable a laboratory's reference toxicant test results are, the wider the limits of acceptability will be. Indeed, it can serve merely to reinforce high variability or bias.

Instead, a process of 'external' quality control is favoured. The accuracy and precision of ecotoxicity tests may be defined and monitored by defining limits on the amount of bias that is considered acceptable and also on the amount of variability between repeat tests. These limits can be estimated from ring-testing involving laboratories that are experienced in the test method and therefore can be regarded as reasonably proficient but are 'imposed' rather than being generated by individual laboratories. Furthermore, they must use the same reference toxicant (sometimes referred to as a 'positive control'). The resulting conclusions are applicable only to this toxicant because it is clear that different toxicants yield different levels of within- and between-laboratory variability (Whitehouse *et al.*, 1996). It follows that subsequent monitoring also must be based on the same toxicant.

Such an approach was trialled in the late 1990s as part of an initiative by the Environment Agency in the UK. It led to proposals for a quality control scheme for use within the Agency's Direct Toxicity Assessment programme for effluent assessment and control. The trial entailed an interlaboratory ring-test of four test methods using two toxicants (3,4-dichloroaniline and zinc sulphate) from which accuracy and precision criteria were estimated, as described below.

2.3.3.7 Defining limits for accuracy

Control limits may be applied to the EC_{50} estimates, defining the range within which a given proportion (say 95%) of values would be expected to fall. This requires an estimate of the underlying toxicity of the toxicant using that test method (the 'consensus' mean) and the variance for reproducibility, which can be estimated from residual maximum likelihood analysis of the ring-test data. This analysis yields an estimate of $\sigma_{reproducibility}$, the sum of the variances attributable to within-test, within-laboratory and between-laboratory sources of error. It is then a simple matter to define the control limits for accuracy as:

95% Control limit for accuracy = 'Consensus' mean $\pm 1.96 \times \sigma_{reproducibility}$

We can then calculate the limits, expressed as concentrations of the reference toxicant, within which we would expect 95% of EC_{50} estimates to lie.

2.3.3.8 Defining limits for precision

Residual maximum likelihood analysis of the ring-test data also provides an estimate of the underlying repeatability of each test method ($\sigma^2_{repeatability}$). As part of a quality control scheme, this can be compared with variances estimated from the EC/LC_{50} values derived from a series of repeat tests carried out within a laboratory using the same toxicant (S^2) and using a conventional χ^2 test, as shown below:

$$S^2/\sigma^2_{repeatability}(n-1) < \chi^2_{(n-1,\,\alpha)}$$

If S^2 is significantly greater than $\sigma^2_{repeatability}$ for that test method, excessive variability between repeat tests is indicated and an investigation of possible sources of error might be prompted to restore within-laboratory variability to an acceptable level.

Once the control criteria have been developed, laboratories can carry out a series of ecotoxicity tests with the reference toxicant. Their data then may be used to monitor both the risks of excessive bias and lack of precision (as illustrated in Figure 2.2) and, if necessary, problems highlighted at an early stage. Formal implementation of this approach might require laboratories to submit reference toxicant data for evaluation and to take remedial action when data fall outside specified limits. Although quality control limits derived in this way are simply based on the results obtained by a group of laboratories in ring-testing, the accumulation of reference toxicant data will help to refine the limits over time.

In this example, each laboratory has performed six tests. Laboratories A, C, D, F and G would meet the criteria for both within- and between-laboratory variability. By contrast, data from laboratory J would be acceptable in terms of the between-laboratory criterion but the poor agreement between EC_{50} values would marginally fail the criterion for within-laboratory variability. In this example, data from laboratories H and I exhibit poor repeatability and also fail the criterion for between-laboratory variability.

2.3.3.9 Test method development and the derivation of quality control criteria

The development of new test guidelines or the revision of existing ones provides an opportunity to address quality control considerations as well as the more conventional considerations concerned with the technical conduct of the tests. Although ring-testing is frequently involved, rarely is it designed in such a way

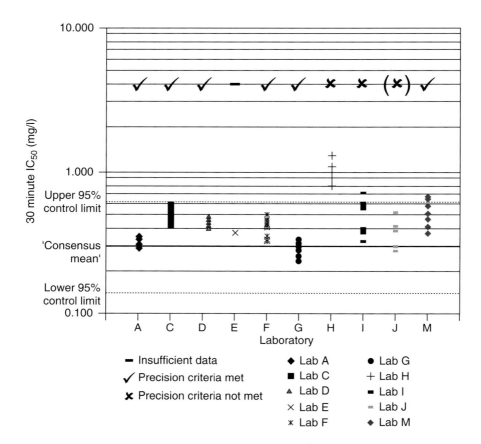

Figure 2.2 Different laboratory performances for 30-min Microtox® tests with 3,4-dichloroaniline: (✓) acceptable precision; (−) marginal acceptability; (×) excessive variability between repeat tests. Values falling between upper and lower 95% confidence intervals indicate acceptable control over between-laboratory variation.

that it will help to define the sort of control criteria outlined here. However, the development of a draft OECD Test Guideline for toxicity testing with aquatic plants (OECD 221) (Sims et al., 2001) and, more recently, the development of an AFNOR standard for a *Brachionus calyciflorus* 48-h toxicity test (Thiebaud and Thomas, 2000) did entail such ring-tests, enabling an examination of the between- and within-laboratory variability associated with these tests.

In our view, such considerations ought to form a more prominent aspect of test guideline development, ideally leading to the inclusion of these important performance characteristics within the test validity criteria. The results thus obtained can then be treated with greater confidence, and the uncertainty in the regulatory decisions based on these data would be reduced.

2.4 Summary

All environmental monitoring tools, including bioassays, must be embedded within a coherent monitoring framework if results are to be related meaningfully to monitoring objectives. Biological, chemical and statistical criteria must be considered before designing such programmes.

The reliability of toxicity estimates that emerge from testing, and therefore the robustness of regulatory decisions made as a result, depends on a number of elements, namely:

- Collection, handling and pretreatment of samples prior to testing all ensure that the toxicity of the sample obtained at source does not change markedly before a test is conducted.
- Standardisation of test protocols to ensure that elements of a procedure that are most critical in terms of its performance are identified and testing is carried out in a consistent manner within a laboratory at different times and also between laboratories.
- Performance characteristics (particularly variability) of bioassays.

References

Alden III, R.W. (1997) Dynamics of an estuarine ecosystem – the Chesapeake Bay experience: statistical approaches and water quality patterns. *Oceanologica Acta*, **20**, 55–69.

Anderson, S.L. and Norberg-King, T.J. (1991) Precision of short-term chronic toxicity tests in the real world. *Environmental Toxicology and Chemistry*, **10**, 143–145.

Ankley, G.T. (1995) Laboratory v. field measurement endpoints: a contaminated sediment perspective. In *Ecological Risk Assessment of Contaminated Sediments*, Ingersoll, C.G., Dillon, T. and Biddinger, G.R. (eds), pp. 115–122. SETAC Press, Pensacola, FL.

Antelo, J.M., Arce, F. and Carballeira, J.L. (1998) Evaluation of sampling frequency for the water quality monitoring network of coastal Galicia (northwest Spain). *Water Environment Research*, **70**, 1327–1329.

ASTM (1994a) *Standard Guide for Collection, Storage, Characterization, and Manipulation of Sediments for Toxicological Testing*, E 1391–94. American Society for Testing and Materials, West Conshohocken, PA.

ASTM (1994b) *Standard Guide for Designing Biological Tests with Sediments*, E 1525–94a. American Society for Testing and Materials, West Conshohocken, PA.

Baird, D.J., Barber, I., Bradley, M. Calow, P. and Soares, A.M.V.M. (1989) The *Daphnia* bioassay: a critique. *Hydrobiologia*, **188/189**, 403–406.

Baird, D.J., Barber, I. and Calow, P. (1990) Clonal variation in general responses of *Daphnia magna* Strauas to toxic stress. I. Chronic life-history effects. *Functional Ecology*, **4**, 399–407.

Baird, D.J., Barber, I., Bradley, M., Soares, A.M.V.M and Calow, P. (1991) A comparative study of genotype sensitivity to acute toxic stress using clones of *Daphnia magna* Straus. *Ecotoxicology and Environmental Safety*, **21**, 257–265.

Bjornestad, E. and Petersen, G. (1992) *Paris Commission Ring Test. Intercalibration and Comparison of Tests and Test Laboratories for Evaluation and Approval of Offshore Chemicals and Drilling Muds*. Water Quality Institute, Horsholm, Denmark.

Botterweg, T. and Rodda, D,W. (1999) Danube river basin: progress with the environmental programme. *Water Science and Technology*, **40**, 1–8.

Boyer, J.N., Sterling, P. and Jones, R.D. (2000) Maximising information from a water quality monitoring network through visualization techniques. *Estuarine Coastal and Shelf Science*, **50**, 39–48.

Bryce, S.A, Larsen, D.P., Hughes, R.M. and Kaufmann, P.R. (1999) Assessing relative risks to aquatic ecosystems: a mid-Appalachian case study. *Journal of the American Water Resources Association*, **35**, 23–36.

Brydges, T. and Lumb, A. (1998) Canada's ecological monitoring and assessment network: where we are at and where we are going. *Environmental Monitoring and Assessment*, **51**, 595–603.

Buikema Jr., A.L. (1983) *Inter- and Intra-laboratory Variation in Conducting Static Acute Toxicity Tests with Daphnia magna Exposed to Effluents and Reference Toxicants*, API Publications 4362. American Petroleum Institute, Washington, DC.

Butler, R., Johnson, I., Horn, K. and Paul, M. (1992) *Discharge Control by Direct Toxicity Assessment (DTA): Evaluation of a Protocol by Case Studies*, WRc Interim Report to the NRA, R&D 049/9/W. National Rivers Authority, Bristol.

Cairns Jr., J., McCormick, P.V. and Niederlehner, B.R. (1993) A proposed framework for developing indicators of ecosystem health. *Hydrobiologia* **263**, 1–44.

Calow, P. (1989) The choice and implementation of environmental bioassays. *Hydrobiologia*, **188/189**, 61–64.

Calow, P. (1993) Seeking standardisation in ecotoxicology. In *Progress in Standardization of Aquatic Toxicity Tests*, Soares, A.M.V.M and Calow, P. (eds), Lewis Publications, Michigan.

Canton, J.H. and Adema, D.M.M. (1978) Reproducibility of short-term and reproduction toxicity experiments with *Daphnia magna* and comparison of the sensitivity of *Daphnia magna* with *Daphnia pulex* and *Daphnia cucullata*, in short-term experiments. *Hydrobiologia*, **59**, 135–140.

Caruso, B.S. and Ward, R.C. (1998) Assessment of nonpoint source pollution from inactive mines using a watershed-based approach. *Environmental Management*, **22**, 225–243.

Casseri, N.A., Ying, W.C. and Sojka, S.A. (1983) Use of a rapid bioassay for assessment of industrial wastewater treatment effectiveness. In *Proceedings of the 38th Purdue Industrial Wastewater Conference*, pp. 867–878. Butterworth Publishers, Woburn, MA.

Champely, S. and Doldedec, S. (1997) How to separate long-term trends from periodic variation in water quality monitoring. *Water Research* **31**, 2849–2857.

Clark, M.J.R, Laidlaw, M.C.A, Ryneveld, S.C. and Ward, M.I. (1996) Estimating sampling variance and local environmental heterogeneity for both known and estimated analytical variance. *Chemosphere*, **32**, 1133–1151.

Comber, S. and Gardner, M. (1999) An assessment of trends in European environmental data for mercury and chlorinated organic compounds in water and biota. *Science of the Total Environment*, **243/244**, 193–201.

Crane, M., Delaney, P., Parker, P., Walker, C. and Watson, S. (1995) The effect of Malathion 60 on *Gammarus pulex* (L.) below watercress beds. *Environmental Toxicology and Chemistry*, **14**, 1181–1188.

Crawford, C.G., Slack, J.R. and Hirsch, R.M. (1983) *Nonparametric Tests for Trends in Water Quality Data using the Statistical Analysis System*. United States Geological Survey Report, 83–550.

CSTEE (1998) *Opinion on Some Practical Implications of the Proposed Modifications to the Water Framework Directive (Annex V)*. Scientific Committee for Toxicity, Ecotoxicity and the Environment Opinion Adopted by Written Procedure, 5 June, 1998 CSTEE, Brussels.

Curtis, C., Lima, A., Lozano, S.J. and Veith, G.D. (1982) Evaluation of a bacterial bioluminescence bioassay as a method for predicting acute toxicity of organic chemicals to fish. In *Aquatic Toxicology and Hazard Assessment: Fifth Conference, ASTM STP 766*, Pearson, J.G., Foster, R.B. and Bishop, W.E. (eds), pp. 170–178. American Society for Testing and Materials, Philadelphia, PA.

Davis, J.C. and Hoos, R.A.W. (1975) Use of sodium pentachlorophenate and dehydroabietic acid as reference toxicants for salmonid bioassays. *Journal of the Fisheries Resources Board of Canada*, **32**, 411–416.

De Zwart, D. and Slooff, W. (1983) 'Microtox®' as an alternative assay in the acute toxicity assessment of water pollutants. *Aquatic Toxicology*, **4**, 129–138.

Dixon, W. and Chiswell, B. (1996) Review of aquatic monitoring program design. *Water Research*, **30**, 1935–1948.

Dixon, D.G. and Sprague, J.B. (1981) Acclimation to copper by rainbow trout (*Salmo gairdneri*) – a modifying factor in toxicity. *Canadian Journal of Fisheries and Aquatic Sciences*, **38**, 880–888.

Dupont, J. (1992) Quebec lake survey: I. Statistical assessment of surface water quality. *Water Air and Soil Pollution*, **61**, 107–124.

Ellis, J.C. (1989) *Handbook on the Design and Interpretation of Monitoring Programmes*, Publication NS 29. Water Research Centre, Medmenham, UK.

El-Shaarawi, A.H. (1993) Environmental monitoring, assessment and prediction of change. *Environmetrics* **4**, 381–398.

Environment Canada (1990) *Guidance Document on Control of Toxicity Test Precision using Reference Toxicants*, Report EPS 1/RM/12. Environment Canada, Ottawa, Ontario.

Esterby, S.R. (1993) Trend analysis methods for environmental data. *Environmetrics*, **4**, 459–481.

Evans, J.C. and Coote, B.G. (1993) Matching sampling designs and significance tests in environmental studies. *Environmetrics*, **4**, 413–437.

Forbes, V.E. and Depledge, M.H. (1992) Predicting population response to pollutants: the significance of sex. *Functional Ecology*, **6**, 376–381.

Fulk, F.A. (1995) Whole effluent toxicity testing variability: a statistical perspective. In *Whole Effluent Toxicity Testing: An Evaluation of Methods and Prediction of Receiving Systems*, Grothe, D.R., Dickson, K.L. and Reed-Judkins, D.K. (eds), SETAC Special Publication Series, pp. 172–179. SETAC Press, Persicola, FL.

Gersich, F.M., Blanchard, F.A., Applegarth, S.L. and Park, C.N. (1986) The precision of daphnid (*Daphnia magna* Straus, 1820) static acute toxicity tests. *Archives of Environmental Contamination and Toxicology*, **15**, 741–749.

Green, R.H. (1979) *Sampling Design and Statistical Methods for Environmental Biologists*. John Wiley, New York.

Hansen, R.J., Courtois, L.A., Espinosa, L.R. and Wiggins, A.D. (1979) *Acute Toxicity Bioassays. Examination of Freshwater Fish Species*, California Water Resources Control Board Publication No. 64.

Harmancioglu, N.B. and Alpaslan, N. (1992) Water quality monitoring network design: a problem of multi-objective decision making. *Water Resources Bulletin*, **28**, 179–192.

Hazleton, C. (1998) Variations between continuous and spot-sampling techniques in monitoring a change in river-water quality. *Journal of the Chartered Institute of Water and Environmental Management*, **12**, 124–129.

Hill, I.R., Matthiessen, P. and Heimbach, F. (1993) *Guidance Document on Sediment Toxicity Tests and Bioassays for Freshwater and Marine Environments*. Society of Environmental Toxicology and Chemistry – Europe Report from the Workshop on Sediment Toxicity Assessment, Renesse, The Netherlands. SETAC Press, Brussels.

Hipel, K.W. and McLeod, A.I. (1994) *Time Series Modelling of Environmental and Water Resources Systems*. Elsevier, Amsterdam.

Hoffmann, A.A. and Parsons, P.A. (1991) *Evolutionary Genetics and Environmental Stress*. Oxford University Press, Oxford.

Horner, R.R., Mar, B.W., Reinhelt, L.E., Richey, J.S. and Lee, J.M. (1986) *Design of Monitoring Programs for Determination of Ecological Change Resulting from Nonpoint Source Water Pollution in Washington State*. Washington State Department of Ecology, Olympia, WA.

Horning, W.B. and Weber, C.I. (1985) *Short-term Methods for Estimating the Chronic Toxicity of Effluents and Receiving Waters to Freshwater Organisms*, EPA 600/4-85-014. United States Environmental Protection Agency, Cincinnati, OH.

Hurlbert, S.H. (1984) Pseudoreplication and the design of ecological field experiments. *Ecological Monographs*, **54**, 187–211.

Hurley, M.A., Currie, J.E., Gough, J. and Butterwick, C. (1996) A framework for the analysis of harmonised monitoring scheme data for England and Wales. *Environmetrics* **7**, 379–390.

ISO (1989) *Water Quality – Determination of the Inhibition of the Mobility of Daphnia magna Straus* (Cladocera, Crustacea), ISO 6341. International Organization for Standardization, Geneva.
ISO (1990) *Water Quality – Sampling – Part 6: Guidance on Sampling of Rivers and Streams*, ISO 5667–6. International Organization for Standardization, Paris.
ISO (1991) *Water Quality – Sampling – Part 2: Guidance on Sampling Techniques*, ISO 5667–2. International Organization for Standardization, Paris.
ISO (1992a) *Water Quality – Sampling – Part 9: Guidance on Sampling from Marine Waters*, ISO 5667–9. International Organization for Standardization, Paris.
ISO (1992b) *Water Quality – Sampling – Part 10: Guidance on Sampling of Waste Waters*, ISO 5667–10. International Organization for Standardization, Paris.
ISO (1994) *Water Quality – Sampling – Part 3: Guidance on the Preservation and Handling of Samples*, ISO 5667–3. International Organization for Standardization, Paris.
ISO (1995) *Water Quality – Sampling – Part 12: Guidance on Sampling of Bottom Sediments*, ISO 5667–12. International Organization for Standardization, Paris.
ISO (1998a) *Water Quality – Sampling – Part 14: Guidance on Quality Assurance of Environmental Water Sampling and Handling*, ISO 5667–14. International Organization for Standardization, Paris.
ISO (1998b) *Water Quality – Sampling – Part 16: Guidance on Biotesting of Samples*, ISO 5667–16. International Organization for Standardization, Paris.
ISO (1999) *Water Quality – Sampling – Part 15: Guidance on Preservation and Handling of Sludge and Sediment Samples*, ISO 5667–15. International Organization for Standardization, Paris.
ISO (2000) *Water Quality – Sampling – Part 19: Guidance on Sampling of Sediments in the Marine Environment*, ISO 5667–19. International Organization for Standardization, Paris.
Kwiatkowski, R.E. (1991) Statistical needs in national water quality monitoring programmes. *Environmental Monitoring and Assessment*, **17**, 253–271.
Lemke, A.E. (1981) *Interlaboratory Comparison. Acute Testing Set*, EPA 600/3-81-005. United States Environmental Protection Agency, Office of Pesticides and Toxic Substances, Washington, DC.
Lewis, P.A. and Horning II, W.B. (1991) Differences in acute toxicity test results of three reference toxicants on *Daphnia* at two temperatures. *Environmental Toxicology and Chemistry*, **10**, 1351–1357.
Lewis, P.A. and Weber, C.I. (1985) A study of the reliability of *Daphnia* acute toxicity tests. In *Aquatic Toxicology and Hazard Assessment: Seventh Symposium, ASTM STP 854*, Cardwell, R.D., Purdy, R. and Bahner, R.C. (eds), pp. 73–86. American Society for Testing and Materials, Philadelphia, PA.
Loftis, J.C., Ward, R.D., Phillips, R.D. and Taylor, C.H. (1989) *An Evaluation of Trend Detection Techniques for Use in Water Quality Monitoring Programs*, EPA/600/3-89/037. US Environmental Protection Agency, Cincinnati, OH.
Loftis, J.C., Taylor, C.H., Newell, A.D. and Chapman, P.L. (1991) Multivariate trend testing of lake water quality. *Water Resources Bulletin*, **27**, 461–473.
Maltby, L. (1999) Studying stress: the importance of organism-level responses. *Ecological Applications*, **9**, 431–440.
Mar, B.W., Mitter, W.S., Palmer, R.N., Carpenter, R.A. (1987) *Cost-effective Data Acquisition. Guidelines for Surveying and Monitoring Watersheds*, Workshop Report 1. East-West Environment and Policy Institute.
Newell, A.D., Blick, D.J. and Hjost, R.C. (1993) Testing trends when there are changes in methods. *Water, Air and Soil Pollution*, **67**, 457–468.
Parkhurst, B.R., Warren-Hicks, W. and Noel, L.E. (1992) Performance characteristics of effluent toxicity tests: summarization and evaluation of data. *Environmental Toxicology and Chemistry*, **11**, 771–791.
Pegram, G.C., Görgens, A.H.M. and Ottermann, A.B. (1997) A framework for addressing the information needs of catchment water quality management. *Water SA*, **23**, 13–20.
Qureshi, A.A., Flood, K.W., Thompson, S.R., Janhurst, S.M., Inniss, C.S. and Rokosh, D.A. (1982) Comparison of a luminescent bacterial test with other bioassays for determining toxicity of pure compounds and complex effluents. In *Aquatic Toxicology and Hazard*

Assessment: Fifth Conference, ASTM STP 766, Pearson, J.G., Foster, R.B. and Bishop, W.E. (eds). American Society for Testing and Materials, Philadelphia, PA.
Ratner, B.A. and Fairbrother, A. (1991) Biological variability and the influence of stress on cholinesterase activity. In *Cholinesterase-inhibiting Insecticides: Their Impact on Wildlife and the Environment*, Mineau, P. (ed.), pp. 89–107. Elsevier, Amsterdam.
Reckhow, K.H. (1990) Bayesian inference in non-replicated ecological studies. *Ecology*, **71**, 2053–2059.
Rosenthal, W.D. and Hoffman, D.W. (1999) Hydrological modelings/GIS as an aid in locating monitoring sites. *Transactions of the American Society of Agricultural Engineers*, **42**, 1591–1598.
Sharp, W.E. (1971) A topographically optimum water sampling plan for rivers and streams. *Water Resources Research*, **7**, 1641–1646.
Sims, I.R., Watson, S. and Holmes, D. (1993) Toward a standard *Daphnia* juvenile production test. *Environmental Toxicology and Chemistry*, **12**, 2053–2058.
Sims, I.R., Whitehouse, P., Forrow, D. and Smrchek, J. (2001) Development and performance characteristics of a draft OECD *Lemna* growth inhibition test. Poster presentation to SETAC Europe 11th Annual Meeting, Leipzig.
Smith, E.P., Orvos, D.R. and Cairns, J. (1993) Impact assessment using the before-after-control-impact (BACI) model: concerns and comments. *Canadian Journal of Fisheries and Aquatic Sciences*, **50**, 627–637.
Spooner, J. and Line, D.E. (1993) Effective monitoring strategies for demonstrating water quality changes from nonpoint source controls on a watershed scale. *Water Science and Technology*, **28**, 143–148.
Stewart-Oaten, A., Bence, J.R. and Osenberg, C.W. (1992) Assessing effects of unreplicated perturbations: no simple solutions. *Ecology*, **73**, 1396–1404.
Stoddard, J.L., Driscoll, C.T., Kahl, J.S. and Kellogg, J.H. (1998) Can site-specific trends be extrapolated to a region? An acidification example for the Northeast. *Ecological Applications*, **8**, 288–299.
Stow, C.A., Carpenter, S.R., Webster, K.E. and Frost, T.M. (1998) Long-term environmental monitoring: some perspectives from lakes. *Ecological Applications*, **8**, 269–276.
Taub, F.B., Kindig, A.C., Conquest, L.L. and Meadow, J.P. (1989) Results of interlaboratory testing of the Standardised Aquatic Microcosm protocol. In *Aquatic Toxicology and Environmental Fate: 11th Volume*, Suterm II, G.W. and Lewis, M.A. (eds), STP 1027, pp. 5–10. American Society for Testing and Materials, Philadelphia, PA.
Thas, O. (1999) Discussion of evaluation of sampling frequency for the water quality monitoring network of coastal Galicia (Northwest Spain). *Water Environment Research*, **71**, 1364.
Thas, O., Van Vooren, L. and Ottoy, J.P. (1998) Nonparametric test performance for trends in water quality with sampling design applications. *Journal of the American Water Resources Association*, **34**, 347–357.
Thiebaud H. and Thomas L. (2000) Inter-laboratory testing of the *Brachionus calyciflorus* 48-hour chronic toxicity test. Poster presentation to Third SETAC World Congress, Brighton.
Timmerman, J.G., Ottens, J.J. and Ward, R.C. (2000) The information cycle as a framework for defining information goals for water-quality monitoring. *Environmental Management*, **25**, 229–239.
UKWIR (2001) *UK Direct Toxicity Assessment Demonstration Programme*, Technical Guidance Report No. 00/TX/02/07. UW Water Industry Research, London.
Underwood, A.J. (1991) Beyond BACI: experimental designs for detecting human environmental impacts on temporal variations in natural populations. *Australian Journal of Marine and Freshwater Research*, **42**, 569–587.
USEPA (1991) *Technical Support Document for Water Quality-based Toxics Control*, EPA 502/2-90-001. US Environmental Protection Agency Office of Water, Washington, DC.
Van Belle, G. and Hughes, J.P. (1983) Monitoring for water quality: fixed stations versus intensive surveys. *Journal of the Water Pollution Control Federation*, **55**, 400–404.
Walker, J. (1988) Relative sensitivity of algae, bacteria, invertebrates and fish to phenol: analysis of 234 tests conducted for more than 149 species. *Toxicity Assessment*, **3**, 415–447.
Ward, R.C. and Loftis, J.C. (1986) Establishing statistical design criteria for water quality monitoring systems: review and synthesis. *Water Resources Bulletin*, **22**, 759–767.

Ward, R.C., Loftis, J.C. and McBride, G.B. (1990) *Design of Water Quality Monitoring Systems*. Van Nostrand Reinhold, New York.

Warren-Hicks, W. and Parkhurst, B.R. (1992) Performance characteristics of effluent toxicity tests: variability and its implications for regulatory policy. *Environmental Toxicology and Chemistry*, **11**, 793–804.

Warren-Hicks, W. and Parkhurst, B.R. (1995) Issues in whole effluent toxicity test uncertainty analysis. In *Whole Effluent Toxicity Testing: An Evaluation of Methods and Prediction of Receiving System*, Grothe, D.R., Dickson, K.L. and Reed-Judkins, D.K. (eds), SETAC Special Publication Series, pp. 180–190. SETAC Press, Pensicola, FL.

Whitehouse, P., van Dijk, P.A.H., Delaney, P.J., Roddie, B.D., Redshaw, C.J. and Turner, C. (1996) The precision of aquatic toxicity tests: its implications for the control of effluents by Direct Toxicity Assessment. In *Toxic Impacts of Wastes on the Aquatic Environment*, Tapp, J.F., Hunt, S.M. and Wharfe, J.R. (eds), Royal Society of Chemistry Special Publication No. 193, pp. 44–53. Royal Society of Chemistry, Cambridge.

Whitfield, P.H. (1988) Goals and data collection designs for water quality monitoring. *Water Resources Bulletin*, **24**, 775–800.

3 The nature and chemistry of toxicants
Ulf Lidman

3.1 Introduction

3.1.1 History

An environmental toxicant can be defined as a substance that, in a given concentration and chemical form, challenges the organisms of the ecosystem and causes adverse or toxic effects.

From a historical and evolutionary point of view, all living organisms are continually exposed to a panorama of toxins of natural origin and in the course of evolution have developed different kinds of defense, e.g. biochemical and physiological mechanisms, in their efforts to handle toxic substances. The toxins produced by different organisms or natural processes differ considerably in nature and mechanism of toxic action, as do the purposes for production. Thus, many organisms, such as bacteria, plants, fungi and animals, produce very potent toxins for defense and protection of integrity (allelopatic substances) or for the capture of prey, whereas in other cases the toxins seems to be just secondary metabolites or waste products from ordinary metabolism. Also, there are natural toxin-producing processes such as fire, volcanic activity and leakage from ores.

With the evolution and development of mankind and modern society, different man-made substances (anthropogenics or xenobiotics) have been added to the toxic panorama. Furthermore, many substances or products (e.g. metals) have been collected and concentrated for medical or industrial purposes and then redistributed in a manner that, today, is causing toxic exposure and harm to the biosphere. These processes have been accelerating during the last three centuries and millions of tons of xenobiotics have been distributed in the environment.

Awareness of the problems with xenobiotics started to emerge in the middle of the 20th century and there is still ongoing discussion today. Much research has been performed, much knowledge has accumulated, various actions have been taken and numerous substances and processes have been banned or restricted. In parallel, however, many new chemicals and products have been developed, in some cases to replace those substances that were phased out as environmental hazards and in other cases as components of new materials or applications. These new substances are continuously added to the environment, in many cases without thorough toxic evaluation or risk assessment, and they pose new risks to individuals and ecosystems. A special problem is the replacement of powerful environmental toxicants with other substances. To fit the process or application, the substituting substance in many cases has to have more or less the same physical and chemical characteristics and thus, in many cases, expresses similar toxicological behaviour to that of the banned substance.

3.1.2 Properties

Most environmental toxins share the properties of being lipid soluble and persistent to metabolism or breakdown, biotic as well as abiotic. Lipophilicity is commonly characterized by the partition coefficient between octanol and water (K_{ow}) and generally the bioavailability or tendency to be absorbed by biotic systems increases with increasing K_{ow}. At very high K_{ow}, however, the uptake might be hindered by the large molecular size or the substance might get stuck in lipophilic structures such as plasma membranes. The ideal value of K_{ow} for absorption is around unity (Walker et al., 2001).

For some substances, special mechanisms involving a recognition mechanism coupled to a transport reaction (active or passive) are available. Such mechanisms are, in most cases, aimed at the uptake and transport of essential substances but might be shared more or less specifically by the xenobiotics.

3.1.3 Exposure

A fundamental prerequisite for a substance to exert action in biological systems is that the system is exposed to the substance in a manner that allows absorption. The main routes for absorption and consequent exposure are the body linings, e.g. skin, respiratory surfaces and alimentary integument. Thus, exposure is through direct skin contact, air, water and food, including food chains and food webs.

3.1.4 Bioavailability

Generally, bioavailability and absorption by organisms are linked to lipophilicity and driven by concentration gradients. If, however, the molecules become too lipophilic, i.e. they possess a K_{ow} value well above unity, they might become stuck in membranes or other lipid-rich structures. Under most such circumstances no toxic action is expressed. Also, many toxicants can use specialized transport mechanisms (passive or active), designed for the uptake of essential substances, for entry into organisms. In the actual physiological stages of an organism, certain transport mechanisms might be induced that enhance the uptake of a toxicant.

A well-known example showing the differences in bioavailability for different forms of an element is mercury. For this substance, the liquid elemental form Hg^0 is poorly absorbed over gut and skin, the gaseous phase is readily absorbed over the lungs, the ions Hg^+ and Hg^{2+} are more readily absorbed over the gut and organic forms such as $(CH_3)_2Hg$ and CH_3Hg^+ are easily absorbed across all body linings (Gochfeld, 2003).

The question of bioavailability is crucial in environmental toxicology and depends, in many cases, on the physicochemical properties of the surrounding environment and the corresponding equilibrium form of the envirotoxicant in question.

3.1.5 Bioaccumulation

After exposure to a bioavailable toxicant, absorption takes place and the substance is distributed throughout the body or to the different compartments of the ecosystem. If the substance has high affinity for a certain structure or is lipophilic and slowly metabolized, then bioaccumulation takes place.

Elimination from the body is generally associated with hydrophilicity and very many envirotoxicants have to be metabolized to more hydrophilic components or derivatives before being excreted because the excretory organs, apart from milk glands and structures in parallel, are not efficient at handling lipophilic substances. From an abiotic point of view, xenobiotics may have settled and accumulated in sediments and, in many cases, exited from the biotic system. Then, however, they may be the target for microbial action and are either broken down or activated or experience a changed physicochemical environment associated with changed equilibriums so that they threaten the ecosystem in a new, more bioavailable form.

After absorption many xenobiotics enter some kind of depot (e.g. adipose tissue, bone tissue, certain proteins or some other tissue components) and are thus excluded from playing a part in leveling the transport gradients. This means that further absorption will take place and the overall levels of the envirotoxicant in the organisms will increase.

3.1.6 Biomagnification

The organisms of an ecosystem constitute, among other things, a food web with different food chains and trophic levels. As a general rule, the organisms on the lower trophic levels are eaten by animals from higher levels. The animals on the higher levels, the top predators, have to eat many times their own body weight to survive, develop and grow. This constitutes the basis for biomagnification, because the natural food components will be used for energy and construction purposes while the persistent xenobiotics will be stored in the body and give rise to very high body burdens and levels of xenobiotics in the depot tissues of the top predators.

3.1.7 Metabolism

The question of metabolism is very complex and extremely important from a toxicological point of view and has been the subject and main topic for extensive research during the last three decades.

One of the main aims of metabolism, and possibly the most important, is to make lipophilic substances more hydrophilic in order to promote excretion and prevent bioaccumulation.

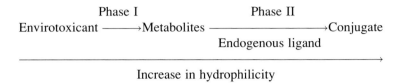

This is, however, a complicated process and from a toxicological point of view two main concerns can be identified: interference of the envirotoxicant metabolism with the endogenous metabolism and turnover of, for example, steroid hormones and other endogenous substances; and the production of many, highly toxic, electrophilic intermediates and metabolites. These metabolites are often very reactive and in many cases act as genotoxicants, inducing specific carcinogenic and teratogenic effects, but they can also cause general cell and tissue necrosis (Figure 3.1).

The metabolism of xenobiotics is commonly divided into phases I and II. Phase I consists of different fundamental chemical reactions (oxidations, reductions, hydratations, hydrolysis, etc.), resulting in slightly more water-soluble products with functional groups such as hydroxyl, amino and sulfhydryl. Phase II involves conjugation with highly water-soluble ligands such as glucuronic or sulphuric acid or glutathione (GSH) for excretion in urine and bile. Most of the highly reactive intermediates, free radicals, etc. are formed during phase I reactions, whereas phase II can be considered a detoxification process. There are, however, exceptions and, in particular, sulphuric acid conjugates may give rise to mutagenic metabolites.

Xenobiotic phase I metabolism can be performed through a vast number of enzymatic pathways, but the most important and most discussed in connection with envirotoxicants is the 'mixed function oxidase' system (MFO), based on a certain class of cytochromes called the cytochrome P450 family. These are membrane bound and localized to the smooth endoplasmic reticulum of liver cells in particular but also to many other cell types, preferably those making up different body linings facing the surrounding environment. Furthermore, the reactions are dependent on the reduced form of NADPH (produced in a membrane-bound NADPH generating system) and molecular oxygen.

Cytochrome P450 exists in many different forms – isozymes – each of which are more or less specific to a certain envirotoxicant or class of envirotoxicants. Some are also inducible; for example, exposure to a certain xenobiotic will cause *de novo* synthesis of a particular cytochrome P450 isozyme and elevated levels and activity of that isozyme, creating a base for synergistic action between substances (Klaassen, 1996).

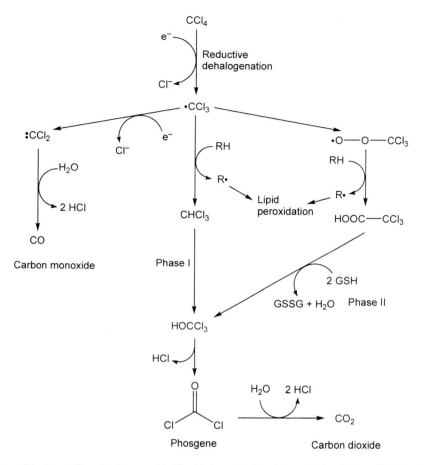

Figure 3.1 Metabolism of carbon tetrachloride with the production of electrophilic radicals and highly toxic metabolites. GSH denotes the important phase II conjugant glutathione. GSSG denotes oxidized glutathione.

The actual isozyme pattern of the MFO system of an animal gives a picture of the exposure situation and history of that animal and thus may serve as an important tool in the characterization of an environment from an envirotoxicological point of view or, more specifically, changes in cytP450 isozyme patterns may be used as a biomarker in bioassays for monitoring the envirotoxicant impact of more or less specified substances. Other phase I xenobiotic metabolizing enzymes are esterases, epoxide hydrolases and reductases, many of which play special and significant roles in toxicology. Phase II reactions are catalyzed by different transferases adding an endogenic polar substance such as glucuronic acid to the substrate, in many cases a phase I metabolite, creating a more polar product for excretion.

3.1.8 Effects of environmental toxicants

A fundamental requirement for a toxic effect to occur is the establishment of chemical contact between the envirotoxicant and some endogenous molecule at the site of action for the toxicant – 'it always starts with a reaction between two molecules'. This, of course, requires exposure, bioavailability and absorption of the substance, as well as distribution within the organism to the site of action.

The endogenous molecule reacting with the envirotoxicant might be of any type, e.g. nucleic acid, protein, lipoprotein, lipid or carbohydrate. Such molecules can constitute DNA, RNA, transport proteins, enzymes, receptors, ion channels, signal molecules, etc. In each case, a more or less severe disturbance of the regulation and internal balance, homeostasis, of the organism will occur, thus activating compensatory biochemical and physiological responses to bring the organism back to equilibrium and optimal function.

The reactions and responses can be very specific and thus serve as biomarkers in ecotoxicological work. Well known examples are: induction of the MFO system by polychlorinated biphenyls (PCBs) and polycyclic aromatic hydrocarbons (PAHs); interference of the DDT metabolite o,p'-DDE in the calcium metabolism of birds, resulting in eggshell thinning; and xenoendocrine effects of synthetic oestrogens and o,p'-DDT, resulting in vitellogenesis (yolk protein synthesis) in male fish. Further information on this subject is given below when discussing the effects of different envirotoxicants in connection with specific substances.

3.1.9 Interactions between envirotoxicants

In nature, organisms are never exposed to and threatened by only one isolated envirotoxicant or substance. In reality, the exposure situation is very complex and consists of a panorama of different substances. Furthermore, there are many interactions between substances and also between substances and organisms, both in the environment and inside the body.

Many products appearing in the environment that are subject to regulatory actions are judged on the basis of one main active compound. In reality, however, the product in most cases consists of a mixture of additives and formulations for optimal functioning and also impurities and by-products from manufacturing. Also, as mentioned, the organisms are not exposed just to one product but rather to many products and different components. Thus, the situation is very complex and to get the full or at least the best picture, testing and evaluation for the ecotoxicity of complete products and mixtures should be considered.

Interactions between substances are expressed in terms of synergism, potentiation and antagonism, sometimes with 'more than additive' and 'less than additive' effects. The terms 'synergism' and 'potentiation' are often used to express the same type of phenomenon, a 'more than additive' effect. Sometimes, however, the term synergism is restricted to the case where one of the compounds has no influence on the toxic effect under study at actual exposure level but strongly enhances the effect of the other in the combined exposure situation. Potentiation is then restricted to describing effects where two or more compounds each contribute to an overall toxicity that is more than additive.

In many cases the toxic contributions are strictly additive, but sometimes the toxicity of a mixture of substances greatly exceeds the sum of the added toxicities of the compounds involved. Whether synergism or potentiation, this will result in effects that are larger than expected, e.g. 'more than additive' effects. This situation can be explained by interactions in the metabolism of the compounds involving metabolic inhibition or activation, resulting in failed or decreased detoxication or accelerated production of a highly toxic metabolite. Antagonism or 'less than additive' effects for corresponding situations then can be explained as increased detoxification or blocked activation (Walker *et al.*, 2001).

3.2 Toxic metals

3.2.1 Introduction

Metals are elements, they are not man-made and they cannot be broken down or created in biological processes. Many metals are essential to living organisms and play fundamental roles in the normal metabolism of the organism. However, for metals and in fact for all substances the general dose–response relationship shows that even essential metals will be toxic at higher exposures, leading to biphasic dose–response curves. Examples of metals and metalloids essential to organisms are Ca, Co, Cu, Fe, K, Mg, Mn, Na, S, Se and Zn, of which Cu, Zn and Se might be of ecotoxicological concern. Among the non-essential metals, often referred to as heavy metals (density $> 5 \,\text{kg}/\text{dm}^3$), are Cd, Hg, Pb and Sn. These metals do not have any known essential biochemical functions but exert a high toxicity to most living organisms and have been recognized as dangerous envirotoxicants for a long time (Fergusson, 1990). Other heavy metals of concern are Au, Ag, and U, which can act as toxic substances and are envirotoxicant candidates but normally have a very low bioavailability (Hogstrand and Wood, 1998).

The metals are elements and, as such, are not produced by man. However, the occurrence of metals and their distribution in nature are often related to anthropogenic activities, resulting in high levels in the surrounding environment and/or bioavailable forms. Of great concern is the natural methylation of Hg, Pb and Sn that takes place in sediments and other deposits. The formation of metallo-organics leads to a large increase in bioavailability and thus increased envirotoxicological impact. Aluminum is not classified as a heavy metal because of its low specific weight. It is not usually considered toxic and is used extensively in household equipment and other trivial applications but it may have a high ecotoxicological impact in connection with acid rain and acidified aquatic ecosystems (Nieboer and Richardson, 1980; Merian, 1991).

Apart from the 'classical' envirotoxic metals (Cd, Hg, Pb) and the others mentioned above, the concept of *new metals* is now emerging as an environmental threat. Modern technology brings into extended use elements such as Am, Bi, In, Ga, La, Mo, Nb, Pd, Pt, Sb, Tc, Te and V, which are or will be distributed in the environment in a more or less uncontrolled manner. Very little is known about the envirotoxicity of these elements and thus toxicological data and risk assessment are a high priority.

3.2.2 Cadmium

Cadmium appears in nature together with zinc and is recovered as a by-product from zinc mining. It is used in different technical applications, such as rechargeable batteries, plating, as pigment and stabilizer in paints and plastic materials, in solders and in many other minor technological applications. It also appears as a natural contaminant in fertilizers and in some soils, and is taken up and accumulated in crops such as wheat and rice and also in fungi. It also appears as an anthropogenic contaminant in sludges from urban wastewater treatment.

Cadmium shows a very high bioconcentration factor (BCF), mainly because of its extremely slow elimination, although the absorption is rather low. The metal is not biomethylated and does not exist in any highly bioavailable organic forms. Its uptake is coupled to calcium and iron transporting mechanisms in which cadmium is mistaken for the essential metal and consequently transported. The uptake of cadmium is thus enhanced in situations of iron deficiency.

Exposure is mainly through food but also a direct uptake from water is also evident for aquatic organisms. This means that organisms exposed to even moderately elevated cadmium levels will bioaccumulate the element very effectively; significantly elevated levels are found, for instance, in kidney and liver tissue from wild herbivorous animals such as the elk (*Alces alces*) and in the hepatopancreas of marine crustaceans such as the lobster (*Homarus gammarus*) and the crab (*Cancer pagurus*).

After uptake, cadmium is tightly bound to a certain depot protein called metallothionein, which is rich in sulfhydryl (SH) groups and stored in depots in kidney and liver tissues, resulting in very high levels of the metal in these organs.

Cadmium is very toxic, with effects on kidney function, nerve and muscle function, ion balance, skeleton performance and circulatory function in organisms.

A common factor for these effects appears to be interference with the very important calcium metabolism of the organisms.

Cadmium is to be considered one of the most threatening heavy metals for the environment because of its very high bioconcentration factor in combination with a very high toxicity. It is also hard to remediate areas or biotopes contaminated with cadmium because it is highly soluble and mobile in the environment.

3.2.3 Mercury

Mercury has been used extensively by man since ancient times, mainly in medical applications and alchemy but also in several technical and industrial applications, including electrical equipment, chlorine production and gold mining. Different organic formulations have been used as pesticides, e.g. fungicides in paper mills and for seed dressing in agriculture. Around the world, action has been taken to phase out mercury in varying applications but it is still in extended use. Also, mercury occurs as a significant contaminant in coal and is released in large amounts when burning coal for energy and electrical power generation. The burning of coal also leads to a high release of sulphur dioxide with concomitant proton generation, giving rise to acidification of the environment, which, in turn, increases the mobility and bioavailability of mercury as well as many other metals.

The bioavailability for mercury depends to a large extent on its actual chemical form. At room temperature, mercury is a heavy liquid with very low bioavailability. The vapour phase, however, is readily absorbed by air-breathing organisms. Mercury ions (Hg^+ and Hg^{2+}) are absorbed mainly across the digestive tract integument, whereas organic forms such as dimethyl mercury [$(CH_3)_2Hg$], monomethyl mercury chloride (CH_3HgCl) and monomethyl mercury (CH_3Hg^+) species are easily absorbed across all body linings. Inside the body, the inorganic forms of mercury are limited to peripheral body compartments whereas the organic forms of mercury penetrate the blood/brain barrier and cause harm to the central nervous system, including the brain. The body burden of mercury is slowly eliminated in the growing hair, fur or other structures made up of fibrous protein.

In ecosystems, the different organic forms of mercury are easily bioaccumulated and effectively biomagnified, especially in aquatic environments and food webs. Thus, birds of prey and large predatory fish show highly elevated levels of total organic mercury compounds, especially in combination with acidification of the environment. Because mercury readily crosses the placental barrier and is a very potent teratogenic substance in mammals, it gives rise to morphological and functional disturbances in the human foetus. Consequently, authorities in many countries have black-listed fish with a total mercury content of more than 1 mg/kg from human consumption, especially with regard to women of child-bearing age.

Besides the teratogenic effects, mercury is also very nephrotoxic, causing damage to the kidneys with accompanying proteinurea and disturbances in ion homeostasis. Furthermore, intoxication in general creates various neural disorders. These disorders are limited to peripheral nervous system functions if the mercury compounds in question are inorganic, whereas absorption of organomercury compounds and elementary mercury in the gaseous phase will affect the brain and other central nervous system functions as well. From an ecotoxicological point of view, the severe neural and other disturbances occurring as a consequence of mercury poisoning might impair the functions of the individual when it comes to capture of prey, avoidance of enemies, reproductive success, etc., and thus, jeopardize the survival of the species and put population under pressure (Walker et al., 2001).

3.2.4 Lead

Lead is often considered the 'symbol' for heavy metals. It has been used since ancient times and its toxicity and harmful effects on organisms, including man, have been known since long before Christ, described among others by Hippocrates in 300 BC. Lead has been and is used in many differing technical applications, such as rechargeable batteries, as a construction material for pipes, cables, roofs, etc., as an anti-knock compound in petrol (although at present this use is being phased out), as bullet material in ammunition, as weights for recreational fishing, as a component in solders, as a pigment and anti-corrosive agent in paints, etc.

The most important overall source of environmental contamination with lead has been the extensive use of leaded petrol all over the world, with the spread of organic lead compounds (e.g. tetraethyl lead, a substance expressing high bioavailability and effectively exposed to the overall ecosystem) dispersed in exhaust gases from automobiles. Other sources of ecotoxicological importance are lead shot from shotguns in wetlands and weights from recreational fishing. These applications expose wetland birds to lead shot and weights through swallowing, grinding and absorption from the gizzard in the upper part of the digestive system, consequently leading to acute lead poisoning.

The bioavailability for lead from inorganic compounds in the environment is very much enhanced by lowered pH, and the ongoing acidification of many ecosystems causes increased exposure and absorption. Further important exposure routes are across respiratory epithelia, e.g. through inhaled air in air-breathing organisms and from the water and across the gills in water-respiring organisms, and also from food.

After uptake, lead is absorbed into red blood cells and effectively distributed throughout the body of the organism, reaching all parts. The metal is then put into depot bound to the metal-binding protein metallothionein and in the crystals of bone tissue. It is excreted slowly, mainly through the bile. The property of being bound in bone structure creates a possibility for analyzing historic material for comparison of lead exposure during different periods and circumstances, with the actual exposure reflected in blood levels.

A very significant and sensitive effect from lead, and a useful biomarker in envirotoxicological studies, is the decrease in the activity of several key enzymes, e.g. δ-aminolevulinic acid dehydratase (ALA-D) and δ-aminolevulinic acid synthetase, (ALA-S), in haemoglobin synthesis. From an overall point of view, these disturbances at the molecular level might result in a pronounced anemia and decreased physical capability and fitness of the organism at the ecosystem level. Further well known results of lead poisoning are neurotoxic effects, involving disturbances at peripheral as well as central levels. The symptoms include motor and mental disturbances, impaired and retarded intellectual development and capacity and behavioural changes.

Overall, lead poisoning will create a wide range of disturbances in the ecosystems affected and further actions for minimizing the impact of lead on ecosystems are needed. The change to unleaded petrol in many countries has improved the exposure situation enormously and is easily quantified. In Stockholm, for example, the mean lead level measured in the blood of the inhabitants has decreased by two-thirds during the last decade and similar improvements are seen all over the world. When planning further actions, however, the question of bioavailability is very important for lead, and usually elementary lead in terrestrial ecosystems is supposed to constitute a low risk. On the other hand, the risk for biomethylation in sediments creates possibilities for increases in bioavailability. Against this background, mitigating actions and alternatives should be evaluated thoroughly before being introduced.

3.2.5 Copper

Copper is a common metal with many technical applications, e.g. electrical equipment, plumbing for drinking water distribution and as roof material, replacement for asbestos in brake clutches for vehicles, bottom paints for ships and other pesticidal paints, etc. It is essential at low levels for most organisms but at the same time it is very toxic to lower life forms such as cyanobacteria, green algae and fungi, and it is frequently used in pesticide formulations. In mammals, absorption from the digestive channel is controlled and fairly large exposures are necessary to achieve toxic effects.

Copper does not seem to pose any general ecotoxicological problem, but the widespread distribution and exposure to dust from car brakes certainly gives rise to local effects structuring the ecosystems. Furthermore, local pollution in connection with industrial activities and in yacht harbours, for example, creates significant changes in habitats.

3.2.6 Tin

Tin is a frequently used metal for plating, soldering and in alloys. It is also produced in different organic forms, such as tri- and tetra-butyltin for use as pesticides or stabilizers in plastic materials. Tin in inorganic form is considered rather non-toxic because of low bioavailability and absorption, whereas organic forms are powerful envirotoxicants with strong anti-fouling properties. They are bioaccumulative and exert androgenic endocrine disturbance in different organisms. The most well known example is imposex, the development of a penis in female marine gastropods in response to tributyltin from anti-fouling paints used to prevent growth on hulls of boats. This phenomenon, first recognized in the late 1960s, occurs at very low levels (around 1 ng/l of seawater for the most sensitive species) (Bryan *et al.*, 1988). It is now recognized worldwide and actions, including bans, have been taken all over the world; the situation is improving in some areas but not in all. At the molecular level, the effect has been shown to depend on elevated levels of the male sex steroid testosterone because of a competitive block of the cytP450-mediated aromatase and the normal conversion of testosterone to estrogens in females. The effect is, however, not an effect of the metal *per se* but rather an effect of the tributyltin molecule as a whole (Matthiessen and Gibbs, 1998).

3.3 Halogenated hydrocarbons

3.3.1 Introduction

The halogenated hydrocarbons are almost exclusively anthropogenic substances, produced and distributed in the environment by man. In this category there are many important products and substances for industrial and technological use but also different types of pesticides and even drugs. Furthermore, many types of very toxic halogenated hydrocarbons, unintentionally produced as contaminants in combustion and certain industrial processes, are polluting the environment.

The hydrocarbons might be aliphatic, cyclic, aromatic or a combined form and the halogens in question are not only chlorine but also bromine and fluorine. The hydrocarbon part gives the substance lipophilic characteristics for absorption and the halogenated group contributes to the persistent, bioaccumulative and biomagnificative properties by making metabolism and biotransformation to a more hydrophilic derivative more difficult, thus preventing excretion or elimination.

Many halogenated hydrocarbons exert toxicity in their native form whereas others are more toxic and exert effects from the intermediates and metabolites formed in the biotransformation for excretion or other purposes. A special group of intermediates or by-products that are formed are different types of very reactive macromolecule and tissue-harming free radicals from carbon, oxygen and chlorine and organic radicals such as CH_3^+, as well as hydrogen peroxide, epoxides and other electrophilic compounds emanating from the oxidative metabolism.

3.3.2 Polychlorinated biphenyls (PCBs)

These substances have been produced and used for differing technical purposes and applications from the end of the 1920s. The molecules consist of a biphenyl skeleton with a varying degree of chlorination and with chlorine in different positions; there are 209 different possible forms or congeners (Figure 3.2). The commercial products, consisting of around 120 congeners, are highly lipophilic, stable, unreactive, viscous liquids of low volatility, the actual physical properties varying with the degree of chlorination.

Well known trademarks are Arochlor 1254 and Clophen A 50, with the last two digits indicating chlorine contents of 54% and 50%, respectively, on a weight basis. The higher the chlorine content, the greater is the relative amount of heavily chlorinated and bioresistant congeners in the mixture.

Among important previous applications of PCB mixtures are hydraulic fluids, insulation dielectrics in transformers and condensers, plasticizers in plastics, outdoor paints and printer's ink and different sealing and construction products for building and mounting purposes.

The PCBs as environmental contaminants were first detected in Baltic Sea seals and herring in the 1960s (Jensen, 1966). Heavy bioaccumulation and biomagnification can result in very high concentrations in the fatty tissues of top predators, especially in aquatic food webs. The PCBs are also subject to long-distance air distribution and today they have contaminated the biosphere worldwide (Skaare et al., 2002).

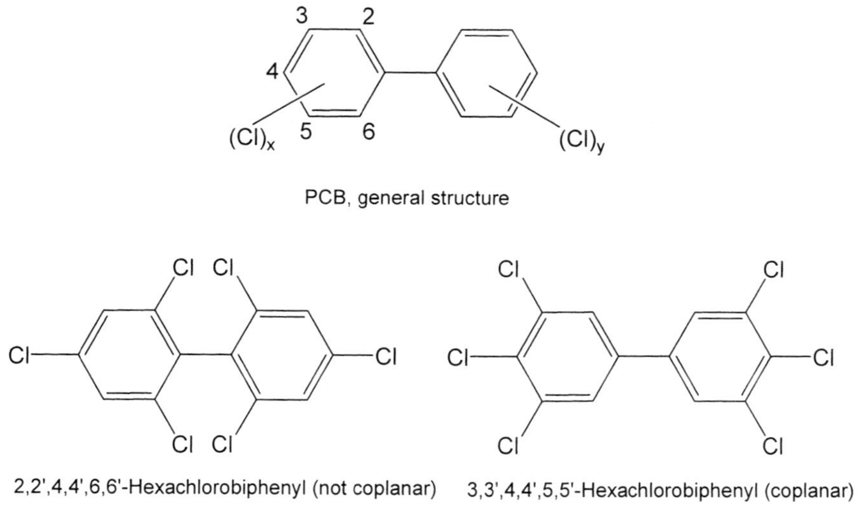

Figure 3.2 General structure of the envirotoxic polychlorinated biphenyls and examples of two congeners, one coplanar with affinity for the Ah-receptor and one not coplanar.

The PCBs are very efficiently bioaccumulated and biomagnified in food webs; in particular, congeners with more than five chlorines are very persistent, with slow or practically absent metabolism. The substances are stored in adipose and fatty tissues and the only effective way for organisms to get rid of the heavier PCBs is through lipid-rich secretions such as mammalian milk, egg yolk, etc. (Skaare *et al.*, 2002). This will, of course, result in serious exposure of embryos and offspring of the following generation.

The acute toxicity exerted by PCBs is generally low. The most prominent effect is the pronounced induction of the MFO system in different tissues, with increased *de novo* synthesis and increased levels in several cytP450 isozymes. The PCBs in the environment occur as a mixture of congeners and every congener exerts specific toxicity, with some PCBs exerting exclusive effects. One such effect is the capability for planar PCBs to bind to and stimulate the cytoplasmatic aryl hydrocarbon receptor in the cell, the same receptor that is stimulated by the highly toxic polychlorinated dibenzodioxins (PCDDs) (see below).

Other effects of different PCBs are anti-oestrogenic (planar or tetrachlorinated dibenzodioxin-like congeners) or estrogenic effects and displacement effects of the thyroid hormone thyroxine and vitamin A from the combined endogenous transport protein, retinol-binding protein/transthyretin, followed by consecutive disturbances in vitamin A and thyroxine-dependent metabolism. The latter effect is caused by a hydroxylated tetrachlorobiphenyl cytP450 metabolite, namely 4'-OH-3,4,3',5'-TCB (Figure 3.3) (Brouwer *et al.*, 1990). Different PCBs also exerts effects on thymus function acting as immunosuppressants (Richter *et al.*, 1994, van Loveren *et al.*, 2000). Other effects are neural disturbances, disturbed testicular development from single prepuberal exposure in the rat (Hsu *et al.*, 2004), disturbed haemoglobin synthesis with increased production of porphyrins, involvement in carcinogenesis, etc. On an integrated level, PCBs cause severe disturbances in reproduction, as for example in Baltic seals (Olsson *et al.*, 1992). In summary, the PCBs are very serious envirotoxicants with many effects at the biomolecular level, giving rise to a significant number of effects at the integrated levels. Even though many actions have been performed and the

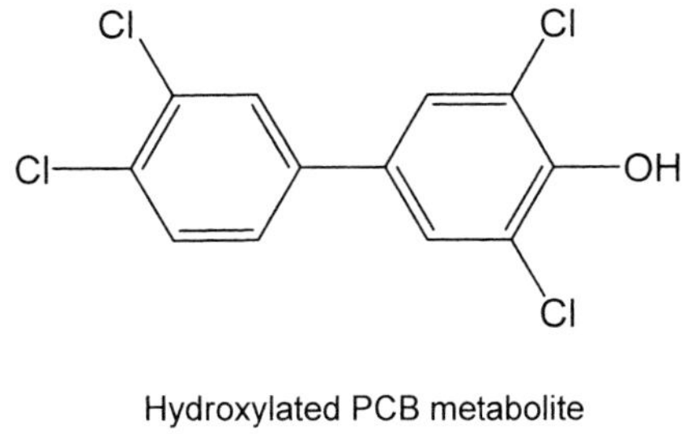

Hydroxylated PCB metabolite

Figure 3.3 Hydroxylated PCB metabolite with affinity for transthyretin.

ecological situation has improved considerably as a result of bans in use and remedial actions, the PCB problem will persist for a long period in the future because of the enormous amounts discharged into the systems and the global motility and the high persistence of especially the more heavily chlorinated congeners.

3.3.3 Polychlorinated dibenzodioxins (PCDDs) and dibenzofurans (PCDFs)

The polychlorinated dioxins and furans are not manufactured or produced for any intentional or commercial purposes. Rather, they occur as by-products from the production of chlorinated phenoxy herbicides, e.g. 2,4,5-trichlorophenoxyacetic acid (2,4,5-T), from production of the fungicide pentachlorophenol (PCP) and from combustion processes, especially metal smelters and the combustion of chlorinated organic substances such as PCBs and also in the burning of industrial and domestic waste. As for the PCBs, the chlorinated dioxins and furans appear in the form of different congeners with different distributions and degrees of chlorination: 75 for the PCDDs and 135 for the PCDFs; altogether 210 different substances (Figure 3.4). Out of these, 17 are considered to be of toxicological significance (Pollitt, 1999). A lot of actions to lower the dioxin pollution have been taken and the known amounts released to the environment have diminished, but so far no decreases in the levels of organisms from the Baltic have been detected (Olsson *et al.*, 2003).

The dioxins, especially 2,3,7,8-tetrachlorodibenzo-*p*-dioxin (TCDD), are extremely toxic and exert effects at submicrogram levels. They are also very lipophilic and persistent and bioaccumulate and biomagnify very effectively. Toxic action is

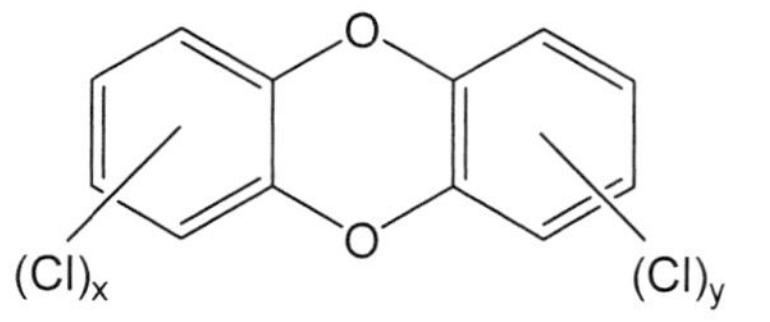

Dibenzodioxin, general structure

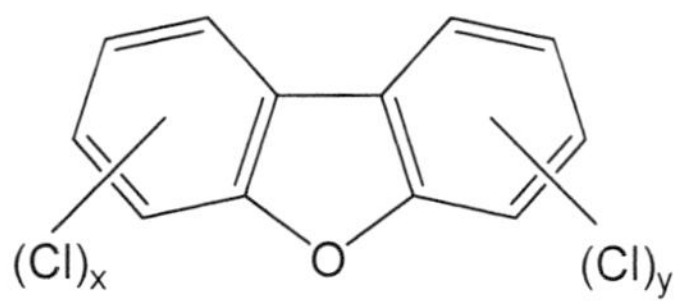

Dibenzofuran, general structure

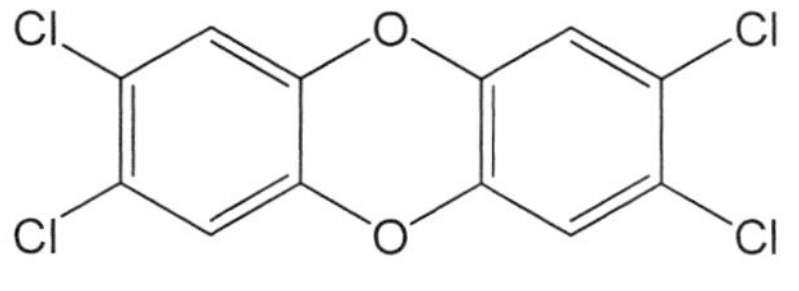

2,3,7,8-Tetrachlorodibenzodioxin (TCDD)

Figure 3.4 General structures for chlorinated dibenzodioxins and dibenzofurans, together with the most toxic dioxin congener TCDD.

exerted by high affinity for the cytoplasmic aryl hydrocarbon (Ah)-receptor and consecutive binding to DNA, with effects on gene expression. Further effects are the blocking of estrogenic receptors, causing anti-estrogenic effects.

Dioxins and other substances (e.g. planar PCB congeners) with affinity for the Ah-receptor give rise to additive toxicity in proportion to the degree of affinity and action on the receptor. Thus, the different congeners are characterized by an affinity factor, the toxic equivalent factor (TEF), which, by multiplication with the actual concentration, gives the toxicity equivalent (TEQ) for the substance. In this context, the TEF of TCDD is set to unity and other substances have TEF < 1. By summing up the contributions or TEQs for all the congeners of PCDDs, PCDFs, PCBs, etc. involved, the total toxic impact on the Ah-receptor can be quantified (Paasivirta, 1991).

Certain PCDDs and PCDFs act as very powerful inducers of the MFO system and give rise to *de novo* synthesis of CYP1A1, a cytochrome P450 isozyme associated with aryl hydrocarbon hydrolase (AHH or Ah), and ethoxyresorufin O-deethylase (EROD), a common biomarker for PCDD exposure.

At the integrated level, PCDDs give rise to an overall 'wasting syndrome', including weight loss, liver damage with porphyria, epithelial changes, etc. Furthermore, the thymus gland is atrophied and the capability of the immunologic system is degraded (Kerkvliet, 1995; de Wit and Strandell, 1999). The PCDDs also affect reproduction and are thought to be carcinogenic (de Wit and Strandell, 1999).

The PCDDs/ PCDFs are highly dangerous envirotoxicants and actions for remediation and for decreasing *de novo* production of the substances are important, e.g. proper combustion methods and care in handling of chlorinated waste, as well as properly controlled manufacturing and use of chlorine-containing products.

3.3.4 Polybrominated flame retardants (PBFRs)

Polybrominated flame retardants are in extensive use in modern society as additives in many products, e.g. computers, other electronic equipment, furniture, cars, construction materials, sealings, etc. They are lipophilic substances with differing K_{ow} and differing persistence, many of them characterized by very high bioavailability and bioconcentration factors.

The increase in use has been dramatic during the last two decades and different classes of PBFRs, as well as many metabolites, are today globally distributed in the environment. The most important products today that are accumulated in abiotic as well as biotic systems, including man, are the polybrominated diphenyl ethers (PBDEs), polybrominated biphenyls (PBBs), tetrabromobisphenol A (TBBPA), pentabromophenol (PBP) and 2,4,6-tribromophenol (TBP). The latter two might be metabolites from the extensively used flame retardant hexabromobenzene (HBB) (Figure 3.5). Out of these, TBBPA is the major PBFR produced but is also the least accumulated in biota. The substance exists in high levels in sediments and sewage sludge but is easily metabolized and excreted and does not biomagnify in animals (de Wit, 2002). The most serious PBFRs from an ecotox-

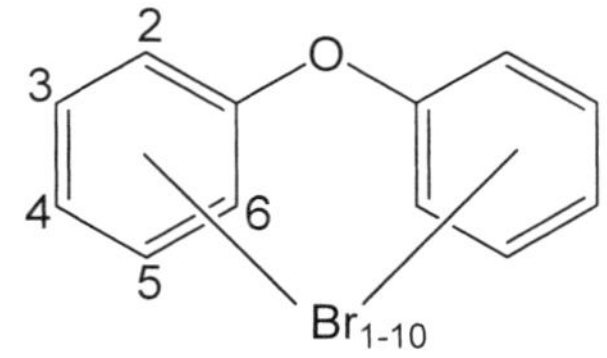

Polybrominated diphenyl ether, general structure

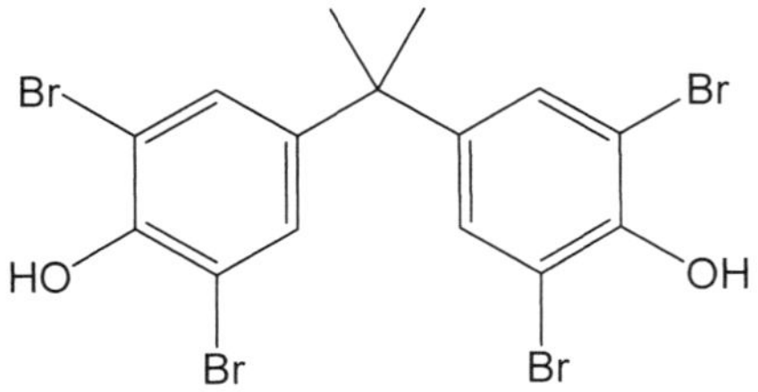

3,3',5,5'-Tetrabromobisphenol A

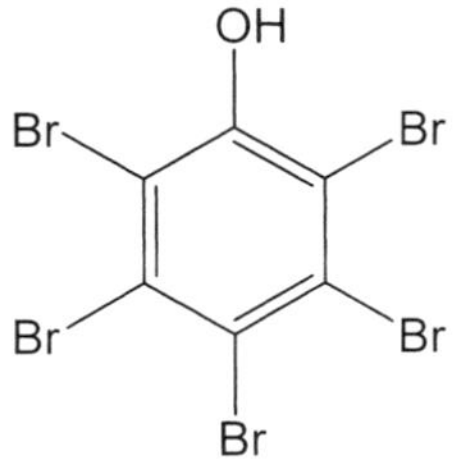

Pentabromophenol

Figure 3.5 Examples of polybrominated flame retardants (PBFRs).

icological point of view appear to be the PBDEs. Like PCBs, these substances theoretically occur in 209 possible congeners with differing toxicological properties. The production, use and distribution of this group of substances have increased exponentially during the last decade and so have the levels in different biota, including man.

The congeners with the highest bioconcentration factors seem to be the tetra- and penta-BDEs (Figure 3.6). Doubling rates in levels for different aquatic animals have been shown to be of the order of 2–3 years and this has resulted in very high levels in many ecologically significant organisms and also in man, with very high levels of the substances in human breast milk, especially in North America.

The toxicology of the PBFRs as a group is similar in many ways to the toxicology of PCBs and the dioxins. PBFRs express a panorama of effects, including low acute toxicity, interference with the Ah-receptor, interference with thyroid functions and transport protein for thyroxine, impairment of neurological development in fetuses, etc. To consider the thyroid disturbances in more detail, hydroxylated metabolites of tetra- and penta-BDEs have been shown to compete with the natural thyroid hormones T_4 and T_3 for the specific transport protein transthyretin (TTR) and to expel the hormones from the protein, resulting in a breakdown of the hormones and xenoendocrine disruption (Figure 3.7) (de Wit and Strandell, 1999; de Boer *et al.*, 2000; Darnerud *et al.*, 2001; Darnerud, 2003; Hakk and Letcher, 2003; Hale *et al.*, 2003).

In a future perspective, the PBFRs impose a great threat to the environment by being produced, used and distributed in large amounts, being globally transported and expressing bioaccumulative and adverse toxic properties of the same kind as

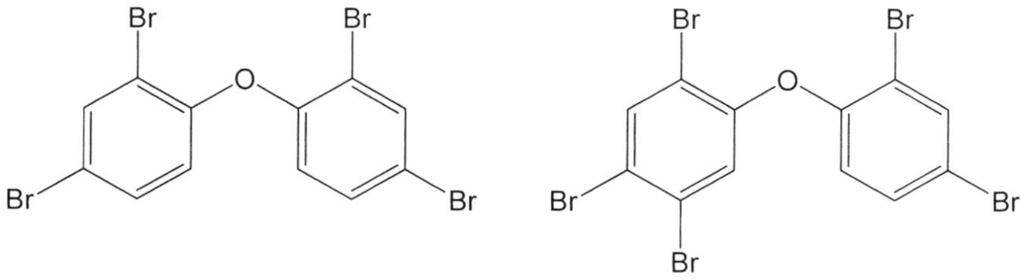

2,2',4,4'-Tetrabromodiphenyl ether 2,2',4,4',5-Pentabromodiphenyl ether

Figure 3.6 The tetra and penta congeners of PBDEs, which are strongly bioaccumulating substances.

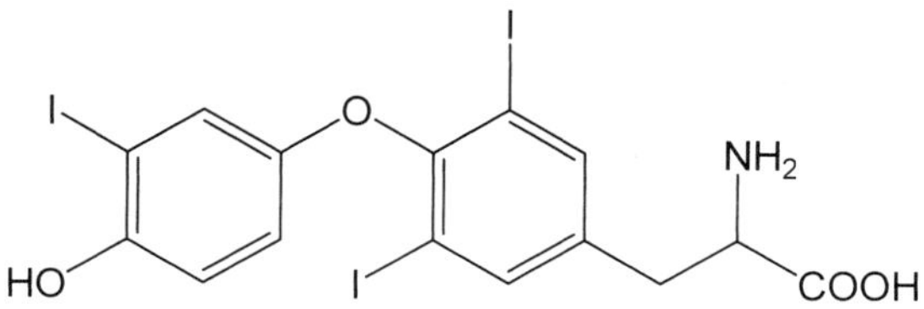

4'-Hydroxy-3,3',5-tri-iodo-L-thyronine (T_3)

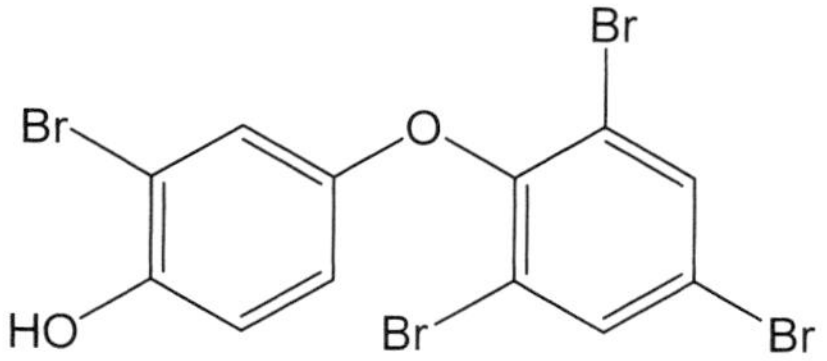

4'-Hydroxy-2,3',4,6-tetra-BDE

Figure 3.7 Structural resemblance between triiodothyronine and a hydroxylated PBDE metabolite.

well known envirotoxicants. Actions are now being taken in parts of the world to limit the use of some of the most threatening products, e.g. tetra- and penta-BDEs, but the problem will continue to develop for a long time on the basis of the huge amount of these substances already present in society and the environment and with continued production approaching the levels of the total PCBs produced and released.

3.3.5 Chlorinated pesticides/insecticides

The production and spread of chlorinated insecticides started in the 1940s and culminated in the 1960s. The group consists of many different and diverse substances, formulations and products, widely used and dispersed in the environment (Brooks, 1974, Hasall, 1990).

The most important product from a commercial as well as a envirotoxicological point of view has been 1,1,1-trichloro-2,2 -bis(4-chlorophenyl) ethane or dichlorodiphenyltrichloroethane (DDT), but also aldrin, dieldrin, lindane (γ-hexachlorocyclohexane, γ-HCH), chlordane and toxaphene are of environmental concern (Figure 3.8).

All products have been used in different applications in agriculture and forestry practices but also for the elimination of malaria-carrying mosquitoes, lice and other sanitary purposes. Features in common are high lipid solubility, high bioavailability and pronounced persistence, either for the parent compound or for the metabolite(s). All substances except for aldrin have a worldwide distribution as a result of persistence and long-distance transportation and ground deposition far away from the source. Aldrin is easily metabolized to dieldrin and thus occurs in the environment in the form of this persistent metabolite (Figure 3.8).

Preparations containing DDT were in general use in many parts of the world as effective insecticides in different applications until the 1970s, when it was banned in many countries and the use and spread declined. As a result of this, levels in biota have declined as well. It is, however, still in use in tropical areas as an effective weapon against malaria and sometime new releases are detected in different parts of

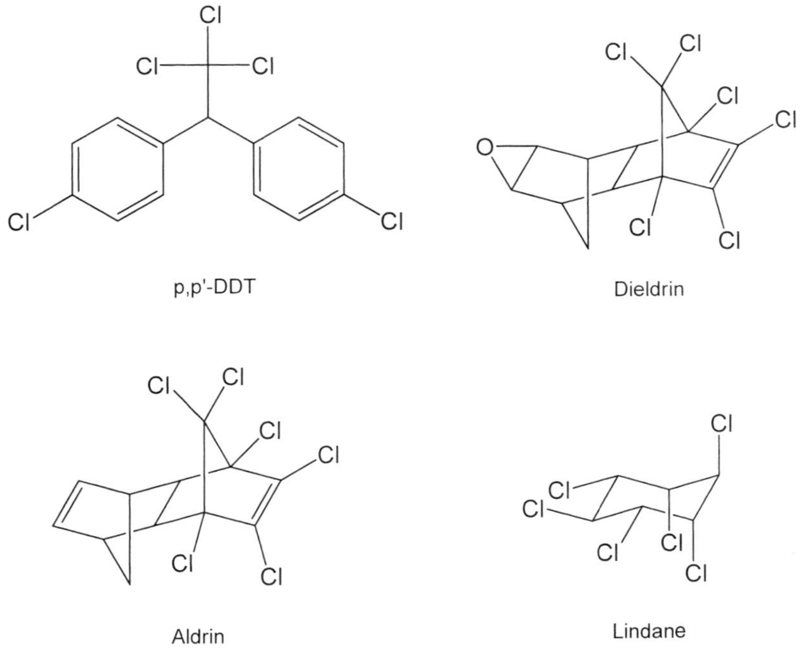

Figure 3.8 Examples of ecotoxicologically important chlorinated insecticides.

the world. DDT expresses a high bioavailability based on lipophilicity and it is easily distributed in the organism. In the technical products, two different isomers p,p'-DDT and o,p'-DDT are formed and recognized. The p,p'-DDT isomer forms the primary insecticide with actions based on neurotoxicity, whereas o,p'-DDT exerts very low neurotoxic activity. In the animal body, these isomers undergo biotransformation or metabolism to p,p'-DDE and o,p'-DDE and further to DDA. Of the metabolites, p,p'-DDE is very stable and persistent and remains in the environment for long periods (Figure 3.9). The acute toxicity of DDT is generally low for mammals and birds but high for poikilothermic fish, invertebrates and insects. The primary insecticidal action as well as effects on poikilothermic animals are based on effects on the nervous system, e.g. interaction with the sodium channels of the nerve axons, especially at low and moderate temperatures (Eldefrawi and Eldefrawi, 1990).

Environmental concern from DDT is based mainly on the blocking of Ca^{2+}-ATPase in the avian oviduct, resulting in impaired Ca^{2+} metabolism in the eggshell gland and eggshell thinning with consequent egg breakage (Peacall, 1993). Also Na^+/K^+-ATPases in different tissues are affected, resulting in impaired ion metabolism. Furthermore, the isomer o,p'-DDT and its metabolite o,p'-DDE are known as xenoestrogenic substances, disturbing sexual differentiation and possibly reproductive success (Janssen et al., 1998). Another class of halogenated insecticides of great environmental concern is the chlorinated

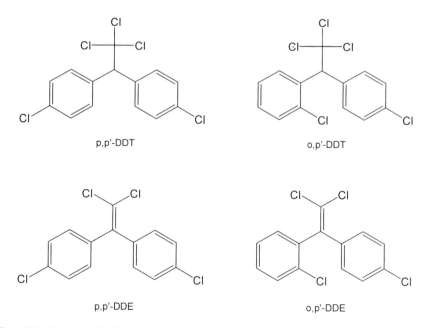

Figure 3.9 Structure of DDT and its metabolites.

cyclodienes represented by aldrin, dieldrin and heptachlor. These substances, characterized by high bioavailability, high toxicity against vertebrates and high persistence, either for the substance *per se* or for the metabolite(s), were produced and spread during the 1950s. Of these, dieldrin has been shown to undergo substantial biomagnification, although this is somewhat limited, because of the high toxicity of the substance, sublethal effects and consequent death of the organism when the levels increase.

Other important cyclodiene insecticides are chlordane, endrin and endosulphan. These substances do not fulfil all the criteria of envirotoxicants previously mentioned and, although causing environmental harm, are considered to be of less environmental concern.

The chlorinated cyclodienes were used in diverse formulations for different purposes such as spraying of crops and vectors of disease, for the control of ectoparasites on animals, for dressing of seed, etc. Their use was more or less phased out during the 1980s and today the uses are very few (Walker *et al.*, 2001).

The different cyclodienes in general and in the formulations are readily bioavailable and thus are absorbed by living organisms and metabolized and bioaccumulated. Aldrin is metabolized to dieldrin, a substance of high toxicity and very high persistence that is still found in significant amounts in the environment. The fundamental mechanism of toxic action for the chlorinated cyclodienes is interference with the GABA receptor functions in the central nervous system, e.g. the brain, resulting in central nervous system disturbances and loss of integration, convulsions and death (Eldefrawi and Eldefrawi, 1990). As an effect of disturbances in brain functions, behavioral and reproductive disturbances might also be suspected (Atchisson *et al.*, 1996). Thus, dieldrin is strongly suspected to be the fundamental cause of the decline in the sparrow hawk (*Accipiter nisus*) and peregrine (*Falco peregrinus*) populations in the UK in the late 1950s and early 1960s, through a combination of acute lethal poisoning and sublethal behavioural effects (Walker and Newton, 1999; Sibly *et al.*, 2000).

Hexachlorocyclohexane has been used as an insecticide for the control of agricultural pests, for dressing of cereal seed and for the control of parasites on farm animals since the 1960s. Its use has been banned in many countries but it still remains in extensive use in others. Hexachlorocyclohexane mainly exists in the form of three different isomers, α-β- and γ-hexachlorocyclohexane, appearing in a crude mixture after manufacture. Only one of these isomers, the γ form, expresses insecticidal activity, and the refined product containing about 99% γ-hexachlorocyclohexane (γ-HCH) is named lindane.

Lindane is a less lipophilic substance in comparison with many other serious envirotoxicants but still is characterized by a high degree of bioavailability. Furthermore, the substance is rather persistent, volatile and subject to long-distance transport. It bioaccumulates to a pronounced degree in lower biota and fish, although mammals and birds biotransform and excrete the metabolites to a great extent. Thus, the levels of lindane in contrast to, for example, PCBs have been shown to be lower in fish-eating birds and seals than in fish edible (Jansson et al., 1993).

The toxicity of lindane is related to the blocking of the Cl⁻ channels and interference with the GABA receptors in the nervous system, including the brain, giving rise to sublethal effects and death at higher levels of exposure (Eldefrawi and Eldefrawi, 1990; Walker et al., 2001).

3.3.6 Other halogenated organic compounds of environmental concern

3.3.6.1 Chlorophenols

Chlorophenols are a class of pesticide substances, e.g. fungicides, used for wood preservation, in pulp production and other miscellaneous applications. The substances were introduced in the 1930s and have been used in very large amounts. Today, the consumption has decreased and the substances are banned in many countries. The main active substance in chlorophenol products is pentachlorophenol (PCP; Figure 3.10). The substance is moderately lipophilic and persistent, yet readily absorbed and accumulated in biota and expresses a rather high acute toxicity. The metabolism and breakdown of this envirotoxicant in biota and in the environment are rather slow, resulting in successively dechlorinated metabolites.

Initially, the main concern with commercial PCP products was the contamination with PCDDs (albeit not TCDD) as by-products from the manufacturing process, and many of the envirotoxic effects originally attributed to PCP have been shown to be effects of the dioxins. Toxic actions directly related to PCP are uncoupling of oxidative phosphorylation resulting in increased metabolism and the breakdown of energy reserves, immunological effects, liver and kidney damage and reproductive disturbances (National Research Council, 1986; Eisler, 1989). Today the use of PCP is rather restricted, but in areas where the preparations are still in extensive use, PCP poses a threat to aquatic environments in particular.

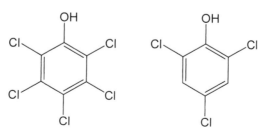

Figure 3.10 Examples of chlorinated phenols.

3.3.6.2 Chlorinated paraffins
Chlorinated paraffins constitute a very large number of different substances, depending on the original paraffin molecule and the degree and positioning of chlorination. They have a lot of different industrial applications and in many cases are used as substitutes for PCBs and other banned envirotoxic products. Examples of use include outdoor paints, hydraulic fluids, plasticizers and flame retardants The chemical and physical properties and the bioavailability, persistence and tendency for bioaccumulation vary a lot between different chlorinated paraffins. In aquatic environments, the chlorinated paraffins bioaccumulate up to the level of fish but diminish at the levels of mammals, indicating an enhanced metabolism and elimination at higher trophic levels. In terrestrial ecosystems, however, the chlorinated paraffins show an increase up to the highest trophic levels.

The toxicology of the chlorinated paraffins as a group is not considered very serious and they are not considered to be one of the most adverse ecological threats. Among all possible chlorinated substances in this group, however, some do exert adverse effects and so the use of chlorinated paraffin products should be restricted (Tomy *et al.*, 1998).

In conclusion, the halogenated hydrocarbons as a group are very powerful envirotoxicants although the potency differs a lot between and in different categories of this immense collection of chemicals. Many are readily bioavailable and persistent and show strong tendencies for bioaccumulation and biomagnification. Some are extremely toxic and exhibit very distinct and specific mechanisms of action, whereas others exert a low and broad toxicity spectrum. Nevertheless, in every case the fundamental toxic action occurs as a reaction between two molecules, and much of the environmental toxicology is concerned with the prerequisites for this to happen. One important factor in this sense is persistence, giving the time for the necessary absorption, distribution and action. Persistence is a common feature for many envirotoxicants; thus persistent substances may act as envirotoxicants, e.g. halogenation of organic molecules is an important step for producing substances with the potential for creating environmental contamination.

3.4 Polycyclic aromatic hydrocarbons (PAHs)

The PAHs are substances mainly formed from incomplete combustion processes (e.g. pyrolysis) of petroleum and coal in connection with energy production, industrial activities, transport, etc. but also from graphite electrodes in industrial processes, from wood preservatives made up from coal tar and from products such as car tyres. The occurrence of PAHs is widespread, although often more or less concentrated to 'hot spots' associated with anthropogenic activities. Of course, PAHs being products from combustion processes, are also formed in natural combustion processes such as forest fires, volcanic activities, etc. and they are also natural constituents in crude oil and coal. However, the sources of PAH production causing environmental harm are almost exclusively anthropogenic (Figure 3.11).

The PAHs are composed of carbon and hydrogen, and different substituted substances can also be assigned to this group of environmental toxicants. The aromatic structure and the absence of polar groups constitute molecules of lipophilic character that are prone to biotransformation or degradation.

For many PAHs the bioavailability is high and they bioaccumulate to a certain degree in living organisms. In higher organisms, e.g. organisms with a well developed xenobiotic metabolism, they are normally readily metabolized and eliminated whereas in abiotic environments, especially under anoxic conditions, many substances exhibit a pronounced persistence with long half-lives, leading to heavy accumulation in sediments and earth (Hoffmann *et al.*, 1995).

The PAHs, like many other groups of envirotoxicants, constitute an immense number of different substances, each one expressing its own toxicity (Bispo *et al.*, 1999). However, certain PAHs express planar structures and affinity for the Ah-receptor and thus interact with certain PCDDs and planar PCBs, adding toxicity equivalents (TEQs) to the overall impact on the receptor. Also, for different PAHs, additive effects towards the Ah-receptor have been documented, as have competition and antagonistic effects from naphthalene in combination with benzo[a]pyrene (White, 2002). Exposure to PAHs gives rise to a panorama of different toxic

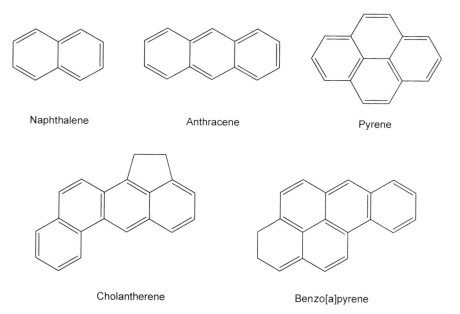

Figure 3.11 Examples of polycyclic aromatic hydrocarbons (PAHs).

effects exerted through different mechanisms. Many PAHs are very efficient inducers of the MFO, leading to an increase in mainly the cytochrome P450 1A1 fraction and corresponding increases in the activity of ethoxyresorufin O-deethylase (EROD), a very efficient biomarker in environmental toxicology (van der Oost *et al.*, 2002). Furthermore, many PAHs exert toxic and genotoxic effects through phase I metabolites from the MFO metabolism, especially epoxides from substances expressing a so-called 'bay region', creating long-lived reactive metabolites. The most pronounced and well known example of such a reactive, genotoxic and carcinogenic metabolite is the benzo[a]pyrene-7,8-diol-9,10-epoxide formed from benzo[a]pyrene (Figure 3.12). The ability to form mutagenic, genotoxic and carcinogenic metabolites in the MFO metabolism is valid also for many other PAHs (Conney, 1982).

Other effects of environmental concern from different PAHs are immunotoxicity, resulting in increased sensitivity towards bacteria and virus. These effects seem to be mediated through the Ah-receptor and expressed at the DNA level (Grundy *et al.*, 1996; Burchiel and Luster, 2001).

The PAHs also give rise to endocrine disruption and can exert estrogenic as well as anti-estrogenic effects on the vitellogenin synthesis in fish liver. Estrogenic effects resulting in an increase in vitellogenin production in male fish are considered to be the result of a direct stimulation of estrogenic receptors by certain hydroxylated PAH metabolites, whereas the anti-estrogenic effects in female fish, registered as a decrease in vitellogenin production, are explained by an increased metabolism and excretion of endogenous estrogens as an effect of the induced MFO system (Nicolas, 1999). Certain PAHs have also been shown to decrease testosterone levels in male fish, which is suggested to give rise to behavioural disturbances and failed reproductive success (Woodhead *et al.*, 1999).

As a whole, the PAHs are a very diverse group of substances produced in nature as well as from anthropogenic activities. Being mainly different substances, with every substance expressing its own toxicity, the toxic effects and environmental risks emanating from the PAHs are very diverse. The overall dominating risks come from PAHs produced and released in high concentrations through human industrial activities and care should be taken not to increase these risks further.

3.5 Medical and veterinary drugs

Quite recently, attention has been drawn to the environmental contamination caused by pharmaceutical products and drugs. These groups constitute an enormous number of different substances, selected and/or manufactured for the purpose of biochemical and physiological activity and interference. By definition, such substances can act as toxicants in elevated concentrations and especially in the wrong situations or places. Furthermore, modern drug design in many cases is directed towards the development of products with high bioavailability and

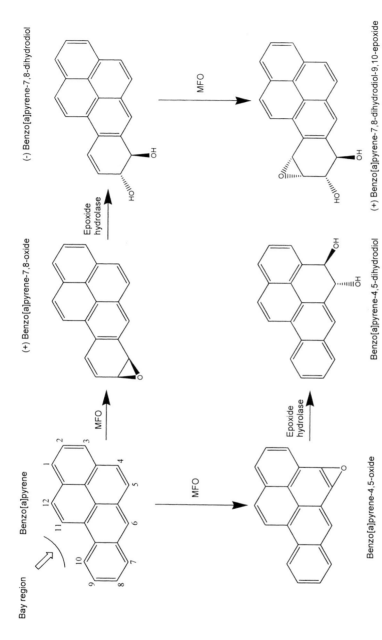

Figure 3.12 Metabolism of the genotoxic and carcinogenic PAH benzo[*a*]pyrene.

persistence in order to keep the dosage as low as possible – an important action for the avoidance of undesirable side-effects. In combination with highly specific physiological effects at very low concentrations, many of the attributes for a harmful environmental toxicant are also fulfilled (Richardson and Bowron, 1985; Halling-Sörensen *et al.*, 1998).

Drugs or medicines are composed of pharmaceutically active substances together with other constituents such as colors, preservatives, fillers, etc. Of these, it is the active substance and on some occasions the preservatives, e.g. mercury, that might be of environmental concern.

The total distribution and consumption of active substance from human drugs in Sweden for the year 2001 amounted to 1000 tonnes, a figure of the same order of magnitude as the total release of pesticides in the agricultural sector (Rosander, 2002). These substances reach the environment through the deposition of unused, surplus drugs in landfills, with consequent leakage, and through the sewage systems, either directly by flushing surplus drugs down the toilet or after passing through the human body in the form of unabsorbed parent substances, metabolites and conjugates released in the urine and faeces. Another important source are the clinical laboratories, which release a lot of drug-contaminated waste into the sewage system (Hubner, 2001). Also, the manufacturing and production of medicines poses a threat, with possible releases to the environment.

In modern society, household and hospital sewage is often transported to and treated in wastewater plants. After treatment the processed water is released into a watercourse and the sludge produced is deposited or spread as fertilizer on agricultural land. The fate of the different drugs and drug metabolites in the wastewater plant is dependent on the actual substance and on the processes and efficiency of the plant. In many cases the substances pass through the plant more or less unchanged. In some cases conjugates are cleaved and free, active substance is released into the environment and in some cases the substance is metabolized, inactivated and destroyed (Figure 3.13). Drug substances suggested or identified to be of environmental concern are synthetic sex steroids, antibiotics, cytostatics (e.g. cyclophosphamide and platinum-containing formulations), drugs for lowering blood lipid levels (e.g. gemfibrozil and bezafibrate), anticonvulsants (e.g. carbamazepine), β-blockers (e.g. metoprolol and propranolol) and fluoxetine formulations, but also other substances might very well appear as candidates in this concern (Ternes, 1998).

Environmental monitoring of drugs has revealed the occurrence of many substances in terrestrial as well as aquatic environments. Thus, antibiotics (e.g. oxytetracycline in the order of mg/kg dry weight) have been detected in sediments and in wild fish in the surroundings of fish farms (Jacobsen and Berglind, 1988; Björklund *et al.*, 1990), river sediments have been shown to contain human as well as veterinary drugs (Zuccato *et al.*, 2000) and so has municipal wastewater (Ternes, 1998). Of special concern in this context are the endocrine-disrupting substances (Tyler *et al.*, 1998).

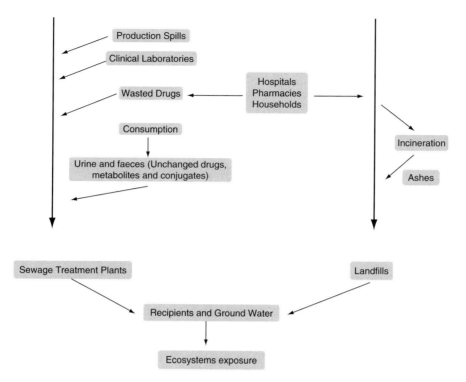

Figure 3.13 An overview of the fate of medical drugs in the environment.

For most of the drug substances the environmental impact is unknown, but for different types of antibiotics or antimicrobials a lot of concern is raised because of the development of bacterial resistance towards the drugs through different molecular mechanisms (Kruse and Sörum, 1994). This poses a threat to animals as well as humans, with the development of multiresistant, sometimes pathogenic, microbial strains (Midtvedt, 1998)

Another group of drugs of serious concern from an environmental point of view is the human female contraceptive based on estrogens and especially ethinylestradiol, a synthetic estrogenic substance used in modern pills. Ethinylestradiol is very persistent in the body and is excreted mainly in an unchanged but conjugated form. In the wastewater treatment plant, the conjugates are broken and the free hormonal substance is liberated to become available to living organisms, e.g. fish, in the receiving water, causing endocrine disruption and feminization of male fish. Measurement of increasing yolk protein synthesis (vitellogenesis) acts as a useful biomarker in this context (Larsson *et al.*, 1999).

The awareness of environmental problems associated with the medical and veterinary use of drugs has arisen recently and it will be an area of much concern for the future. Different protocols for the evaluation of environmental risks from drugs are now being prepared by different organizations (see EMEA, 2003).

3.6 Acid rain and acidification of the environment

A serious environmental problem caused by the combustion of fossil fuels such as oil and coal is the production of sulphur dioxide (SO_2) and nitrogen oxides (NO_x). After oxidation in the atmosphere and local or long-range distribution, precipitation and exposure of the environment to sulphuric acid (H_2SO_4), and nitric acid (HNO_3), will take place. This will result in a titration of buffering substances, e.g. bicarbonate (HCO_3^-) and others, in the terrestrial and aquatic ecosystems and a concomitant decrease in pH, especially in sensitive environments with low buffering capacity. This is the case for Scandinavia, for example, but also in many other areas with bedrocks dominated by acidic types of rock such as granite and gneiss (Schofeld, 1976).

The acid precipitation and concomitant decrease in pH – in most cases from around 6.5 to 4.5, corresponding to an increase in H^+ concentration of 100 times – will exert a heavy impact on the ecosystems with a lot of restructuring. From a toxicological point of view, however, the most striking effects are found in the aquatic ecosystems, with changes in the flora resulting in a predominance of peatmoss (*Sphagnum* spp) and in differing effects on the survival of animals, including fish.

In fish, the primary effects of acidification are exerted on the ion-regulating mechanisms in the gills. In fresh water, the body fluids of fish are more concentrated, i.e. they contain more ions and other dissolved substances than the surrounding water. As a result of this, the fish constantly gain water through osmosis and lose ions through diffusion. These losses and gains have to be compensated through active, energy-demanding transport processes taking place mostly in the gill epithelia. An elevated concentration of H^+, e.g. lowered pH, will impair the essential ion-regulating processes of the gill chloride cells, with concomitant losses of ions and decreased blood levels of Na^+, K^+, Ca^{2+} and Cl^- (Leivestad, 1982; Goss et al., 1996). This, in turn, will disturb the important homeostasis and affect many important physiological mechanisms in the organisms, creating a lot of disturbances at integrated levels.

At very low pH levels, i.e. 3.5–4, the exchange of H^+ produced in the animal metabolism at the gill surfaces is very much impaired, resulting in acidosis and a concomitant Bohr effect with diminished oxygen-carrying capacity of the blood haemoglobin (Leivestad, 1982). This cytotoxic anoxia is exaggerated further by an increased mucus production covering the gills, in part as an effect of increased levels of Al^{3+} (being released in significant amounts from the ground) as part of the natural buffering system at these low pH levels. The aluminum ions also contribute in an additive manner to the H^+ effects on the ion-exchange mechanisms (Potts and McWilliams, 1989). In conclusion, lowered pH in the environment will lead to disturbances in ion homeostasis and impaired oxygen uptake in aquatic animals and, as a consequence, impaired physiological performance and ultimately death. The most sensitive phase is reproduction and the newly hatched fry, and acidification of a lake is often observed through the absence of young generations of different species.

Acid rain as a result of the burning of fossil fuels is an enormous environmental problem in sensitive areas. The situation has improved somewhat during recent years as a result of desulphuration of oil and the installation of scrubbers, etc. for removal of SO_2, but the handling of NO_x (to a great extent produced from different transports) is more difficult. Furthermore, NO_x is produced through incorporation of atmospheric nitrogen during the combustion process.

Another problem with lowered pH in the environment is the mobilization and increased bioavailability of many metals. This will give rise to increased exposure and additional problems from Hg, Cd and Pb, among others (McDonald *et al.*, 1989). In Scandinavia, significant efforts are made to add chalk to the environment. For example, in Sweden around 200 000 tons/year are added (Swedish Environmental Protection Board, 2004), thus giving nature a boost to prevent further acidification.

References

Atchisson, G.J., Sandheinrich, M.B. and Bryan, M.D. (1996) Effects of environmental stressors on interspecific interactions of aquatic animals. In *Ecotoxicology: a Hierarchical Approach*, Newman, M.C. and Jagoe, C.H. (eds), pp. 319–337. Lewis, Chelsea, MI.

Bispo, A., Jourdain, M.J. and Jauzein, M. (1999) Toxicity and genotoxicity of industrial soils polluted by polycyclic aromatic hydrocarbons (PAHs). *Organic Geochemistry*, **8B**, 947–952.

Björklund, H., Bondestam, J. and Bylund, G. (1990) Residues of oxytetracycline in wild fish and sediments from fish farms. *Aquaculture* **86**, 359–367.

Brooks, G.T. (1974) *Chlorinated Insecticides* (2 vol). CRC, Cleveland, OH.

Brouwer, A., Murk, A.J. and Koemann, J.H. (1990) Biochemical and physiological approaches in ecotoxicology. *Functional Ecology*, **4**, 275–281.

Bryan, G.W., Gibbs, P.E. and Burt, G.R. (1988) A comparison of the effectiveness of tri-*n*-butyltin chloride and five other organotin compounds in promoting the development of imposex in the dog whelk *Nucella lapillus*. *Journal of the Marine Biological Association of the United Kingdom*, **68**, 733–744.

Burchiel, S.W. and Luster, M.I. (2001) Short analytical review: signaling by environmental polycyclic aromatic hydrocarbons in human lymphocytes. *Clinical Immunology*, **98**, 2–10.

Conney, A.H. (1982) Induction of microsomal enzymes by foreign chemicals and carcinogenesis by polycyclic aromatic hydrocarbons. *Cancer Research*, **42**, 4875–4917.

Darnerud, P.O. (2003) Toxic effects of brominated flame retardants in man and in wildlife. *Environment International* **29**, 841–853.

Darnerud, P.O., Eriksen, G., Johannesson, T., Larsen, P. and Viluksela, M. (2001) Polybrominated diphenyl ethers: occurrence, dietary exposure and toxicology. *Environmental Health Perspectives*, **109**, 49–68.

de Boer, J., de Boer, K. and Boon, J.P. (2000) Polybrominated biphenyls and diphenylethers. In *The Handbook of Environmental Chemistry* 3. *Part K. New Types of Persistent Halogenated Compounds*, Paasivirta, J. (ed.), pp. 61–95. Springer Verlag, Berlin.
de Wit, C.A. (2002) An overview of brominated flame retardants in the environment. *Chemosphere* **46**, 583–624.
de Wit, C.A. and Strandell, M. (1999) *Levels, Sources and Trends of Dioxins and Dioxin-like Substances in the Swedish Environment – The Swedish Dioxin Survey, vol. 1*, Report no. 5047. Swedish Environmental Protection Agency, Stockholm, Sweden.
Eisler, R. (1989) Pentachlorophenol hazards to fish, wildlife and invertebrates: a synoptic review. *U.S. Department of the Interior, Fish and Wildlife Service Biological Report*, **85**, 1–17.
Eldefrawi, M.E. and Eldefrawi, A.T. (1990) Nervous – system based insecticides. In *Safer Insecticides – Development and Use*, Hodgson, E. and Kuhr, R.J., (eds), pp. 155–207. Marcel Dekker, New York.
EMEA (2003) *Note for Guidance on Environmental Risk Assessment of Medicinal Products for Human Use*. The European Agency for the Evaluation of Medicinal Products, www.e-mea.eu.int.
Fergusson, J.E. (1990) *The Heavy Elements: Chemistry, Environmental Impact and Health Effects*. Pergamon Press, Oxford.
Gochfeld, M. (2003) Cases of mercury exposure, bioavailability, and absorption. *Ecotoxicology and Environmental Safety*, **56**, 174–179.
Goss, G.G., Perry, S.F., Fryer, J.N. and Laurent, P. (1996) Gill morphology and acid – base regulation in freshwater fishes. *Comparative Biochemistry and Physiology*, **119A**, 107–115.
Grundy, M.M., Ratcliffe, N.A. and Moore, M.N. (1996) Immune inhibition in marine mussels by polycyclic aromatic hydrocarbons. *Marine Environmental Research*, **42**, 187–190.
Hakk, H. and Letcher, R.J. (2003) Metabolism in the toxicokinetics and fate of brominated flame retardants – a review. *Environment International*, **29**, 801–828.
Hale, R.C., Alaee, M., Manchester-Neesvig, J.B., Stapleton, H.M. and Ikonomou, M.G. (2003) Polybrominated diphenyl ether flame retardants in the North American environment. *Environment International*, **29**, 771–779.
Halling-Sörensen, B., Nors Nielsen, S., Lanzky, P.F., Ingerslev, F., Holten Lützhof, H.C. and Jörgensen, S.E. (1998) Occurrence, fate and effects of pharmaceutical substances in the environment – a review. *Chemosphere*, **36**, 357–393.
Hassall, K.A. (1990) *The Biochemistry and Uses of Pesticides* (2nd edn). Macmillan, London.
Hoffmann, D. (1995) *Handbook of Ecotoxicology*, pp. 333–347. Lewis Publishers, Boca Raton, FL.
Hogstrand, C. and Wood, C.M. (1998) Toward a better understanding of the bioavailability, physiology and toxicity of silver in fish: implications for water quality criteria. *Environmental Toxicity and Chemistry*, **17**, 547–561.
Hsu, P.-C., Guo, Y.L. and Li, M.-H. (2004) Effects of acute postnatal exposure to 3,3',4,4'-tetrachlorobiphenyl on sperm function and hormone levels. In adult rats. *Chemosphere*, **54**, 611–618.
Hubner, P. (2001) Emissions from clinical – chemical laboratories. In *Pharmaceuticals in the Environment: Sources, Fates, Effects and Risks*, Kümmerer, K. (ed.), pp. 43–48. Springer Verlag, Berlin.
Jacobsen, P. and Berglind, L. (1988) Persistence of oxytetracyclin in sediments from fish farms. *Aquaculture*, **70**, 365–370.
Janssen, P.A.H., Faber, J.H. and Bosveld, A.T.C. (1998) (Fe)male? *IBN Science Contribution*, **13**.
Jansson, B., Andersson, R., Asplund, L., Litzén, K., Nylund, K., Sellström, U., Uvemo, U.-B., Wahlberg, C., Wideqvist, U., Odsjö, T. and Olsson, M. (1993) Chlorinated and brominated persistent organic compounds in biological samples from the environment. *Environmental Toxicology and Chemistry*, **12**, 1163.
Jensen, S. (1966) Report of a new environmental hazard. *New Scientist*, **32**, 612.
Kerkvliet, N.I. (1995) Immunological effects of chlorinated dibenzo-*p*-dioxins. *Environmental Health Perspectives*, **103**, 47–53.
Klaassen, C.D. (1996) *Casarett and Doull's Toxicology: the Basic Science of Poisons* (5th edn). McGraw-Hill, New York.

Kruse, H. and Sörum, H. (1994) Transfer of multiple drug resistance plasmids between bacteria of diverse origins in natural microenvironments. *Applied Environmental Microbiology*, **60**, 4015–4021.

Larsson, D.G.J., Adolfsson-Erici, M., Parkkonen, J., Pettersson, M., Berg, A.H., Olsson, P.-E. and Förlin, L. (1999) Ethinyloestradiol – an undesired fish contraceptive? *Aquatic Toxicology*, **45**, 91–97.

Leivestad, H. (1982) Physiological effects from acid stress on fish. *Proceedings of International Symposium on Acid Rain and Fisheries*, American Fisheries Society, Bethesda, pp. 157–164.

Matthiessen, P. and Gibbs, P.E. (1998) Critical appraisal of the evidence for tributyltin-mediated endocrine disruption in mollusks. *Environmental Toxicology and Chemistry*, **17**, 37–43.

McDonald, D.G., Reader, J.P. and Dalziel, T.R.K. (1989) The combined effects of pH and trace metals on fish ionoregulation. *Seminar Series: Acid Toxicity and Aquatic Animals*, pp. 221–242. Cambridge University Press, Cambridge.

Merian, E. (ed.) (1991) *Metals and their Compounds in the Environment*. VCH, Weinhem.

Midtvedt, T. (1998) The microbial threat – the Copenhagen recommendation. *Microbial Ecology in Health and Disease*, **10**, 65–67.

National Research Council (1986) *Drinking Water and Health*, vol. 6. National Academy of Sciences, Washington, DC.

Nicolas, J.M. (1999) Vitellogenesis in fish and the effects of polycyclic aromatic hydrocarbon contaminants. *Aquatic Toxicology*, **45**, 77–90.

Nieboer, E. and Richardson, D.H.S. (1980) The replacement of the nondescript term 'heavy metals' by a biologically and chemically significant classification of metal ions. *Environmental Pollution*, **1B**, 3–26.

Olsson, M., Andersson, Ö., Bergman, Å., Blomkvist, G., Frank, A. and Rappe, C. (1992) Contaminants and diseases in seals from Swedish waters. *Ambio*, **21**, 561–562.

Olsson, M., Bignert, A., de Wit, C. and Haglund, P. (2003). Dioxiner i Östersjöns fisk – ett hot mot svenskt fiske. Östersjö 2003, Stockholm Marine Research Centre, pp. 3–9 (with English summary).

Paasivirta, J. (1991) *Chemical Ecotoxicology*. Lewis Publishers, Chelsea, MI.

Peacall, D.B. (1993) DDE-induced eggshell thinning: an environmental detective story. *Environmental Review*, **1**, 13–20.

Pollitt, F. (1999) Polychlorinated dibenzodioxins and polychlorinated dibenzofurans. *Regulatory Toxicology and Pharmacology*, **30**, 63–68.

Potts, W.T.W. and McWilliams, P.G. (1989) The effects of hydrogen and aluminium ions on fish gills. *Seminar Series: Acid Toxicity and Aquatic Animals*, pp. 201–220. Cambridge University Press, Cambridge.

Richardson, M.L. and Bowron, J. (1985) The fate of pharmaceuticals in the aquatic environment. *Journal of Pharmacology*, **37**, 1–12.

Richter, C.A., Drake, J.B., Giesy, J.P. and Harrison, R.O. (1994) Immunoassay monitorring of polychlorinated bihenyls (PCBs) in the Great Lakes. *Environmental Science and Pollution Research*, **1**, 69–74.

Rosander, P. (2002) In *Läkemedel i Miljön*, pp. 14–20. Apoteket AB, Stockholm, Sweden.

Sibly, R.M., Newton, I. and Walker, C.H. (2000) Effects of dieldrin on population growth rates of sparrowhawks 1963–1986 *Journal of Applied Ecology*, **37**, 540–546.

Schofeld, C.L. (1976) Acid precipitation: effects on fish. *Ambio*, **5**, 228–230.

Skaare, J.U., Larsen, H.J., Lie, E., Bernhoft, A., Derocher, A.E., Norstrom, R., Ropstad, E., Lunn, N.F. and Wiig, Ö. (2002) Ecological risk assessment of persistent organic pollutants in the arctic. *Toxicology* **181–182**, 193–197.

Swedish Environmental Protection Board (2004) www.environ.se.

Ternes, T. (1998) Occurrence of drugs in German sewage treatment plants and rivers. *Water Research*, **32**, 3245–3260.

Tomy, G.T., Fisk, A.T., Westmore, J.B. and Muir, D.C.G. (1998) Environmental Chemistry and toxicology of polychlorinated n-alkanes. *Review of Environmental Contamination and Toxicology*, **158**, 53–128.

Tyler, C.R., Jobling, S. and Sumpter, J.P. (1998) Endocrine disruption in wildlife: a critical review of the evidence. *Critical Review of Toxicology*, **28**, 319–361.

van der Oost, R., Beyer, J. and Vermeulen, N.P.E. (2002) Fish bioaccumulation and biomarkers in environmental risk assessment: a review. *Environmental Toxicology and Pharmacology*, **13**, 57–149.

van Loveren, H., Ross, P.S., Osterhaus, A.D.M.E. and Vos, J.G. (2000) Contaminant induced immunosuppression and mass mortalities among harbor seal. *Toxicology Letters*, **112**, 319–324.

Walker, C.H. and Newton, I. (1999) Effects of cyclodienes insecticides on raptors in Britain – correction and updating of an earlier paper. *Ecotoxicology*, **8**, 185–189.

Walker, C.H., Hopkin, S.P., Sibly, R.M. and Peakall, D.B. (2001) *Principles of Ecotoxicology*. Taylor & Francis, London.

White, P.A. (2002) The genotoxicity of priority polycyclic aromatic hydrocarbons in complex mixtures. *Mutation Research/Genetic Toxicology and Environmental Mutagenesis*, **515**, 85–98.

Woodhead, R.J, Law, R.J. and Matthiessen, P. (1999) Polycyclic aromatic hydrocarbons in surface sediments around England and Wales, and their possible biological significance. *Marine Pollution Bulletin*, **38**, 773–790.

Zuccato, E., Calamari, D., Natangelo, M. and Fanelli, R. (2000) Presence of therapeutic drugs in the environment. *Lancet*, **355**, 1789–1790.

4 Frameworks for the application of toxicity data

Maria Consuelo Diaz-Baez and Bernard J. Dutka

4.1 Introduction

4.1.1 Background and objectives

With the explosive increase in worldwide industrialization and urbanization over the past five decades coupled with an increasing demand for chemicals (industrial, medical and agricultural), both the developed and developing nations face increasing ecological and toxicological problems from the release of toxic contaminants into the environment. These releases have triggered, internationally, an increasing concern about the impacts of these chemicals. Because there is no instrument devised by man to measure toxicity or genotoxicity, these important properties must be measured with the only suitable material available for this purpose – living organisms (Cairns and Pratt, 1989). This awareness has stimulated an ever growing number of intensive and extensive bioassay investigations into the toxic and genotoxic effects of discarded substances on the environment and on all life-forms.

In response to these increasing stresses on the environment and in the belief that there is, as yet, no single criterion by which we can judge adequately the potential hazard (either to the environment or to man) of a given substance (Draggan and Giddings 1978), a multitude of biological assays and procedures have been developed, proposed and used to assess the impact of toxicants. The increased ability to measure toxic effects has also brought the realization that in any sample or series of samples, collected from the same area, there was no reliable means of establishing whether the observed toxicity/genotoxicity effect was caused by a single constituent, or a combination of constituents or a number of different chemicals, each periodically varying in concentration to cause the observed test response.

Concomitantly it was realized that the tendency to focus all testing and monitoring on a selected group of chemicals, the so-called *priority pollutants*, ignored the thousands of chemicals already in the environment and those being newly created as commercial products or as by-products of biodegradation processes, along with the multitude of synergistic effects potentially occurring in every ecosystem. It must be remembered also that toxicants in air or water recognize no national boundaries, and foods grown in contaminated soils can be/are transported all over the world. Monitoring agencies became aware that the analysis of water, effluent, sediment and other solid-phase samples for every suspected chemical would be very expensive in terms of time and money, and still there would be no assurance that the detection procedures would be sensitive enough or the synergistic responses would be recognized.

One way of addressing these problems was the development of a number of short-term bioassays to screen the samples for an indication of toxic, genotoxic or chronic effects and then to prioritize the samples or sampled areas for chemical analysis and/or more intensive studies. Application of these tests to environmental samples soon revealed that there was no single bioassay that was responsive to all toxicants or mixtures of toxicants. This realization led to the concept of using several bioassays to ascertain environmental water quality and the ecological impacts of effluents, discharges and emissions that became known as the 'battery of tests' approach. A test battery would incorporate a range of trophic levels, life stages, routes of exposure and toxic end-points to approximate more reasonably the real world.

As these short-term screening tests were developed it was found that they could be used to assist chemists and engineers in targeting specific in-plant toxic stream flows and specific extraction fractions of various industrial processes as sources of toxicants/genotoxicants and oestrogenic chemicals. This partnership with chemists could take place by one of the following means. One option is for chemists to quantify the concentrations of expected pollutants and then to check through bioassays whether or not there are other pollutants at risk concentrations. The observed toxicity must be explained on the basis of the chemicals detected. This assessment can be done theoretically by using the toxicity of pure chemicals for the test used or, alternatively, by running the bioassays on a reconstituted sample of similar characteristics spiked with the chemicals at the measured concentrations (Tarazona et al., 1995). However, this option ignores any changes of effects due to synergism or antagonism.

Another option that does not need any kind of prediction on which chemicals could be hazardous starts with the biological screening of a sample using a 'battery of tests' approach. If toxicity is detected, the identification of the toxic substances can be investigated using Toxicity Identification Evaluation (TIE) protocols (USEPA, 1988a,b, 1989a,b, 1991a, 1996, 2001). These methods are useful for any complex mixture, be it an industrial effluent, surface water, sediment pore-water or hazardous waste leachate (Norberg-King et al., 1991; Kszos et al., 1992). The sample is subjected to a set of physical and chemical manipulations and the strength of toxicity before and after each treatment is used to identify the toxic fractions and to guide the chemical analysis. However, neither toxicity interactions (synergistic or antagonistic effects) nor the original bioavailability of chemicals producing the toxic effects can be established.

Because industrial effluents, emissions and toxicants such as pesticides, car exhausts, landfill site leachates, etc. affect life forms at different levels and in many ways, it has become accepted practice that single or multitrophic toxicity tests should be used in monitoring schemes. Researchers are now using a battery of biological tests to estimate the toxicity and mutagenicity of industrial effluents and their impact on receiving waters.

In general, there are two main categories of toxicity screening tests: *in vitro* health effect tests and ecological effect tests. Health effect toxicity tests are based on the use of subcellular components (e.g. enzymes, DNA, RNA), isolated cells (e.g. E-screen, cell cultures, red blood cells), tissue sections or isolated whole organs (Soto *et al.*, 1992). The E-screen test uses human breast estrogen-sensitive MCF7 cells. The assay compares the cell yield after 6 days of incubation in medium supplemented with 5% charcoal–dextran stripped fetal bovine serum in the presence or absence of estradiol (positive and negative controls) to various concentrations of chemicals/samples suspected of being estrogenic. The other tests consist of determining cell viability (vital staining with the dye inclusion test, plating efficiency, colony formation, cell reproduction, DNA unwinding, micronucleus increase or macronuclear biosynthesis). Ecological effect tests are mainly conducted to measure/assess the direct toxicity of chemicals to organisms representing various trophic levels of the food chain. These tests help in the estimation of chemical toxicity in natural and man-modified ecosystems. Bacteria and their enzymes, algae, zooplankton, benthic invertebrates and fish have been used (USEPA, 2002a,b).

Organisms are exposed to a wide range of toxic, organic and inorganic compounds in water systems, sewage treatment processes, industrial wastes and soils. The toxicity of the compounds depends on environmental parameters as well as on the organisms or enzyme systems being tested. The compounds may be altered metabolically to non-toxic metabolites or may exert a direct toxic action on populations. They may be subjected also to synergistic or antagonistic effects between components of toxicant mixtures.

Toxic action is concentration dependent. For example, phenol can be metabolized at low concentrations but becomes toxic at higher concentrations. These types of actions often lead to the occurrence of a hormesis effect, i.e. the stimulating effect of small doses of a toxicant that is known to be inhibitory at larger concentrations. Toxicant action also depends on the presence of other chemicals in solution (Dutka and Kwan, 1982).

There are many known and proposed mechanisms by which toxicants inhibit and eventually kill organisms: halogens, Hg^{2+} and Cd^{2+} may cause damage to the genetic material or their effect may lead to protein denaturation; phenol and quaternary ammonium compounds, may disrupt bacterial cell membranes, the end result being the leakage of DNA, RNA, proteins and other organic materials; and acids and alkalis may displace cations such as Na^+ and Ca^{2+} from adsorption sites on bacterial cells. A more subtle action of toxic pollutants is their ability to block bacterial chemoreceptors (Mitchell and Chet, 1978), which may lead to the inhibition of organic decomposition and self-purification processes in sewage treatment plants and in waters receiving organic materials (Mitchell and Chet,

1978). However, it is believed that one of the most important effects of the toxic action of chemicals is on enzyme activity (Iverson and Brinckman, 1978). Also, in any toxicity study one must take into account the physicochemical factors (presence of other cations, pH, oxidation–reduction potential, temperature, organic matter, clay particles, etc.) that control the toxic action towards organisms (Sprague, 1985; Brezonik et al., 1991).

From this brief overview of toxicity tests, their uses and methods of assessing their responses to toxic substances, it can be seen that this is a very complex topic with many unknowns and no single best way of addressing the problem of bioavailable toxicant estimation. In this chapter, an attempt will be made to describe some applications of toxicological data, how they could be analysed and how toxicity results could be integrated in ecological risk assessment frameworks.

4.2 The purpose of bioassays

4.2.1 Toxicity tests within a triad of techniques

Traditionally, environmental assessment has been directed mainly towards chemical characterization, which has resulted in the establishment of guidelines for regulatory purposes. Such parameters lack the dynamic toxicity information needed to determine the bioavailability of contaminants to the biota residing in an ecosystem (Munawar et al., 1989).

The recognition of the need to base environmental management on the results of toxicological experimentation was an important step in the management of pollution. Several attempts were made to develop rationale and approaches (Giesy et al., 1988; Cairns and Pratt, 1989; Calow, 1989; Giesy and Graney, 1989; Cairns et al., 1995) to measure and deal with environmental perturbation (Blaise et al., 1985; OECD, 1987; Sergy, 1987; Persoone et al., 1989; Scroggins, 1999). As a result of various initiatives, the concept of ecosystem health emerged in environmental toxicology. Under this approach, chemical characterization, toxicity testing and ecoassessment constitute a triad of techniques to deal with whole biological communities in the ecosystem.

As can be seen in the diagram (Figure 4.1), toxicity testing cannot be a substitute for chemical measurements or for surveys of communities of organisms. The strengths of toxicity testing are defined in conjunction with chemical and biological field measurements. These three approaches form a natural triad in which each component enhances the power of the others (Sergy, 1987; Scroggins, 1999). (see also Chapter 9B).

Toxicity tests use biological systems to detect the presence of toxic chemicals in the samples being investigated (e.g. water, effluents, sediments, soil). The term bioassay also is used to describe a toxicity test. The idea is that the response of an organism will be representative of organisms living in the environment where the

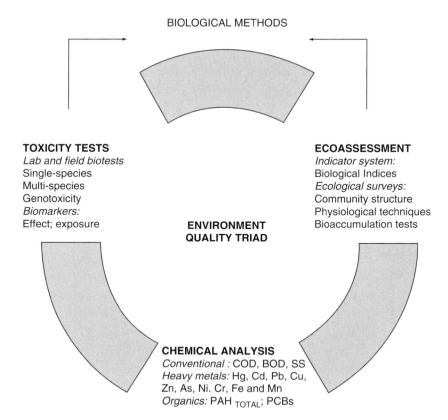

Figure 4.1 Schematic diagram of the triad of techniques that can be used, alone or in combination, in environmental toxicology. COD = chemical oxygen demand; BOD = biochemical oxygen demand; SS = suspended solids.

substance would be released. However, no 'perfect' toxicity test has been developed to represent all biological groups (microbes, plants, invertebrates, fish, etc.) in the ecosystem.

Chemical tests can be used in two different ways. They can be regulatory if a limit is specified for a certain substance in an effluent and they can be predictive by calculating, from the discharge, the concentration expected in the environment; both are achieved by comparison with established quality guidelines for water, sediment or soil (Scroggins, 1999).

Chemical analysis, however, continues to be used to assess the quality of industrial and municipal wastes, as well as to assess the limits of concentration or loading that can be applied to the environment. Such limits have been tied to 'best-treatment' performance of waste treatment plants, also termed *best available treatment* (BAT) and *best available treatment economically achievable* (BATEA).

Biological surveys in the receiving water or in the substrate of the receiving environment are important. These surveys are the definitive check on whether an ecosystem has been affected. Their use is a follow-up audit; however, biotic surveys can hardly serve in a predictive mode to prevent damage, unlike toxicity tests.

Ecoassessments focus directly on populations or communities in order to detect if the exposed biocenosis is changing, how it has changed and whether there will be a change. They cover the methodologies related to analysis of biological communities and can be based on different factors such as the structure of the community, the presence and/or status of sentinel species or the use of biological alterations in individuals related to chemical exposure, e.g. crossed-bills in immature cormorants.

The use of biological indicator systems started more than 100 years ago and is now regarded as an essential tool for river quality monitoring (de Pauw et al., 1992). Several biological indices based on species abundance and diversity are available. These have been reviewed and compared extensively and show good agreement among different assessment methods (Hellawell, 1986; de Pauw et al., 1992). Some physiological or biochemical indices on sentinel populations, i.e. molluscs, have also been proposed (Livingstone, 1991).

Any type of effect produced in an organism due to exposure to a chemical can be considered as a potential indicator for bioassessment. Waldichuck (1989) has categorized the effects that can be studied at the sublethal level, e.g. physiological, biochemical, morphological, pathological, genetical, behavioural and ecological, most of which can be used as potential biomarkers. In an attempt to prevent adverse biological effects in humans and wildlife as a result of contaminant exposure in aquatic ecosystems, a large range of biochemical and molecular indicators of toxicant exposure are being developed. These so-called 'biomarkers' of exposure have been promoted as a first tier in exposure monitoring programs. These early-detection tools can provide sentinel warnings of contaminant exposure in aquatic environments before irreversible, deleterious changes are effected in whole populations of organisms.

The major advantage of using biomarkers is that the results directly answer the important question: are the populations exposed to toxic concentrations of pollutants? Because the effect is seen directly in the population itself, there are fewer concerns with bioavailability, kinetics or factors that modify the toxicity of the media (water, sediment, soil, air). However, the major disadvantage of biomarkers is the lack of information on the biology of wild species (see also Chapter 7).

Environmental agencies and researchers that have embraced toxicity testing to regulate discharges have emphasized the importance of combining toxicity testing with more conventional chemistry and with bioassessments (Blaise et al., 1985; OECD, 1987; Klemm et al., 1990; Hunt et al., 1992; Scroggins, 1999). The three techniques in the triad play different roles and the results should be integrated to provide an early and cost-effective prediction of environmental effects, at an acceptable level of uncertainty, to guide regulatory and industrial decision-making.

4.2.2 Advantages and disadvantages of toxicity testing

There are many applications of toxicity tests but in general they have been used as an environmental tool to provide information about:

- Identification and rank order of toxicity potential of samples
- Definition of the real extent of toxicity
- Identification of the chemical fractions of a complex waste that contribute to toxicity (toxicity identification evaluation, TIE)

From all these uses it is clear that the main advantage of bioassays is that they can provide information about toxicity. However, the strongest disadvantage is that the bioassays cannot identify the substance causing an effect (Scroggins, 1999). This is one of the reasons why toxicity testing should be complemented with chemical characterization.

Associated with this first advantage, toxicity tests are used to measure directly the whole sample toxicity, without identifying and analyzing each pollutant individually. Whole-effluent bioassays are probably more representative of the total toxic effect of the complex effluent mixture. Toxicity tests give an integrated answer of the total effect of diverse toxicants in a given complex effluent.

Toxicity test results expressed in terms of harm to living organisms are more credible or comprehensible than chemical data to non-specialists (Sergy, 1987). They can also be less expensive than comprehensive chemical analysis. However, a toxic finding could be misleading if used in a simplistic way, e.g. failing to allow for dilution or mitigating factors.

Toxicity testing has been a contentious issue. Its opponents argue that toxicity tests are complex, expensive and imprecise. In spite of these arguments, as standardized toxicity tests are developed and their interpretation and regulatory applications are refined, it is clear that toxicity tests will become powerful instruments for pollution monitoring and control. They measure directly the potential environmental effects of complex mixtures and provide information that cannot be obtained by any other means.

4.3 Interpretation of toxicological data

4.3.1 Field validation

The laboratory-based assessment of toxicity using bioassays is limited not only by laboratory conditions but also by the type of species used. As a matter of practice, tests are conducted under controlled and standardized conditions of light, temperature, oxygen, pH, etc. so that the precision of the results obtained is reliable and so that control over variables such as volatilization and biodegradability is maintained (USEPA, 2002a,b).

Even though measurements of toxic effects obtained under laboratory conditions do not always correspond with actual results obtained in the field when toxicants are liberated, they can be used to predict events that could affect populations in a given ecosystem. The difficulty of validating laboratory results with actual field results has been pointed out by several authors over the years. Nevertheless, a considerable number of papers have been published in which the predictive value of toxicity tests has been demonstrated (Stephan *et al.*, 1985; Eagleson *et al.*, 1990; Power and McCarty, 1997; Scroggins, 1999). One of the main difficulties in this context has to do with the risk of making predictions for complex organization levels (populations and communities) based on a single species (individual level). In addition, factors such as the presence and variability of other active contaminants make it even more difficult to deal with this problem. However, specific studies to evaluate the predictive value of toxicity tests have been conducted by governmental agencies (USEPA, Environment Canada) and have shown a positive correlation between field and laboratory results (Geckler *et al.*, 1976; AETE, 1998).

Most of these studies include field validation or the simulation of communities in 'mesocosms'. There are, however, no clear guidelines to carry out these studies and therefore the possibilities for comparison are limited. In order to validate toxicity test results, some authors (Sanders, 1985; Cairns, 1986, 1988) have recommended the need to develop an appropriate experimental design, and the OECD (1992a) has established a number of principles and recommendations for these assays. According to Sprague (1995), there are two levels of validation procedures with varying degrees of complexity:

- Field validation in which the correlation between toxicity tests results and quality measurements in the receiving community is established. The degree of correlation is established by comparison of the results and therefore can be subjective in many cases.
- Scientific field validation, in which strict scientific methods based on formal experiments are used to establish whether toxicity tests do in fact predict the effects of the toxicants on the community. Previously defined acceptability criteria are applied and, consequently, it is possible to establish the level of agreement between laboratory and field results. Under this type of validation it is important to follow a series of steps that can be performed one by one or, in some cases, simultaneously.

The main steps of the scientific field validation are to:

1. Verify the adequacy of the toxicity tests carried out under laboratory conditions.
2. Select the study sites together with a control site.
3. Check existing information about the site, including physical, chemical and biological data.

4. Carry out a preliminary sampling program to establish the magnitude and variability of the site, i.e. wastewater discharge, plume size of the receiving water and the composition of the biological community.
5. Develop a preliminary modeling and analysis method (a phase that can be carried out in parallel with the previous steps).
6. Develop a detailed validation program.
7. Implement the program.

4.3.2 Application factors

Application of toxicity test results was accomplished originally by multiplying median lethal concentration (LC_{50}) values obtained in acute tests by factors of less than unity, in order to predict the concentration in the 'receiving water' at which no negative effects would occur (Henderson and Tarzwell, 1957). These factors were intended to be a realistic representation of the ratio between the LC_{50} and a concentration that just fails to cause sublethal toxicity (i.e. the sublethal threshold or the no-observed-effect concentration NOEC). The application factors have been used to predict the maximum acceptable toxicant concentration (MATC). The MATC is set equal to the product of the application factors and the LC_{50} value (Giesy and Graney, 1989).

The use of application factors was centred on substances that were non-persistent or had non-accumulative effects after mixing with the receiving waters. Recommended factors were designated at 1/10th of the 96-h LC_{50} (as a maximum at any time or place) and at 1/20th or less as a 24-h average concentration after mixing. For toxicants that were persistent or cumulative, the threshold concentrations were required not to exceed 1/20th and 1/100th of the 96-h LC_{50} values.

In some ways a security margin was being sought, because there was a lack of precise knowledge about the sensitivity of the aquatic organisms (Stephan *et al.*, 1985). In addition, a more realistic prediction of the aquatic community concentration threshold was expected to be made. The system was easy and fast because it was based on a lethal concentration obtained from one or several species.

Application factor values have been found to vary by a magnitude of 10 000 and sometimes even as great as 100 000 for different compounds. Moreover, no relationship could be found between the acute lethality (LC_{50}) of the compounds and the application factors. However, when the relationships between the application factors for a pair of species were examined, it was found that the application factor of one species could be predicted from that of another species (Kenaga, 1977).

In spite of the fact that numerical values originally were established intuitively, they became more standardized as new investigations began to generate information on sublethality (Sprague, 1971). During the 1960s, the USA and Canada approved their use for substances of unknown non-effective concentration (NTAC, 1968; OWRC, 1970).

Application factors are still used at the present time, although on a limited scale, particularly when enough knowledge is available regarding the sublethal effects of the toxic substances. For example, application factors of 0.01 and 0.05 have been established in Canada for persistent and non-persistent toxic substances, respectively (CCME, 1991). The potential health risks arising from using these application factors appear to be acceptable until clearer cause and effect dosages are available.

4.3.3 Acute to chronic ratio (ACR)

The definition of the acute to chronic ratio (ACR) has been one of the methods used to predict the threshold concentration at which a toxicant does not produce noticeable effects during a chronic exposure. This ratio is based on the same concept as the application factors, but its numerical value is the inverse (Stephan, 1982). The ACR is the ratio between chemical concentrations exerting a lethal versus sublethal toxic effect and describes the ratio of a lethal to sublethal end-point:

$$ACR = LC_{50}/NOEC \text{ (or } IC_{25})$$

This very simple value has proved to be an effective tool to predict chronic effects. Its main advantage stems from the fact that it can be used to predict, from known ACR values, possible chronic responses in similar species or with similar toxicants. It was possible to use this approach because at least 90% of the substances being evaluated showed that the concentration at which no effect was observed was found to be at least two orders of magnitude below the LC_{50} values (Kenaga, 1982a,b).

Studies aimed at verifying this ratio have shown a 0.88 correlation coefficient for 11 freshwater species and 126 chemicals (Slooff et al., 1983). It was concluded on the basis of these results that compounds that produce acute toxicity also show chronic toxicity. Therefore, with the exception of substances with a very particular toxic action, it is possible to predict the chronic toxicity of a substance on a species based on the acute values obtained for the same compound with similar species (Giesy and Graney, 1989).

In general, the term ACR describes the relation between a lethal and a sublethal end-point (e.g. LC_{50}/IC_{25}), making it possible to estimate the sublethal value known as the threshold observed effect concentration (TOEC). The result, which is given in the form of a whole number, is easy to understand, making the technique quite popular. It is recommended, however, to use this ratio only when the LC_{50} and sublethal values for a given species and a particular toxicant are known. Its use for predictive purposes is similar to that of the application factors, especially when used to predict the TOEC when modifying the environmental conditions for similar industrial discharges and in special cases for other species.

Some environmental protection agencies have indicated that an ACR value of 10 can be used for industrial effluents and common chemicals when no direct measurements are available (Scroggins, 1999). The geometric mean should be used when several toxicity tests for the same species are available. A similar procedure can be used when sublethality test results with the same end-point are known. When two sublethality test results are available (one for reproduction, e.g. IC_{25}, and one for growth, e.g. IC_{50}), the test results that indicate the lowest value (most sensitive) should be used to calculate the ACR (OECD, 1992b).

4.3.4 Toxic units (see also Chapter 11, Section 11.2.5)

As already mentioned, toxicity results obtained in the laboratory cannot always be interpreted directly. Often it is necessary to transform these values into units that are more meaningful in environmental management programs. One way to facilitate the use of toxicity data, whether for regulatory purposes or for modeling of toxicity, is to use *toxic units*.

Toxic units are a measure of toxicity in a sample as determined by the *acute toxic units* (TU_a) or *chronic toxicity units* (TU_c) measured. An acute toxic unit is the reciprocal of the effluent concentration that causes 50% of the organisms to die by the end of the acute exposure period (USEPA, 1991c). It is calculated by expressing the real or true concentration of the effluent as the numerator (which for an effluent will be 100%) and the lethality end-point as the denominator, as follows:

$$\text{Acute toxic units } (TU_a) = 100/LC_{50}$$

However, it should be remembered that for contaminated media such as industrial effluents it is not possible to have TU_a values of less than unity because the lethal concentration LC_{50} cannot be higher than 100% (undiluted effluent). On the other hand, for chemicals found in the environment other units of concentration would be used. If the concentration of a chemical was 5 mg/l within the discharge plume, then this would be the concentration used in the numerator to calculate the toxic unit. If the chemical had an LC_{50} of 0.5 mg/l for a particular species, that would be the denominator, and the toxic strength of the plume would be $5/0.5 = 10\,TU_a$ of the chemical. Because the TU_a was greater than unity, strong lethal effects would be expected (Scroggins, 1999).

Chronic toxic units (TU_c) can be calculated in the same way as with acute exposure (USEPA, 1991c). The TU_c is the reciprocal of the effluent concentration that causes no observable effect on the test organisms by the end of the chronic exposure period (i.e. 100/NOEC). It needs to be remembered, however, that TU_a and TU_c values are different and cannot be mixed in the calculations. Toxic units are dimensionless, so values obtained with other samples and compounds can be compared. Because these values depend on the organism and the end-point used, these data should always be reported (Scroggins, 1999).

4.3.5 Toxicity indices

In general, an index corresponds to expression of the results obtained in different toxicity tests by means of a number that allows the toxicity of the sample to be classified.

There are examples of physical and chemical indices in water quality studies, with which a sample is classified and later compared with control values/sites to establish its own quality. Even though the use of these physicochemical indices is attractive, it should be remembered that they use sets of parameters that, in many cases, may not even include the main contaminant (toxicant). The resulting value may not show any type of hazard because it does not include the substance under investigation (Pratt et al., 1971).

A precursor to a toxicity index is the common use of tests to rank environmental samples or hazardous sites according to the severity of the toxic responses. Toxicity ranking defines priority for action on the most toxic effluents or contaminated sites. Ranking the samples can become complicated if toxicity has been measured with several tests that produced variations in rank. Joining the responses into a toxicity index would express potential hazard in a single number.

An index based on the sum or average of bioassay end-points is the simplest to devise. In some instances it may be desirable to combine tests of acute lethality with sublethal tests in order to include a spectrum of organisms and/or responses. Indices are easier to construct if toxicity end-points are first translated into toxic units. The numerical values then can be summed like the chemical properties of a sample. An alternative would be to classify results on an ordinal scale (e.g. 0–10) based on the observed severity of effect. The approach is more subjective, but at least it incorporates expert judgement that should enter the assessment of data at some point. A ranking scale allows any kind of environmental measurement to be included in the index.

In the design of an index, two decisions must be made about placing weights on components:

- Whether any endpoint is to be considered more important and thus weighted more heavily.
- Whether the extent of toxic response is more important than the strength of effect, or vice versa.

4.3.5.1 Potential ecotoxic effects probe

Based on case studies, many countries have tried to develop similar indices where a number of toxicological parameters are integrated with the aim of quickly measuring the potential danger of the samples, particularly industrial effluents. The potential ecotoxic effects probe (PEEP) is one of the best known. It was developed in Canada as part of the St. Lawrence River Action Plan and was used to rank the impact of 50 industrial effluents discharged daily into the St. Lawrence River (Costan et al., 1993).

The following four bioassays are used to derive the PEEP index:

- Algal growth test using *Selenastrum capricornutum* (Environment Canada, 1992a).
- SOS Chromotest (genotoxicity) using *Escherichia coli* (Quillardet *et al.*, 1982, 1985).
- Luminescence test using *Vibrio fischeri* (formerly *Photobacterium phosphoreum*) (Environment Canada, 1992b).
- Survival and reproduction test using *Ceriodaphnia dubia* (Environment Canada, 1992c).

To apply the PEEP index, the effluent sample is first evaluated using all four bioassays. After biodegradation, the sample is tested using three of the four toxicity tests; the *Ceriodaphnia* test is excluded.

Results of all bioassays are reported as TOEC values, so a value can be entered into the PEEP index even if no inhibitory concentration (IC) or median effective concentration (EC_{50}) was determined. The TOECs are converted to toxic units. The PEEP is calculated as the sum of the toxic units, including both before and after biodegradation, according to the following formula:

$$P = \log_{10}\left[1 + n\left(\frac{\sum_{i=1}^{N} T_i}{N}\right)Q\right]$$

where: P = PEEP value
n = number of end-points exhibiting toxic responses
N = maximum number of obtainable toxic end-points
T_i = toxic units from each test, before and after biodegradation
Q = effluent flow (in m^3/h)

The four tests produce a total of six end-points, because the *Ceriodaphnia* and SOS Chromotest each have two end-points. Without *Ceriodaphnia* there are four end-points after biodegradation, so the value of N is 10. The PEEP formula sums the toxic units from each end-point and divides by N to derive an average toxic unit for the effluent. This is multiplied by n to account for the breadth of toxic response and by the discharge volume to compute the total hourly number of toxic units entering the river or receiving water. The result is expressed as a logarithm (adding 1 in the formula merely ensures that the log is computable). This produces a scale of results that would normally fall between 0 and 10.

The approach is designed to be flexible, and one or more tests can be removed or replaced with another test without preventing the calculation of a value for PEEP. However, indices based on different tests would not be comparable. The PEEP values from different industries are comparable because they are all based on toxicity, and PEEP values also compensate for discharge volume and are thus appropriate for evaluating environmental impact. An example of a PEEP index value calculation is shown in Table 4.1 (Costan *et al.*, 1993, pp. 133–5).

Solution. From Table 4.1, perform the indicated steps:

- Determine the number of bioassays exhibiting geno/toxic responses (n) for each effluent.
- Determine the average toxicity ($\Sigma T_i/N$) for each effluent.
- Determine the toxic print [n ($\Sigma T_i/N$)] for each effluent.
- Determine the toxic charge (effluent flow × toxic print) for each effluent.
- Set up a data summary table with the values obtained (Table 4.2).

Results

- The effluent PEEP values show the wide range of ecotoxic effects.
- Effluent 3 stands out because of its intense toxic print coupled with elevated flow rates. It also highlights the persistence of their toxicity.
- The contribution of effluents 1 and 2 is low compared with that of effluent 3.
- The toxic charge of effluent 4 is negligible.

4.3.5.2 Ranking scheme and 'battery of tests' approach

Another Canadian index combined results from bioassays, biochemical assays and microbial enumerations from both water and sediments. It was used along the St. John River and other Canadian rivers (Dutka *et al.*, 1990). With this approach a point ranking scheme was developed (Dutka *et al.*, 1988) and all data from different rivers were analyzed in an attempt to obtain a representative core group of tests. In the development of the core group, it was believed important that hazards associated with or indicated by microbiological and toxicant pollution should be addressed. Therefore, the core group would have to contain some indicator microbiological tests. From application of Spearman's rank correlation test to the water data, the core group selected was: fecal coliforms, coliphage, SOS Chromotest (genotoxicity), algal ATP and *Daphnia magna* (Dutka, 1988).

In a similar way to the Milli-Q water-extracted sediment data, the following tests were suggested for the core group of the battery of tests: fecal coliforms, *Clostridium perfringens*, SOS Chromotest, algal ATP and *Daphnia magna*. It was expected that the proposed composition of the core battery of tests for water and Milli-Q water-extracted sediments would provide an indication of bacterial hazards, as well as genotoxic and toxic pollutants in the aquatic ecosystem. By using a point scoring scheme based on the degree of positiveness of various tests and by comparing various proposed core groups of tests, it was possible to develop a ranking system for the water and sediment samples collected (Dutka, 1988).

A slightly revised format to award points for specific data values in order to rank the sampled waters and sediments is presented in Table 4.3, which shows the ranking scheme and battery of tests approach used by Dutka *et al.* (1991). Surface water samples were collected at seven sites (three per site) for eight toxicant

Table 4.1 Bioassay results (toxic threshold values in TUs) before (B) and after (A) biodegradation of five effluent samples, each representing one of the following industrial sectors: (1) organic chemical production; (2) inorganic chemical production; (3) pulp and paper; (4) metal plating; (5) textile (Costan et al., 1993, pp. 133–5).

Effluent	Microtox® test		Algal test		Crustacean test		SOS Chromotest (−S9)		SOS Chromotest (+S9)	
	B	A	B	A	L[a] − B	R[b] − B	B	A	B	A
1	17	5	20	18	14	575	8	0	8	0
2	16	2	126	78	2	18	5	0	32	0
3	5	14	160	160	7	5745	0	0	0	6
4	2	0	0	0	6	35	0	0	0	0
5	11	3	1244	1244	0	3	0	0	0	0

[a] Lethality test before biodegradation.
[b] Reproduction inhibition test before biodegradation.

Table 4.2 The PEEP index value calculations based on the Table 4.1 effluent data and the relative toxic charge contribution (%) of each effluent to the total toxic charge of combined effluents (Costan et al., 1993).

Effluent	n	N	Average toxicity	Toxic print	Effluent flow (m³/h)	Toxic charge	PEEP value	%
1	8	10	66.5	532	44	23 408	4.4	0.20
2	8	10	27.9	223.2	281	62 719	4.8	0.55
3	7	10	609.7	4267.9	2568	10 959 967	7.0	95.74
4	3	10	4.3	12.9	3	39	1.6	0.0
5	5	10	250.5	1252.5	321	402 053	6.6	3.51
Total						11 448 186		100

Table 4.3 Point award scheme for sample ranking, based on potential hazards (Dutka et al., 1991).

Daphnia magna		Spirillum volutans		Toxi-Chromotest		Microtox®		ATP-TOX system		Algal ATP		Ceriodaphnia dubia		Mutatox®	
EC%/ml	Points	MI^a/ml	Points	% I^b/ml	Points	% I^b/ml	Points	% I^b/ml	Points	% I^b/ml	Points	% IR^c/ml	Points	Revertants	Points
EC$_{20}$ at 100%	1	Negative	0	0.1–10	1	40	1	1–30	1	50–100	1	100	2	3–6	5
EC$_{40}$ at 100%	2	Positive	10	11–25	3	25–40	3	31–60	3	20–49	3	50	3	7–14	7
EC$_{50}$ at 100%	5			26–50	5	10–24	5	61–90	5	1–19	5	10	5	15–25	10
EC$_{50}$ at 75%	7			51–75	7	1–9	7	91–99	7	0.1–0.9	7	1	10	26–49	15
EC$_{50}$ at 50%	8			76–100	10	<0.9	10	100	10	<0.09	10	0.1	15	50–5000	20
EC$_{50}$ at 25%	10											0.01	20		

^{a} Motility inhibition.
^{b} Inhibition.
^{c} Reproduction inhibition.

Table 4.4 River surface water toxicology data with point scores and ranking (Dutka et al., 1991).

Site and sample	D. magna	S. volutans	Toxi Chromotest	Microtox®	ATP-TOX system	Algal ATP	C. dubia	Mutatox®	Total points
1	2	0	0	1	3	0	0	25	31
2	3.5	0	0	0	9	0	0	0	12.5
3	2	0	0	0	13	0	0	30	45
4	1.5	0	0	0	9	0	0	15	25.5
5	2	0	0	0	9	0	0	30	41
6	4.5	0	0	2.5	11	0	0	15	35
7	2	0	0	0	9	0	0	25	36

screening tests. Table 4.4 summarizes the points awarded to the various genotoxic/toxic results obtained from the water samples tested.

The method of ranking the sites is based on the total accumulated points for only the toxicant/genotoxicant screening tests. In this scheme site 3 would be designated as having the greatest potential hazard load, followed by sites 5, 7, 6, 1, 4 and 2.

4.3.5.3 Toxicity classification system
An example of this procedure is the evaluation/classification system proposed by the Laboratory for Biological Research in Aquatic Pollution (now the Laboratory of Environmental Toxicology and Aquatic Ecology, LETAE) from Ghent University in Belgium (Persoone and Vangheluwe, 2000; Persoone et al., 2003). Two different evaluation/classification systems were proposed: one for gross determination of the degree of toxic contamination of natural aquatic media; and one for quantification of the toxicity of wastes prior to their release into aquatic environments.

Both systems are based on the application of a battery of bioassays with short exposure time (1–3 days). The battery of bioassays is composed of test species belonging to different phylogenetic groups and is comprised of the following bioassays:

- Bacterial luminescence inhibition assay on *Vibrio fischeri* (formerly *Photobacterium phosphoreum*).
- Algaltoxkit: growth inhibition test with the microalgae *Selenastrum capricornutum* (*Raphidocelis subcapitata*), with determination of the 72-h EC_{50} and the NOEC.
- Protoxkit: reproduction inhibition test with the ciliated protozoan *Tetrahymena thermophila*, with determination of the 24-h EC_{50} and the NOEC.
- Rotoxkit: mortality test with the rotifer *Brachionus calyciflorus*, with determination of the 24-h LC_{50}.
- Daphtoxkit: immobilization assay with the crustacean *Daphnia magna*, with determination of the 24-h and 48-h EC_{50}.

- Thamnotoxkit: mortality test with the crustacean *Thamnocephalus platyurus*, with determination of the 24-h LC_{50}.

The principle of the classification systems is a one-step determination of the acute hazard of natural waters on non-diluted samples with the battery of bioassays. The classification system is based on two values:
- A class ranking in five acute hazard categories
- A weight score for each hazard class

Hazard classification of natural media
Waters are ranked into one of five classes on the basis of the highest toxic response shown by at least one of the tests applied. After determination of the percentage effect (PE) obtained with each of the microbiotests, a weight score is calculated for each hazard class to indicate the quantitative importance (weight) of the toxicity in that class. This weight score is expressed as a percentage and ranges from 25% (only one test in the battery has reached the toxicity level of that class) to 93% (all tests but one have reached the toxicity level of that class).

Acute hazard classes
- Class I: no acute hazard = none of the tests shows a toxic effect (i.e. an effect value significantly higher than that in the controls).
- Class II: slight acute hazard = the lowest observed effective concentration (LOEC) is reached in at least one test but the effect level is below 50%.
- Class III: acute hazard = the 50% effect is reached or exceeded in at least one test but the effect level is below 100%.
- Class IV: high acute hazard = the 100% effect is reached in at least one test.
- Class V: very high acute hazard = the 100% effect is reached in all tests.

The weight scores for the effect results of each bioassays are calculated as follows:

No significant toxic effect	=	score 0
Significant toxic effect < PE 50	=	score 1
Toxic effect > PE 50 but < PE 100	=	score 2
PE 100	=	score 3

The class weight score = Σ all test scores/n, where n = number of tests performed. The class weight score as a percentage = (class weight score)/(maximum class weight score) × 100.

The hazard classification system for natural waters is:

Class weight score (%)	Class	Hazard
< 20	I	No acute hazard
20 < PE < 50	II	Slight acute hazard
50 < PE < 100	III	Acute hazard

PE 100 in at least one test	IV	High acute hazard
PE 100 in all tests	V	Very high acute hazard

Classification of wastes

The principle of the toxicity classification system for wastes discharged into the aquatic environment is a two-step determination and quantification of the acute toxicity of the liquid wastes or leachates with the battery of bioassays. In the first step the toxicity is determined on non-diluted samples, and in the second step the toxicity tests are performed on a dilution series of the samples with all bioassays for which more than 50% effect has been found in the non-diluted sample.

The proposed system is also based on two values: an acute toxicity ranking in five classes and a weight score for each toxicity class.

Acute toxicity classes. The effect results obtained with each bioassay are transformed into toxic units (TUs) with the formula $TU = (1/LC_{50}) \times 100$. The samples are classified into one of the following categories on the basis of the highest number of TUs found in one of the tests of the battery:

- Class I: no acute toxicity = none of the tests shows a toxic effect (i.e. an effect value significantly higher than that in the controls)
- Class II: slight acute toxicity = the LOEC is reached with at least one test but the effect level is below 50% (< 1 TU)
- Class III: acute toxicity = the LC/EC_{50} is reached or exceeded in at least one test but in the ten-fold dilution of the sample the effect is lower than 50% (1–10 TUs)
- Class IV: high acute toxicity = the LC/EC_{50} is reached in the tenfold dilution for at least one test but not in the 100-fold dilution (10–100 TUs)
- Class V: very high acute toxicity = the LC/EC_{50} is reached in the 100-fold dilution for at least one test (> 100 TUs).

A weight score is calculated for each toxicity class, to indicate the quantitative importance (weight) of the toxicity in that class. The weight score is expressed as a percentage and ranges from 25% (only one test in the battery has reached the toxicity level of that class) to 93% (all tests but one have reached the toxicity level of that class). The weight scores are calculated as follows:

No significant toxic effect	=	score 0
Significant toxic effect but $< LC/EC_{50}$ (i.e. < 1 TU)	=	score 1
1–10 TUs	=	score 2
10–100 TUs	=	score 3
> 100 TUs	=	score 4

The class weight score and the percentage class weight score are calculated following the scoring method on natural waters. The hazard classification system for wastes discharged into the aquatic environment is:

TU	Class	Toxicity
< 0.4	I	No acute toxicity
0.4 < TU < 1	II	Slight acute toxicity
1 < TU < 10	III	Acute toxicity
10 < TU < 100	IV	High acute toxicity
TU > 100	V	Very high acute toxicity

Example: Toxicity classification system

Natural water

Bioassay	% Effect	Test score
Bacteria	< 10	0
Microalgae	25	1
Protozoa	72	2
Rotifers	55	2
Crustaceans	100	3

Classification: the concerned water sample belongs to class IV (high acute hazard) because the 100% effect level is reached in one test (crustaceans).
Class weight score: $(0 + 1 + 2 + 2 + 3)/5 = 8/5 = 1.6$
Percentage class weight score: $1.6 \times 100/3 = 53\%$

Industrial effluent

Bioassay	% Effect	Test score
Bacteria	12	3
Microalgae	2	2
Protozoa	< 1	1
Rotifers	55	3
Crustaceans	27	3

Classification: the concerned industrial effluent is classified in class IV (high acute toxicity) because the number of TUs in the most sensitive test is situated between 10 and 100.
Class weight score: $3 + 2 + 1 + 3 + 3/5 = 12/5 = 2.4$
Percentage class weight score: $2.4 \times 100/3 = 80\%$

4.3.5.4 The Chimiotox system

The Chimiotox (Scroggins, 1999) is another rating system developed for the St. Lawrence River Action Plan as a companion to the PEEP index. It is not a toxicity index, because toxicity is not measured. Chemical profiles of municipal and industrial effluents and known toxicity data are used to rank the effluents for potential environmental damage. The daily load of each chemical component is multiplied by a toxicity factor, and the sum of toxic loads produces the rating.

Chimiotox provides a physicochemical characterization of substances in an effluent. The substances are weighted for their known toxicity, with the goal of providing a common denominator in terms of potential environmental damage. The toxic weighting factor (F_{tox}) is calculated according to the following formula (Scroggins, 1999):

$$F_{toxi} = 1\,mg/l/MSC_i\,mg/l$$

where: F_{toxi} = toxic weighting factor for substance i
$1\,mg/l$ = arbitrary reference
MSC_i = most stringent water quality criterion (guideline) for substance i

Chimiotox units (UC) for each substance are calculated by multiplying its toxic weighting factor by its polluting load:

$$UC_i = Load_i \times F_{toxi}$$

Chimiotox units for each substance in an effluent are then combined to arrive at the Chimiotox index (IC) for the effluent, using the following formula. If desired, the same formula can be used to integrate the results for individual effluents, or individual substances, into totals according to types of industry, groups of substances, geographic region or overall:

$$IC = \sum_{i}^{n} UC_i$$

where: IC = Chimiotox index for a given effluent
UC_i = Chimiotox units for a given substance
n = number of substances

Based on the physicochemical characterization of an effluent, the toxic weighting used in the Chimiotox model makes it possible to do the following on a predictive basis:

- Compare the results of characterization of different effluents
- Identify the predominant toxic substances
- Combine the results into a single value for compilation into databases
- Obtain an overall picture of the toxic waste situation

Toxicity indices should be used cautiously because there is a loss of information in the calculation of only one summary value.

4.3.6 Structure–activity relationships (SARs)

The structure–activity relationship (SAR) is a means by which the effect of a drug or toxic chemical on an animal, plant or the environment can be related to its molecular structure (Walker *et al.*, 2001). This type of relationship may be assessed by considering a series of molecules and making gradual changes to them, noting the effect of each change upon their biological activity. Alternatively, it may be possible to assess a large body of toxicity data using intelligent tools such as neural networks to try to establish a relationship. Ideally, such relationships can be formulated as quantitative structure–activity relationships (QSARs), in which some degree of predictive capability is present (Walker *et al.*, 2001).

Computational models are used in human health risk assessment to predict the biological activity of potentially health-threatening agents. Structure–activity relationships that relate chemical structure to biological activity are examples of such models. Once a model is developed, it is used to predict the activity of new chemicals. The success or failure of a SAR model lies in its ability to predict correctly the true activity of a chemical.

Hazard assessment of environmental pollutants requires the input of many physicochemical, biomedical and toxicological properties of large numbers of chemicals in the various decision-making steps. Unfortunately, most of the candidate chemicals have very little or no laboratory data, a prerequisite to their proper evaluation. Modern combinatorial chemistry is quickly producing large libraries of real or virtual chemicals for which almost no property is known except their molecular structure.

One solution to this quagmire has been the use of calculated properties estimated from the molecular structure of chemicals instead of their experimental data. Molecular descriptors calculated using different variations of the chemical structure lead to the development of quantitative structure–property/activity relationship (QSPR/QSAR) models.

In general, SARs predict the toxicity of organic compounds based on their physical and chemical properties, particularly in the field of carcinogenesis. Toxicants move from the environment into the organisms, to exert their toxic action within the cells. Both processes depend on the elemental composition and structure of the chemical.

There are diverse uses of SARs:

- To predict the effect of new commercial chemicals
- To rank chemicals for priority in testing
- To develop preliminary water-quality guidelines
- To estimate missing toxicity data
- To select tests

QSARs are used routinely in predicting aquatic toxicity quickly. The USEPA has developed QSARs for all new commercial chemicals evaluated since 1981 and has about 300 QSARs for some 100 classes of compounds, most of which are based on octanol–water coefficient (K_{ow}) and molecular weight data (USEPA, 1994).

In Europe, SARs are coming into use for hazard assessment of new chemicals under the European Economic Community (EEC) (Bol *et al.*, 1993). Many endeavours on the use of SARs as an aid in setting environmental quality criteria have been made in The Netherlands (Verhaar *et al.*, 1992).

In Canada, new chemicals are controlled under the Canadian Environmental Protection Act (CEPA). The regulations promulgated under CEPA require producers of new chemicals to provide toxicological data and, when appropriate, SARs might be used as a substitute for experimental data (Environment Canada and Health and Welfare Canada, 1993).

4.3.7 Toxic emissions

From the point of view of environmental management, the effect of an effluent on receiving water is a function of its toxic concentration as well as of the flow or volume discharged. The product of the concentration and the flow is called load and is usually expressed as the mass of contaminant per day (kg/day). Comparisons of the effect of an effluent therefore should be done on the basis of its concentration and load. In terms of quality, the total mass load of the contaminant discharged per unit of time is usually determined. When the effluent is toxic it is important to know, in addition to its toxic concentration, the total amount of toxicants present in a given volume of discharge. This will make it possible to compare effluent streams of different volumes and toxic levels. These comparisons can be carried out by calculating the *toxic emission rate* (TER), defined in 1972 (Resources Agency, 1972) as:

$$\text{Toxic emission rate} = \text{TU} \times (\text{volume of flow in } m^3/\text{day})$$

The value thus obtained allows the evaluation of the relative quantities of toxicity contributed by various industries to the receiving water, or the contribution from different contaminated streams generated in an industrial process. It is important to note that, even though the TER is expressed in m^3/day, it would be more appropriate to express it as m^3/day of TUs.

Similarly, the *toxic emission factor* (TEF) makes it possible to compare processes in different industrial sites because it is calculated as a function of the industrial product being produced. The TEF value is calculated as follows:

$$\text{Toxic emission factor} = \text{TU} \times (\text{volume of flow in } m^3/\text{t of product})$$

Even though this factor may be considered of little importance, it is used in countries such as Canada for other parameters (e.g. biochemical oxygen demand). This constitutes an adequate way to compare industrial plants with different processes, because the TEF expresses the toxicity corrected for factors such as size of the plant and efficient use of water within the plant.

4.4 Ecological risk assessment

Environmental problems are complex, with multiple causes and diverse ecological effects. Dealing with such problems requires a flexible decision-making process that can accommodate this diversity while providing some measure of the uncertainty associated with decisions that are made (Norton *et al.*, 1995). For these reasons, different environmental protection agencies have an increasing interest in using ecological risk assessment as a basis for environmental decisions (Norton *et al.*, 1995).

Owing to the increased emphasis on ecological issues and the need to provide uniform procedures, the USEPA in 1989 initiated studies to produce ecological risk assessment guidelines based on the same open, peer-reviewed process that was used to develop the human health risk assessment guidelines (USEPA, 1991b). As a result of this work several workshops were organized, and a variety of publications (USEPA, 1990, 1992a,b,c) were produced. To assist the Regions in implementing this policy, the Office of Emergency and Remedial Response (OERR) issued the Superfund Environmental Evaluation Manual (USEPA, 1989c). The manual defines ecological assessment as:

> '... *a qualitative and/or quantitative appraisal of the actual or potential effects of a hazardous waste site on plants and animals other than people or domesticated species.*'

Ecological risk assessment evaluates the likelihood that adverse ecological effects will occur or are occurring as a result of exposure to stressors related to human activities such as dredging or filling of wetlands or release of chemicals. The term *stressor* is used to describe any chemical, physical or biological entity that can induce adverse effects on ecological components, i.e. individuals, populations, communities or ecosystems. Adverse ecological effects encompass a wide range of disturbances, ranging from mortality in an individual organism to a loss in ecosystem function. Thus, the ecological risk assessment process must be flexible while providing a logical and scientific structure to accommodate a broad array of stressors and ecological components (Norton *et al.*, 1995). In general terms, ecological risk assessment comprises four interrelated activities:

1. *Problem formulation*: a qualitative evaluation of contaminant release, migration and fate; identification of contaminants of concern, receptors, exposure pathways and known ecological effects of the contaminants; and selection of end-points for further study. The term *end-point* is used to describe the expected or anticipated effect of a contaminant on an ecological receptor (USEPA, 1989c).
2. *Exposure assessment*: a quantification of contaminant release, migration and fate; characterization of exposure pathways and receptors; and measurement or estimation of exposure point concentrations (USEPA, 1989c).

3. *Ecological effects assessment*: literature reviews, field studies and toxicity tests linking contaminant concentrations to effects on ecological receptors (USEPA, 1989c).
4. *Risk characterization*: a measurement or estimation of both current and future adverse effects (USEPA, 1989c).

The ecological risk evaluation process is illustrated in Figure 4.2. As the diagram indicates, each element in the process can have an impact on others. The components do not always follow one another in a stepwise manner, and sometimes the evaluation can be done on aspects of all four components at the same time (USEPA, 1989c).

Ecological risk may be expressed as a true probabilistic estimate of risk or may be deterministic or even qualitative in nature. The likelihood of adverse effects is expressed through a semi-quantitative or qualitative comparison of effects and exposure.

The approach to ecological risk assessment emphasizes three areas (Norton *et al.*, 1995):

1. Ecological risk assessment can consider effects beyond those on individuals and the populations of a single species and may examine multiple populations, communities and ecosystems.
2. There is no single set of ecological values to be protected (e.g. health effects such as cancer or birth defects) that can be applied generally. Values are selected from a number of possibilities based on both scientific and policy considerations.
3. There is an increasing awareness of the need for ecological risk assessments to consider non-chemical as well as chemical stressors.

Risk assessment and risk management are two processes where the activities can be shared. At the beginning of the risk assessment process, managers can help to ensure that the risk assessment will provide relevant information for making decisions on the issues under consideration, while the risk assessor can ensure that the risk manager addresses all relevant ecological concerns.

Risk assessment deals with data acquisition, verification and monitoring. The collection of data is required and the need for additional data may be identified at any point in the process. Verification and monitoring can help to determine the overall effectiveness of the framework approach, provide feedback concerning any need for future modifications, help to evaluate the effectiveness and practicality of policy decisions and point out the need for new or improved risk assessment techniques (USEPA, 1992a).

4.4.1 Risk characterization

Risk characterization for ecological risk assessments could be performed by weight of evidence (Risk Assessment Forum, 1992). In this process ecological

FRAMEWORKS FOR THE APPLICATION OF TOXICITY DATA

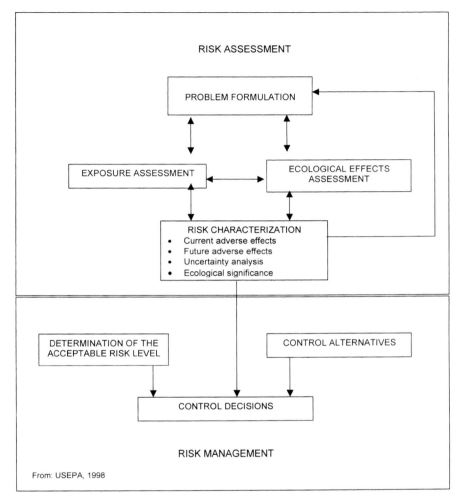

Figure 4.2 The framework for ecological risk assessment and risk management processes. The chart incorporates procedures and outlines.

risk assessors examine all available data from chemical analysis, toxicity tests, biological surveys, and biomarkers to estimate the likelihood that significant effects are occurring or will occur, and they describe the nature, magnitude and extent of effects on the designated assessment end-points (Suter, 1996).

The weight of evidence strategy is used to evaluate problems of toxicity by combining data obtained from three sources: chemical measurements, ecoassessment and toxicity tests. The concomitant analysis of data from these three sources will allow for a precise interpretation of the environmental impact produced by the discharge of toxic substances. In addition, more information on the presence, potency, nature and effects of the toxicants present can be obtained. However, it is necessary to check carefully the information obtained because equal weight can be given to a set of different answers that might not have enough validity or applicability (Munkittrick and McCarty, 1995).

The approach for estimating risk on the weight of evidence is based on individual lines of evidence that are combined through a process of weighting. The lines are integrated independently so that the implications of each are presented explicitly. For each line of evidence it is necessary to evaluate the relationship of the measurement end-point to the assessment end-point, the quality of the data and the relationship of the exposure metrics in the exposure–response data to the exposure metrics for the site (Suter et al., 1995).

Because each technique has a different function and should be used for a specific purpose, each will contribute one of the following types of data:

4.4.1.1 Chemical toxicity line
Chemical techniques may be able to identify the various chemicals/compounds that are producing the toxic effect as well as their concentrations. With this information one can predict if there is the possibility of an effect, but toxic activity cannot be proved.

The chemical toxicity line of evidence uses analysis of individual chemicals in individual media to estimate exposure and uses literature values for effects of individual chemicals to estimate effects. These are combined in two steps.

1. The chemicals are screened against ecotoxicological benchmarks, against background exposures and against characteristics of the source to determine which are the chemicals of potential ecological concern (COPECs).
2. A more definitive characterization is performed by comparing the distributions of exposure and the effects for each COPEC.

The integration of exposure with single chemical toxicity data is expressed as a quotient of the environment exposure concentration (EEC) divided by the toxicologically effective concentration (TEC):

$$\text{Hazard quotient (HQ)} = \text{EEC}/\text{TEC}$$

The TEC may be a test end-point, a test end-point corrected by a factor or other extrapolation model or a regulatory criterion or other benchmark value. A hazard quotient (HQ) greater than unity is treated as evidence that the chemical is worthy of concern. Suter (1996) also suggests that, if numerous chemicals occur at potentially toxic concentrations, an index of total toxicity could be calculated by the sum of toxic units (ΣTUs). This permits a comparison of COPECs and examines their distribution across areas within a site. The TUs are quotients of the concentration of a chemical in a medium divided by the standard test end-point concentration for that chemical.

4.4.1.2 Toxicity tests line
Toxicity tests will tell whether chemicals, waste, surface water, sediments or soil exhibit an adverse effect to the test organism. However, with this information it will not be possible to identify the toxicant causing the observed effect.

Risk characterization for the toxicity line of evidence begins by determining whether the tests show significant toxicity. Toxicity is not significant if the effects relative to controls are less than 20% (e.g. less than 20% mortality) (Suter, 1996) and the effects are not statistically significantly different from the controls. Effects are considered significant if:

- The hypothesis of no difference between responses in contaminated media and in either reference media or control media is rejected with 95% confidence (i.e. statistical significance).
- An effect of 20% or greater in survival, growth or reproduction relative to either reference media or control media is observed (i.e. biological significance).

If no significant toxicity is found, the risk characterization consists of determining the likelihood that the result constitutes a false negative. If significant toxicity occurs in the tests, the risk characterization should describe the nature and magnitude of the effects and the consistency of effects among tests conducted with different species in the same medium (Suter, 1996).

If significant toxicity is found at one site, then the relationship of toxicity to exposure must be characterized. This may be done by analyzing the relation between toxicity and concentration of chemicals in the media. When sources of toxicants have been identified and tests have been performed on dilution series, the transport and fate of toxicity can be modeled like that of individual chemicals (Suter, 1996). Nevertheless, combined toxic effects of pollutant mixtures are not considered with this approach.

4.4.1.3 Ecoassessment line
Biological monitoring in the receiving ecosystem makes it possible to confirm whether or not there was an effect on the exposed community.

When biological survey data are available for an end-point species or community, then it has to be established if significant effects are occurring. Normally, for some groups of organisms (fish, invertebrates, algae, bacteria) there are abundant data from reference sites. However, in order to produce more reliable results, it would be useful to obtain information from contaminated and reference sites directly. In addition, it is necessary to take into account that for some taxa, such as birds, traditional survey data are not useful for estimating risks from toxic contaminants because factors such as mobility and territoriality do not permit the clear establishment of demographic effects (Suter, 1996).

Suter (1996) also suggests considering the sensitivity of field data to toxic effects relative to other lines of evidence. Biological surveys may be very sensitive, moderately so or insensitive. This characteristic is important, especially when it is compared with other lines of evidence. For instance, if chemical analysis shows a high concentration of one toxicant suggesting that a medium should be highly toxic, that toxicity tests of the same medium do not show toxicity and that the community has not been affected or modified, it could be an indication that the chemical is present in a non-bioavailable form. Similarly, when a highly modified community is found and chemical analysis does not detect high levels of toxic chemicals, possible explanations may include combined toxic effects, toxic levels of unanalysed contaminants or episodic contamination (Suter, 1996).

When biological survey data show a reduction in abundance, production or diversity, it is necessary to analyze the potential causes to see if there is an association with the apparent effects. The following procedure is used to carry out this analysis:

- Compare the distribution of the apparent effects in space and time with the distribution of sources of contaminants.
- Compare the distribution of apparent effects with the distribution of habitat factors that may affect the organisms.
- Examine the natural variability of the populations and communities and the accuracy of the survey methods to estimate the probability that the effects are due to chance.

4.4.1.4 Biomarkers line

Biomarkers are physiological or biochemical measurements that may be indicative of exposure to contaminants. Although they are not used to estimate risks, they can be used to support other lines of inference (Suter, 1996). Because the effect is seen directly in the population itself, there is no problem with bioavailability, kinetics or other factors that modify the toxicity of the media (water, sediment, soil). A positive response means that the population is exposed to toxicants. The inference is established when the levels of biomarkers from contaminated and reference sites are significantly different. When the levels of biomarkers are characteristic of contaminant exposures, the distribution and frequency of elevated levels should be compared with the distribution and concentrations of contaminants. In addition, the implications of the observed biomarker levels for populations or communities should be estimated (Suter, 1996).

The integration of results obtained from each line and for each end-point will permit an overview of the conditions at a particular site or ecosystem being studied (Table 4.5), which then can be compared with the values found at the reference site.

The weight of evidence approach uses information from all sources, but particularly from the chemical, toxicity, ecoassessment and biomarker characterization. These approaches are compatible, and together they reveal more about the presence, strength, nature and effects of toxicants than could be discovered from any single approach. Each technique serves to both confirm and extend the understanding of environmental contamination in surface waters, sediments or soil (Scroggins, 1999).

In the sediment triad approach, sites are studied by monitoring benthic invertebrates, environmental chemistry and a battery of bioassays on whole sediment, porewater, elutriates or extracts. Results are compared with parallel measurements at nearby reference sites. Results for a given polluted site can be summarized as *ratio-to-reference values*, the ratio of a measurement at the test site to that at the reference site. The ratios can be presented on a triaxial graph with confidence limits (Alden, 1992; Canfield *et al.*, 1994; Scroggins, 1999). As an alternative, data can be scaled from zero (reference site) to 100 (worst site) and be presented as triangles on axes with those scales (Figure 4.3). This analysis has been applied for sediments but in some cases interpretation may be difficult. The main problem resides in the fact that the axes are not equivalent because different parameters are measured with each of the techniques, which also have different purposes and sensitivities. In order to compare exposure and effect distributions to obtain a more complete picture of the nature of possible effects in the environment, weights are determined based on

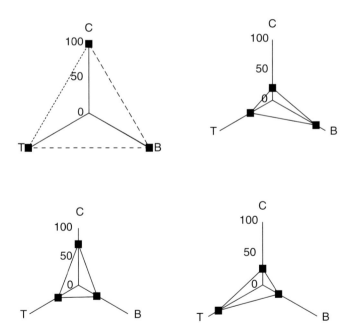

Figure 4.3 Use of triaxial diagrams to illustrate results from the quality triad (Canfield *et al.*, 1994). Axes are scaled from zero (reference value) to 100 (worst value measured). T = toxicity; C = chemistry; B = ecoassessment.

Table 4.5 'Weight of evidence' conclusions.

Chemical guideline(s) exceeded	Evidence results				Possible conclusion
	Biotic indices impaired	Toxicity (field or lab.)	Tissue residues elevated	Habitat impairment	
+	+	+	+	+	Contaminant and habitat stress
−	−	−	+	−	Potential food chain effects
−	+	−	−	−	Unknown chemical or other unknown stressor
−	−	+	−	−	Chemical including potential degradation
+	−	+	−	−	Chemical contamination: monitor biota closely
+	−	−	−	−	Chemicals not bioavailable or tests insensitive

the following aspects (Suter, 1993; Menzie *et al.*, 1996): relevance, exposure/response, temporal scope, spatial scope, quality and quantity.

These and other considerations are used in forming an expert judgment or consensus about which way the weight of evidence tips the balance. Table 4.5 summarizes possible conclusions from different lines of evidence.

Risk characterization in ecological assessment is a process of applying professional judgment to determine whether adverse effects are occurring or will occur as a result of contamination associated with a site. Available methods (both quantitative and qualitative) seek to answer the following questions:

- Are ecological receptors currently exposed to site contaminants at levels capable of causing harm, or is future exposure likely?
- If adverse ecological effects are observed or predicted, what are the types, extent and severity of the effects?
- What are the uncertainties associated with risk characterization?

Risk characterization concludes with a risk description that:

- Includes a summary of the risks and uncertainties
- Interprets the ecological significance of the observed or predicted effects

Risk description is a key step in communicating ecological risks to site managers and decision makers. Description, interpretation and communication of the ecological significance of the risk should include the nature and magnitude of the effects and their spatial and temporal distribution, as well as potential for recovery. A significant risk is sufficient to prompt consideration of remedial actions, but the nature, magnitude and distribution of its effects determine whether remediation is justified, given the remedial costs and risks (Suter, 1996).

References

AETE (Aquatic Effects Technology Evaluation) (1998) *Summary of Cost-effectiveness Evaluation of Aquatic Effects Monitoring Technologies Applied in the 1997 AETE Field Evaluation Program*. Beak International and Golder Associates, Ottawa, Ontario.

Alden III, R.W. (1992) Uncertainty and sediment quality assessments: I. Confidence limits for the triad. *Environmental Toxicology and Chemistry*, **11**, 637–644.

Blaise, C., Bermingham, N. and van Coillie, R. (1985) The integrated ecotoxicological approach to assessment of ecotoxicity. *Water Quality Bulletin*, **10**(1), 3–10, 60–61.

Bol, J., Verhaar, H.J.M., van Leeuwen, C.J. and Hermens, J.L.M. (1993) *Predictions of the Aquatic Toxicity of High-production-volume-chemicals. Part A: Introduction and Methodology. Part B: Predictions*, No. 1993/9A/B. VROM, The Hague. Netherlands Ministerie van Volkshuisvesting, Ruimtelijke Ordening en Milieubeheer, Diretoraat-Generaal Milieubeheer.

Brezonik, P.L., King, S.O. and Mach, C.E. (1991) The influence of water chemistry on trace metal bioavailability and toxicity to aquatic organisms. In *Metal Ecotoxicology: Concepts and Applications*, Newman, M.C. and McIntosh, A.W. (eds), pp. 1–31. Lewis Publishers, Chelsea, MI.

Cairns Jr., J. (1986) What is meant by validation of predictions based on laboratory toxicity tests? *Hydrobiologia*, **137**, 271–278.

Cairns Jr., J. (1988) What constitutes field validation of predictions based on laboratory evidence? In *Aquatic Toxicology and Hazard Assessment*, Adams, W.J., Chapman, G,A. and Landis, W.G. (eds), ASTM STP 971, pp. 361–368. American Society for Testing and Materials, Philadelphia, PA.

Cairns Jr., J. and Pratt, J.R. (1989) The scientific basis of bioassays. *Hydrobiologia*, **188/189**, 5–20.

Cairns Jr., J., Niederlehner, B.R. and Smith, E.P. (1995) The emergence of functional attributes as endpoints in ecotoxicology. In *Sediment Toxicity Assessment*, Burton, GA, (ed.), pp. 111–128. Lewis Publishing, Boca Raton, FL.

Calow, P. (1989) The choice and implementation of environmental bioassays. *Hydrobiologia*, **188/189**, 61–64.

Canfield, T.J., Kemble, N.E., Brumbaugh, W.G., Dwyer, F.J., Ingersoll, C.G. and Fairchild, J.F. (1994) Use of benthic invertebrate community structure and the sediment quality triad to evaluate metal contaminated sediment in the Upper Clark Fork River, Montana. *Environmental Toxicology and Chemistry*, **13**, 1990–2012.

CCME (Canadian Council of Ministers of the Environment) (1991) *Interim Canadian Environmental Quality Criteria for Contaminated Sites*, Report CCME EPC-CS34. CCME, Subcommittee on Environmental Quality Criteria for Contaminated Sites, Winnipeg, Manitoba.

Costan, G., Bermingham, N., Blaise, C. and Ferard, J.F. (1993) Potential ecotoxic effects probe (PEEP): a novel index to assess and compare the toxic potential of industrial effluents. *Environmental Toxicology and Water Quality: An International Journal*, **8**, 115–140.

de Pauw, N., Ghetti, P.F., Manzini, P. and Spaggiari, R. (1992) Biological assessment methods for running water. In *River Water Quality: Ecological Assessment and Control*, Newman, P., Piavaux, A. and Sweeting, R. (eds), EUR 14606 EN-FR. Commission of the European Communities, Brussels.

Draggan, A. and Giddings, J.M. (1978) Testing toxic substances for protection of the environment. *Science of the Total Environment*, **9**(1), 63–74.

Dutka, B.J. (1988) Priority setting of hazards in waters and sediments by proposed ranking scheme and battery of tests approach. *Angewandte Zoologie*, **20**, 303–316.

Dutka, B.J. and Kwan, K.K. (1982) Application of four bacterial screening procedures to assess changes in the toxicity of chemicals in mixtures. *Environmental Pollution Ser. A*, **29**, 125–138.

Dutka, B.J., Jones, K., Kwan, K.K., Bailey, H. and McInnis, R. (1988) Use of microbial and toxicant screening tests for priority site selection of degraded areas in water bodies. *Water Research*, **22**, 503–510.

Dutka, B.J., Seidli, P. and Munro, D. (1990) Using microbial and toxicant screening-test data to prioritize water bodies. In *Encyclopedia of Fluid Mechanics, Volume 10, Surface and Groundwater Flow Phenomena*, Cheremisinoff, N.P. (ed.), pp. 429–452. Gulf Publishing, Houston, Tx.

Dutka, B.J., Kwan, K.K., Rao, S.S., Jurkovic, A. and Liu, D. (1991) River Evaluation using ecotoxicological and microbiological procedures. *Environmental Monitoring and Assessment*, **16**, 287–313.

Eagleson, K.W., Lenat, D.L., Ausley, L.W. and Winborne, F.B. (1990) Comparison of measured in stream biological responses with responses predicted using *Ceriodaphnia dubia* chronic toxicity test. *Environmental Toxicology and Chemistry*, **9**(5), 1019–1028.

Environment Canada (1992a) Biological test method: growth inhibition test using the freshwater alga, *Selenastrum capricornutum*, Report EPS 1/RM/25. Environment Canada, Environmental Protection Service, Ottawa, Ontario.

Environment Canada (1992b) Biological test method: toxicity test using luminescent bacteria *Photobacterium phosphoreum*, Report EPS 1/RM/24. Environment Canada, Environmental Protection Service, Ottawa, Ontario.

Environment Canada (1992c) Biological test method: test of reproduction and survival using the cladoceran *Ceriodaphnia dubia*, Report EPS 1/RM/21. Environment Canada, Environmental Protection Service, Ottawa, Ontario.

Environment Canada and Health and Welfare Canada (1993) *Guidelines for the Notification and Testing of New Substances: Chemicals and Polymers*. Pursuant to the New Substances Notification Regulations of Canadian Environmental Protection Act. Environment Canada, Conservation and Protection, and Health and Welfare Canada, Health Protection Branch, Ottawa, Ontario.

Geckler, J.R., Horning, W.B., Neiheisel, T.M., Pickering, Q.H., Robinson, E.L. and Stephan, C.E. (1976) Validity of laboratory tests for predicting cooper toxicity in streams, EPA/600/3-76/116. USEPA Ecological Research Service, Washington, DC.

Giesy, J.P. and Graney, R.L. (1989) Recent development in and intercomparisons of acute and chronic bioassays and bioindicators. *Hydrobiologia*, **188/189**, 21–60.

Giesy, J.P., Versteeg, D.L. and Graney, R.L. (1988) A review of selected clinical indicators of stress-induced changes in aquatic organisms. In *Toxic Contaminants and Ecosystem Health: A Great Lakes Focus*, Evans, M.S. (ed.), pp. 169–200. John Wiley and Sons, New York.

Hellawell, J.M. (1986) *Biological Indicators of Fresh Water Pollution and Environmental Management*. Elsevier Applied Science, London.

Henderson, C. and Tarzwell, C.M. (1957) Bioassays for control of industrial effluents. *Sewage Industrial Wastes*, **29**(10), 1002–1017.

Hunt, D.T.E., Johnson, I. and Milne, R. (1992) The control and monitoring of discharges by biological techniques. *Journal of the Institution of Water and Environmental Management*, **6**, 269–277.

Iverson, W.P. and Brinckman, F.E. (1978) Microbial metabolism of heavy metals. In *Water Pollution Microbiology*, vol. 2, Mitchell, R. (ed.), John Wiley and Sons, New York, pp. 201–232.

Kenaga, E.E. (1977) Aquatic test organisms and methods useful for assessment of chronic toxicity of chemicals. In *Analyzing the Hazard Evaluation Process*, Dickson, K.L., Maki, A.W. and Cairns Jr., J. (eds), pp. 101–111. American Fisheries Society, Bethesda, MD.

Kenaga, E.E. (1982a) Review: the use of environmental toxicology and chemistry data in hazard assessment: progress, needs, challenges. *Environmental Toxicology and Chemistry*, **1**, 69–79.

Kenaga, E.E. (1982b) Predictability of chronic toxicity from acute toxicity of chemicals in fish and aquatic invertebrates. *Environmental Toxicology and Chemistry*, **1**, 347–358.

Klemm, D.J., Lobring, L.B. and Horning, W.H. (1990) *Manual for the Evaluation of Laboratories Performing Aquatic Toxicity Tests*, EPA/600/4-90/031, p. 109. USEPA Environmental Monitoring Systems Laboratory, Cincinnati, OH.

Kszos, L.A., Stewart, A.J. and Taylor, P.A. (1992) An evaluation of nickel toxicity to *Ceriodaphnia dubia* and *Daphnia magna* in contaminated stream and in laboratory tests. *Environmental Toxicology and Chemistry*, **11**(7), 1001–1012.

Livingstone, D.R. (1991) Towards a specific index of impact by organic pollution in marine invertebrates. *Comparative Biochemistry and Physiology, Part C: Comparative Pharmacology and Toxicology*, **100**(1/2), 151–155.

Menzie, C., Henning, M.H., Cura, J., Finkelstein, K., Gentile, J., Maughan, J., Mitchell, D., Petron, S., Potocki, B. and Svirsky, S. (1996) A weight-of-evidence approach for evaluating ecological risks: report of the Massachusetts Weight-of-evidence Work Group. *Human and Ecological Risk Assessment*, **2**(2), 277–304.

Mitchell, R. and Chet, I. (1978) Indirect ecological effects of pollution. In *Water Pollution Microbiology* Vol. 2, Mitchell, R. (ed.), pp. 177–199. John Wiley and Sons, New York.

Munawar, M., Munawar, I.F., Mayfield, C.I. and McCarthy, L.H. (1989) Probing ecosystem health: a multi-disciplinary and multi-trophic assay. *Hydrobiologia*, **188/189**, 93–116.

Munkittrick, K.R. and McCarty, L.S. (1995) An integrated approach to aquatic ecosystem health: top-down, bottom-up or middle-out? *Journal of Aquatic Ecosystem Health*, **4**, 77–90.

Norberg-King, T.J., Durham, E.J. and Ankley, G.T. (1991) Application of toxicity identification procedures to the ambient waters of the Colusa Basin Drain, California. *Environmental Toxicology and Chemistry*, **10**(6), 891–900.

Norton, S.B., Gentile, J.H., Landy, R.B. and van der Schalie, W. (1995) The EPA's framework for ecological risk assessment. In *Handbook of Ecotoxicology*, Hoffman, D.J., Rattner, B.A., Burton, G.A. and Cairns Jr., J. (eds), pp. 703–716. Lewis Publishers, Boca Raton, FL.

NTAC (National Technical Advisory Committee) (1968) Water quality criteria. Report of the National Technical Advisory Committee to the Secretary of the Interior. US Federal Water Pollution Control Administration, Washington, DC.

OECD (Organization for Economic Co-operation and Development) (1987) Utilisation des tests biologiques pour l'evaluation et le contrôle de la pollution de l'eau. Monographies sur l'Environnement no. 11. OECD, Paris.

OECD (Organization for Economic Co-operation and Development) (1992a) Report of the OECD workshop on the extrapolation of laboratory aquatic toxicity data to the real environment, OECD/GD (92) 169, OECD Environmental Monographs no. 59. OECD, Paris.

OECD (Organization for Economic Co-operation and Development) (1992b) Report of the OECD workshop on quantitative structure–activity relationships (QSARS) in aquatic effects assessment, OECD/GD (92) 168, OECD Environmental Monographs no. 58. OECD, Paris.

OWRC (Ontario Water Resources Commission) (1970) *Guidelines and Criteria for Water Quality Management in Ontario*. OWRC, Toronto, Ontario.

Persoone, G., Van de Vel, A., Van Steertegem, M. and De Nayer, B. (1989) Predictive value of laboratory tests with aquatic invertebrates: influence of experimental conditions. *Aquatic Toxicology*, **14**(2), 149–166.

Persoone, G. and Vangheluwe, M.L. (2000) Toxicity determination of the sediments of the river Seine in France by application of a battery of microbiotests. In *New Microbiotests for Routine Toxicity Screening and Biomonitoring*, Persoone, G, Janssen, C. and De Coen, W. (eds), pp. 427–439. Kluwer Academic/Plenum Publishers, New York.

Persoone, G., Marsalek, B., Blinova, I., Törökne, A., Zarina, D., Manusadzianas, L., Nalecz-Jawecki, G., Tofan, L., Stepanova, N., Tothova, L. and Kolar, B. (2003) A practical and user-friendly toxicity classification system with microbiotests for natural waters and wastewaters. *Environmental Toxicology*, **18**(6), 395–402.

Power, M. and McCarty, L.S. (1997) Fallacies in ecological risk assessment practices. *Environmental Science and Technology/News and Research Notes*, **331**, 370A–375A.

Pratt, L., Pavanello, R. and Pesarin, F. (1971) Assessment of surface water quality by a single index of pollution. *Water Research*, **5**(9), 741–751.

Quillardet, P., Huisman, O., D'Ari, R. and Hofnung, M. (1982) SOS Chromotest, a direct assay of induction of an SOS function in *Escherichia coli* K-12 to measure genotoxicity. *Proceedings of the National Academy of Sciences*, **79**, 5971–5975.

Quillardet, P., de Bellecombe, C. and Hofnung, M. (1985) The SOS Chromotest, a colorimetric bacterial assay for genotoxins: validation study with 83 compounds. *Mutation Research*, **147**, 79–95.

Resources Agency (1972) *Water Quality Control Plan for Ocean Waters of California*. (Unnumbered report). The Resources Agency, State Water Resources Control Board, Sacramento, CA.

Risk Assessment Forum (1992) *Framework for Ecological Risk Assessment*, EPA/630/R-92/001. United States Environmental Protection Agency, Washington, DC.

Sanders, W.M. (1985) Field validation. In *Fundamentals of Aquatic Toxicology: Methods and Applications*, Rand, G.M. and Petrocelli, S.E. (eds), pp. 601–618. Hemisphere Publishing Corporation, Washington, DC.

Scroggins, R.P. (1999) *Guidance Document on Application and Interpretation of Single-species Tests in Environmental Toxicology*, Report EPS 1/RM/34. Environmental Technology Centre, Environment Canada, Ottawa, Ontario.

Sergy, G.A. (1987) *Recommendations on Aquatic Biological Tests and Procedures for Environmental Protection*. C & P DOE, Environment Canada, Environmental Protection, Technology Development and Technical Services Branch, Edmonton, Alberta.

Slooff, W., Canton, J.H. and Hermens, J.L.M. (1983) Comparison of the susceptibility of 22 freshwater species to 15 chemical compounds. I (Sub) acute toxicity tests. *Aquatic Toxicology*, **4**(2), 113–128.

Soto, A.M., Lin, M.T., Justica, H., Silvia, R.M. and Sonnenschein, C. (1992) An in culture bioassay to assess the estrogenicity of xenobiotics. In *Chemically Induced Alterations in Sexual Development the Wildlife/Human Connection*, Colburn, T. and Clement, C. (eds), pp. 295–309. Princeton Scientific Publishing, Princeton, NJ.

Sprague, J.B. (1971) Measurement of pollutant toxicity to fish: III Sub-lethal effects and safe concentrations. *Water Research*, **5**, 245–266.

Sprague, J.B. (1985) Factors that modify toxicity. In *Fundamentals of Aquatic Toxicology: Methods and Applications*, Rand, G.M. (ed.), pp. 124–163. McGraw-Hill, New York.

Sprague, J.B. (1995) Planning a field validation program. In *Review of Methods for Sub-lethal Aquatic Toxicity Tests Relevant to the Canadian Metal Mining Industry, and Design of Field Validation Programs*, Chapter 3. Prepared for Natural Resources Canada, Canada Centre for Mineral and Energy Technology, Aquatic Effects Technology Evaluation Program (AETE), Ottawa, Ontario by Sprague Associates, Salt Spring Island, BC.

Stephan, C.E. (1982) Increasing the usefulness of acute toxicity tests. In *Aquatic Toxicology and Hazard Assessment: Fifth Conference*, Pearson, J.G., Foster, R.B. and Bishop, W.E. (eds), ASTM STP no. 766, pp. 69–81. American Society for Testing and Materials, Philadelphia, PA.

Stephan, C.E., Mount, D.I., Hanson, D.J., Gentile, J.H., Chapman, G.A. and Brungs, W.A. (1985) *Guidelines for Deriving Numeric National Water Quality Criteria for the Protection of Aquatic Organisms and their Uses*. PB85-227049. USEPA, National Technical Information Service, Duluth, MN.

Suter II, G.W. (1993) *Ecological Risk Assessment*. Lewis Publishers, Boca Raton, FL.

Suter II, G.W. (1996) *Risk Characterization for Ecological Risk Assessment of Contaminated Sites*, ES/ER/TM-200. Environmental Restoration Division, US Department of Energy, Oak Ridge, TN.

Suter II, G.W., Sample, B.E., Jones, D.S., Ashwood, T.L. and Loar, J.M. (1995) *Approach and Strategy for Performing Ecological Risk Assessments for the Department of Energy's Oak Ridge Reservation: 1995 Revision*, ES/ER/TM-33/R2. Environmental Restoration Division, Oak Ridge, TN.

Tarazona, J.V., Carballo, M., Castaño, A. and Muñoz, M.J. (1995) The role of cell biology on the application of toxicology to environmental sciences: biological assessment and biomarkers. In *Cell Biology in Environmental Toxicology*, Cajaraville, M.P. (ed.), pp. 15–28. University of the Basque Country Press, Bilbao.

USEPA (United States Environmental Protection Agency) (1988a) *Toxicity Reduction Evaluation Protocol for Wastewater Treatment Plants*. EPA/600/2-88/062. USEPA, Office of Research and Development, Risk Reduction Engineering Laboratory, Cincinnati, OH.

USEPA (United States Environmental Protection Agency) (1988b) *Generalized Methodology for Conducting Industrial Toxicity Reduction Evaluations*, EPA/600/2-88/070. USEPA, Office of Research and Development, Risk Reduction Engineering Laboratory, Cincinnati, OH.

USEPA (United States Environmental Protection Agency) (1989a) *Methods for Aquatic Toxicity Identification Evaluations, Phase II, Toxicity Identification Procedures*, EPA/600/3-88/035. USEPA, Office of Research and Development, Environmental Research Laboratory, Duluth, MN.

USEPA (United States Environmental Protection Agency) (1989b) *Methods for Aquatic Toxicity Identification Evaluations, Phase III, Toxicity Confirmation Procedures*, EPA/600/3-88/036. USEPA, Office of Research and Development, Environmental Research Laboratory, Duluth, MN.

USEPA (United States Environmental Protection Agency) (1989c) *Risk Assessment Guidance for Superfund. Volume II: Environmental Evaluation Manual*, EPA/540-1-89/001. Office of Emergency and Remedial Response, US Environmental Protection Agency, Washington, DC.

USEPA (United States Environmental Protection Agency) (1990) *Reducing Risk: Setting Priorities and Strategies for Environmental Protection*, Science Advisory Board SAB-EC-90–021. US Environmental Protection Agency, Washington, DC.

USEPA (United States Environmental Protection Agency) (1991a) *Methods for Aquatic Toxicity Identification Evaluations, Phase I, Toxicity Characterization Procedures*, EPA/600/6-91/003. USEPA, Office of Research and Development, Environmental Research Laboratory, Duluth, MN.

USEPA (United States Environmental Protection Agency) (1991b) *Summary Report on Issues in Ecological Risk Assessment*, Risk Assessment Forum, EPA/625/3-91/018. US Environmental Protection Agency, Washington, DC.

USEPA (United States Environmental Protection Agency) (1991c) *Technical Support Document for Water Quality Based Toxics Control*, Office of Water, EPA/505/2-90–001, PB91–127415. US Environmental Protection Agency, Washington, DC.

USEPA (United States Environmental Protection Agency) (1992a) *A Framework for Ecological Risk Assessment*, Risk Assessment Forum, EPA/630/R-92.001. US Environmental Protection Agency, Washington, DC.

USEPA (United States Environmental Protection Agency) (1992b) *Ecological Risk Assessment Guidelines Strategic Planning Workshop*, Risk Assessment Forum, EPA/625/3-91/018. US Environmental Protection Agency, Washington, DC.

USEPA (United States Environmental Protection Agency) (1992c) *Peer Review Workshop Report on a Framework for Ecological Risk Assessment*, Risk Assessment Forum, EPA/630/R-92/001. US Environmental Protection Agency, Washington, DC.

USEPA (United States Environmental Protection Agency) (1994) *Estimating Toxicity of Industrial Chemicals to Aquatic Organisms using SAR* (2nd edn), EPA/748/R/93/001. US Environmental Protection Agency, Office of Pollution Prevention and Toxics, Washington, DC.

USEPA (United States Environmental Protection Agency) (1996) *Marine Toxicity Identification Evaluation (TIE): Phase I, Guidance Document*, EPA/600/R-96/054. USEPA, Cincinnati, OH.

USEPA (United States Environmental Protection Agency (1998) *Guidelines for Ecological Risk Assessment*, Risk Assessment Forum, EPA/630/R095/002F. US Environmental Protection Agency, Washington, DC.

USEPA (United States Environmental Protection Agency) (2001) *Clarifications Regarding Toxicity Reduction and Identification Evaluations in the National Pollutant Discharge Elimination System Program, March 27, 2001*. US Environmental Protection Agency, Office of Wastewater Management, Office of Regulatory Enforcement, Washington, DC.

USEPA (United States Environmental Protection Agency) (2002a) *Short-term Methods for Estimating the Acute Toxicity of Effluents and Receiving Waters to Freshwater and Marine Organisms*, EPA/821/R-02-012. USEPA, Office of Water, Washington, DC.

USEPA (United States Environmental Protection Agency) (2002b) *Short-term Methods for Estimating the Chronic Toxicity of Effluents and Receiving Waters to Freshwater Organisms*, EPA/821/R-02/013. USEPA, Office of Water, Washington, DC.

Verhaar, H.J.M., van Leewen, C.J. and Hermens, J.L.M. (1992) Classifying environmental pollutants. 1: structure–activity relationships for prediction of aquatic toxicity. *Chemosphere*, **25**, 471–491.

Waldichuck, M. (1989) Aquatic toxicology in management of marine environmental quality: present trends and future prospects. In *Aquatic Toxicology and Water Quality Management*, Nriagu, J.O. and Lakshminarayama, J.S.S. (eds), pp. 7–22. John Wiley & Sons, New York.

Walker, C.H., Hopkin, S.P., Sibly, R.M. and Peakall, D.B. (2001) *Principles of Ecotoxicology*, 2nd edn. Taylor and Francis, London.

5 The aquatic environment
William L. Goodfellow Jr.

5.1 Introduction

The overall quality of the aquatic environment has been increasingly scrutinized by regulatory agencies over the last few decades. However, the concept of using biological organisms in the measurement of potential impacts on aquatic systems was documented considerably earlier. For example, Hart *et al.* (1945) presented various techniques for evaluating wastewater toxicity and the associated effects on the receiving water, wastewater variability of toxicity, and multiple dischargers.

Prior to the 1970s, protection of aquatic systems in the USA was focused principally on operations of sewage treatment plants and the eutrophication of lakes, streams, rivers, and estuaries. Identification of pollution problems throughout the 1960s and 1970s led State and Federal agencies to hold workshops or 'Enforcement Conferences' to develop pollution abatement programs (Tebo, 1986). Within the USA, these working groups and associated activities led to the development of various water quality criteria incorporating toxicity assessment with chemical specific data. These data were incorporated into national water quality criteria with water quality criteria books such as the *Green Book* (USEPA, 1972), *Blue Book* (USEPA, 1976), *Red Book* (USEPA, 1977), and *Gold Book* (USEPA, 1986), which were published periodically.

During the 1970s, it became increasingly obvious that approaches other than only chemical-specific water quality criteria were needed for controlling pollutants. Engineering technology with the goal of developing *best practical technology* (BPT) and a longer term goal of developing *best available technology* (BAT) were being used in regulatory settings. An initiative to have all point sources directly discharging to wastewater treatment plants (WWTP) or to install specific wastewater treatment as part of the discharger's operations began to be required. Through this initiative, it was believed that adverse pollutants to aquatic systems would be controlled or eliminated. However, once the BPT was completed in the USA by the end of 1974, studies demonstrated that many wastewaters remained toxic. For example, in 1974 roughly 80% of the WWTPs were still acutely toxic, and during 1975–1982, 62% still exhibited acute toxicity (Tebo, 1986).

The Clean Water Act (CWA) was enacted in 1972 for the further protection of rivers, lakes, and coastal waters of the USA. The goal of the CWA was to restore and maintain the chemical, physical and biological integrity of the Nation's waters.

Additionally, the CWA established a permitting process as a management tool to regulate the discharge of pollutants into the waters of the USA. The National Pollutant Discharge Elimination System (NPDES) permitting process marked the shift from controlling 'conventional' pollutants to controlling 'toxic discharges'.

Currently the USA has more than 400 000 NPDES permitted facilities (www.epa.gov/ow) with more than 50 industrial classifications containing several hundred thousand businesses and more than 16 000 publicly operated treatment works (POTWs). Each facility is required to comply with limits in the NPDES permits. Permits are typically expressed as an amount, load or concentration of chemicals that can be discharged safely to the aquatic system. The primary objective of these permits is to identify controls to protect the receiving water for its 'designated uses'.

In 1984 the US Environmental Protection Agency (USEPA) issued a policy designated to reduce further or eliminate toxic wastewaters and to achieve the objectives of the CWA (USEPA, 1984). This integrated toxics control strategy applies both chemical-specific and biological methods to manage the discharge of toxic pollutants. The resulting USEPA publication, *Technical Support Document for Water-Quality-based Toxics Control* (often referred to as the TSD or Technical Support Document) (USEPA, 1984, 1991b), is the guidance for this approach. This process uses whole effluent toxicity (WET) testing as the principal tool to protect the Nation's receiving waters. This is accomplished by establishing water quality standards and limits using NPDES permits.

Whole effluent toxicity testing relies on acute and chronic aquatic toxicity tests to monitor discharges to receiving waters (Grothe *et al.*, 1996; USEPA, 2000). Acute toxicity tests are short-term tests designed to measure effects during a short exposure duration and is typically expressed as effects on survival over a 24-h to 96-h period (USEPA, 2000). Chronic toxicity tests are designed to evaluate the toxic effects on survival, growth, reproduction, and other sublethal effects over a significant or sensitive portion of the organism's life cycle (USEPA, 1995; 2002b,c).

Similar approaches that incorporate WET tests and toxicity-based controls have been or are being developed throughout the world. For example, the UK (Wharfe, 1996), Sweden (Svenson *et al.*, 1992), France (Vandevelde and Fauchon, 1998), as well as throughout the European Union (Van Leeuwen and Herman, 1995), Canada (Sergy, 1987), and much of South America (Barra *et al.*, 2002) have adopted management processes that not only use chemical-specific monitoring of aquatic systems but also incorporate biological assessment of the wastewater using a 'biological detector' to indicate the presence of potential pollutants.

Although the majority of regulatory issues have focused on the water in aquatic systems, recently sediment contamination and its effect on aquatic organisms have received greater interest in both the research and regulatory arenas. In the USEPA's 1998 Report to Congress on contaminated sediments (USEPA, 1997), the agency reported that 26% of surveyed watersheds in the continental USA, including streams, lakes, harbors, near-shore areas, and oceans, are sufficiently contaminated with toxic pollutants to pose potential risks to aquatic life and to human health. As part of this assessment, the agency went further in identifying 96 watersheds that were deemed to be areas of probable concern. These watersheds represent approximately 10% of the sediment underlying US surface waters. Roughly 70% of these watersheds have fish consumption advisories already in place. The sediment contaminants most frequently identified to be associated with causing adverse effects include polychlorinated biphenyls (PCBs), mercury, organophosphate pesticides, and polycyclic aromatic hydrocarbons (PAHs).

Although most data on contamination are largely characterizations of chemicals in the water and sediment/soil samples, the fraction that is biologically available (or bioavailable) can be considerably less than the total or even dissolved fraction. Bioavailability is the portion of the chemical that directly influences or interacts with the biota. It is difficult to 'truly evaluate' bioavailability, thus surrogate parameters such as acute and chronic toxicity and bioaccumulation are also used as direct evidence for whether a contaminant in a sediment is biologically available. There are many factors that determine the bioavailability of contaminants in sediments, including the chemistry of the water, the geochemistry of the sediment, the influence of the organism on the environment, the physical conditions of the water and sediment, and the wide variety of biochemical reactions that may occur once an organism has come in contact with the contaminant (Forbes *et al.*, 1998). The response of the organism to the contaminant is directly related to the complexity of the exposure route (e.g. ingestion, dermal exposure, inhalation/respiration), which will influence the amount of the contaminant that enters the organism or cell of the organism. Additionally, the portion of the contaminant that has the potential to be bioavailable influences how the organism reacts to the contaminant.

Bioavailability of contaminants in waters and sediments also changes in temporal and spatial scales. For example, because sediments are at best mixtures, rather than solutions like most water-borne pollutants, the conditions that influence the bioavailability of a contaminant in two similar sites may change from day to day as well as be considerably different even though they are in close proximity to each other. In general, the characteristics of the water above the sediment (overlaying water) and the interstitial water (water in-between the particles of the sediment, often referred to as pore water) most influence the bioavailability of the contaminants in the sediment. It is believed that pore water has the greatest influence on the bioavailability of contaminants in sediments (Suter *et al.* 2000). One way to think of contaminants in water is that they are today's pollution issue, whereas contaminants in sediments are both today's and yesteryear's issues.

5.2 Aquatic toxicity

Theoretically, any species can be used in a toxicity test if it meets the toxicological and other objectives of the assessment. However, several aspects of the studies are important to consider when selecting a species. Additionally, several different types of tests can be used as part of traditional aquatic toxicity testing. As discussed earlier, the two standard types of tests are acute and chronic toxicity tests. Furthermore, these types of test are divided into testing based on the handling of the test solutions, such as static testing without solution renewal, static-renewal testing, and flow-through testing. The amount of sample available for carrying out the toxicity test may be limited, especially in the case of wastewater treatability studies and pore water (sediment testing). Most often the samples are either hand-carried to the laboratory or shipped via overnight express courier service. As a result, tests that need very large amounts of samples may require on-site testing. This is not always practical and is often expensive. To facilitate the utilization of reduced volume tests, tests using species or early life stages of traditionally tested organisms have been developed that require limited sample volume, e.g. acute tests using daphnids or fathead minnows, and mysid tests requiring only 15–500 ml of sample per test chamber (USEPA, 2002a). Chronic toxicity tests often use larger test volumes because the sensitive end-points of growth and reproduction require larger volumes to minimize density-dependent variables in the testing (USEPA, 2002b,c). Many of these tests may require 30–1000 ml of solution per test chamber (USEPA, 2002b,c). Other factors to consider when selecting a species for testing are: organism availability and sensitivity, relevance to study (e.g. pelagic vs. benthic, freshwater vs. marine), availability of toxicity data for the species and ease of test (e.g. feeding, temperature, light regime, water quality requirements) (USEPA, 2002a,b,c). The species most commonly used for both freshwater and marine testing in the USA are summarized below.

5.2.1 Freshwater species

A summary of the freshwater species that have been shown to be suitable for testing in the USA are presented in Table 5.1. Upon review of the available literature it quickly becomes apparent that the majority of toxicity tests performed in the USA use *Ceriodaphnia dubia* (water flea), *Daphnia pulex* (water flea), and *D. magna* (water flea) followed by *Pimephales promelas* (fathead minnow). The reason for this is that the vast majority of the toxicity tests carried out in the USA are in freshwater. In states (e.g. Washington) where protection of salmon population or other coldwater species is an issue, *P. promelas* may be replaced by *Oncorhynchus mykiss* (rainbow trout). Testing carried out on ambient water and sediments to identify toxicants from agricultural run-off often include the green algae species *Selenastrum capricornutum* (recently renamed, see Table 5.1) (Ankley *et al.*, 1990, 1992; SETAC, 1998) as part of the toxicity assessment.

5.2.2 Estuarine and marine species

A summary of the estuarine and marine species that have been used most consistently in the USA is presented in Table 5.2. The species used in these

Table 5.1 Commonly used freshwater species in the USA.

Species	Common name
Pelagic	
Ceriodaphnia dubia	Water flea
Daphnia magna	Water flea
Daphnia pulex	Water flea
Gammarus pulex	Amphipod
Oncorhynchus mykiss	Rainbow trout
Pimephales promelas	Fathead minnow
Benthic	
Chironomus tentans	Midge larvae
Hyalella azteca	Side-swimmer (Amphipod)
Lumbriculus variegatus	Oligochaete
Plant	
Selenastrum capricornutum (renamed *Raphidocelis subcapitata* and more recently *Pseudokirchneriella subcapitata*)	Green algae

Table 5.2 Commonly used estuarine/marine species in the USA.

Species	Common name
Pelagic	
Atherinops affinis	Topsmelt (fish)
Cyprinodon variegatus	Sheepshead minnow
Menidia beryllina	Inland silverside
Americamysis bahia	Mysid shrimp
Benthic	
Ampelisca abdita	Amphipod
Arbacia punctulata	Sea urchin
Corophium volutator	Mud shrimp
Crassostrea gigas	Portuguese oyster
Dendraster excentricus	Sand dollar
Haliotus rufescens	Abalone
Mulinia lateralis	Dwarf surf clam
Mytilus galloprovincialis	Mussel
Mytilus californianus	California sea mussel
Psammechinus miliaris	Shore urchins
Strongylocentrotus purpuratus	Purple urchin
Plant	
Champia parvula	Red macroalgae
Macrocystis pyrifera	Giant kelp

studies are much more diverse than those used for freshwater. One reason for this is that both Atlantic and Pacific species are used. The most commonly used pelagic species for toxicity testing are *Americamysis bahia* (formally *Mysidopsis bahia)*, *Menidia beryllina*, and *Cyprinodon variegatus*. Plant species used for biomonitoring tests include *Champia parvula* and *Macrocystis pyrifera*. Benthic species commonly used for testing are *Ampelisca abdita*, *Arbacia punctulata*, and *Strongylocentrotus purpuratus*. Other species that have been used in toxicity testing are also included in Table 5.2.

5.2.3 Other organisms used for toxicity testing

In addition to the species listed above, other test systems have been used successfully for aquatic toxicity testing. Although not presented in this chapter in detail, one example of a test system in use is Microtox®. Microtox® is discussed in this chapter because of its regular use in Europe. Microtox® uses the luminescent marine bacteria *Vibrio fischeri* as the test species, with light output as the endpoint. The main advantages of Microtox® are the minimal exposure durations needed for responses (15–30 min) and the low sample volume required. Evidence exists implying that Microtox® does not always model the results of traditional species used in some of the compliance tests and is often not sensitive to secondary treatment effluent (Goodfellow *et al.*, 1989). Examples of the investigators that have used Microtox® as the primary test include Guzzella *et al.* (1996), Hoke *et al.* (1992), Mazidi *et al.* (1992), and Svenson *et al.* (1992, 1998).

Although there are reservations about using Microtox® as the primary test in the USA, it is considered useful in species sensitivity comparisons (Ankley *et al.*, 1990, 1992; Bleckmann *et al.*, 1996; Gupta and Karuppiah, 1996a,b). The Microtox® test is sensitive to certain toxicants but relatively insensitive to other toxicants, including ammonia. Ammonia is a common toxicant in water and sediment samples (Schot *et al.*, 1995). Microtox® often fails to identify ammonia as the toxicant at concentrations toxic to most other aquatic organisms. As a result, when Microtox® is used in combination with other species tests, its relative insensitivity to ammonia may lead to results that provide information on toxicants that may be masked by ammonia when only traditional species are used.

Other species that have been used for rapid detection of toxicants include rotifers, copepods, and brine shrimp. Additionally, organisms that have been fed chemical indicators as well as genetically engineered organisms have started to be explored as alternatives to traditional testing. For example, a product developed by Aqua Survey (Flemington, NJ, USA) that evaluates the toxicity of water samples has shown great promise as an early warning detection system with regard to monitoring for Homeland Security in a round-robin testing program.

Recently a move towards the assessment of cellular damage, chemical effects on enzyme or stress proteins and the relationship towards environmental assessments of contaminants have become more accepted within the regulatory community. Traditionally, histopathological evaluations of fish or bivalves, as they related to the commercial or recreational value of stock, were the extent of the evaluations performed with regard to regulatory situations. In the realm of human health-related toxicology, cellular-level assessments have long been the norm. However, as the breadth of experience in correlating environmental conditions and contamination with specific cellular damage or related enzymes and/or stress proteins knowledge increases, more and more regulatory situations will be evaluated at the cellular level rather than only at the organism level.

In an excellent summary of biochemical mechanisms with discussions on metabolism, adaptation and toxicity, Di Guilio *et al.* (1995) provided the following insight. They believe that biochemical assessments have played a lesser role in aquatic toxicology for the following reasons. First, traditional toxicology is only concerned with one species (humans) and thus considerable information can be obtained with regard to the biochemical and cellular impacts to humans from specific chemicals or mixtures. Although models such as mice, rats and monkeys are used for investigative purposes, because they relate to humans, the number of organisms required to assess the toxicological impacts of contaminants in the aquatic environment are in the hundreds if not thousands. Di Guilio *et al.* (1995) also offered another reason for the minimal use of biochemical- or cellular-level assessment in that the principal concern of environmental contamination is not an individual organism but is rather at the population, community, or ecosystem level.

Environmental scientists are applying more modern medical toxicology tools and theory to the assessment of the aquatic environment. Investigators have begun to employ biochemically based tool or biomarkers for monitoring environmental quality. Biomarkers are the assessment of biochemical responses measured in test organisms and they offer a potential diagnostic tool or early warning parameter with regard to environmental concerns. Specific biochemical assessment parameters that have been used by investigators include: enzyme systems that metabolize or biotransform organic contaminants; and protective and toxic responses associated with oxyradicals, of which the production can be enhanced by specific chemicals, metal-binding proteins, and stress proteins.

5.2.4 *Multiple species assessments*

Using multiple species may provide the investigator with better insight into multiple toxicants or provide additional toxicant information based on differing species sensitivities. For example, inclusion of more than one species as part of the testing program (such as *P. promelas*, which is more sensitive to surfactants than *C. dubia*, while *C. dubia* is more sensitive to organophosphate pesticides) would help to resolve toxicants that are masked in their response when both are found at toxic concentrations in an aquatic sample. Additionally, any changes in the proportional relationship of toxicity between the species over time may indicate that the toxicants present in the sample are variable (USEPA, 2002a).

When the toxicity of two species is compared in order to reduce uncertainties due to fluctuations of various parameters such as pH and oxygen saturation, it may be informative and cost-efficient to test two species in the same test chamber. Species that have been tested successfully together include:

- *Ceriodaphnia dubia* and *Pimephales promelas* (USEPA, 2002a)
- *C. dubia* and *Daphnia magna* (USEPA, 2002a)
- *C. dubia* and *Daphnia pulex* (USEPA, 2002a)
- *Ampelisca abdita* and *Americamysis bahia* (USEPA, 1994a)

5.2.5 Dilution water

5.2.5.1 Freshwater

Several water types and strategies are available for the performance of toxicity testing in freshwater discharge situations. The investigator can use natural receiving water for the discharge situation, synthetic laboratory-prepared dilution water such as reconstituted freshwater following USEPA (2002a) or Standard Methods (APHA *et al.*, 1998), dilute mineral water (USEPA, 2002a) or other appropriate surface water that can be a surrogate. Laboratory-prepared dilution waters such as reconstituted freshwater or dilute mineral water offer the investigator a water source that can be prepared easily, does not require collection, has characteristics that are consistent and has a composition that is known and controllable. Laboratory-prepared dilution waters can be prepared readily using the desired hardness and alkalinity to mimic the effluent or receiving water in order to assist the investigator in performance of the toxicity test. The hardness of the dilution water can be prepared to allow for the concentration of the wastewater's hardness during testing (i.e. maintaining the same hardness for the experiment, regardless of the test concentration), which is beneficial during evaluation of divalent metals, which are characterized by having a toxic response that is affected directly by hardness. However, laboratory waters are not equivalent to mature freshwater sources and are often devoid of the microelements and nutrients, organic acids or bacteria and micro food particles that may be necessary for the long-term survival and reproduction of some test organisms.

The use of receiving waters and other surface waters offers the investigators an opportunity to mimic the physical/chemical characteristics of the wastewater's receiving water in an effort to model what would happen to the test organisms in the lake, stream or river after discharge of the effluent. The advantage of using receiving waters is that they are more chemically and biological mature compared with laboratory-prepared dilution waters. The major disadvantage of using receiving waters is that the water could contain unknown toxicant(s) that could adversely affect the results of the investigation. In addition, chemicals such as organic acids could reduce the toxicity of the wastewater being evaluated if metals are the toxicants. Because the receiving water composition is often largely unknown, certain physical/chemical characteristics may fluctuate greatly during the overall study, necessitating the use of appropriate test controls. In situations where the receiving water is unacceptable for testing and the test organism requires natural water for adequate testing, other surface water sources can be used.

5.2.5.2 Marine and estuarine waters
For performing toxicity tests on wastewaters discharged to marine receiving waters, it is frequently desirable to test with marine species. If the sample wastewater is non-saline it will be necessary to adjust the salinity of the wastewater to a level that is acceptable to the test organisms. The need to adjust the sample raises several issues with regard to water type. First, how should the salinity be adjusted? There are two primary methods: addition of hypersaline brine or addition of reagent salts. Each method has advantages and disadvantage (Burgess *et al.*, 1997). Hypersaline brine is prepared from natural seawater and thus is composed of the essential constituents needed for marine organisms. However, addition of hypersaline brine does result in a dilution of the wastewater sample. For mildly toxic wastewater samples, the dilution may result in the wastewater being inferred as non-toxic. Addition of reagent salts to adjust sample salinity does not involve a dilution effect but does mean incorporating artificial constituents into a sample, which may themselves cause artifactual toxic effects. The reagent salt recipe described as GP2 consists of the major ions in marine waters (Burgess *et al.*, 1997; USEPA, 2002a). This recipe has been used successfully for marine/estuarine assessments. There are several commercially available reagent salt formulas to select from, including Forty Fathoms™ which is commonly used with Atlantic and Gulf coast species, and Tropic Marin,™ which is frequently used with Pacific coast species (Nicely *et al.*, 2000). These recipes contain both the macro and microelements representing marine/estuarine waters.

A second critical issue raised by salinity adjustment is how to determine the final salinity value. Should the final salinity be calibrated to the receiving water salinity or a standard test salinity (i.e. 20‰)? This issue is most dependent upon the objectives of the investigation being performed. For routine testing, using a standard test salinity may be most appropriate; however, for studies relating wastewater toxicity to receiving water effects, calibrating salinity to the receiving water conditions may be critical.

5.2.6 Selection of end-points

Typically, selection of the major test end-point to be used in the investigation is based on the compliance needs or the objective of the investigation. For acute exposures, survival is often the end-point selected for performance of the toxicity assessment. Although chronic exposures often use survival, growth expressed as dry weight, or some measure of reproductive potential (i.e. fecundity, offspring production) is the major test end-point. Survival of the test organisms (expressed as mortality of the test organism) is probably the most utilized test end-point because it is easy to measure, may not require long test exposures, often is the least costly of all of the tests and most single chemical testing data exists for survival as the end-point in toxicology databases. However, the major disadvantage of using survival as the end-point is that it is not a sensitive end-point.

Growth and/or reproduction in concert with survival are the typical end-points for chronic toxicity assessments. These end-points are usually more sensitive to toxicants than survival alone but are often more variable in test response because they may be influenced by test conditions that are less controllable than for tests evaluating only survival. In addition, tests evaluating growth and/or reproduction are often performed at temperatures higher than survival-only testing in order to maximize the organism's response to a toxicant, which may influence interpretation of the testing. Chronic exposure testing typically incorporates at least one additional end-point other than survival as the major test end-point.

Other end-points that might have diagnostic benefits that are typically not used during compliance testing include end-points such as symptoms or the organism's behavior resulting from exposure to the wastewater, avoidance, and time to lethal or impairment response. The behavior of the test organism may lead the investigator to a specific chemical class, or be a diagnostic fingerprint for tracking the specific toxicant within the sewer system without identification of the toxicant. For example, an organism may be influenced by the effluent, causing a spinning behavior response that could be tracked upstream of the WWTP without specifically identifying the toxicant but identifying the source of the toxicant. Time to lethality or impairment has been used successfully by several investigators to assist in the assessment of specific toxicants. McCulloch *et al.* (1993) used time to lethality as one of their diagnostic end-points for a steel mill effluent. The time to lethality end-point offers the investigator another tool that can assist in developing the specific 'weight of evidence' response of the wastewater to the test organisms. Toxicants that affect respiration or the gill tissue of fish and invertebrates often influence the organism's response to the toxicant more quickly than a toxicant that has an impact on specific organs or systems of the organism. Thus, the time to lethality end-point may provide additional information to the investigator than only survival data.

5.3 Wastewater toxicity identification evaluations

The presence and potency of the toxicants in the effluent and other wastewater samples are detected by performing various manipulations on the sample and by using aquatic organisms to track the changes in the toxicity. This toxicity-tracking step is the basis of toxicity identification evaluation (TIE). The methodology/strategy most often employed was developed by USEPA as three phases (characterization, identification, and confirmation) and currently five guidance documents cover the use of freshwater and estuarine/marine organisms. These methods are intended to provide techniques and approaches to those who need to characterize, identify or confirm the cause of toxicity in effluents.

The general approach for TIE is described in the document *Methods for Aquatic Toxicity Identification Evaluations: Phase I Toxicity Characterization Procedures* (Mount and Anderson-Carnahan, 1989; Norberg-King *et al.*, 1991). Norberg-King *et al.* (1992) describe methods that are applied to toxicity tests that estimate the chronic toxicity. The characterization manipulations include aeration, filtration, C_{18} SPE extraction and chromatography, chelation with EDTA, oxidant reduction and/or complexation with sodium thiosulfate, and toxicity testing at different pH values. The main differences between the acute and chronic techniques are that the concentrations of additives must be lower and the test conditions must be less severe in a chronic TIE because the longer exposure durations result in the test organisms being more sensitive to these conditions.

In some instances, not all TIE treatments and/or tests must be performed (i.e. a full Phase I characterization) when the TIE can be applied for specific types of dischargers, i.e. dischargers that have high levels of ammonia and cationic metals may need to home in on the techniques of Phases I and II, which will help in characterizing the toxicity sufficiently for toxicity reduction evaluation (TRE). Another example would be when a discharger needs to evaluate the toxicity resulting from a specific substance, e.g. if ammonia is causing the toxicity in a wastewater or whether chemicals other than ammonia are involved. Conversely, eliminating tests may result in erroneous conclusions or unnecessary efforts, especially during the early stages of Phase I.

After conducting the Phase I characterization tests, it is recommended that the investigator concentrates on the steps that are most clean-cut and have the major effect of reducing or eliminating the toxicity in the effluent. If toxicity in every effluent sample is not caused by the same toxicant(s), the characterization tests should indicate if the type of toxicant(s) is the same or different. Once identification is initiated, and suspect substances identified, the varying causes of toxicity can be evaluated because the concentrations of toxicants should correlate with the toxicity.

Toxicity, and particularly chronic toxicity, must be present sufficiently frequently so that an adequate number of toxic samples can be obtained. Enough routine toxicity testing should be done on each effluent before a TIE is initiated to ensure that toxicity is consistently present. It is not important that the same amount of toxicity is present in each sample, rather the changing levels of toxicity can assist in determining the cause of toxicity. Also, once the acute toxicant is removed, it cannot be assumed that the sublethal toxicity that may occur is due to the same compound.

Classes of chemicals characterized in Phase I include: (1) volatile toxicants such as organic solvents (e.g. xylene, benzene); (2) particulate-associated toxicants present in sample suspended matter, which may become bioavailable to a testing organism upon particle ingestion; examples of particulate toxicants include some pesticides and metals; (3) oxidants such as chlorine, commonly added to municipal effluents to eliminate pathogens before discharge; (4) non-polar organics, which include the vast suite of compounds such as solvents, hydrocarbons, fuels, some PAHs and some pesticides; (5) cationic metals, which are those elements commonly associated with toxicity to aquatic organisms [e.g. cadmium, copper, lead, nickel and zinc this group of toxicants can expand to include less toxic metals such as calcium, strontium, barium, manganese, cobalt and indium (Garvan, 1964)]; and (6) pH-dependent toxicants, which are substances that show differing toxicity as a function of pH. Examples of these include ammonia, which generally is most toxic at high pH ranges (USEPA, 2002a), and hydrogen sulfide, which is most toxic at low pH ranges (National Academy of Sciences, 1974).

Although Phase I of a TIE *characterizes* the types of toxicants suspected of being active in a sample, Phase II is designed to *identify* the specific toxicant(s) active. Methods for accomplishing this objective are described for freshwater samples in Durhan *et al*. (1993). Specific Phase II marine TIE methods are not available but Phase II of a marine TIE can be performed based on Durhan *et al*. (1993) methods. The procedures used to identify active toxicants characterized in Phase I are specifically designed to demonstrate the role of non-polar organic toxicants, ammonia, cationic metals, oxidants and filterable toxicants.

The final part of a TIE is Phase III. Phase III has the objective of *confirming* that toxicants characterized and identified in Phases I and II are the actual causes of toxicity. Phase III is described in great detail in Mount and Norberg-King (1993) for freshwater TIEs and, as noted above, specific Phase III marine TIE methods do not exist but general approaches in Mount and Norberg-King (1993) can be substituted. Phase III is essential in the conduct of a TIE. Owing to the inherent complexity of the toxicity tests, analytical chemistry and other TIE manipulations in Phases I and II, it is often necessary to reduce the statistical power of the TIE design (e.g. often a limited number of replicates are used due to the large number of treatments). Clearly, this compromise in experimental design can lead to erroneous or ambiguous results. Several Phase III procedures are available, including correlation, symptom, species sensitivity, spiking, mass balance and deletion methods. Another approach (a variation of spiking) is to prepare a serial dilution of a mock sample from control water that contains identical quantities of the identified toxicant(s). Statistically robust toxicity tests are then used with the mock sample and the actual sample; in principle, the two samples should demonstrate similar dose–response curves if the toxicant has been identified correctly. Other Phase III approaches include assessing the individual toxic unit contribution of each manipulation to ensure that they are equivalent to the original sample (mass balance) and relating quantities of identified toxicant to known toxicity values (e.g. single chemical toxicity data) to determine if enough toxicant is present to have the observed effects (correlation).

5.4 Sediment toxicity

Although the majority of regulatory issues have focused on the water in aquatic systems, many investigators have attempted to correlate toxicity with measured concentrations in sediment. However, most studies do not measure the toxicity as an internal dose to the organism vs. potential exposures by the contaminant in the sediment. Thus, the major assumption of these studies is that the retrospective observation of organism response (reduced survival, reduced growth, etc.) is directly correlated to chemical characterization of the sediment, without direct data on cause and effect of the chemical in the sediment to the organism. Often, water column toxicity test results are applied to the sediment samples to evaluate the potential toxicity of various chemicals to aquatic organisms. The use of water quality criteria applied to the sediment concentrations determined in the sediment can lead to biased assessment of the potential risks of sediments to aquatic organisms. Additionally, these tests are determined in very controlled laboratory systems that typically have maximum bioavailability of the chemical being evaluated. The presence of a contamination does not indicate the potential for adverse effects. A contaminant can have toxic effects only if it occurs in a bioavailable form (Suter *et al.*, 2000) and the compound can enter the cell of the animal or influence the normal processes of the organism. As discussed earlier, contaminants that are more bioavailable pose greater risk to the environment and potentially to human health than similarly toxic but less available contaminants. It is necessary to understand the factors and processes that influence the bioavailability of contaminants in sediments and to integrate this concept and information into the regulatory framework. The general factors influencing bioavailability predictions include:

- Organic geochemistry
  — water and particulate organic matter
  — chemical speciation (this can be difficult to determine)
- Organism interaction or niche variations
- Biological variability/diversity of biochemical responses
- Body burden or thermodynamic equilibrium

The most common technique for assessing the impacts due to the bioavailable fraction of a contaminant in sediments is the use of toxicity testing. By exposing the aquatic organisms directly to the contaminant of concern or the matrix under investigation, the fraction that is bioavailable can be determined. Principally, the most common exposure regime used for toxicity testing is short-term exposures designed to evaluate the acute toxicity of a toxicant. Longer term chronic exposures that evaluate the survival/mortality of the test organisms, as well as growth and reproductive potential, are also employed. Less routine in the assessment of bioavailability, but gaining momentum in the evaluation of stressors and toxicants, is the use of other end-points such as enzymes or stress proteins, as well as other biochemical end-points.

The methodology for sediment toxicity testing varies, depending upon the regulatory requirement, but is fairly uniform. For detailed method descriptions, the reader should refer to USEPA (1994a,b), APHA *et al.* (1998), ASTM (2001) for water column, bulk sediment and bioaccumulation testing. The USEPA/USACE (1994) is a good starting point to better understand toxicity testing for dredged material/sediment assessments that include elutriate, bulk sediment and bioaccumulation testing as well as water-only testing. Table 5.3 summarizes the major toxicity test types, matrices and species most commonly utilized in the USA.

As discussed earlier, toxicity testing is generally classified into three categories. Acute toxicity tests refer to relatively short exposure durations with an end-point of survival/mortality or other significant impact to the organism, such as immobilization. Acute tests have exposure durations that range from 48 to 96 h for water column/pore water exposures and up to 10 days for bulk sediment exposures. Chronic toxicity tests are longer exposure tests that evaluate the impacts on the aquatic organisms using end-points such as growth and reproductive potential as well as survival/mortality. The duration of exposure for chronic tests is highly variable, ranging from several days to years, depending on the objective of the test. Typically, the exposure durations used in most chronic sediment toxicity tests are 7–35 days for water column/pore water exposures and 28 days for bulk sediment testing. Chronic test durations are often determined by the amount of time necessary to achieve an adequate end-point response. For example, the shorter durations are used with growth and survival/mortality end-points and longer durations are used with reproductive potential. Bioaccumulation testing evaluates the body burden (or tissue concentration) of the contaminant(s) of concern from exposure to the sediment. The organisms are held for exposure durations typically ranging from 28 to 56 days. Following exposure, the organisms are held for an additional time (1+ days, depending on organism and testing strategy) to remove sediment from the digestive system, which is termed depuration. Immediately following this step the organism's tissue is evaluated analytically to determine the uptake of the contaminant by the organism.

Similar to the aquatic-only testing discussed earlier, the type of test is also classified by the renewal frequency of the test. Typically the renewal frequency refers to the water used in the test, with the sediment sample typically not replaced during testing. Static testing denotes testing where the water is not replaced during the toxicity test. This type of test is often used when the water quality characteristics are not expected to change during exposure and is typically used only for short-duration exposures. Static renewal testing is used when the water quality characteristics are anticipated to change during the exposure period or the test has a longer duration of exposure. This is the most typical type of test employed for sediment testing. Flow-through testing is used when the water quality characteristics are anticipated to change considerably during testing. Flow-through tests are

Table 5.3 Summary of major toxicity test types, matrices and species commonly utilized in the USA.

Test type	Matrices	Species commonly used[a]
Acute toxicity tests (typically 48–96 h of exposure)	Water, pore water, elutriate	Freshwater – water flea (*Daphnia magna*); water flea (*Daphnia pulex*); water flea (*Ceriodaphnia dubia*); fathead minnow (*Pimephales promelas*); bluegill (*Lepomis macrochirus*); channel catfish (*Ictalurus punctatus*); and rainbow trout, (*Oncorhynchus mykiss*) Marine/estuarine – mysid shrimp (*Americamysis bahia* and *Neomysis americana*); grass shrimp (*Palaemonetes* sp.); oyster (*Crassostrea* sp.); mussel (*Mytilus edulis*); sea urchin (*Strongylocentrotus* sp.); inland silverside (*Menidia beryllina*); and sheepshead minnow (*Cyprinodon variegatus*)
Chronic toxicity tests (typically 7–35 days of exposure)	Water, pore water, elutriate	Freshwater – water flea (*C. dubia*); fathead minnow (*P. promelas*); and rainbow trout (*O. mykiss*) Marine/estuarine – mysid shrimp (*A. bahia*); inland silverside (*M. beryllina*); and sheepshead minnow (*C. variegatus*)
Acute toxicity tests (typically 96 h to 10 days of exposure)	Whole sediment	Freshwater – amphipod (*Hyalella azteca*); mayfly (*Hexagenia limbata*); and midge (*Chironomus tentans* and *C. riparius*) Marine/estuarine – amphipods (*Ampelisca abdita, Rhepoxynius abronius, Leptocheirus plumulosus, Eohaustorius estuarius*); mysid shrimp (*A. bahia*); and polychaete (*Nenathes arenaceodentata*)
Chronic toxicity tests (typically 28 days of exposure)	Whole sediment	Freshwater – amphipod (*H. azteca*); midge (*C. tentans*) Marine/estuarine – amphipod (*L. plumulosus*)
Bioaccumulation (typically 28–56 days of exposure)	Water, whole sediment	Freshwater – amphipod (*H. azteca*); fathead minnow (*P. promelas*); rainbow trout (*O. mykiss*); channel catfish (*I. punctatus*); mayfly (*H. limbata*); and oligochaetes (*Lumbriculus variegatus*) Marine/estuarine – polychaetes (*N. arenaceodentata* and *Neris virens*); bivalves (*Macoma nasuta* and *M. edulis*); sheepshead minnow (*C. variegatus*); and inland silverside (*M. beryllina*)

[a] Species commonly used in US Environmental Protection Agency (USEPA), American Testing and Materials (ASTM) and US Army Corps of Engineers (USACE) testing programs.

often employed for dredged material testing and bioaccumulation testing. The major limitation of this test is the large water requirement necessary for the test and the generation of a substantial volume of wastewater during testing.

Sediment toxicity testing strategies are used to evaluate the potential toxicity of the sediment. The characteristics of the sediment, such as the water column or elutriate, pore water and bulk sediment, are evaluated using different testing regimes to tease out the impacts of various aspects of the complex sediment toxicity. Each of these aspects in concert with unknown contaminants or site conditions provides a weight of evidence for the overall impact of sediment toxicity on aquatic organisms.

5.4.1 Water column or elutriate testing

Water column or elutriate testing considers the effects on aquatic organisms that are only in direct contact with the water. Typically, the organisms are exposed to the overlying water or a prepared elutriate. The overlying water is the water that is in direct contact with the sediment but is not in the interstitial waters around the sediment particles. This testing is the easiest and most straightforward of all sediment-related testing. However, the toxicity associated with the interstitial water or bulk sediment is not fully considered during these assessments. Additionally, the overlying water often is not in equilibrium with the sediment. Thus, overlying water assessments alone can be biased low for the overall toxicity or bioavailability of the sediment.

Elutriate testing attempts to consider the toxicity impacts related more directly to the sediment. The test most often employed is the elutriate or modified elutriate test, referenced in the US Army Corps of Engineers dredged material management programs (USEPA/USACE, 1994). This procedure uses four parts of water and one part of sediment, which is mixed for a period of time and allowed to settle. The supernatant is decanted and evaluated toxicologically. This test attempts to evaluate the effects related to dredging and in water placement of sediment. The advantage of this test is that a large volume of test solutions can be prepared. The major disadvantage of this test is that the water chemistry related to the sediment/water interface can be altered drastically and influence the bioavailability due to the oxidation of chemicals that are in a reduced state in the sediments.

5.4.2 Pore water

Pore water assessments evaluate the toxicity of the interstitial water of the sediment to the aquatic organism (Mayer, 1993). The interstitial water is removed from the sediment by sediment compression or centrifugation. The advantage of this assessment is that the equilibrium of the sediment/water interface is more closely evaluated toxicologically, which allows more confidence in the bioavailability assessment. However, toxicity artifacts such as ammonia and sulfide toxicity that influence the observed sediment toxicity may overestimate the assigned sediment toxicity in the pore water assessments. A disadvantage of pore water testing is the difficulty in obtaining larger volumes of pore water from the sediment.

5.4.3 Bulk sediment

Bulk sediment testing is the second most common strategy for the assessment of toxicity and the bioavailability of contaminants related to the sediment. In bulk sediment testing the organisms are exposed directly to the sediment and their responses are evaluated. The advantage of this type of testing is that the organisms are allowed to interact directly with the sediment and interstitial water. However, it is difficult to isolate the causative agent in bulk sediment tests when toxicity is observed. Bioaccumulation testing is also performed on bulk sediments. As for all laboratory testing strategies, the bioavailability of a toxicant from the sediment may be altered from what is available in the environment. Caution must be employed in the interpretation of any laboratory studies and their integration to 'real life conditions'. Often, laboratory studies are best considered as worst-case situations because the organisms are exposed continuously to the toxicant or at greatly elevated concentration compared with the actual environmental conditions available in the field.

5.4.4 In situ testing or ambient testing

In situ or ambient testing attempts to integrate the field conditions into the toxicological assessment of the sediment. Many investigators have used cages or other devices to ensure that organisms are exposed to the sediment in relation to the other natural site conditions (e.g. site water quality impacts, temperature, ambient light levels, native food conditions, etc.). The advantage of this type of testing is that an attempt is made to integrate the field conditions with exposure to the sediment. However, artifacts resulting from the cage or other devices can be subject to damage or tampering, greatly influencing interpretation of the test results.

5.5 Assessment of bioavailability

Several tools have been investigated for assessing the bioavailability of contaminants in addition to the use of toxicity testing. Typically each of these methods is developed for specific purposes and may not be applicable directly outside of the original purpose of their development. It is also important to keep in mind that the total amount of a compound that can be extracted from sediment using analytical techniques does not correlate to the exposure or bioavailability of a toxicant.

5.5.1 Alternative extraction methods

Several investigators have used digestive fluid extraction to measure the bioavailability of contaminants in sediments. Mayer *et al.* (1996) and Weston and Mayer (1998a,b) investigated the impacts of ingested sediment on deposit-feeding aquatic organisms by determining the role that the digestive system plays on the absorption of contaminants from the sediment. This bioavailability assessment incubates sediments in digestive fluids of the polychaete, *Arenicola brasiliensis*. Unlike the low pH of the digestive fluids of vertebrates, invertebrates have digestive fluids that are close to neutral. Weston and Mayer (1998a,b) found that the digestive fluid extraction yielded only 12–50% of PAHs in spiked sediment, thus assessment based solely on total PAHs would have overestimated the potential risk of the sediments by a factor of two to eight times. Furthermore, they found that the extraction of PAHs in the digestive fluid was dependent indirectly on the organic carbon concentration and contaminant concentration in the sediment (e.g. a higher organic carbon level yielded lower bioavailability of PAH). Weston and Mayer (1998a,b) found that holding time affected digestive fluid extraction efficiency. When sediments spiked with PAHs were extracted immediately by the digestive fluids, approximately 70% of the PAHs were extracted and bioavailable. However, storage of the sediment for 3 weeks reduced the bioavailability of the PAHs in the sediments by 35%, as demonstrated by lower digestive fluid extractability. Sediment ageing or weathering has been demonstrated by other studies to decrease the amount of contaminants that are bioavailable or bioaccumulative (Luthy *et al.*, 1994; Ghoshal and Luthy, 1996; Forbes *et al.*, 1998) and the use of digestive system extraction models provides further evidence for the reduced bioavailability due to sediment ageing.

Lawrence *et al.* (1999) used digestive fluids as a way to evaluate the bioavailability of mercury and methylmercury with the lugworm, *Arenicola marina*. They found that digestive fluid solubilized these compounds to a greater level than seawater alone. Digestive fluid extracted the monomethylmercury more readily than mercury. The investigators found that an indirect relationship of organic content in the sediment to bioavailability existed, as evidenced from the digestive system model. Kelsey and Alexander (1997) evaluated the extraction of atrazine and phenanthrene using a variety of organic solvents and solvent combinations to mineralization by *Pseudomonas*. The bioavailability of atrazine was predictable for the bacteria with a methanol–water mixture, whereas phenanthrene bioavailabilty was best predicted with a butanol–water mixture. They demonstrated further that a relative decrease in bioavailability was related directly to a decrease in extractability of the contaminant from the sediment. Wood *et al.* (1997) found similar results in the assessment of PAH and PCB bioaccumulation patterns in sediment-exposed *Chironomus tentans* larvae, as did Thomann and Komlos (1999) for PAHs with crayfish and sunfishes.

5.5.2 Membrane analog methods

Semipermeable membrane devices (SPMDs), Tenax TA and polyethylene tube dialysis (PTD) systems have been used to provide a better understanding of the bioavailability of many contaminants in the environment, including PAHs in sediments. Macrae and Hall (1998) found that for all of the methods listed above they had similar extraction amounts of PAHs, with PTDs being slightly more efficient. They also found that absorption of the PAHs increased with the log octanol/water partition factor (K_{ow}) for two to four aromatic rings but decreased with more than four aromatic rings (probably due to the lower permeability of larger PAHs by the membranes). However, the researchers found complications (damage) in the use of the membranes after contact with the sediment for more than 24 h and recommended that the membranes only be used for short-term exposures. Macrae and Hall (1998) reported that for the Tenax TA method, because there is no membrane in this method to limit diffusion, it is most likely more representative of total dissolved PAH rather than simply the bioavailability of PAHs in the sediment. They reported also that as the organic matter increases, the amount of PAH desorbed decreases using the SPMD method.

Other investigators have used SPMDs for estimations of bioavailability. For example, Utvik and Johnsen (1999) used SPMDs and blue mussels to determine the bioavailability of PAHs from oil/natural gas exploration and the resulting 'produced waters'. They found that the SPMDs reflected the water-soluble fraction of the PAHs, which they believed represented the exposure route for lower trophic levels of aquatic organisms. In their hypothesis, the tissue residues represent both the water-soluble fraction and the bioavailable particle-bound fraction.

In addition to direct membrane analogs, many researchers have used C_8, C_{18}, and XAD resins to simulate the bioavailability of organic compounds. Lake *et al.* (1996) used C_{18} resins as a surrogate for benthic organism bioaccumulation of hydrophobic compounds from sediments. Gustafson and Dickhut (1997) and Ankley *et al.* (1991) also utilized resins to sorb compounds for the assessment of toxicity (and bioavailability) of contaminants in sediments.

5.5.3 Pore water assessments

Characterization of the pore water of sediment samples is a simple way of assessing the potential bioavailability of contaminants in sediments and is one of the principal matrices used in the performance of sediment TIEs (USEPA, 1991a). Ankley *et al.* (1991) found that pore water is a better predictor of bioavailability of contaminants than sediment elutriates. Pore water typically has the toxicant freely dissolved, which is the most readily bioavailable fraction. The contaminants are partitioned between the pore water and sediment particles. This partitioning was described by Landrum and Robbins (1990) as a multikinetic process involving both a reversible portion and a resistant portion. However, pore water assessments alone do not consider the role of sediment ingestion on the total bioavailability of contaminants in sediments. The use of elutriates, although helpful for assessing the impacts on bioavailability due to disburances of the sediment (e.g. dredging), have limited use for assessing the normal bioavailabililty of contaminants. This is primarily because of substantial changes in sediment/water chemistry during preparation of the elutriate (Ankley *et al.*, 1991; USEPA/USACE, 1994). It should be cautioned that changes in water/sediment chemistry during sampling and assessment of the pore water will occur and these changes need to be considered as part of the assessment of bioavailability.

Kosnian *et al.* (1999) observed that pore waters strongly correlated with bioavailability determined using Ambersorb resin. This observation was based on reduced bioavailability of fluoranthene in sediment by reducing the pore water concentration and the overall toxicity to *Lumbriculus variegatus*. Zwiernik *et al.* (1999) also reported that the bioavailability of PCBs to bacteria correlated with PCB pore water concentrations. They found that the presence of PAHs reduced the pore water concentration of PCBs, thus reducing their bioavailability. This is an important result, because it is direct evidence that contaminants can influence the bioavailability of other contaminants in sediments. Carvalho *et al.* (1999) compared the bioavailability of free and colloidal-bound metals, including barium, cadmium, cobalt, iron, manganese, mercury, silver, tin, and zinc. They reported that generally the kinetics were similar for both free and colloidal forms, however, barium, tin and zinc free ions accumulated to a greater extent than colloidal-bound metals, suggesting that they were more bioavailable.

5.6 Factors controlling bioavailability

Current research suggests that the availability of contaminants to aquatic organisms that do not ingest sediments is controlled by the concentrations that dissolve in the sediment pore waters, or water in the case of effluent toxicity testing (Boucher and Watzin, 1999; Kosnian *et al.*, 1999). Tomson and Pignatello (1999) suggest that pollutant mass transfer regulates bioavailability. However, others are currently testing these theories. Two key factors that have been demonstrated clearly to control bioavailability in sediment are acid-volatile sulfide and organic carbon for both sediment and wastewater toxicity.

5.6.1 Acid-volatile sulfide

Acid-volatile sulfide has been studied extensively and incorporated widely into models for assessing sediment contamination (DiToro *et al.*, 1990, 1992; Carlson *et al.*, 1991; Berry *et al.*, 1996; Ankley *et al.*, 1996; Wang and Chapman, 1999). In marine waters where the concentration of sulfate is approximately 28 mM (Stumm and Morgan, 1996), sulfate-reducing bacteria convert sulfate to sulfide during the decomposition of organic matter (Wang and Chapman, 1999). Although concentrations of sulfate are much lower in freshwater streams and rivers (typically 0.12 mM; Stumm and Morgan, 1996), under conditions of large amounts of organic sedimentary matter measurable amounts of sulfide may be present (Wang and Chapman, 1999). In anoxic sediments, sulfide forms iron and manganese sulfide solids, including amorphous iron sulfide, and other metal complexed solids. Thus, in these sediments the metals are not bioavailable owing to the insoluble metal sulfides. All of these complexes are easily dissolved in acid, which is why they are termed acid-volatile sulfide.

5.6.2 Organic carbon

As noted previously, many studies have correlated increased sediment and waterborne organic carbon concentrations with decreased bioavailability (Kelsey and Alexander, 1997; Macrae and Hall, 1998; Lawrence *et al.*, 1999). Organic carbon is widely recognized as an important determinant of bioavailability. Weston and Mayer (1998a,b) found that the extractability of PAH in digestive fluids depends on the sediment's organic carbon level. Baptista Neto *et al.* (2000) claimed a strong correlation between sediment organic carbon and iron, manganese, nickel, lead, zinc, and chromium levels. Davies *et al.* (1998) developed an assay for copper availability in freshwater that describes the strong role of organic carbon in reducing the bioavailability of copper. Standley (1997) found that sediment organic matter reduced the bioavailability of the organochlorine pesticide dieldrin to the oligochaete, *Lumbricus variegatus*. Baptista Neto *et al.* (2000) also found that sediment size (in addition to organic carbon content) influences the abundance and distribution of metals in surface sediments.

In addition to sediment organic matter, dissolved organic carbon (DOC) affects contaminant concentrations through sorption, reducing bioavailability (Servos *et al.*, 1989; Schubauer-Bergan and Ankley, 1991; Landrum *et al.*, 1996; Lawrence *et al.*, 1999). Breault *et al.* (1996) found that 84–99% of copper present in sediment was bound to dissolved fulvic acid and EDTA (added to sediment during the resulting laboratory investigations) in seven stream samples. They also found that the fraction of copper bound to DOC decreased with increasing water hardness, suggesting a role in availability, which is influenced by other metals such as calcium and magnesium.

Not only the quantity but also the quality of organic matter has been shown to influence bioavailability (Karickhoff *et al.*, 1979). Sediment organic carbon provides a primary food source for benthic organisms. DeWitt *et al.* (1992) found that sediment pore water partitioning and toxicity of fluoranthene to the amphipod, *Rheopyxinius abronius*, were affected by sediment organic matter quality.

Gunnarsson *et al.* (1999) found that the polychaete, *Neries diversicolor*, preferentially fed on enriched sediment organic matter that also contained higher concentrations of the contaminant trichlorobenzene (TCB). Phillips *et al.* (2004) found that N:C ratios provide a reasonable indicator of bacterial bioavailability for complex, recalcitrant carbon moieties typical of many aquatic systems. Gunnarsson *et al.* (1999) examined the impact of sediment organic carbon quality on contaminant uptake and benthic organism growth. Reactivity of the organic carbon was measured by respiration and dissolved inorganic nitrogen flux. Growth rates of the brittle star and TCB uptake rates and steady-state TCB concentrations differed significantly between treatments and were correlated to the quality of the sediment organic carbon, as measured by amino acid, lipid, C, N, and polyphenolic compounds (Gunnarsson *et al.*, 1999). They further suggested that organic matter quality should be studied and considered when developing sediment quality criteria. DeWitt *et al.* (1992) suggested that such a consideration is of limited utility because they found that organic matter quality affected partitioning and toxicity but the absolute range of toxicities was small. Schlekat *et al.* (2000) found that silver and cadmium attached to labile polymeric organic carbon were significantly more bioavailable for digestive uptake than metal complexed with recalcitrant organic carbon, mineralogical features such as iron oxides, or even foods such as phytoplankton.

5.7 Processes effecting bioavailability in sediments

Many toxicants are sequestered in sediments (such as sediments contaminated by wastewater) or they are naturally the source of the toxicant (such as for metal toxicity). In both of these examples the sediment would be termed the toxicant reservoir. Thus, an understanding of the processes affecting the bioavailability of contaminants in sediments is important for a comprehensive understanding of bioavailability in the aquatic environment.

5.7.1 *Ageing or weathering*

Although bioavailability depends on the target organism, there are several trends or governing factors. For instance, it has been shown that bioavailability decreases with increasing sediment-contaminant contact time for many organisms (Landrum, 1989). Ageing or weathering of sediments appears to make the contaminant less available to aquatic organisms (Macrae and Hall, 1998).

As compounds interact with sediment over a period of time – a process termed ageing or weathering – it is generally accepted that there are two fractions, a rapidly desorbed fraction and a fraction, that is more slowly desorbed. This fraction potentially may not be desorbed at all, depending on environmental conditions (Reid *et al.*, 2000). Thus, aged or weathered sediments may be absent or virtually void of the rapidly desorbed fraction, leaving only the slowly desorbed fraction, which is considerably less bioavailable.

5.7.2 Sorption

For contaminants in sediments as well as waters, the portion of the contaminant in the vapour or solution phase primarily determines bioavailability. Therefore, it is important to understand not only the thermodynamic forces that govern equilibrium distribution, but also the rates of exchange of contaminants between the fluid and solid phases. Sorption of contaminants to sediments reduces bioavailability (Chen et al., 1999). Sorption is influenced by both sediment characteristics and contaminant characteristics, most notably hydrophobicity for organic contaminants (as measured by K_{ow}). Many researchers have implicated organic matter as the primary characteristic governing bioavailability (Cornelissen et al., 1998). Equilibrium sorption and desorption processes at solid–liquid interfaces in subsurface-saturated systems are referred to generally as either abortion or adsorption processes. Absorption relates to the partitioning of a contaminant into the natural organic matter or mineral components of the aquatic system. Adsorption refers to the physical or chemical binding at surfaces or interfaces between a solution and a sorbent. In general there are two distinct kinetic phases of desorption of chemicals from sediments. Cornelissen et al. (1998) indicated that the fast desorption represented the fraction of a compound that is bioavailable, whereas the slowly desorbing fraction is less available biologically.

5.7.3 Seasonality

Researchers have found that bioavailability changes seasonally (Mendoza et al., 1996; Balras, 1999). In an assessment of various metals and their bioavailability, Mendoza et al. (1996) found that the bioavailabilty of cadmium and lead peaked in July, whereas cobalt peaked in early spring for organisms in their study area. Balras (1999) reported that fish tissue and sediment concentrations varied seasonally, with lead, cadmium and cobalt concentrations increasing in sediment samples in August–October samples. In general, it is believed that the seasonality of the bioavailable fractions is linked directly to increased temperature-influenced reaction kinetics and increased organism activity as temperature increases, whereas reduced temperatures typically reduce the bioavailability of a contaminant.

5.7.4 Bioturbation and sediment resuspension

It has been demonstrated that water movement through bioturbation, irrigation, wave action and resuspension from dredging activities, etc. can increase bioavailability (Aller and Aller, 1992) and enhance mixing and redistribution of sediment-associated contaminants (Karickhoff and Morris, 1985). Bioturbation is the mixing of sediment particles, redistribution within the sediment column and sediment resuspension in overlying water as a consequence of burrowing, feeding, defecation and tube-building activities by benthic organisms (Ciarrelli et al., 1999). Such redistribution could prevent or delay the establishment of the sediment water equilibrium of contaminants, significantly increasing the potential for their accumulation by food or suspended sediment. Similarly, dredging activities or other factors such as storms, etc., may cause considerable redistribution of sediments within the aquatic system, which will influence the equilibrium of contaminants and alter the predicted bioavailability of the contaminant.

5.8 Estimating bioavailability

Traditionally, strong acids or highly non-polar organic solvents are used in a chemical analysis of metals and organic compounds to determine the total rather than the bioavailable amount of the contaminant. However, harsh extraction procedures such as Soxhlet extraction or metal digestion procedures often do not correlate with the amount of a compound that is available to an aquatic organism (Bosma et al., 1997) or the associated environmental risk (Suter et al., 2000). Recently, several additional selective extraction procedures have been used to quantify the bioavailable fraction, but none have been generally accepted or broadly adopted within the regulatory community (Suter et al., 2000). Cornelissen et al. (1997), Kelsey and Alexander (1997) and Reid et al. (1999) have attempted to use more representative measures for bioavailable compounds. However, until the complex relationships are understood that control exposure, uptake and effect, no chemical method will predict accurately the bioavailability in the environment (Reid et al., 2000). Thus, the risk associated with contaminants often will be overestimated.

The concentration of a contaminant can be used to estimate bioavailability for non-ionic organic compounds and metals using an equilibrium partitioning approach. The water only or free pore water concentration of non-ionic compounds, most often considered the bioavailable fraction, can be estimated by normalizing the concentration to organic carbon content. This is the preferred method by the USEPA (1993). The dissolved organic carbon concentration and the partition coefficient must be known because a high proportion of the chemical can be complexed to dissolved organic matter. For metals, one can measure the fraction of the concentration that is bound to sulfide. This method is considered valid for the five divalent metals, cadmium, copper, lead, nickel, and zinc (DiTorro et al., 1990). Extracted pore water has been shown to give a conservative estimate of sediment toxicity for metals (Ankley et al., 1991), however, researchers have demonstrated that toxicity in pore water does not necessarily indicate toxicity in the overlying waters.

Biological methods measuring toxicity or bioaccumulation are currently used widely to indicate bioavailability, yet interpretation of the results can be confounded by factors unrelated to bioavailability. For example, estimates of toxicity can be modified by the organism's general health, acclimation or adaptation. Bioaccumulation as a measure of bioavailability is confounded by behavior-affected exposure (such as feeding and respiration rates) as well as by metabolism of the contaminant of interest (USEPA, 1998).

The use of models for the prediction of metal toxicity using acid-volatile sulfide (DiTorro *et al.*, 1990, 1992) and equilibrium partitioning for hydrophobic compounds (DiTorro *et al.*, 1991) has been very useful for the prediction of bioavailability and associated toxicity of contaminants in sediments. Recently the Biotic Ligand model has been developed to relate metal bioavailability (and its potential toxicity) using the most recent chemical and physiological effects information on metals in aquatic environments (Paquin *et al.*, 1999; DiTorro *et al.*, 2001; Santore *et al.*, 2001).

Although thousands of anthropogenic compounds exist in surface waters, sediments, and other ambient media, results of previous studies have tended to group compounds in a few chemical classes. Many of these chemical groups have distinguishing characteristics that can be used to aid in their identification. Trace metal toxicants have been identified in sediments (Burgess *et al.*, 2000), pore waters (Anderson *et al.*, 2000), as well as in urban highway and industrial runoffs (Hunt *et al.*, 1998). The ability to chelate trace metals and manipulate their bioavailability through pH adjustments aids in their identification. Non-polar organic compounds such as pesticides, PAHs and PCBs are often found in ambient samples and sediments, and their association with toxicity is usually determined by exploiting their different affinities, bioavailability, and relative toxicity. Organophosphate pesticides, being relatively water-soluble, may enter surface waters more readily than higher K_{ow} compounds such as organochlorine or pyrethroid pesticides, which tend to enter surface water only through erosion and transport of associated soil particles (Phillips *et al.*, 2004). These more hydrophobic compounds (including PAHs and PCBs) tend to accumulate in sediments, where they may occur at toxic concentrations (Phillips *et al.*, 2004). Ammonia occurs in the environment naturally, but its concentrations are often enhanced by anthropogenic activities associated with livestock operations, sewage treatment plants, and storm drains that concentrate organic matter in poorly mixed aquatic environments. Ammonia is commonly identified as the cause of sediment and pore water toxicity (Ankley *et al.*, 1991), as is hydrogen sulfide, another naturally occurring and anthropogenically enriched compound. Hydrogen sulfide is not usually an issue in surface waters, however, because it seldom persists in aerobic media.

5.9 Summary and conclusions

It is important to understand that the bioavailability of a chemical in aquatic systems must be considered before the assessment can be complete. The fact that an elevated total chemical concentration is observed does not necessarily mean that the contaminant has any impact on the aquatic organisms. A contaminant must be biologically available before it will have a toxicological impact or risk to the environment. The bioavailability of a contaminant is influenced by numerous complexities related to the environment and site conditions. However, the use of toxicity testing procedures coupled with analytical characteristics of the water or sediment will provide an estimate of the potential risk associated with the contamination when used as a 'weight of evidence' approach. By exposing the aquatic organisms directly to the contaminant of concern or the matrix under investigation, the fraction that is biologically available can be determined. Principally, the most common exposure regimes used for toxicity testing are short-term exposures designed to evaluate the acute toxicity of a toxicant, typically expressed as survival/mortality. Longer term exposures that evaluate the survival/mortality of the test organisms after exposure, as well as the effect of the toxicant on organism growth and reproductive potential, are termed chronic toxicity assessments. These tests provide more detail concerning the toxicity and bioavailability of contaminants but are more expensive and resource-dependent.

Several tools have been investigated for measuring bioavailable compounds in addition to the use of toxicity testing. Methods such as alternative extraction methods, membrane analogs, and pore water assessments have been employed. Also, factors that control bioavailability, such as acid-volatile sulfide, organic carbon, the Biotic Ligand model, ageing or weathering of sediments, sorption, seasonality, and bioturbation, all greatly influence the bioavailability of contaminants. Typically each of these methods is developed for specific purposes and may not be applicable directly outside the original purpose of their development. It is important to keep in mind that the total amount of a compound that can be extracted from water or sediment using analytical techniques does not correlate to the exposure or bioavailability of a toxicant. Thus, for the assessment of water and sediments and their impacts on the environment, it is extremely important to consider all data in an integrated approach rather than using brief investigative 'snap shots' to understand the environmental condition and determine mitigation responses where appropriate.

References

Aller, R.C. and Aller, J.Y. (1992) Meiofauna and solute transport in marine muds. *Limnol. Oceanogr.*, **37**, 1018–1033.

American Public Health Association (APHA), American Water Works Association, and Water Environment Federation (1998) *Standard Methods for the Examination of Water and Wastewater* (20th edn). APHA, Washington, DC.

American Society for Testing and Materials (ASTM) (2001) *Annual Book of ASTM Standards, Section 11, Water and Environmental Technology*. vol. 11.05. ASTM, Philadelphia, PA.

Anderson, B.S., Hunt, J.W., Phillips, B.M., Stoelting, M., Becker, J., Fairey, R., Puckett, H.M., Stephenson, M., Tjeerdema, R.S. and Martin, M. (2000) Ecotoxicology change at a remediated superfund site in San Francisco, California, USA. *Environ. Toxicol. Chem.*, **19**, 879–887.

Ankley, G.T., Katko, A. and Arthur, J.W. (1990) Identification of ammonia as an important sediment associated toxicant in the Lower Fox River and Green Bay, Wisconsin. *Environ. Toxicol. Chem.*, **9**, 313–322.

Ankley, G.T., Schubauer-Berigan, M.K. and Dierkes, J. R. (1991) Predicting the toxicity of bulk sediments to aquatic organisms with aqueous test fractions: pore water vs. elutriate. *Environ. Toxicol. Chem.*, **10**, 1359–1366.

Ankley, G.T., Schubauer-Berigan, M.K. and Hoke, R.A. (1992) Use of toxicity identification evaluation techniques to identify dredged material disposal options: a proposed approach. *Environ. Manage.*, **16**, 1–6.

Ankley, G.T., DiTorro, D.M., Hansen, D.J. and Berry, W.J. (1996) Technical basis and proposal for deriving sediment quality criteria. *Environ. Toxicol. Chem.*, **15**, 2056–2066.

Balras, N. (1999) A pilot study of heavy metal concentration in various environments and fishes in the Upper Sakarya River Basin, Turkey. *Environ. Toxicol.*, **14**, 367–373.

Baptista Neto, J.A., Smith, B.J. and McAllister, J.J. (2000) Heavy metal concentrations in surface sediments in a near shore environment. Jurujuba Sound, Southeast Brazil. *Environ. Pollut.*, **109**, 1–9.

Barra, R., Colombo, J.C., Eguren, G. and Jardim, N.G.W. (2002) Regionally based assessment of persistent and toxic substances (PTS) in South America. *VII Congresso Brasileiro de Ecotoxicologia. V. Reuniao da SETAC Latino-America*, October 2002, Vitoria-Espirito Santo, Brazil.

Berry, W.J., Hansen, D.J., Boothman, W.S., Mahoney, J.D., Robson, D.L., DiToro, D.M., Shipley, B.P., Rodgers, B. and Corbin, J.M. (1996) Predicting the toxicity of metal-spiked laboratory sediments using acid-volatile sulfide and interstitial water normalizations. *Environ. Toxicol. Chem.*, **15**, 2067–2079.

Bleckmann, C.A., Rabe, B., Edgmon, S.J. and Fillingame, D. (1996) Aquatic toxicity variability for fresh- and saltwater species in refinery wastewater effluent. *Environ. Toxicol. Chem.*, **14**, 1219–1223.

Bosma, T.N.P., Middledorp, P.J.M., Schraa, G. and Zehnder, A.J.B. (1997) Mass transfer limitations of biotransformation: quantifying bioavailability. *Environ. Sci. Technol.*, **31**, 248–252.

Boucher, A.M. and Watzin, M.C. (1999) Toxicity identification evaluation of metal-contaminated sediment using artificial pore water containing dissolved organic carbons. *Environ. Toxicol. Chem.*, **18**, 509–518.

Breault, R.F., Colman, J.A., Aiken, G.R. and McKnight, D. (1996) Copper speciation and binding by organic matter in copper-contaminated stream water. *Environ. Sci. Technol.*, **30**, 3477–3486.

Burgess, R.M., Ho, K.T., Morrison, G.E., Chapman, G. and Denton, D.L. (1997) *Marine Toxicity Identification Evaluation (TIE): Phase I Guidance Document*, EPA 600/R-96/054. USEPA, Office of Research and Development, Washington, DC.

Burgess, R.M., Cantwell, M.G., Pelletier, M.C., Ho, K.T., Serbst, J.R., Cook, H.F. and Kuhn, A. (2000) Development of a toxicity identification evaluation procedure for characterizing metal toxicity in marine sediments. *Environ. Toxicol. Chem.*, **19**, 982–991.

Carlson, A.R., Phipps, G.L., Mattson, V.R., Kosnian, P.A. and Cotter, A.M. (1991) The role of acid-volatile sulfide in determining cadmium bioavailability and toxicity in freshwater sediments. *Environ. Toxicol. Chem.*, **10**, 1309–1319.

Carvalho, R.A., Benfield, M.C. and Santschi, P.H. (1999) Comparative bioaccumulation studies of colloidally complexed and free-ionic heavy metals in juvenile brown shrimp *Penaeus aztecus*. *Limnol. Oceanogr.*, **44**, 403–414.

Chen, W., Kan, A.T., Fu, G., Vignona, L.C. and Tomson, M.B. (1999) Adsorption–desorption behaviors of hydrophobic organic compounds in sediments of Lake Charles, Louisiana, USA. *Environ. Toxicol. Chem.*, **18**, 1610–1616.

Ciarelli, S., van Straalen, N.M., Klap, V.A. and van Wezel, A.P. (1999) Effects of sediment bioturbation by the estuarine amphipod *Corophium volutator* on fluoranthene resuspension and transfer into the mussel (*Mytilus edulis*). *Environ. Toxicol. Chem.*, **18**, 318–328.

Cornelissen, G., Rigterink, H., Ferdinandy, M.M.A. and Van Noort, P.C.M. (1998) Rapidly desorbing fractions of PAHs in contaminated sediments as a predictor of the extent of bioremediation. *Environ. Sci. Technol.*, **32**, 966–970.

Davies, C.M., Apte, S.C. and Johnstone, A.L. (1998) A bacterial bioassay for the assessment of copper bioavailability in freshwaters. *Environ. Toxicol. Water Qual.*, **13**, 263–271.

DeWitt, T.H., Ozretich, R.J., Swartz, R.C., Lamberson, J.O., Schultz, D.W., Ditsworth, G.R., Jones, J.K.P., Huselton, L. and Smith, L.M. (1992) The effects of organic matter quality on the toxicity and partitioning of sediment-associated fluoranthene to the infaunal amphipod, *Rhepoxynius abronius*. *Environ. Toxicol. Chem.*, **11**, 197–208.

Di Guilio, R.T., Benson, W.H., Sanders, B.M. and Van Veld, P.A. (1995) Biochemical mechanisms: metabolism, adapation, and toxicity. In *Fundamentals of Aquatic Toxicology: Effects, Environmental Fate, and Risk Assessment* (2nd edn), Rand, G.M. (ed.), Chapter 17. Taylor & Francis, Washington, DC.

DiToro, D.M., Mahoney, J.D., Hansen, D.J. and Berry, W.J. (1990) Toxicity of cadmium in sediments: the role of acid volatile sulfide. *Environ. Toxicol. Chem.*, **9**, 1487–1502.

DiToro, D.M., Zarba, C.S., Hansen, D.J., Berry, W.J., Swartz, R.C., Cowan, C.E., Paulon, S.P., Allen, H.E., Thomas, N.A. and Parquin, P.R. (1991) Technical basis for establishing sediment quality criteria for non-ionic organic chemicals using equilibrium partitioning. *Environ. Toxicol. Chem.*, **10**, 1541–1583.

DiToro, D.M., Mahoney, J.M., Hansen, D.J., Scott, K.J., Carlson, A.R. and Ankley, G.T. (1992) Acid volatile sulfide predicts the acute toxicity of cadmium and nickel in sediments. *Environ. Sci. Technol.*, **26**, 96–101.

DiToro, D.M., Allen, H.E., Bergman, H.L., Meyer, J.S., Paquin, P.R. and Santore, R.C. (2001) A Biotic Ligand Model of the acute toxicity of metals. I. Technical basis. *Environ. Toxicol. Chem.*, **20**, 2383–2396.

Durhan, E.J., Norberg-King, T.J. and Burkhard, L.P. (1993) *Methods for Aquatic Toxicity Identification Evaluations: Phase II Toxicity Identification Procedures for Acutely and Chronically Toxic Samples*, EPA-600/R-92/080. US Environmental Protection Agency, Office of Research and Development, Environmental Research Laboratory, Duluth, MN.

Forbes, T.L., Forbes, V.E., Giessing, A., Hansen, R. and Kure, L.K. (1998) Relative role of pore water versus ingested sediment in bioavailability or organic contaminants in marine sediments. *Environ. Toxicol. Chem.*, **17**, 2453–2462.

Garvan, F.L. (1964) Metal chelates of ethlyenediaminetetraacetic acid and related substances. *Chelating Agents and Metal Chelates*, In Dwyer, F.P. and D.P. Mellor (eds), pp. 283–333. Academic Press, New York.

Ghosal, S. and Luthy, R.G. (1996) Bioavailability of hydrophobic organic compounds from non-aqueous-phase liquids: the biodegradation of naphthalene from coal tar. *Environ. Toxicol. Chem.*, **15**, 1894–1900.

Goodfellow Jr, W.L., McCulloch, W.L., Botts, J.A., McDearmon, A.G. and Bishop, D.F. (1989) Long-term multispecies toxicity and effluent fractionation study at a municipal wastewater treatment plant. In *Aquatic Toxicology and Environmental Fate*, Suter II, G.W. and Lewis, M.A. (eds), vol. 11. STP 1007. pp. 139–158. American Society for Testing and Materials, Philadelphia, PA.

Grothe, D.R., Dickson, K.L. and Reed-Judkins, D.K. (1996) *Whole Effluent Toxicity Testing: an Evaluation of Methods and Prediction of Receiving System Impacts*. SETAC Press, Pensacola, FL.

Gunnarsson, J.S., Granberg, M.E., Nielsson, H.C., Rosenberg, R. and Hellman, B. (1999) Influences of sediment-organic matter quality on growth and polychlorobiphenyl bioavailability in echinodermata (*Amphiura filiformis*). *Environ. Toxicol. Chem.*, **18**, 1534–1543.

Gupta, G. and Karuppiah, M. (1996a) Toxicity identification of Pocomoke River porewater. *Chemosphere*, **33**, 939–960.

Gupta, G. and Karuppiah, M. (1996b) Toxicity study of a Chesapeake Bay tributary – Wicomico River. *Chemosphere*, **32**, 1193–1215.

Gustafson, K.E. and Dickhut, R.M. (1997) Distribution of polycyclic aromatic hydrocarbons in southern Chesapeake Bay surface water: evaluation of three methods for determining freely dissolved water concentrations. *Environ. Toxicol. Chem.*, **16**, 452–461.

Guzzella, L., Gartone, C., Ross, P., Tartari, G. and Muntau, H. (1996) Toxicity identification evaluation of Lake Orta (Northern Italy) sediments using the Microtox® system. *Ecotoxicol. Environ. Saf.*, **35**, 231–235.

Hart, W.B., Doudoroff, P. and Greenbank, J. (1945) *The Evaluation of the Toxicity of Industrial Wastes, Chemicals and Other Substances to Freshwater Fishes*. Atlantic Refining Co., Philadelphia, PA.

Hoke, R.A., Giesy, J.P. and Kreis Jr, R.G. (1992) Sediment pore water toxicity identification in the lower Fox River and Green Bay, Wisconsin using the Microtox® assay. *Ecotoxicol. Environ. Saf.*, **23**, 343–354.

Hunt, J.W., Anderson, B.A., Phillips, B.P., Newman, J., Tjeerdema, R.S., Taberski, K., Wilson, C.J., Stephenson, M., Pudcett, H.M., Fairey, R. and Oakden, J. (1998) *Sediment Quality and Biological Effects in San Francisco Bay: Bay Protection and Clean Up Program*, Final Technical Report. State Water Resources Board, Sacramento, CA.

Karickhoff, S.W. and Morris, K.R. (1985) Impact of tubificid oligochaetes on pollutant transport in bottom sediments. *Environ. Sci. Technol.*, **19**, 51–56.

Karickhoff, S.W., Brown, D.S. and Scott, T.A. (1979) Sorption of hydrophobic pollutants on natural sediments. *Water Res.*, **13**, 241–248.

Kelsey, J.W. and Alexander, M. (1997) Declining bioavailability and inappropriate estimation of risk of persistent compounds. *Environ. Toxicol. Chem.*, **16**, 582–585.

Kosian, P.A., West, C.W., Pasha, M.S., Cox, J.S., Mount, D.R., Huggett, R.J. and Ankley, G.T. (1999) Use of nonpolar resin for reduction of fluoranthene bioavailability in sediment. *Environ. Toxicol. Chem.*, **18**, 201–206.

Lake, J.L., McKinney, R., Osterman, F.A. and Lake, C.A. (1996) C-18 coated silica particles as a surrogate for benthic uptake of hydrophobic compounds from bedded sediment. *Environ. Toxicol. Chem.*, **15**, 2284–2289.

Landrum, P.F. (1989) Bioavailability and toxicokinetics of polycyclic aromatic hydrocarbons sorbed to sediments for the amphipod *Pontoporeia hoy. Environ. Sci. Technol.*, **23**, 588–595.

Landrum, P.F. and Robbins, J.A. (1990) Bioavailability of sediment-associated contaminants to benthic invertebrates. In *Sediments: Chemistry and Toxicity of In-Place Pollutants*, Baudo, R., Geisy, J.P. and Muntau, H. (eds), pp. 237–263. Lewis Publishers, Ann Arbor, MI.

Landrum, P.F., Harkey, G.A. and Kukkonen, J. (1996) Evaluation of organic contaminant exposure in aquatic organisms: the significance of bioconcentration and bioaccumulation, In *Ecotoxicology: A Hierarchical Treatment*, Newman, M.C. and Jagoe, C.H. (eds), pp. 85–131. Lewis Publishers, Boca Raton, FL.

Lawrence, A.L., McAloon, K.M., Mason, R.P. and Mayer, L.M. (1999) Intestinal solubilization of particle-associate organic and inorganic mercury as a measure of bioavailability to benthic invertebrates. *Environ. Sci. Technol.*, **33**, 1871–1876.

Luthy, R.G., Dzombak, D.A., Peters, C.A., Ramaswami, A., Roy, S.B., Nakles, D. and Nott, B.R. (1994) Remediating tar contaminated soils at manufactured gas sites: Technological challenges. *Environ. Sci. Technol.* **28**, 266A–277A.

Macrae, J.D. and Hall, K.I. (1998) Comparison of methods to determine the availability of polycyclic aromatic hydrocarbons in marine sediment. *Environ. Sci. Technol.*, **32**, 3809–3815.

Mayer, L.M. (1993) Organic matter at the sediment-water interface, In *Organic Geochemistry: Principles and Applications*, Engel, M.H. and Macko, S.A. (eds), pp. 171–184. Plenum Press, New York.

Mayer, L.M., Chen, Z., Findlay, R.H., Fang, J., Sampson, S., Self, R.F.L., Jumars, P.A., Quetel, C. and Donard, O.F.X. (1996) Bioavailability of sedimentary contaminants subject to deposit-feeder digestion. *Environ. Sci. Technol.*, **27**, 1719–1728.

Mazidi, C.N., Koopman, B., Bitton, G. and Neita, D. (1992) Distinction between heavy metal and organic toxicity using EDTA chelation and microbial assays. *Environ. Toxicol. Water Qual.*, **7**, 339–353.

McCulloch, W.L., Goodfellow, W.L. and Black, J.A. (1993) Characterization, identification and confirmation of total dissolved solids as effluent toxicants. In *Environmental Toxicology and Risk Assessment*, Grouch, J.W., Dyer, F.J., Ingersoll, C.G. and LaPoint, T.W. (eds), vol. 2, STP 1216, pp. 213–227. ASTM, Philadelphia, PA.

Mendoza, C.A., Cortes, G. and Munoz, D. (1996) Heavy metal pollution in soils and sediments of rural developing district 063, Mexico. *Environ. Toxicol. Water Qual.*, **11**, 327–333.

Mount, D.I. and Anderson-Carnahan, L. (1989) *Methods for Aquatic Toxicity Identification Evaluations: Phase II Toxicity Identification Procedures*, EPA-600/3-88/035. US Environmental Protection Agency, Office of Research and Development, Environmental Research Laboratory, Duluth, MN.

Mount, D.I. and Norberg-King, T.J. (1993) *Methods for Aquatic Toxicity Identification Evaluations: Phase I Toxicity Characterization Procedures*, EPA-600/3-88-034. US Environmental Protection Agency, Office of Research and Development, Environmental Research Laboratory, Duluth, MN.

National Academy of Sciences (1974) *Ammonia: Nitrate–Nitrite. Water Quality Criteria 1972*, EPA Ecol. Res. Ser. EPA-R3-73-033. US Environmental Protection Agency, Washington, DC.

Nicely, P.A., Phillips, B.M., Anderson, B.S., Hunt, J.W., Huntley, S.A., Tjeerdema, R.S., Palmer, F.H. and Carley, S. (2000) *Tolerance of Several Marine Toxicity Test Organisms to Ammonia and Artificial Salts* (Abstract Book). Society of Toxicology and Chemistry, Nashville, TN.

Norberg-King, T.J., Mount, D.I., Durhan, E.J., Ankley, G.T., Burkhard, L.P., Amato, J.R., Lukasewycz, M.T., Schubauer-Berigan, M.K. and Anderson-Carnahan, L. (1991) *Methods for Aquatic Toxicity Identification Evaluations: Phase I Toxicity Characterization Procedures* (2nd edn). US Environmental Protection Agency, Washington, DC.

Norberg-King, T.J., Mount, D.I., Amato, J.R., Jensen, D.A. and Thompson, J.A. (1992) *Toxicity Identification Evaluations: Characterization of Chronically Toxic Effluents, Phase I*, EPA-600/6-91/005F. US Environmental Protection Agency, Washington, DC.

Paquin, P.R., DiToro, D.M., Santore, R.C., Trivedi, D. and Wu, K.B. (1999) A Biotic Ligand Model of the acute toxicity of metals. III. Application to fish and *Daphnia* exposure to sediments. In *Integrated Approach to Assessing the Bioavailability and Toxicity of Metals in Surface Waters and Sediments*, Section 3 pp. 3-59–3-102. USEPA Science Advisory Board, Office of Water, Office of Research and Development, Washington, DC.

Phillips, B.M., Anderson, B.S., Hunt, J.W., Nicely, P.A., Kosaka, R.A., Tjeerdema, R.S., deVlaming, V. and Richard, N. (2004) *In situ* water and sediment toxicity in an agricultural watershed. *Environ. Toxicol. Chem.*, **23**, 435–442.

Reid, B.J., Jones, K.C. and Semple, K.T. (1999) Can bioavailability of PAHs be assessed by chemical means? In *Proceedings of the Fifth In Situ and On-Site Bioremediation International Symposium*, Leeson, A. and Alleman, B. (eds), vol. 8, pp. 253–258. Battelle Press, Columbus, OH.

Reid, B.J., Jones, K.C. and Semple, K.T. (2000) Bioavailability of persistent organic pollutants in soils and sediments – a perspective on mechanisms, consequences, and assessment. *Environ. Pollut.*, **108**, 103–112.

Santore, R.C., DiToro, D.M., Paquin, P.R., Allen, H.E. and Meyer, J.S. (2001) A Biotic Ligand Model of the acute toxicity of metals. II. Application to acute copper toxicity in freshwater fish and *Daphnia*. *Environ. Toxicol. Chem.*, **20**, 2397–2402.

Schlekat, C.E., Decko, A.W. and Chandler, G.T. (2000) Bioavailability of particle-associated silver, cadmium, and zinc to the estuarine amphipod, *Leptocheirus plumulosus* through dietary ingestion. *Limnol. Oceanogr.*, **45**, 11–21.

Schot, M., Schout, P., Jol, J., Vonck, W., Ho, K., Spronk, G. and Stonkhorst, J. (1995) Vergelijking van *Mysidopsis bahia* en *Corophium volutator* in een mariene toxicity identification evaluation (TIE), RIKZ Werkdocument RIKZ/OS-95.839X. RIKZ, The Hague.

Schubauer-Berigan, M.K. and Ankley, G.T. (1991) The contribution of ammonia, metals, and non-polar organic compounds to the toxicity of sediment interstitial water from an Illinois River tributary. *Environ. Toxicol. Chem.*, **10**, 925–939.

Sergy, G. (1987) *Recommendations on Aquatic Biological Tests and Procedures for Environmental Protection*, C & P, DOE, Environment Canada, Conservation and Protection. Environment Canada, Edmonton, Alberta.

Servos, M.R., Muir, D.C.G. and Webster, G.R.B. (1989) The effect of dissolved organic matter on the bioavailability of polychlorinated dibenzo-*p*-dioxins. *Aquat. Toxicol.*, **14**, 169–184.

SETAC (1998) *WET (Whole Effluent Toxicity): Toxicity of Complex Effluents*, short course developed by SETAC Foundation for Environmental Education. Society for Environmental Toxicology and Chemistry, Pensacola, FL.

Standley, L.J. (1997) Effect of sedimentary organic matter composition on the partitioning and bioavailability of dieldrin to the oligochaete *Lumbriculus variegatus*. *Environ. Sci. Technol.*, **31**, 2577–2583.

Stumm, W. and Morgan, J.J. (1996) *Aquatic Chemistry* (3rd edn). John Wiley and Sons, New York.

Suter, G.M., Efroymson, R.A., Sample, B.E. and Jones, D.S. (2000) *Ecological Risk Assessment for Contaminated Sites*. Lewis Publishers, Washington, DC.

Svenson, A., Linlin, Z. and Kaj, L. (1992) Primary chemical and physical characterization of acute toxic components in wastewaters. *Ecotoxicol. Environ. Saf.*, **24**, 234–242.

Svenson, F., Viktor, T. and Remberger, M. (1998) Toxicity of elemental sulfur in sediments. *Environ. Tox. Water Qual.*, **13**, 217–224.

Tebo Jr, L.B. (1986) Effluent monitoring: historic perspective. *Pellston Environmental Workshop: Environmental Hazard Assessment of Effluents*, SETAC Special Publication, Bergman, H.L., Kimerle, R.A. and Maki, A.W. (eds), pp. 13–31. Pergamon Press, New York.

Thomann, R.V. and Komlos, J. (1999) Model of biota-sediment accumulation factor for polycyclic aromatic hydrocarbons. *Environ. Toxicol. Chem.*, **18**, 1060–1068.

Tomson, M.B. and Pignatello, J.J. (1999) Causes and effects of resistant sorption in natural particles. *Environ. Toxicol. Chem.*, **18**, 1609–1619.

USEPA (1972) *Water Quality Criteria 1972*, EPA-R-3-003. US Environmental Protection Agency, Washington, DC (referred to as the 'Green Book').

USEPA (1976) *Quality Criteria for Water*. Report of the Committee on Water Quality Criteria. National Academy of Sciences, National Academy of Engineering. US Environmental Protection Agency, Washington, DC (referred to as the 'Blue Book').

USEPA (1977) *Quality Criteria for Water and Hazardous Materials*, EPA-440/9-76-023. US Environmental Protection Agency, Washington, DC (referred to as the 'Red Book').

USEPA (1984) *Technical Support Document for Water Quality-based Toxics Control*. US Environmental Protection Agency, Washington, DC (referred to as the Technical Support Document or TSD).

USEPA (1986) *Quality Criteria for Water, 1986*, EPA 400-5-86-001. US Environmental Protection Agency, Office of Water, Washington, DC (referred to as the 'Gold Book').

USEPA (1991a) *Sediment Toxicity Identification Evaluation: Phase I (Characterization), Phase II (Identification) and Phase III (Confirmation) Modification of Effluent Procedures*, Technical Report 08-91. USEPA, National Effluent Toxicity Assessment Center, Duluth, MN.

USEPA (1991b) *Technical Support Document for Water Quality-based Toxics Control*, EPA-505/2-90-001. US Environmental Protection Agency. Office of Water, Washington, DC.

USEPA (1993) *Technical Basis for Deriving Sediment Quality Criteria for Non-ionic Contaminants for the Protection of Benthic Organisms by Using Equilibrium Partitioning*, EPA 822-R-93-011. US Environmental Protection Agency, Office of Water, Washington, DC.

USEPA (1994a) *Methods for Measuring the Toxicity and Bioaccumulation of Sediment-associated Contaminants with Freshwater Invertebrates*, EPA 600-R-94-024. US Environmental Protection Agency, Office of Research and Development, Washington, DC.

USEPA (1994b) *Methods for Assessing the Toxicity of Sediment-associated Contaminants with Estuarine and Marine Amphipods*, EPA 600-R-94-025. US Environmental Protection Agency, Office of Research and Development, Washington, DC.

USEPA (1995) *West Coast Marine Species WET Test Methods: Short-term Methods for Estimating the Chronic Toxicity of Effluents and Receiving Waters to West Coast Marine and Estuarine Organisms*, EPA-600-R-95-136. US Environmental Protection Agency, Washington, DC.

USEPA (1997) *Incidence and Severity of Sediment Contamination in Surface Waters of the United States. Volume I: National Sediment Quality Survey*, EPA 823-R-97-006. US Environmental Protection Agency, Washington, DC.

USEPA (1998) UC Berkeley researchers use *in vitro* techniques to measure bioavailability of sediment-associated contaminants. *Contaminated Sediments News 1998(22)*, EPA 823-N-98-007. US Environmental Protection Agency, Washington, DC.

USEPA (2000) *Stressor Identification Guidance Document*, EPA 822-B-00-025. US Environmental Protection Agency, Office of Water, Office of Research and Development, Washington, DC.

USEPA (2002a) *Methods for Measuring the Acute Toxicity of Effluents and Receiving Waters to Freshwater and Marine Organisms* (5th edn), EPA-821-R-02-012. US Environmental Protection Agency, Washington, DC.

USEPA (2002b) *Short-term Methods for Estimating the Chronic Toxicity of Effluents and Receiving Waters to Freshwater Organisms* (4th edn), EPA-821-R-02-013. US Environmental Protection Agency, Washington, DC.

USEPA (2002c) *Short-term Methods for Estimating the Chronic Toxicity of Effluents and Receiving Waters to Marine and Estuarine Organisms* (3rd edn), EPA-821-R-02-014. US Environmental Protection Agency, Washington, DC.

USEPA/USACE (1994) *Evaluation of Dredged Material Proposed for Discharge in Waters of the US-Testing Manual. Inland Testing Manual*, EPA-823-B-94-002. US Environmental Protection Agency and US Army Corps of Engineers. Office of Water, Washington, DC.

Utvik, T.I.R. and Johnsen, S. (1999) Bioavailability of polycyclic aromatic hydrocarbons in the North Sea. *Environ. Sci. Technol.*, **33**, 1963–1969.

Vandevelde, T. and Fauchon, N. (1998) Intergrated management and water resources protection: the case of the Paris (France) region. In *Watershed Management: Practice, Policies, and Coordination*, Reimold, R.J. (ed.), Chapter 17. McGraw-Hill, New York.

Van Leeuwen, C.J. and Herman, J.L.M. (1995) *Risk Assessments of Chemicals: An Introduction*. Kluwer, Dordrecht, The Netherlands.

Wang, F. and Chapman, P.M. (1999) Biological implications of sulfide in sediment – a review focusing on sediment toxicity. *Environ. Toxicol. Chem.*, **16**, 2526–2532.

Weston, D.P. and Mayer, L.M. (1998a) *In vitro* digestive fluid extraction as a measure of the bioavailability of sediment-associated polycyclic aromatic hydrocarbons: sources of variation and implications for partitioning models. *Environ. Toxicol. Chem.*, **17**, 820–829.

Weston, D.P. and Mayer, L.M. (1998b) Comparison of *in vitro* digestive fluid extraction and traditional *in vivo* approaches as measures of polycyclic aromatic hydrocarbon bioavailability from sediments. *Environ. Toxicol. Chem.*, **17**, 830–840.

Wharfe, J.R. (1996) Toxicity based criteria for the regulatory control of waste discharges and for environmental monitoring and assessment in the United Kingdom. In *Toxic Impacts of Wastes on the Aquatic Environment*, Tapp, J.F., Hunt, S.M. and Wharfe, J.R. (eds), pp. 26–35. The Royal Society of Chemistry, Cambridge.

Wood, L.W., O'Keefe, P. and Bush, B. (1997) Similarity analysis of PAH and PCB bioaccumulation patterns in sediment-exposed *Chironomus tentans* larvae. *Environ. Toxicol. Chem.*, **16**, 283–292.

Zwiernik, M.J., Quensen III, J.F. and Boyd, S.A. (1999) Residual petroleum in sediments reduces the bioavailability and rate of reductive dechlorination of Aroclor 1242. *Environ. Sci. Technol.*, **33**, 3574–3578.

6 Biological methods for assessing potentially contaminated soils

David J. Spurgeon, Claus Svendsen and Peter K. Hankard

6.1 Why biological testing

Biological approaches offer a number of advantages that complement the detective capability of chemical analysis in the assessment of contaminated soils. These include:

- Direct measurement of effects on biota, rather than inferring these from comparisons of residue data and the results of laboratory toxicity tests.
- Responding to all contaminants, rather than only those in a predefined analytical suite.
- Accounting for contaminant interactions with soil properties.
- Integrating the combined effects of simple and complex mixtures.
- Providing powerful tools for risk communication by demonstrating the presence/absence of the components and functions of a healthy ecosystem.

Taken together, these arguments make a convincing case for the inclusion of biological tools within any soil quality assessment framework. Recognising the need for biological testing raises the question of how biological methods should best be used within any assessment framework and which of the available indicators are suitable? That biological tools are available that could be used for assessing soil quality is beyond doubt. A great deal of research has been undertaken that has contributed to knowledge of the effects of contamination on the ecological status of soil ecosystems. From this work, a number of adapted and new methods have been developed. These tests, which measure parameters ranging from ecosystem functions to molecular genetic responses, can be categorised as follows (Van Gestel and Van Brummelen, 1996):

- *Ecological indicators.* These are assays that measure ecosystem functional-level responses.
- *Bioindicator.* The bioindicator concept is based on the presence/absence of a species behaving in its normal manner, thereby indicating an acceptable set of environmental conditions.
- *Bioassays.* These are laboratory tests in which the toxicity of a sample is measured by exposing a specific organism and measuring life-cycle responses.

- *Biomarkers.* These have had a number of definitions in the context of ecotoxicology, which have ultimately agreed that a biomarker is '*any biological response to an environmental chemical, at the individual level or below, demonstrating a departure from normal status*'.
- *Biosensors.* These are analytical devices composed of a biological recognition element interfaced to a signal transducer that relate pollutant concentration to a measurable response.

Included within each of these categories is a range of procedures in various stages of development. Some tests are well established, with standardised protocols published by national and international organisations; other tests are established in the academic literature but have not been standardised yet; a final series of procedures are still in development at the research horizon and can be considered as still 'emerging'. The range of biological tests that are available within these categories and examples of their application are detailed below. At the end of the chapter, the most suitable tests for current use and the most likely methods for future application are discussed.

6.2 Standardised procedures

A number of bodies issue standardised guidelines for ecotoxicity testing, of which those produced by the Organization for Economic Cooperation and Development (OECD) and the International Organization for Standardization (ISO) are regarded as the gold standard. Standardised tests include ecological indicators, bioassays and biosensors using soil microbial communities, individual microbial species, plants and invertebrates. As part of the process of producing a standardised guideline, each test has been subjected to interlaboratory comparison and ring-testing. This has improved aspects of test reproducibility, responsiveness and robustness.

In all cases, the tests were originally developed for the regulatory approval of single compounds (usually plant protection products) and occasionally for effluent testing. This means that they were not intended for use in assessing contaminated soils. Despite this, these tests are now being more widely used for this purpose. Some of the tests are more readily adaptable than others. In particular, those that use the response of an aquatic animal or plant require sample manipulation that limits their relevance for contaminated soils.

6.2.1 *Ecological indicators*

6.2.1.1 *Soil functional assessments*

There are two standardised procedures for measuring the functional activity of the soil microbial community. These provide an indication of the size and activity of the microbial community by measuring carbon production rate following breakdown of organic compounds and nitrate and ammonia formation rate, respectively.

The OECD/ISO carbon transformation/mineralisation (C-min) and substance-induced respiration (SIR) tests. The C-min and SIR tests have had a fairly wide application in examining the effects of selected chemicals on soil microbial function. This includes fungicides (Jones and Ananyeva, 2001), hydrocarbons (Brohon *et al.*, 2001) and 2,4,6-trinitrotoluene (TNT) (Gong *et al.*, 2000). In an examination of a range of responses to copper, SIR was found to be more sensitive than microbial nitrification, earthworm growth, soil enzyme activity and nematode abundance (Bogomolov *et al.*, 1996). In contrast to these positive results, Murray *et al.* (2000) found that SIR was not affected by cadmium, copper, nickel, lead and zinc in urban soils, whereas SIR was found to increase with zinc concentrations (Aceves *et al.*, 1999). These contrary results suggest that there may be a limited sensitivity of the C-min/SIR end-point to particular chemicals under some soil conditions that may limit the usefulness of the test.

The OECD/ISO nitrogen mineralisation (N-min) test. Because of the fundamental role of nitrogen in soil ecosystems, measurement of mineralisation has been used widely in soil ecology. Studies of the effects of a range of contaminants have also been made. These include fluoride (Pomazkina *et al.*, 2001), heating oil (Weissman and Kunze, 1994), metal-containing sewage sludge (Chander *et al.*, 1995) and polychlorinated biphenyls (PCBs) (Dusek, 1995a). Among the studies that have been conducted to date, a number have found that nitrification measurements do not produce dose-responsive results. A further issue is that soil and climatic factors, such as soil moisture content and temperature, affect mineralisation rate. The presence of these co-varying factors and the absence of clear dose – response relationships in some cases does not support the widespread application of nitrification tests for contaminated soil assessment.

6.2.2 Bioassays

6.2.2.1 Aquatic plant tests

The OECD Algal Toxicity Test and the OECD Draft and ISO Draft Growth Inhibition Test using Lemna *spp.* Although intended for testing the toxicity of contaminated waters, these two aquatic plant tests can be adapted for use with soil solutions or soil extract. Examples of applications of the algal test include extracts containing polycyclic aromatic hydrocarbons (PAHs) (Barich *et al.*, 1987), heavy metals (Greene *et al.*, 1988; Barich *et al.*, 1992), pesticides and chlorinated solvents (Linz and Nakles, 1996). An examination of the correlation between the results of the aquatic and terrestrial plant tests found good agreement for pentachlorophenol, which is easily soluble, but a poor correlation for the moderately soluble compound lindane. These results suggest that it may be difficult to obtain representative samples for these aquatic tests, particularly when the contaminant is poorly soluble in water. With the *Lemna* test, Clement and Bouvet (1993) determined the toxicity of water-soluble leachates from municipal landfills by using serial dilutions. As outlined above, a concern was that because the test is aquatic it was first necessary to extract a representative sample. This issue means that the use of both of these tests for soil assessment should be limited, except in cases where risks to groundwater are a concern.

6.2.2.2 Terrestrial plant tests
The OECD/ISO seedling emergence, growth and vegetative vigour tests. Plants have inherent properties that make them good candidates for toxicity testing. They interact intimately with the soil, have a large root (and mycorrhizal) surface area (which actively and passively absorbs contaminants) and are static. Additionally, they are essential to a healthy ecosystem.

Despite the OECD and ISO plant tests originally being designed for the regulation of plant protection products, both have been successfully used for the assessment of contaminated field soils (Linz and Nakles, 1996). In a comparative study of 13 bioassays, Saterbak *et al.* (1999) recommended the seed germination test with dicotyledon (mustard and lettuce) and monocotyledon (maize and wheat) for the ecological risk assessment of hydrocarbon-contaminated sites. Barud-Grasset *et al.* (1993) successfully used the germination of lettuce, oat and millet to assess the effectiveness of bioremediation of a PAH-contaminated site, and Linder *et al.* (1990) used the germination and growth of cucumber and radish to screen soils from military sites. The success of all these studies clearly demonstrates the potential for using plants for contaminated soil assessment. To date, the majority of plant tests in contaminated field soils have used crop species. Although potentially suitable for testing plant protection products, these species are of course less relevant for contaminated soils. As an alternative to crop species, a few studies have used wild plants.

6.2.2.3 Aquatic invertebrate tests
The OECD Daphnia *spp acute immobilisation and reproduction tests.* A mainstay in aquatic toxicity testing, *Daphnia* tests have been used also to evaluate the toxicity of contaminated groundwaters and leachates (Kross and Cherryholmes, 1992). As with any of the aquatic tests, the principal problem with the *Daphnia* test is the need to extract a suitable aqueous sample. This problem is illustrated by Kross and Cherryholmes (1992), who compared *D. magna* and Microtox® assay results in leachates but found a poor correlation between the two methods.

6.2.2.4 Terrestrial invertebrate tests
Tests with three soil invertebrate groups, lumbricid earthworms, enchytraeids and springtails, have been ratified. All three tests were developed originally for testing the toxicity of plant protection products in an artificial soil consisting of 10% peat,

20% kaolin clay and 70% sand, with the pH adjusted to 6.0 and the moisture content to 33% wet weight. For testing, groups of animals are incubated in dosed or undosed soil for 28 days and supplied with suitable food. On termination of exposure, the survival and reproduction are determined either by wet sieving soil and counting the number of survivors and cocoons (for earthworms and enchytraeids) or by counting adults and hatched juveniles (springtails) after flotation or tullgren extraction.

The ISO and OECD Draft Earthworm Reproduction Test. Earthworms are suitable sentinel organisms for assessing the impact of anthropogenic stresses to soil. This is generally accepted by academics, regulators and the general public. Although most early work in earthworm ecotoxicology focused on lethal endpoints, sublethal test methods are now available. These allow the measurement of reproduction (Van Gestel *et al.*, 1989; Kula, 1992), growth and development (Spurgeon *et al.*, 2004) and behaviour (Stephenson *et al.*, 1998). So far, however, only the reproduction test has been adopted for standardisation.

Among the many published studies that have used the earthworm reproduction test to measure chemical toxicity, only a few have assessed contaminated field soils. Two studies have been conducted along the contamination gradients from two smelting works (Spurgeon and Hopkin, 1995; Posthuma *et al.*, 1998). Both studies found that earthworm reproduction was affected by elevated metal concentrations in soil close to the point source. Studies in contaminated soils have indicated that reproduction can be influenced also by the soil characteristics (Saterbak *et al.*, 1999). This makes the choice of suitable control soils vital for the assessment.

The earthworm species recommended for the OECD/ISO reproduction test is *Eisenia fetida*. Other species have, however, also been used successfully. These include *Lumbricus rubellus* (Spurgeon *et al.*, 2003), *Aporrectodea caliginosa* (Khalil *et al.*, 1996), *Lumbricus terrestris* (Spurgeon *et al.*, 2000) and *Eudrilus eugeniae* (Reinecke and Reinecke, 1997). Comparisons of species sensitivity generally have found that *E. fetida* is less susceptible (Edwards and Bohlen, 1992; Spurgeon *et al.*, 2000). Despite this, the use of *E. fetida* as the test species has continued, although because this species does not occur naturally in soil (instead inhabiting manure heaps) it would seem more relevant to use soil-dwelling species for field soils.

The ISO springtail reproduction test. The standard laboratory-based springtail reproduction test using *Folsomia candida* or *Folsomia fimetaria* has been used for the toxicity assessment of TNT, phenols, PAHs, PCBs and fungicides and adapted to field soils, where it has been used successfully for the assessment of zinc-, copper- and mineral oil-contaminated soils (Smit and Van Gestel, 1996; Scott-Fordsmand *et al.*, 2000; Van Gestel *et al.*, 2001). Of the parameters that can be measured during the test, reproduction has been shown to be more sensitive than adult toxicity. In some cases, effects on reproduction are direct, although reproduction can be affected indirectly by reductions in growth.

A disadvantage of the existing springtail test is that toxic effects are not known until the Collembola are extracted from the soil at the end of the exposure period (e.g. Sandifer and Hopkin, 1996). Hence, parameters such as growth, oviposition and hatching times cannot be monitored (Scott-Fordsmand et al., 2000). A second issue is that soil factors such as organic matter content and pH are known to affect springtail reproduction. This means that the choice of a suitable control soil is vital. Given that a suitable control soil can be identified, the springtail reproduction test clearly has potential for the assessment of contaminated soils.

The OECD and ISO Draft Enchytraeid Reproduction Test. Enchytraeids are one of the most important invertebrate families in European soils. For this reason they have been recommended for use in a now standardised reproduction test. Despite the apparent suitability of the enchytraeid test for contaminated soil assessment, the method has not been as widely used as either the earthworm or springtail tests. This is due to the relative unfamiliarity of the enchytraeid family. From the results of the few studies that have been undertaken to date, mostly with spiked soils, responsiveness, robustness and reliability are similar to those of the earthworm reproduction test (Collado et al., 1999; Kuperman et al., 1999). Latterly, the test was used in an innovative study to understand the role of soil factors in modifying the toxicity of metals (Lock et al., 2000; Lock and Janssen, 2001). These studies illustrated the importance of pH and soil organic content in moderating toxic effects. A further similarity to the earthworm test is that the recommended test species *Enchytraeus albidus* is rarely found in meadows and other agricultural habitats, inhabiting instead organic-rich sites (compost heaps). As an alternative, the forest, grassland and peat-dwelling species *Cognottia sphagnorum* has been recommended (Augustsson and Rundgren, 1998).

6.2.3 Biosensors

The OECD Ames Test. This measures both chemical mutagenicity and carcinogenicity. The test uses a strain of *Salmonella typhimurium* that carries a mutant gene, making it unable to synthesize the amino acid histidine from the ingredients of a standard culture medium. In the presence of mutagens, this mutation can be reversed, making the bacteria able to survive and form colonies on histidine-free culture plates. Use of the Ames test to assess the mutagenicity of environmental samples is more limited than in routine chemical screening. The assay has been applied to samples from various industrial wastes as well as contaminated sediments and air-borne particulates (Linz and Nakles, 1996). These studies suggest that there is value in the method (and similar systems such as Mutatox$^{\circledR}$) for detecting mutagens.

The ISO Draft Microtox® Toxicity Test. The Microtox® system (which uses the strain of *Vibrio fischeri* NRRL B-11177) and other similar luminescent bacteria-based assays can be used to measure the toxicity of environmental samples. Over 500 papers and reports pertaining to the use of such bacterial systems have been produced. These include a number describing environmental applications, including wastewater monitoring and treatment plant influent and effluent testing. To conduct the test as originally developed with soils, there is a need to extract a suitable aqueous phase sample. A variety of methods are available for this, each extracting a different fraction of pollutants. As outlined earlier, the need to extract an aqueous sample means that the test is less well suited for soils contaminated with hydrophobic compounds. As a result of this possible limitation, the procedure has been modified for direct testing of solid matrices such as soils (Kwan and Dutka, 1992).

When testing contaminated soils, a perceived drawback of Microtox® is that the organism is a marine bacterium. This means that in order to conduct the assay it is essential to add salt to the sample. One of the implications of this is that the ecological relevance of the assay for the niches present in soils is reduced, especially with respect to metals. Although this does not necessarily preclude use as a screening tool, the development of genetic engineering technology has allowed the gene that codes for bioluminescence to be transferred to naturally soil-dwelling bacteria. These may have greater relevance to soil systems and thereby could prove more suitable for soil assessment.

6.3 Academically established methods

These are tests that are established within the academic community as monitors of chemical exposure and chemical effect. Unlike the procedures discussed in section 6.2, standards do not exist for these tests. Development instead can be tracked via a series of scientific papers. Some methods are long established in soil ecology and ecotoxicology. Examples include invertebrate bioassays, soil enzyme assays and litterbags. In contrast, a number of methods, such as bait lamina and some biochemical assays (e.g. lysosomal membrane stability), have been developed more recently but have passed quickly into widespread use.

6.3.1 Ecological indicators

6.3.1.1 Invertebrate feeding activity using bait lamina strip
The bait lamina test measures the feeding activity of soil organisms by assessing the removal of a series of bait material pellets embedded in plastic strips (Törne, 1990a,b). Larink and Lübben (1991) used the bait lamina test to study the influence of heavy metals in arable fields amended with sewage sludge. Reduced feeding activity was found in the more contaminated plots. Kula and Rombke

(1998) used four functional assays (bait lamina, litterbags, minicontainers and cotton strips) to assess the effects of diflubenzuron in a deciduous woodland. Results indicated strong effects of the pesticide on invertebrate feeding activity measured using bait lamina. These changes were in accordance with changes in the abundance of different springtail species, confirming the ability of the assay to detect community-level changes. In addition to direct effects of chemicals on feeding, soil factors such as moisture content and pH are also known to influence bait lamina results. These effects mean that bait lamina studies should be timed carefully and suitable controls used.

6.3.1.2 Community-level physiological profiling (CLPP) using BIOLOG plates

The BIOLOG multiwell plate system for assessing carbon substrate utilisation by microorganisms was designed originally for the identification of bacterial isolates but has been adopted for environmental samples. Chapman et al. (2000) reviewed the studies that have used the method for soil bacterial community assessment. For contaminant monitoring, these included: the effects of metals in smelter-contaminated soils, for which Kelly and Tate (1998) found altered community metabolic profiles close to a factory; depleted uranium, for which Meyer et al. (1998) reported effects on functional diversity; and hydrocarbons, for which the pattern of utilisation corresponded to changes in the abundance of hydrocarbon-utilising strains (Wunsche et al., 1995). In a study of CLPP responses to pesticides, Engelen et al. (1998) found that one pesticide, dinoterb, inhibited substrate usage but responses for others were less clear. This demonstrates that there can be variability in the CLPP response and, as a result, a clear dose–effect relationship cannot always be established. This problem can compromise the utility of the assay with respect to complexly polluted soils.

A particular application of the BIOLOG system is analysis of pollution-induced community tolerance (PICT) in microbial communities. This is the process by which exposure to a chemical alters the structure of microbial assemblages through adaptation and extinction so that the resulting community becomes more resilient to the effects of that chemical. For PICT analysis, a range of concentrations of the chemical is spiked into the BIOLOG plates. These are then used to test the substrate utilisation of each community. Rutgers et al. (1998) used the PICT approach to study the effects of zinc in 2.5-year-old contaminated field plots. For most substrates, the metabolic activities showed an increased community tolerance with increasing zinc concentration, indicating that PICT had evolved. The method proved sensitive because PICT could be demonstrated at soil zinc concentrations close to the Dutch soil protection guideline value.

6.3.1.3 Litterbags

Litterbags are used to assess rates of organic decomposition by detritivorous soil fauna. Litterbags are constructed at least in part from mesh, the size of which can be selected to allow the activity of particular invertebrate groups to be assessed. The low cost of litterbags means that they have been used very extensively to assess the impact of numerous soil factors, seasonal effects and land uses changes. They have had widespread use also in studying the effects of chemicals. Some studies have found a clear negative effect of exposure, e.g. for carbendazim (Forster *et al.*, 1996) and captan (De Jong, 1998). In contrast, Huuselaveistola *et al.* (1994) found no effect of two pyrethroids on decomposition rate, indicating that either the method could be insensitive to some exposures or, alternatively, that the approach is accurately predicting a minimal effect on community composition.

6.3.1.4 Minicontainers
Minicontainers have been developed from litterbags. Each minicontainer has a central body over which gauze discs of variable mesh sizes are placed. The mesh allows access to selected soil fauna that facilitate litter breakdown. Since first published by Eisenbeis (1993), the system has been used in different areas of soil ecology. Only a few studies have used the minicontainer method to assess chemical effects on organic matter decomposition. One study by Paulus *et al.* (1999) compared three methods for assessing biological activity in a field experiment with two levels of diflubenzuron. The methods were litterbags, minicontainers and bait lamina. No visible effect on decomposition was seen using litterbags, whereas minicontainers showed lower decomposition in both biocide-treated plots compared with controls. This study suggests that there may be value in using minicontainers for site assessment, although care should be taken to include suitable controls.

6.3.2 Microbial assemblage functions

Microbial assemblage activities, including elements of the nitrogen cycle and enzyme functions, have the potential to provide information concerning the effects of chemicals on soil systems.

6.3.2.1 Nitrification
Nitrification is the second step in the mineralisation of nitrogen from organic matter and is characterised by the oxidation of ammonium to nitrite and nitrate. Only a small number of bacteria are capable of carrying out this function, with *Nitrosomonas* and *Nitrobacter* being the most common. Because only a few bacteria undertake nitrification, it has been suggested that this parameter could be particularly sensitive to soil pollutants. For organic compounds, Dusek (1995b) demonstrated an effect of PCBs on nitrification. Remde and Hund (1994) successfully used the activity of nitrifying bacteria as an indicator of the potential toxicity of anthracene oil and dicyandiamide. For metals, effects on soil nitrification are inconsistent. Dusek (1995a) found that cadmium disrupted nitrification in spiked soils, however, the fact that effects were found only at 100 and 500 mg/kg suggests

that the parameter was not particularly sensitive. Murray *et al.* (2000) found no effect of cadmium, copper, nickel and lead, whereas Kandeler *et al.* (1992) found that nitrification was influenced more by the supply of organic substrate in sludge than the heavy metal content. These complex results and the possible low sensitivity do not, at present, support regulatory use of this test for contaminated soil assessment.

6.3.2.2 Nitrogen fixation
Nitrogen is fixed in soil both symbiotically and asymbiotically by a number of bacterial species. Domsch *et al.* (1983) concluded that nitrogen fixation under certain conditions can be sensitive to pollutants. Population sizes of nitrogen-fixers, nodulation, nitrogen fixation rate and growth of legumes can be limited by pesticide application (Gonzalezlopez *et al.*, 1993; Martineztoledo *et al.*, 1993). Additionally, symbiotic nitrogen-fixers have been found to be sensitive to metals (Simon, 2000), and in the clover *Rhizobium leguminosarum* bv *trifolii* association both reduced nitrogen fixation and clover yield have been associated with poor survival of the bacteria in contaminated soils (McGrath, 1994). However, despite these promising results, other studies have not always seen clear impacts of contamination on fixation. This is true particularly if fixation is already low, as would be likely at many industrial sites (Lorenz *et al.*, 1992).

6.3.2.3 Soil enzyme activity
Enzymes such as dehydrogenase, glucose oxidase, catalase, peroxidase, phosphatases, cellulase, proteinase and urease carry out important functions in soil and inhibition of one or more of these could have important consequences for soil sustainability. Chemical effects on soil enzyme activities have been measured in a number of studies. Vink and Van Straalen (1999) studied the effects of a fungicide (benomyl), which is expected to inhibit microflora, and an insecticide (diazinon), which is expected to inhibit arthropod feeding in soil microcosms containing isopods. Benomyl hardly affected dehydrogenase and respiration but nitrification did decrease at high concentrations. Diazinon reduced respiration, dehydrogenase activity and nitrification. This study thus indicated that soil functions can be disrupted by effects on both micro- and macroorganisms. In the study by Tu (1990), effects on enzyme activity due to four insecticides were found. Strong temporal variations, however, made it difficult to draw conclusions on the effects of each pesticide. Soil enzyme responses to metals have also been analysed. Chander and Brookes (1991) found that both dehydrogenase and phosphatase activity was reduced by copper, with the former enzyme being more sensitive. In industrially contaminated soils containing hydrocarbons and metals, Brohon *et al.* (2001) found that urease and dehydrogenase were inhibited in the most contaminated soil. Trasar-Cepeda *et al.* (2000), however, found that activities of individual enzymes were apparently unaffected in three soils contaminated by tanning effluent, hydrocarbons or landfill effluent. Taken together, these results

give a somewhat inconsistent picture for chemical effects on soil-borne enzymes and, as a result, the methods should be used with caution.

6.3.3 Bioassays in non-standardised species

The most direct tool for assessing the toxicity of soil species is direct measurement of life-cycle parameters following exposure. For higher terrestrial plants, a single generic toxicity test has been recommended that can be adapted for a variety of species. For soil invertebrates where life-cycles and habitats vary, the development of such a species general protocol is not feasible and instead tests for a few representative species have been designed. So far only the springtail and two oligochaete reproduction tests have been developed and ring-tested. Among remaining species groups, a series of toxicity tests have been developed and proposed as potential bioassays for contaminated soils. These include Oribatid mites, Staphylinid beetles, woodlice, nematodes and snails. Competition (between two nematode species) and predation (of a mite on a collembolan) tests also have been developed. A full description of the methods for these tests is beyond the scope of this chapter and the reader is instead referred to the reviews of Van Gestel and Van Straalen (1994) and Løkke and Van Gestel (1998). Overall, some of these tests could have potential for contaminated soil and litter assessment but many will need further development before they can be used for this purpose.

6.3.4 Biomarkers

6.3.4.1 Tissue/cellular histopathological changes
Histological and ultrastructural alterations in cells, tissues or organs can afford good biomarkers of pollutant stress. Early histological work developed approaches for fish and marine invertebrates. Only more recently have techniques been transferred to soil species. The most detailed histopathological change studies in terrestrial invertebrates are available for the metal-containing granules found in the digestive and excretory organs of most species (see comprehensive review by Hopkin, 1989). At the subcellular level of ultrastructure, Köhler and Triebskorn (1998) have revealed that cell organelles possess different susceptibilities to pollutants and show different spectra of reactions. By using the totality of all organelle reactions to build up a syndrome of intoxication, the influence of stress factors may be interpreted. Köhler and Triebskorn (1998) developed this idea into a protocol for evaluating ultrastructural stress reactions. The suitability of this protocol was demonstrated in slugs, woodlice, millipedes and springtails exposed to metals under laboratory conditions. The data showed different susceptibilities for different organelles in each species, for the investigated species to the particular metals and for the monitored tissue. The system is, however, somewhat complex and is usable only by experts.

6.3.4.2 Lysosomal membrane stability
One of the characteristic cellular changes occurring following chemical exposure is the increased fragility of the lysosomal membrane. Lowe *et al.* (1992) developed an *in vitro* technique to measure the stability of the lysosomal membrane in isolated liver cells in fish caught in 'clean' and contaminated sites. The technique was based on the ability of cell lysosome to retain neutral red dye. On the basis of this technique, further modifications were implemented by Weeks and Svendsen (1996) to enable use with terrestrial invertebrates. This method (the neutral red retention time assay) has since been applied widely for measuring lysosomal damage.

Chemical effects on lysosomal membrane in earthworms were demonstrated initially in *L. rubellus* and *E. andrei* exposed to copper (Weeks and Svendsen, 1996), in earthworms sampled from an arsenic-contaminated gold mine (Weeks and Williams, 1994) and in worms at an industrial accident site (Svendsen *et al.*, 1996). Compared with ecologically relevant sublethal parameters (e.g. reproduction, feeding rate, weight change), the neutral red retention time was decreased at lower soil copper concentrations (Svendsen and Weeks, 1997). Since this initial work, lower lysosomal membrane stability in earthworms has been demonstrated after exposure to nickel (Scott-Fordsmand *et al.*, 1998), mercury (Stubberud, 1998), TNT (Robidoux *et al.*, 2002), zinc, lead, cadmium, chloropyrifos and carbendazim (Eason *et al.*, 1999; Svendsen, 2000). So far, only the carbamate pesticide methiocarb has been shown not to elicit the response (Grafton, 1995). The ubiquitous nature of the response and the fact that it has been found to be sensitive to chemicals but not influenced by environmental factors (such as temperature) make the neutral red retention time assay one of the most suitable biomarker methods for detecting chemical exposure of invertebrates.

6.3.4.3 Immune system activity
Immune systems are responsible for providing organisms with the ability to resist infections from various sources. It is well known that interactions of environmental chemicals with the immune system can both suppress and enhance immune activity (Luster *et al.*, 1988). This means that immune responses have potential as biomarkers of chemical exposure and effect. To date, immune system studies in terrestrial ecosystems have concentrated on earthworms. This is because the immunobiology of this group is well researched and understood. Activities or competencies of the cellular immune system have been measured through a range of parameters, such as:

- The ability to reject allo- and xenografts and perform wound healing
- Phagocytosis of rosette formation of foreign cells adhering to coelomocytes
- Production of reactive oxygen species (H_2O_2 and O^{2-})
- Spontaneous cytotoxicity (natural killing)
- Elimination of non-pathogenic bacteria

Compounds used in tests with earthworm immunity include PCBs (e.g. Goven et al., 1994b; Cikutovic et al., 1999), metals (e.g. Goven et al., 1994a; Fugere et al., 1996) and the organic pesticides carbaryl and 2,4-D (e.g. Ville et al., 1997). Although not all immunological parameters have been measured for each compound, some distinct response patterns have been observed. Most compounds caused some effect on the respective immune parameters at the concentrations used, the exceptions being some of the pesticides and lead. It was apparent that different compounds had different mechanisms, as coelomocytes demonstrated different levels of phagocytotic activity (Ville et al., 1997; Giggleman et al., 1998). Importantly, cellular immunity usually was depressed following exposure to any compound, whereas the various humoral parameters in general were enhanced, with only a few becoming depressed and some remaining unaffected (Ville et al., 1997). The complexity of these responses mean that further work would be needed prior to use of immunity in soil assessment, although there does appear to be some promise in the approach.

6.3.4.4 DNA alterations
Some pollutants are genotoxic or are converted into genotoxins during metabolism. Thus, DNA damage has potential as a biomarker. The actual damage and alterations to DNA occur in four basic forms, the detection of which employs specific techniques:

1. Pollutants, or metabolites, bind covalently to DNA to form 'DNA adducts'. The DNA adducts are usually detected by the non-specific ^{32}P-post-labelling technique.
2. Strand breaks can be measured by two methods, the alkaline unwinding technique and the single-cell gel electrophoresis or comet assay.
3. Change in DNA minor base composition can be observed as a decreased level of methylation of the base 5-methyl deoxycytidine.
4. Increased DNA synthesis due to repair can be measured by recording tritium-labelled thymidine incorporation.

Among the terrestrial invertebrates, increased numbers of strand breaks were found when earthworms were exposed to a mixture of dioxins, X-rays, mitocin C (a known genotoxin) and soil contaminated with a range of organic compounds such as coke, benzene and aniline (Verschaeve et al., 1993; Verschaeve and Gilles, 1995; Salagovic et al., 1996). The level of DNA adducts was studied in *E. andrei* exposed to increasing concentrations of benzo[*a*]pyrene (B[*a*]P) for different periods of time (Saint-Denis et al., 2000). The level of DNA adducts formed was dose-dependent for short-term exposure and significant at 50 μg B[*a*]P/kg. At the highest doses, the formation of DNA adducts reached a steady state.

In terrestrial plants, the comet assay can be used in a range of species for *in situ* monitoring (Gichner et al., 1999). As an example, strand breaks in tobacco plant cells showed a concentration–response relationship after exposure to ethyl methanesulphonate (Gichner and Plewa, 1998; Stavreva et al., 1998). These studies demonstrate that there is potential for the measurement of DNA alteration in cases where exposure of plants or animal species to genotoxins is suspected.

6.3.4.5 Enzyme activity/induction
Phase 1 detoxification enzymes. The response of the mixed-function oxidase system can be measured via several of its components, the most common being cytochrome P450 levels and the activities of ethoxyresorufin *O*-deethylase (EROD), ethoxycoumarin *O*-dealkylase (ECOD), aryl hydrocarbon hydroxylase (AHH) and NADPH cytochrome *c* reductase.

As is commonly the case for biomarker systems, most of the work with cytochrome P450 has been undertaken in mammals and fish, but there are examples in invertebrates and in particular earthworms. The presence of cytochrome P450 monooxygenases in earthworms was documented by isolation and characterisation in *L. terrestris* by Berghout *et al.* (1990) and from whole-body microsomes of *E. fetida* by Achazi *et al.*, (1998). Although induction of EROD is one of the most commonly used measures for P450 activity, no studies yet have observed this response in earthworms but ECOD activity has been observed in *L. terrestris* (Stenersen, 1984), *L. rubellus, A. caliginosa* and *E. andrei* (Eason *et al.*, 1998). The presence of cytochrome P450 has also been demonstrated in *Enchytraeus crypticus*, where short-term exposure to B[*a*]P reduced EROD activity significantly but did not affect pentotyresorufin *O*-deethylase (PentROD). Long-term (8 weeks) exposure to B[*a*]P, however, had no effect on EROD activity but PentROD decreased to zero (Achazi *et al.*, 1998).

In other soil invertebrates, microsomal activity of cytochrome P450 and accompanying conjugation enzymes has been demonstrated through the formation of pyreneglucoside and pyrenesulphate in the isopods *Porcellio scaber* and *Oniscus asellus* exposed to pyrene (de Knecht *et al.*, 2001). When the isopods were given food containing B[*a*]P and 3-methyl-cholanthrene, the majority of the enzyme activities appeared to be non-inducible. This non- or low inducibility agrees with earlier cytochrome P450 1A (CYP1A) measurements in *O. asellus* (Zanger *et al.*, 1997). In this animal, therefore, as in earthworms, the picture regarding the link between exposure and P450 enzyme activity is yet to be established. This limits the application of these biomarkers for contaminant detection.

Glutathione S-transferase (GST). The induction and activities of GST, like the phase 1 enzymes above, have been measured as a potential biomarker of exposure to organic contaminants. In earthworms, Stokke and Stenersen (1993) found no increase of GST in *E. andrei* following exposure to three classic inducers (*trans*-stilbene oxide, 3-methylcholanthrene and 1,4-bis[2(3,5-dichloropyridoxyl)]benzene) for 3 days, a result confirmed by Borgeraas *et al.* (1996) for both *E. andrei* and *Eisenia veneta*. In contrast, Hans *et al.* (1993) found induction of GST in another earthworm *Pheretima posthuma* following exposure to aldrin, endosulphan and lindane. However, this induction was transient, which may explain some of the differences between the studies. In other invertebrates, the activity of GST increased in larva of the lacewing *Micromus tasmaniae* exposed to sublethal doses of cypermethrin and decreased when exposed to fenoxycarb, whereas GST activity in another lacewing species remained unaffected by all exposures (Rumpf *et al.*, 1997). These inconsistencies, like those for the phase 1 enzymes, currently limit the potential of GST measurements for contaminated soil assessment.

Antioxidant enzymes. Toxic derivatives of O_2, such as superoxide (O_2^-) radical, are scavenged by protective enzymes called antioxidants, which all the organisms analysed so far possess. The key enzymes are superoxide dismutase (SOD), catalase (CAT), peroxidase and glutathione reductase (see Gagné and Blaise, Chapter 7 of this volume). Increased antioxidant enzyme activities have been found to be a good indicator of solid or leachate phase toxicity in three terrestrial plants, oats, Chinese cabbage and lettuce (Ferrari *et al.*, 1999), exposed to different concentrations of the herbicide sodium trichloroacetate. These studies showed that increases in the activities of SOD, CAT, peroxidase and glutathione reductase were more sensitive than effects on biomass and germination rate (Radetski *et al.*, 2000). The use of these enzyme activities thus may permit the detection of early injury in plants, although further development is required.

The utility of antioxidants in terrestrial animals has received relatively little attention so far. In earthworms, inhibition of CAT and glutathione peroxidase but not SOD have been demonstrated in *E. fetida* exposed to lead and uranium (Labrot *et al.*, 1996). However, SOD and CAT activity were not induced in *E. veneta* and *E. fetida* exposed to Zn, Cu and Hg and the herbicide paraquat (Honsi *et al.*, 1999). Antioxidant enzyme measurement therefore cannot be considered a reliable biomarker of exposure in soil invertebrates.

Isozymes. Several studies have demonstrated that various chemicals can cause differences in the quantities of isoenzymes. So far, responses have been measured in aquatic species. For example, the effect of mercury and cadmium on marine gastropods was investigated, with phosphoglucomutase, glucosephosphate isomerase and aminopeptidase showing a clear response (reviewed in Nevo *et al.*, 1987).

In addition to pollutants, several studies have shown a correlation between various alleles and environmental factors (Guttman, 1994). Temperature has been shown to effect allozymes of glucosephosphate isomerase in the isopod *Porcellio laevis*; the allele expressed in organisms with the slower migration rate was favoured in normal 'unstressed' conditions, but the allele expressed in organisms that migrate faster was favoured in stressed environments (McCluskey *et al.*, 1993). Another factor that has been shown to be important is nutrition. For example, a change in the diet of the caterpillar *Helicoverpa zea* had an effect on representation of the enzyme esterase (Salama *et al.*, 1992). Such studies suggest that the diagnosis of isoenzyme difference in wild populations may be complex and as a result the methods will require further development.

6.3.4.6 Enzyme inhibition

Cholinesterase (ChE). Inhibition of ChE has been studied in various species such as lacewing (Rumpf et al., 1997), carabid beetles (Jensen et al., 1997) and grasshoppers (Schmidt and Ibrahim, 1994) but, as for many biochemical assays, most studies in soil invertebrates are with earthworms. In the initial study of ChE activity in earthworms, Stenersen et al. (1973) investigated the effects of seven organophosphate and three carbamate compounds on *L. terrestris*. The results showed that the organophosphates caused a more severe inhibition ($>99\%$) than the carbamates (30–80%). Booth et al. (2000) examined ChE inhibition following exposure of *A. caliginosa* to chlorpyrifos and diazinon. Both pesticides reduced enzyme activity but the work confirmed that the link between ChE inhibition and mortality observed in vertebrates does not appear to translate to earthworms. As an example, Stenersen (1979) found that earthworms could survive severe (99%) ChE inhibition that would result in vertebrate mortality, although behavioural changes were found. Similar links between ChE inhibition and behaviour have been found in carabids (Jensen et al., 1997).

Although it is often thought that ChE measurements are specific for organophosphates and carbamate pesticides, some research has challenged this. Inhibition of ChE has been found following cadmium, mercury and lead exposure in grasshoppers (Schmidt and Ibrahim, 1994). Similarly, lead and uranium inhibited ChE activity in *E. andrei* after both *in vitro* and *in vivo* exposures (Labrot et al., 1996). However, lead was, not found to cause ChE inhibition in *E. fetida* (Scaps et al., 1997) or *E. andrei* (Saint-Denis et al., 2001). This work has caused some uncertainty and at present the situation is far from clear. However, this non-specificity may enhance the applicability of ChE for mixtures, as demonstrated in exposure of the marine copepod *Tigriopus brevicornis* to binary combinations of metals (arsenic, copper, cadmium) and pesticides (carbofuran, dichlorvos, malathion) (Forget et al., 1999).

δ-*Aminolevulinic acid dehydratase (ALAD).* The best example of an enzyme that is specifically inhibited by a single pollutant is ALAD (for a review, see Mayer et al., 1992). This enzyme is present in the haem biosynthetic pathway and is specifically inhibited by lead (other metals being 1000 times less effective). Inhibition of ALAD is a standard assay for detecting lead exposure in humans. In soil invertebrates, application is limited to those species possessing haemoglobin. Earthworms are one such group and inhibition of ALAD has been demonstrated in *L. terrestris* sampled from roadside soils high in lead (Rozen and Mazur, 1997).

6.3.4.7 Protein-based biomarkers

Stress/heat shock proteins (see Gagné and Blaise, Chapter 7 of this volume). These constitute a set of protein families of different molecular weights that generally have been referred to as 'heat shock proteins' (HSPs), classified into different protein families on the basis of their molecular weight. Among the families, a number are stress inducible, with HSP60 and HSP70 in particular being widely studied.

The HSP70s are some of the most well-conserved proteins known and they comprise two major types, namely a 73 kDa protein and a 72 kDa protein, each in multiple isoforms. In terrestrial invertebrates, the first use of HSP70 induction in toxicity assessment was for the isopod *Oniscus asellus* (Köhler *et al.*, 1992), with studies since then covering most common taxonomic groups and contaminants. For example, isopod HSP70 induction has been found to respond to metal mixtures, cadmium, lead, zinc, lindane, pentachlorophenol, PCB52 and B[*a*]P (Eckwert *et al.*, 1994, 1997; Köhler and Eckwert, 1997; Köhler *et al.*, 1999). Investigations of HSP70 in two springtail species along a metal gradient from a brass mill showed that individuals from the middle of the gradient had 40% higher HSP70 levels than populations from the least and most contaminated sites (Köhler *et al.*, 1999). This study also demonstrated that the induction of HSP70 was reversible.

The HSP60s have been observed in all investigated organisms and, after heat shock, induction can cause HSP60 levels to increase to about 0.3% of the total protein (Langer and Neupert, 1990). Measurement of induction in terrestrial invertebrates indicated that metal exposure caused only slightly increased levels of HSP60 in the supernatant of homogenates of slugs, millipedes and woodlice. Comparing this response to that for HSP70 indicated that the induction of HSP60 was much lower (Eckwert *et al.*, 1997). In the nematode *Plectus acuminatus*, an induction of HSP60 has been demonstrated following exposure to heat, copper and cadmium (Kammenga *et al.*, 1998). For the two metals, HSP60 induction was three (copper) and one (cadmium) order of magnitude more sensitive than the EC_{20} for reproduction.

To date, field studies of HSPs in invertebrates exposed to pollutants have been scarce, but some aspects show promise. Köhler *et al.* (1992) found elevated HSP70 levels in *O. asellus* living on a smelter spoil bank. This study suggests there may be some value in measuring HSP levels as a biomarker, although it must be recognised that there is still a need for further development. In particular, the fact that some studies have suggested that there may be a non-sigmoidal response of HSPs to chemical exposure, with highest induction found at intermediate concentrations (Köhler *et al.*, 1999), presents a possible problem. Additionally, many environmental factors are known to alter HSP levels and these effects would need to be taken into account during any assessment.

Metallothionein, metal-binding proteins and phytochelatins (see Gagné and Blaise, Chapter 7 of this volume). Metallothioneins are low molecular weight, cysteine-rich proteins with a high affinity for transition metals. After they were first discovered in the kidney cortex of the horse they have been detected in a variety of animal species. It is widely accepted that metallothioneins are multi-functional proteins primarily involved in the homeostasis of essential trace metals, zinc-mediated gene regulation, and in the protection of cells against oxidative stress. Traditional methods for measuring metallothionein concentrations are based on quantitative protein assays (see review by Stegeman et al., 1992). As an alternative, advances in molecular biology have provided tools for measuring metallothionein gene expression.

The nature and responses of metallothioneins in terrestrial organisms have been gaining increasing attention. Four groups of invertebrates have been studied in detail, nematodes, insects, snails and earthworms, with most studies addressing the latter two groups. Snails possess distinct metallothionein isoforms involved in different metal-specific tasks (Dallinger et al., 2000). In *Helix pomatia*, one isoform is rapidly induced in response to cadmium, becoming nearly exclusively loaded with this metal (Berger et al., 1995). The second isoform, which is non-inducible by cadmium, binds nearly exclusively to copper and is probably involved in homeostatic regulation (Dallinger et al., 1997). This strong differential responses gives the snail metallothionein system potential as a multiple biomarker system for detecting metal exposure (Dallinger, 1996).

In a detailed analysis of earthworm metallothioneins, Stürzenbaum et al. (1998c) identified two complementary DNAs (cDNAs) encoding metallothionein-like proteins. Both isoforms were induced by copper and cadmium and also a number of mine site soils. This work for both snails and earthworms confirms that metallothioneins are inducible by metal exposure and are thus useful biomarkers of exposure.

In addition to metallothioneins, some species of terrestrial invertebrates possess non-metallothionein metal-binding proteins. As examples, a glycoprotein involved in metal storage and/or detoxification has been isolated from isopods (Dallinger, 1993), and a non-metallothiorein metal-binding protein that is induced by cadmium was found in enchytraeids (Willuhn et al., 1996). Other species also shown to possess non-metallothionein metal-binding proteins include the nematode *Caenorhabditis elegans*, the grasshopper *Aiolopus thalassinus* and the stonefly *Pteronarcys californica* (Clubb et al., 1975; Slice et al., 1990; Schmidt and Ibrahim, 1994). It remains to be demonstrated whether such proteins have the same biomarker potential as metallothioneins.

In plants, two kinds of metal-binding peptides or proteins are synthesized. Plant metallothioneins are inducible cysteine-rich entities very like those found in animals. Differential expression (induction) of metallothionein genes can be due to both variation of external heavy metal concentrations and the influence of various environmental factors. The principle role of plant metallothioneins seems to be in homeostasis rather than in metal detoxification. Plants are also known to have so-called phytochelatins, which are non-protein thiols specifically induced upon exposure to heavy metals. A close positive relationship between the concentrations of cadmium and phytochelatins in the plant shoot material has been observed and linked to the degree of growth inhibition (Keltjens and Van Beusichem, 1998). These observations make the use of phytochelatins promising for the assessment of heavy metal effect on plants.

6.4 'Emerging' techniques with future potential

There are a number of research areas with potential to produce tools for assessing the effects of chemicals on organisms and populations inhabiting contaminated soils. These are:

- Lipid and nucleic acid analysis approaches for assessing changes in the composition of the soil microbial community.
- Use of life-cycle theory to predict the long-term fate of exposed populations.
- Measurement of single gene biomarkers using high-throughput systems.
- Profiling changes in activities of all genes, proteins and/or metabolites using the 'omic' technologies.
- Transgenic organism biosensors.

These areas are all fast developing and already have produced a number of new techniques that could have future uses in risk assessment and monitoring. In many cases the techniques are still in development, although a few already have demonstrated clear potential.

6.4.1 Bioindicators

6.4.1.1 Microbial community profiling

Signature lipid biomarker analysis. Signature lipid biomarker analysis uses phospholipid fatty acid (PLFA), glycolipids, esterified lipids or lipopolysaccharide hydroxy fatty acid to identify the species and strain composition of the microbial community. Analysis of PLFA was first used to investigate the composition of microbial communities in soils by Thompson *et al.* (1993). The method is suitable for fungi, bacteria and actinomycetes. Since its development, PLFA has become widely used in studies of the effects of environmental perturbations. Changes in phospholipid fatty acids indicating a shift in the bacterial species composition have been found in soil polluted by alkaline dust (Bååth *et al.*, 1998), sewage sludge (Witter *et al.*, 2000), metals (Kelly *et al.*, 1999; Khan and Scullion, 2000) and organic compounds (Fuller and Manning, 1998; Thompson *et al.*, 1999). Most studies have been conducted in spiked soils, although limited data are available for contaminated field soils. These have suggested that the approach could be developed for contaminated soil assessment, assuming that suitable baseline data are available.

Nucleic acid analysis. Methods that analyse the microbial community from nucleic acid composition are based on use of the polymerase chain reaction (PCR). Universal forward and reverse primers are used in combination with PCR to amplify species-specific DNA fragments (usually the 16S subunit of ribosomal DNA) from samples isolated directly from soil. Samples then are separated according to sequence by a range of methods such as: denaturing and temperature gradient gel electrophoresis (DGGE/TGGE); single-strand confirmation polymorphism (SSCP); amplified ribosomal DNA restriction analysis (ARDRA) and terminal-restriction fragment length polymorphism (T-RFLP). Separation is followed then by analysis, such as image capture of gel banding. If needed, the PCR products can be collected for sequencing and thereby species identification.

Both DGGE and TGGE have been applied to examine the soil bacterial community responses to pollutants. For metals, Kozdroj and van Elsas (2001) found that the microbial community in an industrially contaminated soil differed in 'richness' and structure from clean soils. Muller *et al.* (2001) found altered community structure and decreased diversity along a mercury gradient and Rasmussen and Sorensen (2001) also found a negative effect of mercury on microbial diversity. For organic compounds, Duarte *et al.* (2001) found a reduction in the numbers of detected bands with increasing soil hydrocarbon content. In contrast to these results, Kandeler *et al.* (2000) found that addition of high concentrations of zinc, copper, nickel and cadmium did not change the DGGE profiles.

The methods SSCP, ARDRA and T-RFLP have not been as widely applied as DGGE for assessing microbial community change in contaminated soils. Cho and Kim (2000) used SSCP to identify shifts in bacterial communities in soils subject to two remediation treatments. Beaulieu *et al.* (2000) showed reduced complexity of SSCP profiles in the pentachlorophenol-treated soil slurry but no change in a combined pentachlorophenol and hydrocarbon wood preservative-amended soil. Using ARDRA, Sandaa *et al.* (2001) found higher bacterial diversity in the high-sludge/high-metal-amended soil compared with low-sludge, low-metal treatments. This later study indicates that soil microbial community diversity may not always be linked to soil contaminant status, but instead can be dependent on other soil factors (e.g. organic content). These studies all suggest that there is potential value for the use of microbial community profiling methods based on nucleic acid analysis. Clearly care must be taken because profiles can be influenced by a range of soil factors as well as exposure to high contaminant concentrations.

6.4.2 Bioassays

Life-cycle tests combined with demographic modelling. Toxicity tests have been standardised for plants and a number of soil invertebrates. A prevailing view in these is that estimation of critical effect levels for a single sensitive trait has high predictive power for population effects. In reality, however, this is not always the case. It is only possible to predict the long-term consequences of chemical exposure by considering the effects of pollutants on all life-cycle parameters (Van Straalen and Kammenga, 1998). This is the principle that underlies demographic methods. Because demographic studies require measurement of chemical effects throughout the full life-cycle, it is unsurprising that most studies of this type have used short-lived species (Van Straalen and Kammenga, 1998). In invertebrates, favourites are cladocerans and copeopods in freshwater and springtails and nematodes in soil. For the latter group, Kammenga *et al.* (1996) optimised a life-cycle test for *P. acuminatus* that can be adapted for other species, such as *C. elegans*, and Crommentuijn *et al.* (1997) have developed a life-cycle-based approach with the springtail *F. candida*. For longer lived invertebrates, practical difficulties mean that demographic studies are rare. Laskowski and Hopkin (1996) combined data from a 3-month metal exposure with the snail *Helix aspersa* and information on age-specific survival in the field to construct transition matrices. The approach proved useful, suggesting that with further development this targeted testing approach could prove useful (Spurgeon *et al.*, 2003).

Plants also have potential for demographic testing. Sheppard *et al.* (1993) proposed a 35-day test with *Brassica rapa*. Comparison of the results with a suite of assays in metal-spiked soils indicated that bloom initiation was more sensitive than lettuce emergence or earthworm survival for zinc but not mercury. Saterbak *et al.* (1999) used this test for the assessment of hydrocarbon-contaminated soil. The study, which used *Brassica rapa* selectively bred to reduce generation time, proved unsuccessful due to low germination in the control soil, suggesting that there is still a need for further method development.

6.4.3 Biomarkers

6.4.3.1 Molecular genetic assays

Genome mutation analysis using PCR-based methods. Exposure to genotoxins can result in the formation of DNA adducts, the faulty repair of which can result in point sequence mutations. Two molecular genetic techniques, randomly amplified polymorphic DNA (RAPD) and arbitrarily primed PCR (AP-PCR), are available for detection of point mutations. The RAPD technique has been used to investigate DNA mutation in parallel with life-cycle responses in *D. magna* exposed to B[*a*]P (Atienzar *et al.*, 1999). Results indicated that RAPD was more sensitive than co-measured fitness parameters. However, although there was a negative correlation with B[*a*]P concentration, the shape of the relationship was complex. Savva (2000b) used AP-PCR to detect DNA damage in rats and shore crabs exposed to B[*a*]P in the laboratory and in crabs from control and polluted areas. Results indicated differences in fingerprints between treatment in both the laboratory and field samples. Although not conducted in soil species, these results suggest that the method could be used to detect genotoxin exposure in soil organisms.

An introduction to single gene transcript quantification methods. A number of methods are now available for the detection of gene activity via messenger RNA (mRNA). In all protocols, the first step is to isolate total RNA. This can be probed directly or used for reverse transcriptase polymerase chain reaction (RT-PCR) analysis after conversion to complementary DNA (cDNA). Once prepared, cDNA samples can be quantified by a series of analytical techniques. These include Northern, dot and slot blotting, quantitative competitive PCR and fluorescence-based protocols.

The key aspect of transcript quantification is selecting suitable gene targets. In reality, the range and complexity of contaminants present in soils mean that even for a single species it is unlikely that a single gene suitable for quantitative assessment of contaminants in all situations will ever be identified. The best candidate genes can be categorised into three primary groups (although Gagné and Blaise, Chapter 7 this volume, separate early biological effect and defence from exposure to give four categories). These are biomarkers of exposure, physiological compensation and effect.

- *Exposure*. This category includes genes encoding proteins such as metallothionein, cytochrome P450 and GST that are known to be involved in pollutant handling and detoxification.
- *Physiological compensation*. This category includes genes involved in adaptive biochemical pathways, such as HSPs, antioxidant enzymes, mitochondrial genes, lysosomal genes and metalloenzymes, for which toxic metals compete for binding in the active site.
- *Effect*. These genes are linked to toxicity following exposure. The most well known is vitellogenin following xenoestrogen exposure, but other reproductively linked genes (e.g. the zona pellucida proteins), hormones and apoptotic genes are also potential candidates.

Northern and dot/slot blotting. These hybridisation techniques were among the first methods to be developed for RNA quantification. Both use a radiolabelled oligonucleotide probe for detection of the gene of interest. Use of these techniques for the quantitative assessment of gene expression is routine in the medical and toxicology sciences and application in ecotoxicology is also established. For the enchytraeid worm *Enchytraeus buchholzi*, Willuhn et al. (1994) used Northern blotting to investigate expression of a novel cysteine-rich non-metallothionein protein. Analysis reveals that the gene was not expressed in untreated worms but was rapidly induced by cadmium.

Fluorescence-based quantitative PCR. Initially, a major obstacle to the use of PCR for gene expression quantification was that the final level of product that could be detected was derived from a limitation not related to the starting quantity of template. This means that parallel reactions with vastly different template inoculations resulted in near identical final product levels. To overcome this problem, techniques have been developed that take a snap shot of product levels during the PCR reaction. The most commonly used of these are based on fluorescence *in situ* monitoring of PCR progress using two main detection methods.

1. SYBR® Green dye, which fluoresces when bound inside the double helix of DNA.
2. The fluorogenic 5' nuclease assay, which utilises the 5' nuclease activity inherent as a secondary function of Taq DNA polymerase to cleave a gene-specific probe, thereby releasing a fluorescent molecule for detection.

For both detection systems, the inclusion of a series of calibration standards containing cloned copies of the target gene at known concentrations can be used to obtain the relationship between transcript frequency and the number of cycles required to obtain a specific threshold. This standard curve is then used to determine gene concentration in samples.

Since inception, fluorogenic detection has been used to monitor expression of a range of genes in different media. Following development in the medical field, work has begun to adapt the methods for environmental diagnostics. An early application of quantitative PCR by Stürzenbaum (1997) and Stürzenbaum et al. (1998a,b,c, 1999) was to detect the transcript level of metal-responsive genes in L. rubellus exposed to metal-spiked and contaminated field soils. Results indicated increased expression for many of the analysed genes, with, in some cases, transcript levels increased over 100-fold in the exposed animals. This demonstrates the ability of fluorescence detection RT-PCR to detect gene transcription changes and suggests that the method could have a wide scope for assessing the exposure and toxic response of organisms inhabiting contaminated soils.

Chemiluminescent RNA hybridisation. This method offers an alternative to quantitative RT-PCR for the detection of mRNA. Already applied in aquatic ecotoxicology to measure vitellogenin and zona radiata protein expression in fish exposed to xenoestrogens, the assay has broad potential. For the assay, labelled probes complementary to the target gene are added to an RNA sample. This hybridises to the target and a reaction occurs from which light is given off. The light generated is proportional to the number of gene transcripts present.

6.4.3.2 The 'omic' technologies
Transcriptomics using oligonucleotide and cDNA microarrays. Over the past few years the genomes of over 50 organisms have been sequenced, with another 100 or so in progress. To take advantage of this increasing volume of sequence information, new technologies have been developed that measure changes in the expression of multiple genes simultaneously, the best known of which are microarrays.

Microarrays consist of at least many hundreds and usually many thousands of oligonuceolide sequences or cloned fragment or whole cDNAs printed onto a solid support (slide or membrane). Gene expression is analysed by labelling total mRNA from different treatments, such as pollutant exposed and control organism, with either a red or green fluorescent dye. Equal volumes of the labelled cDNA are mixed and hybridised to the DNA spots on the array slide. If the gene is expressed differentially between the two conditions, one colour dominates and the spot appears red or green. If there is no difference, colours merge and the spot appears yellow.

To date, the use of microarrays for analysis of environmentally derived changes in global gene expression has been limited, with applications in functional genomics and medical diagnostics dominating. Bartosiewicz *et al.* (2001) used a limited (148 expressed sequence tag) array to examine gene expression in response to β-naphthoflavone, cadmium, B[*a*]P and trichloroethylene in mice. The upregulated genes found were those predicted from prior knowledge of metabolism (e.g. metallothionein, P450, HSPs). This confirmed the suitability of the array for detecting differentially expressed genes. In a more comprehensive microarray study of 6000 yeast genes, Alexandre *et al.* (2001) found that 3.1% of the genes encoded in the yeast genome were up-regulated by at least a factor of three after 30 min of ethanol stress, whereas 3.2% of genes were down-regulated by a similar factor. This indicates the number of genes that can be involved in chemical acclimatisation.

Another detailed study of the effects of chemicals on the transcriptome was conducted by Kitagawa *et al.* (2003). This study investigated the relationship between gene expression profiles and chemical structure for a range of pesticides. These results suggest that DNA microarrays had potential for predicting which major chemicals will cause environmental toxicity and therefore they could be developed for use in risk assessment and monitoring. The power of the microarray approach means that in future it may be possible to elucidate simultaneously the combined toxic effects of various chemicals on a species and determine which chemicals are causing the observed effect.

Proteomics. Proteomics has been made possible by the advent of two-dimensional gel electrophoresis and protein identification using time-of-flight mass spectrometry. Two-dimensional gel electrophoresis is used to separate protein on a sodium dodecyl sulphate polyacrylamide gel. The first dimension separation is by isoelectric focusing and the second dimension separation is by size. Following separation, gels are visualised and image analysis is used to identify differentially expressed proteins. These can be excised for characterisation and ultimate identification by time-of-flight mass spectrometry. This latter area is rapidly developing and a number of new mass spectroscopy-based approaches to proteome analysis are being developed that circumvent the need for the two-dimensional separation step.

Initial studies of proteome responses to environmental chemicals in soil-dwelling animals or plants are currently under way. Kuperman *et al.* (2004) have used the approach to identify differentially expressed proteins in earthworms exposed to chemical warfare agents. Toxicological studies also have been undertaken. Vido *et al.* (2001) analysed yeast cells exposed to an acute cadmium stress: 54 proteins were induced and 43 repressed. Finally, Bradley (2000) used two-dimensional gel electrophoresis to demonstrate differences in protein expression signatures from oysters collected along a chemical gradient. Although these results are preliminary, they suggest that proteomics may become a valued tool in future ecotoxicological assessment.

Metabonomics (e.g. using nuclear magnetic resonance spectroscopy). As one of a range of potential detection methods for low molecular weight metabolites (e.g. chromatography mass spectrometry and spectroscopy methods), nuclear magnetic resonance (NMR) spectroscopy is based on the fact that atomic nuclei orientated by a strong magnetic field absorb radiation at characteristic frequencies. In different atomic environments, nuclei of the same element give rise to distinct spectral lines. This makes it possible to observe and measure signals from individual atoms in complex macromolecules and, from these, to interpret molecular structure. There are a number of NMR active nuclei: ^{1}H is the most sensitive and stable nucleus, ^{13}C also may be used but has low sensitivity and abundance, and ^{31}P and ^{19}F can be used for specific applications.

The use of NMR spectroscopy in ecotoxicology is undeveloped compared with soil science. Here, it has been used to study both the nature of soil organic matter and also how pollutants bind to soil constituents (Knicker, 1999; Xiong et al., 1999; Kohl et al., 2000). In (eco)toxicology an application of ^{1}H-NMR spectroscopy is to produce biochemical profiles, or fingerprints, of the small molecules in a complex matrix such as a biofluid. This non-selective detection can be combined with multivariate pattern recognition in order to probe toxic effects at the small molecule level (Nicholson et al., 1999). ^{1}H-NMR has been used for initial characterisations of metabolic profiles in a range of invertebrates (Gibb et al., 1997a). In earthworms, the method has been used to identify increasing free histidine in tissue extracts of *L. rubellus* (Gibb et al., 1997b; Bundy et al., 2001) and model compound (substituted anilines) effects on metabolite profiles in *E. veneta* (Warne et al., 2000; Bundy et al., 2001). NMR spectroscopy is also a well-established tool in plant biology and has been used to study physiological stress (Fan et al., 1992). It would be easy to extend these studies to investigate toxic effects at contaminated sites.

6.4.4 Biosensors

Transgenic bacterial biosensors. Systems such as the Microtox® assay detailed earlier use the marine species *Vibrio fischeri* as the sensor. Because it uses a marine bacterium, Microtox® must be conducted in saline solution, which is ecologically irrelevant for most soils. Because no naturally luminescent soil bacteria are known that could be used as an alternative, one solution is to fuse the genes responsible for bioluminescence into soil-dwelling strains using recombinant technology (Paton et al., 1997). Two approaches can be used:

1. In an approach analogous to Microtox®, light output can be linked to metabolic activity so that any chemical that disrupts metabolism decreases light output.
2. The *lux* gene can be fused to a functional gene linked to pollutant homeostasis.

Soil-dwelling bacteria marked with the *lux* gene cassette include *Pseudomonas fluorescens*, *R. leguminosarum* bv *trifolii*, *Escherichia coli* and *Burkholderia* sp. Of these, *P. fluorescens* has been used most widely for the metabolic system. Light output has been found to be reduced by heavy metals (singly and in combination) (Paton *et al.*, 1995; Palmer *et al.*, 1998; Chaudri *et al.*, 1999), organometals (Bundy *et al.*, 1997), chlorophenols (Shaw *et al.*, 2000; Boyd *et al.*, 2001) and benzene (Boyd *et al.*, 1997b). There are a few examples in which a change in bioluminescence has not been found, namely the PAHs phenanthrene, pyrene and B[*a*]P (Reid *et al.*, 1998).

The increased ecological relevance of engineered soil bacteria over Microtox® is clear but comparative studies with the two systems are relatively rare. Boyd *et al.* (1997a) screened contaminated groundwater using both biosensors and the results indicated that the reproducibility of *lux*-marked *P. fluorescens* and Microtox® was similar. For chlorobenzenes and chlorophenols, response profiles for the *lux* system agreed with fathead minnow, ciliate and diatom toxicity test results but not with Microtox® (Boyd *et al.*, 1998, 2001). Bundy (1997) also found different response profiles between the two systems for organotins, with Microtox® being more sensitive to tri-organotins and *P. fluorescens* to di-organotins.

An alternative to linking *lux* to metabolism is fusion to the promoters for functionally active genes such as HSPs and enzymes involved in degradation or resistance. Van Dyk *et al.* (1994) fused *lux* and heat shock genes in *E. coli*. This system was used to establish the toxicity of a range of organic and inorganic pollutants. A further application of the *lux* gene used by Heitzer *et al.* (1992) is insertion in the genes responsible for naphthalene and salicylate catabolism into *P. fluorescens*. This approach could have application in the *in situ* monitoring of hydrocarbon remediation.

Transgenic nematode biosensors. A number of transgenic strains of the nematode *C. elegans* have been developed in which a fluorescence or *lacZ* reporter gene has been linked to the HSP16 promoter (Power *et al.*, 1998; Guven *et al.*, 1999). Because of the relative novelty of transgenic nematode technology, the system has not yet been applied widely in contaminated soil assessment. Power and De Pomerai (1999) examined responses to cadmium, copper and zinc. High concentrations (250 mg/kg) of cadmium induced a stress response but the response to copper and zinc was minimal. Another study found that the surfactant Pluronic F-127 enhanced the stress response of transgenic nematodes to cadmium, mercury, copper, manganese and zinc (Dennis *et al.*, 1997). As an alternative to HSP, Cioci *et al.* (2000) linked a β-galactosidase reporter to the *C. elegans* metallothionein-2 promoter. A comparative study indicated that the system was more sensitive for cadmium, mercury, zinc, and nickel than either a 24-h LC_{50} assay or the HSP strain.

6.4.5 Further methods

The discussion in this chapter has been limited to techniques where there is some evidence of the potential of the method in ecotoxicology, ecology or toxicology. Outside these fields, there are methods that could also have potential for contaminated soil assessment in the future. One example is the use of remote sensors to detect contaminant-induced stress in terrestrial plants. Such techniques, although not capable of differentiating small-scale site heterogeneity, could be used to identify large-scale physicochemical boundaries and gradients of contamination. Another area where research may cross over into ecotoxicology in the future is medical diagnostics. Here, research in cell biology, cell signalling, apoptosis and molecular genetics focused on disease diagnosis and drug discovery may yield methods with potential for environmental investigations.

6.5 Community census analysis using macrofauna/flora

The pollution-induced absence of species from contaminated areas where they would be expected to occur normally is strong evidence of effects on ecosystem properties. The ecological importance of changes in community for soil sustainability has lead to interest in using census data to assess contaminated ecosystems since the 1950s. Despite this, and the fact that such procedures are well established in non-terrestrial ecosystems (e.g. freshwaters), the analysis of higher organism communities for assessing changes in soil quality in terrestrial systems remains an 'emerging' area.

6.5.1 Selection of groups for census studies

Community studies for assessing contaminated ecosystems have been undertaken in a range of habitats. Macroinvertebrates are the best studied group because they are present over a range of diverse habitats, are relatively easy to collect and identify (when good keys are available), are functionally important as ecosystem engineers and have relatively low mobility, meaning that they are representative of the local environment. These benefits outweigh some disadvantages such as the uncertain taxonomic status of selected groups, the limitations of some sampling methods to provide a representative sample and the time demands for sample processing and identification. Another group with potential for community-based analysis are plants. Plants are sedentary, have established taxonomy, can be sampled readily and are of course functionally important. Despite these advantages, the number of studies that have used plant community structure to assess contaminated sites is far below that for macroinvertebrates.

6.5.2 Community census analysis with macroinvertebrates

In soil, numerous studies have assessed the effects of chemical pollutants on soil communities. These have quantified community changes for earthworms, springtails, ants, ground beetles, molluscs, Oribatid mites and spiders along contamination or land-use gradients. The changes observed included reduced community size (including extinctions), lower diversity, altered composition and changes in species dominance/structure. In considering the use of invertebrate communities for detecting pollution effects, it is difficult to ignore the body of work that has been conducted around the Avonmouth smelter in south-west England. As a case study, this site provides a long and defined contamination gradient across semi-natural land or pasture. Studies undertaken include assessing the distribution of major invertebrate groups (Martin *et al.* 1982; Hopkin and Martin, 1985), a carabid beetle study (Read *et al.*, 1987) and earthworm community profiling (Spurgeon and Hopkin, 1996, 1999). By far the most detailed study of the effects of metal deposition on macroarthropods at this site was conducted by Sandifer (1996). Results indicated severe impacts on decomposers such as earthworms, isopods, molluscs, myriapods, springtails and mites. These groups were all absent or reduced in abundance and/or diversity at the two sites closest to the smelter and in some cases also at more distant sites.

The feasibility of using community census as a tool to monitor the impact of chemical contamination and land-use change on soil ecosystems has been addressed in two studies:

1. *The 'SOILPACS' study (Weeks et al. 1998).* This study investigated the feasibility of the community census approach for soils. The study concluded that a system could be constructed but substantial development was needed. In particular, full faunistic, physical, geological and chemical surveys of many (diverse) reference sites would be needed to develop a system with predictive capability. For development of the UK river invertebrate prediction and classification scheme, over 400 sites were sampled in this manner. The fact that this number would need to be matched for SOILPACS presents a substantial hurdle to development.
2. *The BBSK project.* This was launched in Germany as an initiative to develop a soil community-based monitoring scheme, but currently the method is not in day-to-day use, for a number of reasons. Some are political, but other scientific issues, including a lack of data from which to establish the normal biocoenosis, taxonomic problems for some groups and the absence of internationally standardised sampling methods, are also hindering development.

An invertebrate-based system worth considering separately is the nematode maturity index (MI) (Bongers, 1990). The system is based on the fact that nematode families adapted for rapid colonisation dominate soil nematode communities in disturbed ecosystems. At any site, recording of the families of nematodes present can be used to derive an MI value, with high values indicating undisturbed soils and low values those subject to perturbation. Reduced MI values have been found after nitrogen addition (Sarathchandra *et al.*, 2001) and pesticide use (Neher and Olson, 1999; Ruess *et al.*, 2001). For metals, Korthals *et al.* (1996) found a lower MI in copper, nickel and zinc (but not cadmium) spiked soils. Not all uses of the nematode MI have shown clear responses to contamination. Nagy (1999) found that of 13 metals only selenium and chromium consistently reduced the MI and Fuller *et al.* (1997) found no effect of toluene on MI in two soils and an effect of trichloroethylene in only one soil. Despite these occasional anomalies, the vast majority of published work does support the fact that pollution or land-use change decreases the MI whereas recovery coincides with an increase in MI. The approach is thus, at present, probably the best invertebrate community-based approach for soil quality assessment.

6.5.3 Community census analysis with plants

Examples of community-based surveys of the effects of chemical contamination on plants are relatively limited. Salemaa *et al.* (2001) recorded under-storey vegetation along a gradient from a copper/nickel smelter and studies with sewage sludge and mine wastes have also been conducted (Stoughton and Marcus, 2000; Vasseur *et al.*, 2000). These all showed clear effects on coverage, community composition and biodiversity. Overall, however, considering the frequency with which plant surveys are used to analyse long-term change in habitat structure and land use, full plant community surveys appear to have been underused for evaluating chemical effects. Instead the focus has been on bioindicator species such as metallophytes.

6.6 Summary and selection of suitable assays

The discussion above shows that there are a vast array of biological methods with potential for assessing contaminated soils. Outputs from a series of recent research programmes have now made it possible to envisage the use of a suite of biological assessment tools in temporal and spatial surveys of chemical effects in terrestrial ecosystems. Some methods are already standardised, others are established and further assays are in development. In parallel with the research to adapt or develop biological assessment methods, there has been increased regulatory emphasis on contaminated land and the link between this and ecosystem status. This is being driven by the increased awareness of the need to evaluate the ecological effects caused by contaminated land and the potential impacts of point source and diffuse pollution on vulnerable habitats.

The field of ecological risk assessment for contaminated soils is still developing and as a result the precise frameworks and conceptual models being used are evolving. This means that it is difficult to make precise recommendations concerning the most suitable assessment tools. Choice will be dependent on data requirements. Despite these limitations, biological methods can provide specific information on the effects of industrial, agricultural and diffuse pollution on the soil ecosystem. Among the methods, assays suitable for a range of assessment applications can be identified. For example, after a desk study of site historic use and chemical analyses have indicated potential risk, the most likely biological approach would be as follows:

1. Use a rapid, easy to interpret and preferably sensitive biological system to provide initial quantitative spatial data concerning the likelihood of exposure. When possible, use should allow risk visualisation via geographical information systems. By overlaying this information onto chemical analysis data, contamination hot spots can be identified.
2. Apply biological methods capable of accurate assessment of ecological effects that integrate mixture and bioavailability effects.

The diverse nature of soil contamination means that it is currently not possible to recommend a single assay for soil quality assessments in all circumstances. A suite of assays is likely to be the most useful approach. Within the suite, complementary methods applicable for a range of contaminants at different concentration levels and in a range of soils are needed. These include bioassay/biomarker/biosensor methods that:

- Detect both generic effects and also selected priority contaminant groups.
- Measure responses in a range of organisms, including plants, bacteria, fungi and subsurface and surface-dwelling soil invertebrates.
- Measure responses at a range of biological levels of organisation, thus ecological and biological indicators will show that changes to ecosystem functions have occurred; bioassays indicate that there may be significant impacts on individuals, whereas biomarkers may provide early warning of exposure and effect.

In recommending assays, the existence of an internationally standardised protocol is an important factor for the credibility of a given test. Coupled to this, however, is the recognition that there can be a considerable time delay before standardisation is attained and in any case current standard tests have not been fully developed for contaminated soil assessment. Even when tests become standardised, developments do not cease and methods can change on the basis of improved knowledge. For example, a revision of the OECD terrestrial plant growth test (OECD, 2000) first published in 1984 (OECD, 1984) has been released. That said, some of the internationally standardised tests are among the most suitable assays for use in contaminated soil assessment. These include Microtox® (and similar systems), which has the potential as a screening tool, and toxicity tests with plants and the three invertebrate groups (earthworm, enchytraieid and springtail), which can be used for detailed effect assessment. The remaining standardised assays, however, appear either to be less relevant for use with soil systems (in the case of the aquatic test) or insufficiently sensitive to the effects of chemical disturbance (in the case of carbon and nitrogen mineralisation).

Accepting that the absence of an international guideline should not be a barrier to the use of developing methods, further useful methods can be identified. Some, such as the bait lamina (Törne, 1990a) and earthworm lysosomal membrane stability assay (Weeks and Svendsen, 1996), are now well established in the scientific literature and appear particularly suited to assessing exposure and effects. Some of the biomarker techniques, such as ChE inhibition, metallothionein induction and possibly immune function assessment, can be used to provide valuable information regarding the nature of the chemicals present and the exposure of soil species.

For all these tests, interpretation and use would be easier if data on baseline responses were available. Currently, the quantity and quality of baseline data vary between tests. In the cases of standardised tests, such as the earthworm reproduction bioassay, performance criteria exist that can be used to discern if an observed effect is significant. For other assays, information of this type is not available. One area that any work to establish baselines should focus on is the interplay between soil factors, ecotypes and measured responses. This will give baselines for variation within the tests and will help to derive performance criteria that can be used as part of an algorithm to discern the status of a given location.

Among the developing techniques there are clearly some assays that have potential for ecotoxicological studies and contaminated soil assessment. New approaches for microbes, such as PLFA and nucleic acid-based analysis, appear to offer an improved resolution for the effects of chemical exposure on the composition of the microbial community. Demographic approaches offer an apparent means by which to improve our ability to predict the long-term effects of chemical exposure. One of the most rapidly developing and promising areas is molecular genetics and biochemistry. Here, developments in medical diagnostics are the principal driving force behind the development of a suite of new technologies. The pace of development in this field means that regulatory agencies should remain informed on current developments (Kille et al., 2003). Methods such as microarrays, proteomics and metabolic profiling in particular appear to offer a real potential for elucidating the exposure of species to chemicals in contaminated soils and a platform from which to predict toxic effects.

Acknowledgement

This chapter has been condensed from a report of R & D project PS-063, funded by the UK Environment Agency.

References

Aceves, M.B., Grace, C., Ansorena, J., Dendooven, L. and Brookes, P.C. (1999) Soil microbial biomass and organic C in a gradient of zinc concentrations in soils around a mine spoil tip. *Soil Biology & Biochemistry*, **31**, 867–876.

Achazi, R.K., Flenner, C., Livingstone, D.R., Peters, L.D., Schaub, K. and Scheiwe, E. (1998) Cytochrome P450 and dependent activities in unexposed and PAH-exposed terrestrial annelids. *Comparative Biochemistry and Physiology B*, **121**, 339–350.

Alexandre, H., Ansanay-Galeote, V., Dequin, S. and Blondin, B. (2001) Global gene expression during short-term ethanol stress in *Saccharomyces cerevisiae*. *FEBS Letters*, **498**, 98–103.

Atienzar, F.A., Conradi, M., Evenden, A.J., Jha, A.N. and Depledge, M.H. (1999) Qualitative assessment of genotoxicity using random amplified polymorphic DNA: comparison of genomic template stability with key fitness parameters in *Daphnia magna* exposed to benzo[a]pyrene. *Environmental Toxicology and Chemistry*, **18**, 2275–2282.

Augustsson, A.K. and Rundgren, S. (1998) The enchytraeid *Cognettia sphagnetorum* in risk assessment: advantages and disadvantages. *Ambio*, **27**, 62–69.

Bååth, E., Diaz-Ravina, M., Frostegard, A. and Campbell, C.D. (1998) Effect of metal-rich sludge amendments on the soil microbial community. *Applied and Environmental Microbiology*, **64**, 238–245.

Barich, J.J., Greene, J. and Bond, R. (1987) Soil stabilization treatability study at the Western Processing Superfund site. Superfund '87: *Proceedings of the 8th National Conference on Management of Uncontrolled Hazardous Waste Sites*, Washington, DC.

Barich, J.J., Peterson, S.A. and Greene, J.C. (1992) The toxicological assessment of remedial and restoration techniques. International Seminar on Nuclear War and Planetary Emergencies, Erice, Italy.

Bartosiewicz, M., Penn, S. and Buckpitt, A. (2001) Applications of gene arrays in environmental toxicology: fingerprints of gene regulation associated with cadmium chloride, benzo(a)pyrene, and trichloroethylene. *Environmental Health Perspectives*, **109**, 71–74.

Baud-Grasset, F., Baud-Grasset, S. and Safferman, S.I. (1993) Evaluation of the bioremediation of a contaminated soil with phytotoxicity tests. *Chemosphere*, **26**, 1365–1374.

Beaulieu, M., Becaert, V., Deschenes, L. and Villemur, R. (2000) Evolution of bacterial diversity during enrichment of PCP-degrading activated soils. *Microbial Ecology*, **40**, 345–355.

Berger, B., Dallinger, R. and Thomaser, A. (1995) Quantification of metallothionein as a biomarker for cadmium exposure in terrestrial gastropods. *Environmental Toxicology and Chemistry*, **14**, 781–791.

Berghout, A., Buld, J. and Wenzel, E. (1990) The cytochrome P 450 dependent monooxygenase system of the midgut of the earthworm *Lumbricus terrestris*. *European Journal of Pharmacology*, **183**, 1885–1886.

Bogomolov, D.M., Chen, S.K., Parmelee, R.W., Subler, S. and Edwards, C.A. (1996) An ecosystem approach to soil toxicity testing: a study of copper contamination in laboratory soil microcosms. *Applied Soil Ecology*, **4**, 95–105.

Bongers, T. (1990) The maturity index – an ecological measure of environmental disturbance based on nematode species composition. *Oecologia*, **83**, 14–19.

Booth, L.H., Hodge, S. and O'Halloran, K. (2000) Use of cholinesterase in *Aporrectodea caliginosa* (Oligochaeta; Lumbricidae) to detect organophosphate contamination: comparison of laboratory tests, mesocosms, and field studies. *Environmental Toxicology and Chemistry*, **19**, 417–422.

Borgeraas, J., Nilsen, K. and Stenersen, J. (1996) Methods for purification of glutathione transferases in the earthworm genus *Eisenia*, and their characterization. *Comparative Biochemistry and Physiology C*, **114**, 129–140.

Boyd, E.M., Killham, K., Wright, J., Rumford, S., Hetheridge, M., Cumming, R. and Meharg, A.A. (1997a) Toxicity assessment of xenobiotic contaminated groundwater using lux modified *Pseudomonas fluorescens*. *Chemosphere*, **35**, 1967–1985.

Boyd, E.M., Meharg, A.A., Wright, J. and Killham, K. (1997b) Assessment of toxicological interactions of benzene and its primary degradation products (catechol and phenol) using a *lux*-modified bacterial bioassay. *Environmental Toxicology and Chemistry*, **16**, 849–856.

Boyd, E.M., Meharg, A.A., Wright, J. and Killham, K. (1998) Toxicity of chlorobenzenes to a *lux*-marked terrestrial bacterium, *Pseudomonas fluorescens*. *Environmental Toxicology and Chemistry*, **17**, 2134–2140.

Boyd, E.M., Killham, K. and Meharg, A.A. (2001) Toxicity of mono-, di- and tri-chlorophenols to lux marked terrestrial bacteria, *Burkholderia* species *Rasc c2* and *Pseudomonas fluorescens*. *Chemosphere*, **43**, 157–166.

Bradley, B.P. (2000) Environmental proteomics. SETAC Third World Congress, Brighton, UK.

Brohon, B., Delolme, C. and Gourdon, R. (2001) Complementarity of bioassays and microbial activity measurements for the evaluation of hydrocarbon-contaminated soils quality. *Soil Biology & Biochemistry*, **33**, 883–891.

Bundy, J.G., Wardell, J.L., Campbell, C.D., Killham, K. and Paton, G.I. (1997) Application of bioluminescence-based microbial biosensors to the ecotoxicity assessment of organotins. *Letters in Applied Microbiology*, **25**, 353–358.

Bundy, J.G., Osborn, D., Week, J.M., Lindon, J.C. and Nicholson, J.K. (2001) NMR-based metabonomic approach to the investigation of coelomic fluid biochemistry in earthworms under toxic stress. *FEBS Letters*, **500**, 31–35.

Chander, K. and Brooks, P.C. (1991) Is the dehydrogenase assay invalid as a method to estimate microbial activity in copper-contaminated soils? *Soil Biology & Biochemistry*, **23**, 909–915.

Chander, K., Brookes, P.C. and Harding, S.A. (1995) Microbial biomass dynamics following addition of metal-enriched sewage sludges to a sandy loam. *Soil Biology and Biochemistry*, **27**, 1409–1421.

Chapman, S.J., Campbell, C.D., Edwards, A.C. and McHenery, J.G. (2000) *Assessment of the Potential of Biotechnology Environmental Monitoring Techniques*, SR (99) 10F. January 2000. Macaulay Research and Consultancy Services. Aberdeen.

Chaudri, A.M., Knight, B.P., BarbosaJefferson, V.L., Preston, S., Paton, G.I., Killham, K., Coad, N., Nicholson, F.A., Chambers, B.J. and McGrath, S.P. (1999) Determination of acute Zn toxicity in pore water from soils previously treated with sewage sludge using bioluminescence assays. *Environmental Science and Technology*, **33**, 1880–1885.

Cho, J.C. and Kim, S.J. (2000) Computer-assisted PCR-single-strand-conformation polymorphism analysis for assessing shift in soil bacterial community structure during bioremediational treatments. *World Journal of Microbiology and Biotechnology*, **16**, 231–235.

Cikutovic, M.A., Fitzpatrick, L.C., Goven, A.J., Venables, B.J., Giggleman, M.A. and Cooper, E.L. (1999) Wound healing in earthworms *Lumbricus terrestris*: a cellular-based biomarker for assessing sublethal chemical toxicity. *Bulletin of Environmental Contamination and Toxicology*, **62**, 508–514.

Cioci, L.K., Qiu, L. and Freedman, J.H. (2000) Transgenic strains of the nematode *Caenorhabditis elegans* as biomonitors of metal contamination. *Environmental Toxicology and Chemistry*, **19**, 2122–2129.

Clement, B. and Bouvet, Y. (1993) Asessment of landfill leachate toxicity using the duckweed (*Lemna minor*). *Science of the Total Environment*, Suppl., 1179–1190.

Clubb, R.W., Lords, J. and Gaufin, A. (1975) Isolation and characterisation of a glycoprotein from stonefly, *Pteronarcys californica*, which binds cadmium. *Journal of Insect Physiology*, **21**, 53–60.

Collado, R., Schmelz, R.M., Moser, T. and Rombke, J. (1999) Enchytraeid Reproduction Test (ERT): sublethal responses of two Enchytraeus species (Oligochaeta) to toxic chemicals. *Pedobiologia*, **43**, 625–629.

Crommentuijn, G.H., Doodeman, C.J.A.M., Doornekamp, A. and Van Gestel, C.A.M. (1997) Life-table study with the springtail *Folsomia candida* (Wilem) exposed to cadmium, chlorpyrifos and triphenyl tin hydroxide. In *Ecological Principles for Risk Assessment of Contaminants in Soil*, pp. 275–291. Chapman and Hall, London.

Dallinger, R. (1993) Strategies of metal detoxification in terrestrial invertebrates. In *Ecotoxicology of Metals in Invertebrates*. Dallinger, R. and Rainbow, P.S. (eds), pp. 245–89. Lewis Publishers, Boca Raton FL.

Dallinger, R. (1996) Metallothionein research in terrestrial invertebrates: synopsis and perspectives. *Comparative Biochemistry and Physiology C*, **113**, 125–133.

Dallinger, R., Berger, B., Hunziker, P.E. and Kägi, J.H.R. (1997) Metallothionein in snail Cd and Cu metabolism. *Nature*, **388**, 237–238.

Dallinger, R., Berger, B., Gruber, C., Hunziker, P. and Sturzenbaum, S. (2000) Metallothioneins in terrestrial invertebrates: structural aspects, biological significance and implications for their use as biomarkers. *Cellular and Molecular Biology*, **46**, 331–346.

De Jong, F.M.W. (1998) Development of a field bioassay for the side effects of pesticides on decomposition. *Ecotoxicology and Environmental Safety*, **40**, 103–114.

de Knecht, J.A., Stroomberg, G.J., Tump, C., Helms, M., Verweij, R.A., Commandeur, J., van Gestel, C.A.M. and van Straalen, N.M. (2001) Characterization of enzymes involved in biotransformation of polycyclic aromatic hydrocarbons in terrestrial isopods. *Environmental Toxicology and Chemistry*, **20**, 1457–1464.

Dennis, J.L., Mutwakil, M., Lowe, K.C. and de Pomerai, D.I. (1997) Effects of metal ions in combination with a non-ionic surfactant on stress responses in a transgenic nematode. *Aquatic Toxicology*, **40**, 37–50.

Domsch, K.H., Jagnow, G. and Anderson, T.-H. (1983) An ecological concept for the assessment of side-effects of agrochemicals on soil microorganisms. *Residue Reviews*, **86**, 65–105.

Duarte, G.F., Rosado, A.S., Seldin, L., de Araujo, W. and van Elsas, J.D. (2001) Analysis of bacterial community structure in sulfurous-oil-containing soils and detection of species carrying dibenzothiophene desulfurization (dsz) genes. *Applied and Environmental Microbiology*, **67**, 1052–1062.

Dusek, L. (1995a) Activity of nitrifying populations in grassland soil polluted by polychlorinated-biphenyls (PCBs) *Plant and Soil*, **176**, 273–282.

Dusek, L. (1995b) The effect of cadmium on the activity of nitrifying populations in two different grassland soils. *Plant and Soil*, **177**, 43–53.

Eason, C.T., Booth, L.H., Brennan, S. and Ataria, J. (1998) Cytochrome P450 activity in 3 earthworm species. In *Advances in Earthworm Ecotoxicology. Proceedings from the Second International Workshop on Earthworm Ecotoxicology, 2–5 April 1997*, Sheppard, S. Bembridge, J., Holmstrup, M. and Posthuma, L. (eds), pp. 191–198. SETAC Press, Pensacola, FL.

Eason, C.T., Svendsen, C., O'Halloran, K. and Weeks, J.M. (1999) An assessment of the lysosomal neutral red retention test and immune function assay in earthworms (*Eisenia andrei*) following exposure to chlorpyrifos, benzo-a-pyrene (BaP), and contaminated soil. *Pedobiologia*, **43**, 641–645.

Eckwert, H., Zanger, M., Reiss, S., Musolff, H., Albert, G. and Köhler, H.-R. (1994) The effect of heavy metals on the expression of hsp70 in soil invertebrates. *Verhandlungen der Deutschen Zoologischen Gesellschaft*, **87**, 325.

Eckwert, H., Alberti, G. and Köhler, H.-R. (1997) The induction of stress proteins (hsp) in *Oniscus asellus* (Isopoda) as a molecular marker of multiple heavy metal exposure .1. Principles and toxicological assessment. *Ecotoxicology*, **6**, 249–262.

Edwards, C.A. and Bohlen, P.J. (1992) The effects of toxic chemical on earthworms. *Review of Environmental Contamination and Toxicology*, **125**, 23–99.

Eisenbeis, G. (1993) Zersetzung im Boden. *Inf. Natursch. Landschaftspfl.*, **6**, 53–76.

Engelen, B., Meinken, K., von Wintzingerode, F., Heuer, H., Malkomes, H.P. and Backhaus, H. (1998) Monitoring impact of a pesticide treatment on bacterial soil communities by metabolic and genetic fingerprinting in addition to conventional testing procedures. *Applied and Environmental Microbiology*, **64**, 2814–2821.

Fan, T.W.M., Lane, A.N. and Higashi, R.M. (1992) Hypoxia does not affect rate of ATP synthesis and energy-metabolism in rice shoot tips as measured by P-31 NMR *in vivo*. *Archives of Biochemistry and Biophysics*, **294**, 314–318.

Ferrari, B., Radetski, C.M., Veber, A.M. and Ferard, J.F. (1999) Ecotoxicological assessment of solid wastes: a combined liquid- and solid-phase testing approach using a battery of bioassays and biomarkers. *Environmental Toxicology and Chemistry*, **18**, 1195–1202.

Forget, J., Pavillon, J.F., Beliaeff, B. and Bocquene, G. (1999) Joint action of pollutant combinations (pesticides and metals) on survival (LC_{50} values) and acetylcholinesterase activity of *Tigriopus brevicornis* (Copepoda, Harpacticoida). *Environmental Toxicology and Chemistry*, **18**, 912–18.

Forster, B., Eder, M., Morgan, E. and Knacker, T. (1996) A microcosm study of the effects of chemical stress, earthworms and microorganisms and their interactions upon litter decomposition. *European Journal of Soil Biology*, **32**, 25–33.

Fugere, N., Brousseau, P., Krzystyniak, K., Coderre, D. and Fournier, M. (1996) Heavy metal-specific inhibition of phagocytosis and difference in *in vitro* sensitivity of heterogeneous coelomocytes from *Lumbricus terrestris* (Oligochaeta). *Toxicology*, **109**, 157–166.

Fuller, M.E., Scow, K.M., Lau, S. and Ferris, H. (1997) Trichloroethylene (TCE) and toluene effects on the structure and function of the soil community. *Soil Biology & Biochemistry*, **29**, 75–89.

Fuller, M.E. and Manning, J.F. (1998) Evidence of differential effects of 2,4,6-trinitrotoluene and other munitions compounds on specific sub-populations of soil microbial communities. *Environmental Toxicology and Chemistry*, **17**, 2185–2195.

Gibb, J.O.T., Holmes, E., Nicholson, J.K. and Weeks, J.M. (1997a) Proton NMR spectroscopic studies on tissue extracts of invertebrate species with pollution indicator potential. *Comparative Biochemistry and Physiology C*, **118**, 587–598.

Gibb, J.O.T., Svendsen, C., Weeks, J.M. and Nicholson, J.K. (1997b) H-1 NMR spectroscopic investigations of tissue metabolite biomarker response to Cu(II) exposure in terrestrial invertebrates: identification of free histidine as a novel biomarker of exposure to copper in earthworms. *Biomarkers*, **2**, 295–302.

Gichner, T. and Plewa, M.J. (1998) Induction of somatic DNA damage as measured by single cell gel electrophoresis and point mutation in leaves of tobacco plants. *Mutation Research – Fundamental and Molecular Mechanisms of Mutagenesis*, **401**, 143–152.

Gichner, T., Ptacek, O., Stavreva, D.A. and Plewa, M.J. (1999) Comparison of DNA damage in plants as measured by single cell gel electrophoresis and somatic leaf mutations induced by monofunctional alkylating agents. *Environmental and Molecular Mutagenesis*, **33**, 279–286.

Giggleman, M.A., Fitzpatrick, L.C., Goven, A.J. and Venables, B.J. (1998) Effects of pentachlorophenol on survival of earthworms (*Lumbricus terrestris*) and phagocytosis by their immunoactive coelomocytes. *Environmental Toxicology and Chemistry*, **17**, 2391–2394.

Gong, P., Gasparrini, P., Rho, D., Hawari, J., Thiboutot, S., Ampleman, G. and Sunahara, G.I. (2000) An *in situ* respirometric technique to measure pollution-induced microbial community tolerance in soils contaminated with 2,4,6-trinitrotoluene. *Ecotoxicology and Environmental Safety*, **47**, 96–103.

Gonzalezlopez, J., Martineztoledo, M.V., Rodelas, B. and Salmeron, V. (1993) Studies on the effects of the insecticides phorate and malathion on soil-microorganisms. *Environmental Toxicology and Chemistry*, **12**, 1209–1214.

Goven, A.J., Chen, S.C., Fitzpatrick, L.C. and Venables, B.J. (1994a) Lysozyme activity in earthworm (*Lumbricus terrestris*) coelomic fluid and coelomocytes – enzyme assay for immunotoxicity of xenobiotics. *Environmental Toxicology and Chemistry*, **13**, 607–613.

Goven, A.J., Fitzpatrick, L.C. and Venables, B.J. (1994b) Chemical toxicity and host-defense in earthworms – an invertebrate model. *Annals of the New York Academy of Sciences*, **712**, 280–300.

Grafton, M.J. (1995) The effects of application of methiocarb using drilled and broadcast techniques on the earthworm *Lumbricus terrestris*. *MSc thesis*, University of Reading, UK.

Greene, J.C., Miller, W.E., Debacon, M., Long, M.A. and Bartels, C.L. (1988) Use of *Selenastrum capricornatum* to assess the toxicity potential of surface and groundwater contamination caused by chromium waste. *Environmental Toxicology and Chemistry*, **7**, 35–39.

Guttman, S.I. (1994) Population genetic structure and ecotoxicology. *Environmental Health Perspectives*, **102**, 97–100.

Guven, K., Power, R.S., Avramides, S., Allender, R. and de Pomerai, D.I. (1999) The toxicity of dithiocarbamate fungicides to soil nematodes, assessed using a stress-inducible transgenic strain of *Caenorhabditis elegans*. *Journal of Biochemical and Molecular Toxicology*, **13**, 324–333.

Hans, R.K., Khan, M.A., Farooq, M. and Beg, M.U. (1993) Glutathione S transferase activity in an earthworm (*Pheretima posthuma*) exposed to 3 insecticides. *Soil Biology & Biochemistry*, **25**, 509–511.

Heitzer, A., Webb, O.F., Thonnard, J.E. and Sayler, G.S. (1992) Specific and quantitative assessment of napthalene and salicylate bioavailability using a bioluminescent catabolic reporter bacterium. *Applied and Environmental Microbiology*, **58**, 1839–1846.

Honsi, T.G., Hoel, L. and Stenersen, J.V. (1999) Non-inducibility of antioxidant enzymes in the earthworms *Eisenia veneta* and *E. fetida* after exposure to heavy metals and paraquat. *Pedobiologia*, **43**, 652–657.

Hopkin, S.P. (1989) *Ecophysiology of Metals in Terrestrial Invertebrates*. Elsevier Applied Science, London.

Hopkin, S.P. and Martin, M.H. (1985) Assimilation of zinc, cadmium, lead, copper and iron by the spider *Dysdera crocata*, a predator of woodlice. *Bulletin of Environmental Contamination and Toxicology*, **34**, 183–187.

Huuselaveistola, E., Kurppa, S. and Pihlava, J.M. (1994) Effects of fenvalerate and permethrin on soil arthropods and on residues in and decomposition of barley straw. *Agricultural Science in Finland*, **3**, 213–223.

Jensen, C.S., Garsdal, L. and Baatrup, E. (1997) Acetylcholinesterase inhibition and altered locomotor behavior in the carabid beetle *Pterostichus cupreus*. A linkage between biomarkers at two levels of biological complexity. *Environmental Toxicology and Chemistry*, **16**, 1727–1732.

Jones, W.J. and Ananyeva, N.D. (2001) Correlations between pesticide transformation rate and microbial respiration activity in soil of different ecosystems. *Biology and Fertility of Soils*, **33**, 477–483.

Kammenga, J.E., Busschers, M., Van Straalen, N.M., Jepson, P.C. and Bakker, J. (1996) Stress-induced fitness reduction is not determined by the most sensitive life-cycle trait. *Functional Ecology*, **10**, 106–111.

Kammenga, J.E., Arts, M.S.J. and OudeBreuil, W.J.M. (1998) HSP60 as a potential biomarker of toxic stress in the nematode *Plectus acuminatus*. *Archives of Environmental Contamination and Toxicology*, **34**, 253–258.

Kandeler, E., Luftenegger, G. and Schwarz, S. (1992) Soil microbial processes and *Testacea* (Protozoa) as indicators of heavy-metal pollution. *Zeitschrift fur Pflanzenernahrung und Bodenkunde*, **155**, 319–322.

Kandeler, E., Tscherko, D., Bruce, K.D., Stemmer, M., Hobbs, P.J., Bardgett, R.D. and Amelung, W. (2000) Structure and function of the soil microbial community in microhabitats of a heavy metal polluted soil. *Biology and Fertility of Soils*, **32**, 390–400.

Kelly, J.J. and Tate, R.L. (1998) Use of BIOLOG for the analysis of microbial communities from zinc-contaminated soils. *Journal of Environmental Quality*, **27**, 600–608.

Kelly, J.J., Haggblom, M. and Tate, R.L. (1999) Changes in soil microbial communities over time resulting from one time application of zinc: a laboratory microcosm study. *Soil Biology & Biochemistry*, **31**, 1455–1465.

Keltjens, W.G. and Van Beusichem, M.L. (1998) Phytochelatins as biomarkers for heavy metal stress in maize (*Zea mays* L) and wheat (*Triticum aestivum* L): combined effects of copper and cadmium. *Plant and Soil*, **203**, 119–126.

Khalil, M.A., AbdelLateif, H.M., Bayoumi, B.M. and Van Straalen, N.M. (1996) Analysis of separate and combined effects of heavy metals on the growth of *Aporrectodea caliginosa* (Oligochaeta; Annelida), using the toxic unit approach. *Applied Soil Ecology*, **4**, 213–219.

Khan, M. and Scullion, J. (2000) Effect of soil on microbial responses to metal contamination. *Environmental Pollution*, **110**, 115–125.

Kille, P., Blaxter, M., Tyler, C.R., Spurgeon, D.J., Small, G. and Snape, J. (2003) *Genomics – An Introduction*. Environment Agency, Wallingford.

Kitagawa, E., Momose, Y. and Iwahashi, H. (2003) Correlation of the structures of agricultural fungicides to gene expression in *Saccharomyces cerevisiae* upon exposure to toxic doses. *Environmental Science and Technology*, **37**, 2788–2793.

Knicker, H. (1999) Biogenic nitrogen in soils as revealed by solid-state carbon-13 and nitrogen-15 nuclear magnetic resonance spectroscopy. *Journal of Environmental Quality*, **29**, 715–723.

Kohl, S.D., Paul, T.J. and Rice, J.A. (2000) Solid-state NMR investigation of dual-mode sorption to soil organic matter. *Abstracts of Papers of the American Chemical Society*, **220**, 2-ENVR.

Köhler, H.-R. and Triebskorn, R. (1998) Assessment of the cytotoxic impact of heavy metals on soil invertebrates using a protocol integrating qualitative and quantitative components. *Biomarkers*, **3**, 109–127.

Köhler, H.R., Triebskorn, R., Stöcker, W. and Kloetzal, P.M. (1992) The 70kD heat shock protein (HSP70) in soil invertebrates: a possible tool for monitoring environmental toxicants. *Archives of Environmental Toxicology and Chemistry*, **22**, 334–338.

Köhler, H.-R., Knodler, C. and Zanger, M. (1999) Divergent kinetics of HSP70 induction in *Oniscus asellus* (Isopoda) in response to four environmentally relevant organic chemicals (B[a]P, PCB52, gamma-HCH, PCP): suitability and limits of a biomarker. *Archives of Environmental Contamination and Toxicology*, **36**, 179–185.

Korthals, G.W., Alexiev, A.D., Lexmond, T.M., Kammenga, J.E. and Bongers, T. (1996) Long-term effects of copper and pH on the nematode community in an agroecosystem. *Environmental Toxicology and Chemistry*, **15**, 979–985.

Kozdroj, J. and van Elsas, J.D. (2001) Structural diversity of microbial communities in arable soils of a heavily industrialised area determined by PCR-DGGE fingerprinting and FAME profiling. *Applied Soil Ecology*, **17**, 31–42.

Kross, B.C. and Cherryholmes, K. (1992) Toxicity screening of sanitary landfill leachates: a comparative evaluation with Microtox® analyses, chemical and other toxicity screening methods. Ecotoxicology Monitoring: International Symposium, London.

Kula, C. and Rombke, J. (1998) Evaluation of soil ecotoxicity tests with functional endpoints for the risk assessment of plant protection products – state of the art. *Environmental Science and Pollution Research*, **5**, 55–60.

Kula, H. (1992) Measuring effects of pesticides on earthworms in the field: test design and sampling methods. In *Ecotoxicology of Earthworms*, Greig-Smith, P.W., Becker, H., Edwards, P.J. and Heimbach, F. (eds), pp. 90–99. Intercept Ltd, Andover, Hants.

Kuperman, R.G., Simini, M., Phillips, C.T. and Checkai, R.T. (1999) Comparison of malathion toxicity using enchytraeid reproduction test and earthworm toxicity test in different soil types. *Pedobiologia*, **43**, 630–634.

Kuperman, R.G., Simini, M., Phillips, C.T. and Checkai, R.T. (2004) A proteome based assessment of the earthworm *Eisenia fetida* response to chemical warfare agents (CWA) in a sandy loam soil. *Pedobiologia*, **57**, 48–53.

Kwan, K.K. and Dutka, B.J. (1992) Evaluation of Toxi-chromotest direct sedimentary toxicity testing procedure and Microtox® solid-phase testing procedure. *Bulletin of Environmental Contamination and Toxicology*, **49**, 656–662.

Labrot, F., Ribera, D., SaintDenis, M. and Narbonne, J.F. (1996) *In vitro* and *in vivo* studies of potential biomarkers of lead and uranium contamination: lipid peroxidation, acetylcholinesterase, catalase and glutathione peroxidase activities in three non mammalian species. *Biomarkers*, **1**, 21–28.

Langer, T. and Neupert, W. (1990) Heat shock proteins hsp60 and hsp70: their roles in folding, assembly and membrane translocation of proteins. In *Heat Shock Proteins and Immune Response, Current Topics in Microbiology and Immunology*, Kaufmann, S.H.E. (ed.), Springer Verlag, Berlin.

Larink, O. and Lübben, B. (1991) Bestimmung der biologischen aktivität von böden mit dem köderstreifen-test nach v. Törne. *Mitteilungen der Deutschen Bodenkundlichen Gesellschaft*, **66**, 551–554.

Laskowski, R. and Hopkin, S.P. (1996) Effects of Zn, Cu, Pb and Cd on fitness in snails (*Helix aspersa*). *Ecotoxicology and Environmental Safety*, **34**, 59–69.

Linder, G., Green, J.C., Ratsch, H., Nwosu, J., Smith, S. and Wilborn, D. (1990) Seed germination and root elongation toxicity tests in hazardous waste site evaluation: methods development and applications. In *Plants for Toxicity Assessment*, ASTM STP 1091, Lower, W.R. (ed.), pp. 177–187. American Society for Testing and Materials, Philadelphia, PA.

Linz, D.G. and Nakles, D.V. (1996) *Environmentally Acceptable Endpoints in Soil: Risk-based Approach to Contaminated Site Management Based on Availability of Chemicals in Soil*. American Academy of Environmental Engineers.

Lock, K. and Janssen, C.R. (2001) Test designs to assess the influence of soil characteristics on the toxicity of copper and lead to the oligochaete *Enchytraeus albidus*. *Ecotoxicology*, **10**, 137–144.

Lock, K., Janssen, C.R. and de Coen, W.M. (2000) Multivariate test designs to asses the influence of zinc and cadmium bioavailability in soils on the toxicity to *Enchytraeus albidus*. *Environmental Toxicology and Chemistry*, **19**, 2666–2671.

Løkke, H. and Van Gestel, C.A.M. (1998) Soil toxicity tests in risk assessment of new and existing chemicals. In *Handbook of Soil Invertebrate Toxicity Tests*, Løkke, H. and Van Gestel, C.A.M. (eds), pp. 3–19. John Wiley & Sons, Chichester.

Lorenz, S.E., McGrath, S.P. and Giller, K.E. (1992) Assessment of free-living nitrogen-fixation activity as a biological indicator of heavy-metal toxicity in soil. *Soil Biology & Biochemistry*, **24**, 601–606.

Lowe, D.M., Moore, M.N. and Evans, B.M. (1992) Contaminant impact on interactions of molecular probes with lysosomes in living hepatocytes from dab *Limanda limanda*. *Marine Ecology Progress Series*, **91**, 135–140.

Luster, M.I., Munson, A.E., Thomas, P.T., Holsapple, M.P., Fenters, J.D., White, K.L., Lauer, L.D., Germolec, D.R., Rosenthal, G.J. and Dean, J.H. (1988) Development of a testing battery to assess chemical-induced immunotoxicity – National Toxicology Programs Guidelines for immunotoxicity evaluation in mice. *Fundamental and Applied Toxicology*, **10**, 2–19.

Martin, M.H., Duncan, E.M. and Coughtrey, P.J. (1982) The distribution of heavy metals in a contaminated woodland ecosystem. *Environmental Pollution (Series B)*, **3**, 147–157.

Martineztoledo, M.V., Salmeron, V., Rodelas, B. and Gonzalezlopez, J. (1993) Response of soil microflora to the insecticides fonofos and parathion. *Toxicological and Environmental Chemistry*, **39**, 139–145.

Mayer, F.L., Versteeg, D.J., McKee, M.J., Folmar, L.C., Graney, R.L., McCume, D.C. and Rattner, B.A. (1992) Physiological and nonspecific biomarkers. In *Biomarkers. Biochemical, Physiological, and Histological Markers of Anthropogenic Stress*, Huggett, R.J., Kimerle, R.A., Mehrle, Jr P.M. and Bergman, H.L. (eds), pp. 125–153. Lewis Publishers, Boca Raton, FL.

McCluskey, S., Mather, P.B. and Hughes, J.M. (1993) The relationship between behavioural responses to temperature and genotype at a PGI locus in the terrestrial isopod *Porcellio laevis*. *Biochemical Systematics and Ecology*, **21**, 171–179.

McGrath, S.P. (1994) Effects of heavy metals from sewage sludge on soil microbes in agricultural ecosystems. In *Toxic Metals in Soil Plant Systems*, Ross, S.M. (ed.), pp. 247–274. John Wiley & Sons, New York.

Meyer, M.C., Paschke, M.W., McLendon, T. and Price, D. (1998) Decreases in soil microbial function and functional diversity in response to depleted uranium. *Journal of Environmental Quality*, **27**, 1306–1311.

Muller, A.K., Westergaard, K., Christensen, S. and Sorensen, S.J. (2001) The effect of long-term mercury pollution on the soil microbial community. *FEMS Microbiology Ecology*, **36**, 11–19.

Murray, P., Ge, Y. and Hendershot, W.H. (2000) Evaluating three trace metal contaminated sites: a field and laboratory investigation. *Environmental Pollution*, **107**, 127–135.

Nagy, P. (1999) Effect of an artificial metal pollution on nematode assemblage of a calcareous loamy chernozem soil. *Plant and Soil*, **212**, 35–43.

Neher, D.A. and Olson, R.K. (1999) Nematode communities in soils of four farm cropping management systems. *Pedobiologia*, **43**, 430–439.
Nevo, E., Lavie, B. and Noy, R. (1987) Mercury selection of allozymes in marine gastropods: Prediction and verification in nature revisited. *Environmental Monitoring and Assessment*, **9**, 233–238.
Nicholson, J.K., Lindon, J.C. and Holmes, E. (1999) 'Metabonomics': understanding the metabolic responses of living systems to pathophysiological stimuli via multivariate statistical analysis of biological NMR spectroscopic data. *Xenobiotica*, **29**, 1181–1189.
OECD (1984) *Guideline for the testing of chemicals:208:Terrestrial Plants, Growth Test*. Organisation for Economic Cooperation and Development, Paris.
OECD (2000) *Proposal for Updating Guideline 208: Terrestrial (Non-Target) Plant Test:208 A: Seedling Emergence and Seedling Growth Test 208 B: Vegetative Vigour Test*. Organisation for Economic Cooperation and Development, Paris.
Palmer, G., McFadzean, R., Killham, K., Sinclair, A. and Paton, G.I. (1998) Use of *lux*-based biosensors for rapid diagnosis of pollutants in arable soils. *Chemosphere*, **36**, 2683–2697.
Paton, G.I., Campbell, C.D., Glover, L.A. and Killham, K. (1995) Assessment of bioavailability of heavy-metals using *lux* modified constructs of *Pseudomonas fluorescens*. *Letters in Applied Microbiology*, **20**, 52–56.
Paton, G.I., Palmer, G., Burton, M., Rattray, E.A.S., McGrath, S.P., Glover, L.A. and Killham, K. (1997) Development of an acute and chronic ecotoxicity assay using *lux*-marked *Rhizobium leguminosarum* biovar *trifolii*. *Letters in Applied Microbiology*, **24**, 296–300.
Paulus, R., Rombke, J., Ruf, A. and Beck, L. (1999) A comparison of the litterbag-, minicontainer- and bait-lamina-methods in an ecotoxicological field experiment with diflubenzuron and BTK. *Pedobiologia*, **43**, 120–133.
Pomazkina, L.V., Radnaev, A.B.D., Kotova, L.G. and Petrova, I.G. (2001) The transformation and balance of nitrogen in fluoride-polluted agroecosystems on gray forest soils of the Baikal region. *Eurasian Soil Science*, **34**, 645–650.
Posthuma, L., Van Gestel, C.A.M., Smit, C.E., Bakker, D.J. and Vonk, J.W. (1998) *Validation of Toxicity Data and Risk Limits for Soils: Final Report*, No. 607505004. Rijksinstituut voor Volksgezondheid en Milieu, Bilthoven, The Netherlands.
Power, R.S. and De Pomerai, D.I. (1999) Effect of single and paired metal inputs in soil on a stress-inducible transgenic nematode. *Archives of Environmental Contamination and Toxicology*, **37**, 503–511.
Power, R.S., David, H.E., Mutwakil, M.H.A.Z., Fletcher, K., Daniells, C., Nowell, M.A., Dennis, J.L., Martinelli, A., Wiseman, R., Wharf, E. and dePomerai, D.I. (1998) Stress-inducible transgenic nematodes as biomonitors of soil and water pollution. *Journal of Biosciences*, **23**, 513–526.
Radetski, C.M., Cotelle, S. and Ferard, J.F. (2000) Classical and biochemical endpoints in the evaluation of phytotoxic effects caused by the herbicide trichloroacetate. *Environmental and Experimental Botany*, **44**, 221–229.
Rasmussen, L.D. and Sorensen, S.J. (2001) Effects of mercury contamination on the culturable heterotrophic, functional and genetic diversity of the bacterial community in soil. *FEMS Microbiology Ecology*, **36**, 1–9.
Read, H.J., Wheater, C.P. and Martin, M.H. (1987) Aspects of the ecology of Carabidae. *Environmental Pollution*, **48**, 61–76.
Reid, B.J., Semple, K.T., Macleod, C.J., Weitz, H.J. and Paton, G.I. (1998) Feasibility of using prokaryote biosensors to assess acute toxicity of polycyclic aromatic hydrocarbons. *FEMS Microbiology Letters*, **169**, 227–233.
Reinecke, A.J. and Reinecke, S.A. (1997) Uptake and toxicity of copper and zinc for the African earthworm *Eudrilus eugeniae* (Oligochaeta). *Biology and Fertility of Soils*, **24**, 27–31.
Remde, A. and Hund, K. (1994) Response of soil autotrophic nitrification and soil respiration to chemical pollution in long-term experiments. *Chemosphere*, **29**, 391–404.
Robidoux, P.Y., Svendsen, C., Sarrazin, M., Hawari, J., Thiboutot, S., Ampleman, G., Weeks, J.M. and Sunahara, G.I. (2002) Evaluation of tissue and cellular biomarkers to assess 2,4,6-trinitrotoluene (TNT) exposure in earthworms: effects-based assessment in laboratory studies using *Eisenia andrei*. *Biomarkers*, **7**, 306–321.

Rozen, A. and Mazur, L. (1997) Influence of different levels of traffic pollution on haemoglobin content in the earthworm *Lumbricus terrestris*. *Soil Biology & Biochemistry*, **29**, 709–711.

Ruess, L., Schmidt, I.K., Michelsen, A. and Jonasson, S. (2001) Manipulations of a microbial based soil food web at two arctic sites – evidence of species redundancy among the nematode fauna? *Applied Soil Ecology*, **17**, 19–30.

Rumpf, S., Hetzel, F. and Frampton, C. (1997) Lacewings (Neuroptera: Hemerobiidae and chrysopidae) and integrated pest management: Enzyme activity as biomarker of sublethal insecticide exposure. *Journal of Economic Entomology*, **90**, 102–108.

Rutgers, M., van't Verlaat, I.M., Wind, B., Posthuma, L. and Breure, A.M. (1998) Rapid method for assessing pollution-induced community tolerance in contaminated soil. *Environmental Toxicology and Chemistry*, **17**, 2210–2213.

Saint-Denis, M., Pfohl-Leszkowicz, A., Narbonne, J.F. and Ribera, D. (2000) Dose-response and kinetics of the formation of DNA adducts in the earthworm *Eisenia fetida/andrei* exposed to B(a)P-contaminated artificial soil. *Polycyclic Aromatic Compounds*, **18**, 117–127.

Saint-Denis, M., Narbonne, J.F., Arnaud, C. and Ribera, D. (2001) Biochemical responses of the earthworm *Eisenia fetida andrei* exposed to contaminated artificial soil: effects of lead acetate. *Soil Biology & Biochemistry*, **33**, 395–404.

Salagovic, J., Gilles, J., Verschaeve, L. and Kalina, I. (1996) The comet assay for the detection of genotoxic damage in the earthworms: a promising tool for assessing the biological hazards of polluted sites. *Folia Biologica*, **42**, 17–21.

Salama, M.S., Schouest Jr, L.P. and Miller, T.A. (1992) Effect of the diet on the esterase patterns in hemolymph of the corn earworm and the tobacco budworm (Lepidoptera: Noctuidae). *Journal of Economic Entomology*, **85**, 1079–1087.

Salemaa, M., Vanha-Majamaa, I. and Derome, J. (2001) Understorey vegetation along a heavy-metal pollution gradient in SW Finland. *Environmental Pollution*, **112**, 339–350.

Sandaa, R.A., Torsvik, V. and Enger, O. (2001) Influence of long-term heavy-metal contamination on microbial communities in soil. *Soil Biology & Biochemistry*, **33**, 287–295.

Sandifer, R.D. (1996) The effects of cadmium, copper, lead and zinc contamination on arthropod communities in the vicinity of a primary smelting works. PhD thesis, University of Reading, UK.

Sandifer, R.D. and Hopkin, S.P. (1996) Effects of pH on the toxicity of cadmium, copper, lead and zinc to *Folsomia candida* Willem, 1902 (Collembola) in a standard laboratory test system. *Chemosphere*, **33**, 2475–2486.

Sarathchandra, S.U., Ghani, A., Yeates, G.W., Burch, G. and Cox, N.R. (2001) Effect of nitrogen and phosphate fertilisers on microbial and nematode diversity in pasture soils. *Soil Biology & Biochemistry*, **33**, 953–964.

Saterbak, A., Toy, R.J., Wong, D.C.L., McMain, B.J., Williams, M.P., Dorn, P.B., Brzuzy, L.P., Chai, E.Y. and Salanitro, J.P. (1999) Ecotoxicological and analytical assessment of hydrocarbon-contaminated soils and application to ecological risk assessment. *Environmental Toxicology and Chemistry*, **18**, 1591–1607.

Savva, D. (2000) AP-PCR fingerprinting for detection of genetic damage caused by polycyclic aromatic hydrocarbons. *Polycyclic Aromatic Compounds*, **20**, 291–303.

Scaps, P., Grelle, C. and Descamps, M. (1997) Cadmium and lead accumulation in the earthworm *Eisenia fetida* (Savigny) and its impact on cholinesterase and metabolic pathway enzyme activity. *Comparative Biochemistry and Physiology C*, **116**, 233–238.

Schmidt, G.H. and Ibrahim, N.M.M. (1994) Heavy-metal content (Hg^{2+}, Cd^{2+}, Pb^{2+}) in various body parts – its impact on cholinesterase activity and binding glycoproteins in the grasshopper *Aiolopus thalassinus* adults. *Ecotoxicology and Environmental Safety*, **29**, 148–164.

Scott-Fordsmand, J.J., Weeks, J.M. and Hopkin, S.P. (1998) Toxicity of nickel to the earthworm and the applicability of the neutral red retention assay. *Ecotoxicology*, **7**, 291–295.

Scott-Fordsmand, J.J., Krogh, P.H. and Weeks, J.M. (2000) Responses of *Folsomia fimetaria* (Collembola: Isotomidae) to copper under different soil copper contamination histories in relation to risk assessment. *Environmental Toxicology and Chemistry*, **19**, 1297–1303.

Shaw, L.J., Beaton, Y., Glover, L.A., Killham, K., Osborn, D. and Meharg, A.A. (2000) Bioavailability of 2,4-dichlorophenol associated with soil water-soluble humic material. *Environmental Science and Technology*, **34**, 4721–4726.
Sheppard, S.C., Evenden, W.G., Abboud, S.A. and Stephenson, M. (1993) A plant life-cycle bioassay for contaminated soil, with comparison to other bioassays – mercury and zinc. *Archives of Environmental Contamination and Toxicology*, **25**, 27–35.
Simon, T. (2000) The effect of nickel and arsenic on the occurrence and symbiotic abilities of native rhizobia. *Rostlinna Vyroba*, **46**, 63–68.
Slice, L.W., Freedman, J.H. and Rubin, C. (1990) Purification, characterization, and cDNA cloning of a novel metallothionein-like, cadmium-binding protein from *Caenorhabditis elegans*. *Journal of Biological Chemistry*, **265**, 256–263.
Smit, C.E. and Van Gestel, C.A.M. (1996) Comparison of the toxicity of zinc for the springtail *Folsomia candida* in artificially contaminated and polluted field soils. *Applied Soil Ecology*, **3**, 127–136.
Spurgeon, D.J. and Hopkin, S.P. (1995) Extrapolation of the laboratory based OECD earthworm toxicity test to metal contaminated field sites. *Ecotoxicology*, **4**, 190–205.
Spurgeon, D.J. and Hopkin, S.P. (1996) The effects of metal contamination on earthworm populations around a smelting works – quantifying species effects. *Applied Soil Ecology*, **4**, 147–160.
Spurgeon, D.J. and Hopkin, S.P. (1999) Seasonal variation in the abundance, biomass and biodiversity of earthworms in soils contaminated with metal emissions from a primary smelting works. *Journal of Applied Ecology*, **36**, 173–183.
Spurgeon, D.J., Svendsen, C., Rimmer, V.R., Hopkin, S.P. and Weeks, J.M. (2000) Relative sensitivity of life-cycle and biomarker responses in four earthworm species exposed to zinc. *Environmental Toxicology and Chemistry*, **19**, 1800–1808.
Spurgeon, D.J., Svendsen, C., Weeks, J.M., Hankard, P.K., Stubberud, H.E. and Kammenga, J.E. (2003) Quantifying copper and cadmium impacts on intrinsic rate of population increase in the terrestrial oligochaete *Lumbricus rubellus*. *Environmental Toxicology and Chemistry*, **22**, 1465–1472.
Spurgeon, D.J., Svendsen, C., Kille, P., Morgan, A.J. and Weeks, J.M. (2004) Responses of earthworms (*Lumbricus rubellus*) to copper and cadmium as determined by measurement of juvenile traits in a specifically designed test system. *Ecotoxicology and Environmental Safety*, **57**, 54–64.
Stavreva, D.A., Ptacek, O., Plewa, M.J. and Gichner, T. (1998) Single cell gel electrophoresis analysis of genomic damage induced by ethyl methanesulfonate in cultured tobacco cells. *Mutation Research – Fundamental and Molecular Mechanisms of Mutagenesis*, **422**, 323–330.
Stegeman, J.J., Brouwer, M., Di Giulio, R.T., Förlin, L., Fowler, B.A., Sanders, B.M. and Van Veld, P.A. (1992) Molecular responses to environmental contamination: enzyme and protein systems as indicators of chemical exposure and effect. In *Biomarkers. Biochemical, Physiological, and Histological Markers of Anthropogenic Stress*, Huggett, R.J., Kimerle, R.A., Mehrle Jr P.M. and Bergman, H.L. (eds), pp. 125–153. Lewis Publishers, Boca Raton, FL.
Stenersen, J. (1979) Action of pesticides on earthworms. Part I: The toxicity of cholinesterase-inhibiting insecticides to earthworms as evaluated by laboratory tests. *Pesticide Science*, **10**, 66–74.
Stenersen, J. (1984) Detoxication of xenobiotics by earthworms. *Comparative Biochemistry and Physiology C*, **78**, 249–252.
Stenersen, J., Gilman, A. and Vardanis, A. (1973) Carbofuran: its toxicity to and metabolism by earthworm (*Lumbricus terrestris*). *Journal of Agricultural and Food Chemistry*, **21**, 166–171.
Stephenson, G.L., Kaushik, A., Kaushik, N.K., Solomon, K.R., Steele, T. and Scroggins, R.P. (1998) Use of an avoidance-response test to assess the toxicity of contaminated soils to earthworms. In *Advances in Earthworm Ecotoxicology*, Sheppard, S., Bembridge, J., Holmstrup, M. and Posthuma, L. (eds). pp. 67–81. SETC, Pensecola, FL.
Stokke, K. and Stenersen, J. (1993) Non inducibility of the glutathione transferases of the earthworm *Eisenia andrei*. *Comparative Biochemistry and Physiology C*, **106**, 753–756.

Stoughton, J.A. and Marcus, W.A. (2000) Persistent impacts of trace metals from mining on floodplain grass communities along Soda Butte Creek, Yellowstone National Park. *Environmental Management*, **25**, 305–320.

Stubberud, H.E. (1998) Lysosomal fragilitet som biomarkør for tungmetall-forurensning i meitemark. Cand. Scient (MSc) Thesis, University of Oslo, Oslo, Norway.

Stürzenbaum, S. (1997) Molecular genetic responses of earthworms to heavy metals. PhD, University of Wales, Swansea.

Stürzenbaum, S.R., Kille, P. and Morgan, A.J. (1998a) Identification of heavy metal induced changes in the expression patterns of the translationally controlled tumour protein (TCTP) in the earthworm *Lumbricus rubellus*. *Biochimica et Biophysica Acta*, **1398**, 294–304.

Stürzenbaum, S.R., Kille, P. and Morgan, A.J. (1998b) Identification of new heavy-metal-responsive biomarker in the earthworm. In *Advances in Earthworm Ecotoxicology*, Sheppard, S.C., Bembridge, J.D., Holmstrup, M. and Posthuma, L. (eds), pp. 215–224. SETAC, Pensacola, FL.

Stürzenbaum, S.R., Kille, P. and Morgan, A.J. (1998c) The identification, cloning and characterization of earthworm metallothionein. *FEBS Letters*, **431**, 437–442.

Stürzenbaum, S.R., Morgan, A.J. and Kille, P. (1999) Characterisation and quantification of earthworm cyclophilins: identification of invariant and heavy metal responsive isoforms. *Biochimica et Biophysica Acta*, **1489**, 467–473.

Svendsen, C. (2000) Earthworm biomarkers in terrestrial ecosystems. PhD Thesis, University of Reading, UK.

Svendsen, C., Meharg, A.A., Freestone, P. and Weeks, J.M. (1996) Use of an earthworm lysosomal biomarker for the ecological assessment of pollution from an industrial plastics fire. *Applied Soil Ecology*, **3**, 99–107.

Svendsen, C. and Weeks, J.M. (1997) Relevance and applicability of a simple earthworm biomarker of copper exposure 1. Links to ecological effects in a laboratory study with *Eisenia andrei*. *Ecotoxicology and Environmental Safety*, **36**, 72–79.

Thompson, I.P., Bailey, M.J., Ellis, R.J. and Purdy, K.J. (1993) Subgrouping of bacterial-populations by cellular fatty–acid composition. *FEMS Microbiology Ecology*, **102**, 75–84.

Thompson, I.P., Bailey, M.J., Ellis, R.J., Maguire, N. and Meharg, A.A. (1999) Response of soil microbial communities to single and multiple doses of an organic pollutant. *Soil Biology & Biochemistry*, **31**, 95–105.

Törne, E.V. (1990a) Assessing feeding activities of soil living animals. *Pedobiologia*, **34**, 89–101.

Törne, E.V. (1990b) Schätzungen der freßaktivitäten bodenlebender tiere. II. Mini köder test. *Pedobiologia*, **34**, 269–279.

Trasar-Cepeda, C., Leiros, M.C., Seoane, S. and Gil-Sotres, F. (2000) Limitations of soil enzymes as indicators of soil pollution. *Soil Biology & Biochemistry*, **32**, 1867–1875.

Tu, C.M. (1990) Effect of 4 experimental insecticides on enzyme-activities and levels of adenosine-triphosphate in mineral and organic soils. *Journal of Environmental Science and Health Part B*, **25**, 787–800.

Van Dyk, T.K., Majarain, W.R., Konstantinov, K.B., Young, R.M., Dhurjati, P.S. and Larossa, R.A. (1994) Rapid and sensitive pollutant detection by induction of heat shock gene bioluminescece gene fusions. *Applied and Environmental Microbiology*, **60**, 1414–1420.

Van Gestel, C.A.M. and Van Brummelen, T.C. (1996) Incorporation of the biomarker concept in ecotoxicology calls for a redefinition of terms. *Ecotoxicology*, **5**, 217–225.

Van Gestel, C.A.M. and Van Straalen, N.M. (1994) Ecotoxicological test systems for terrestrial invertebrates. In *Ecotoxicology of Soil Organisms*, Donker, M.H., Eijsackers, H. and Heimbach, F. (eds), pp. 205–228. CRC Press, Boca Raton FL.

Van Gestel, C.A.M., Van Dis, W.A., Van Breemen, E.M. and Sparenburg, P.M. (1989) Development of a standardized reproduction toxicity test with the earthworm species *Eisenia andrei* using copper, pentachlorophenol and 2,4-dichloroaniline. *Ecotoxicology and Environmental Safety*, **18**, 305–312.

Van Gestel, C.A.M., Van der Waarde, J.J., Derksen, J.G.M., van der Hoek, E.E., Veul, M., Bouwens, S., Rusch, B., Kronenburg, R. and Stokman, G.N.M. (2001) The use of acute and chronic bioassays to determine the ecological risk and bioremediation efficiency of oil-polluted soils. *Environmental Toxicology and Chemistry*, **20**, 1438–1449.

Van Straalen, N.M. and Kammenga, J.E. (1998) Assessment of ecotoxicity at the population level using demographic parameters. In *Ecotoxicology*, Schüürmann, G. and Markert, B. (eds), pp. 621–644. John Wiley and Sons and Spektrum Akademischer Verlag, London.

Vasseur, L., Cloutier, C. and Ansseau, C. (2000) Effects of repeated sewage sludge application on plant community diversity and structure under agricultural field conditions on podzolic soils in eastern Quebec. *Agriculture Ecosystems and Environment*, **81**, 209–216.

Verschaeve, L. and Gilles, J. (1995) Single-cell gel-electrophoresis assay in the earthworm for the detection of genotoxic compounds in soils. *Bulletin of Environmental Contamination and Toxicology*, **54**, 112–119.

Verschaeve, L., Gilles, J., Schoeters, J., Van Cleuvenbergen, R. and de FrJ, R. (1993) The single cell gel electrophoresis technique or comet test for monitoring dioxin pollution and effects. In *Organohalogen Compounds II*, Fiedler, H., Frank, H., Hutzinger, O., Pazzefal, W., Riss, A. and Safe, S. (eds), pp. 213–216. Federal Environmental Agency, Vienna.

Vido, K., Spector, D., Lagniel, G., Lopez, S., Toledano, M.B. and Labarre, J. (2001) A proteome analysis of the cadmium response in *Saccharomyces cerevisiae*. *Journal of Biological Chemistry*, **276**, 8469–8474.

Ville, P., Roch, P., Cooper, E.L. and Narbonne, J.F. (1997) Immuno-modulator effects of carbaryl and 2,4-D in the earthworm *Eisenia fetida andrei*. *Archives of Environmental Contamination and Toxicology*, **32**, 291–297.

Vink, K. and Van Straalen, N.M. (1999) Effects of benomyl and diazinon on isopod-mediated leaf litter decomposition in microcosms. *Pedobiologia*, **43**, 345–359.

Warne, M.A., Lenz, E.M., Osborn, D., Weeks, J.M. and Nicholson, J.K. (2000) An NMR-based metabonomic investigation of the toxic effects of 3-trifluoromethyl-aniline on the earthworm *Eisenia veneta*. *Biomarkers*, **5**, 56–72.

Weeks, J.M. and Svendsen, C. (1996) Neutral red retention by lysosomes from earthworm (*Lumbricus rubellus*) coelomocytes: a simple biomarker of exposure to soil copper. *Environmental Toxicology and Chemistry*, **15**, 1801–1805.

Weeks, J.M. and Williams, T.M. (1994) *Preliminary Field Trial of a Method for the Rapid Assessment of Invertebrate Stress from Mining-related Arsenic and Heavy Metal Contamination: Wanderer Gold Mine, Zimbabwe*. Overseas Development Administration, London.

Weeks, J.M., Wright, J.F., Eversham, B.C., Hopkin, S.P., Roy, D., Svendsen, C. and Black, H. (1998) A Demonstration of the Feasibility of SOILPACS. UK Environment Agency, London.

Weissman, S. and Kunze, C. (1994) Microbial activity in heating oil contaminated soil under field and controlled conditions. *Angewandte Botanik*, **68**, 137–142.

Willuhn, J., Schmittwrede, H.P., Greven, H. and Wunderlich, F. (1994) cDNA cloning of a cadmium inducible messenger RNA encoding a novel cysteine rich, nonmetallothionein 25 KDa protein in an Enchytraeid earthworm. *Journal of Biological Chemistry*, **269**, 24688–91.

Willuhn, J., Otto, A., SchmittWrede, H.P. and Wunderlich, F. (1996) Earthworm gene as indicator of bioefficacious cadmium. *Biochemical and Biophysical Research Communications*, **220**, 581–5.

Witter, E., Gong, P., Baath, E. and Marstorp, H. (2000) A study of the structure and metal tolerance of the soil microbial community six years after cessation of sewage sludge applications. *Environmental Toxicology and Chemistry*, **19**, 1983–91.

Wunsche, L., Bruggemann, L. and Babel, W. (1995) Determination of substrate utilization patterns of soil microbial communities – an approach to assess population changes after hydrocarbon pollution. *FEMS Microbiology Ecology*, **17**, 295–305.

Xiong, J.C., Lock, H., Chuang, I.S., Keeler, C. and Maciel, G.E. (1999) Local motions of organic pollutants in soil components, as studied by H-2 NMR. *Environmental Science & Technology*, **33**, 2224–33.

Zanger, M., Graff, S., Braunbeck, T., Alberti, G. and Kohler, H.R. (1997) Detection and induction of cytochrome P4501A (CYP 1A) like proteins in *Julus scandinavius* (Diplopoda) and Oniscus asellus (Isopoda): a first analysis. *Bulletin of Environmental Contamination and Toxicology*, **58**, 511–7.

7 Review of biomarkers and new techniques for *in situ* aquatic studies with bivalves

Francois Gagné and Christian Blaise

7.1 Introduction

The social and economic expansion of society has contributed to the release of xenobiotics into aquatic environments that could impede the long-term survival of feral populations. These substances are either by-products of technology, including biotechnology, or of geological alterations by mining and land modifications. In addition to global changes, such as depletion of the ozone layer, climate change, loss of natural habitats and the reduced availability of drinking water, environmental degradation due to the release of thousands of chemicals poses a threat to the long-term survival of organisms and the maintenance of biodiversity.

The impacts of environmental contaminants on wildlife are usually determined by resorting to the measurement of biomarkers of exposure and effects. Biomarkers can be defined as any biochemical, physiological or morphological measurement that provides information about an interaction between a particular living entity and its immediate environment. These interactions can involve biological, chemical and/or physical agent(s). Although the domain of biochemical ecology or systematics deals with these interactions, especially in the context of signalling and communication with the environment, the domain of ecotoxicology deals with deleterious interactions between a chemical or physical agent and target tissues. Biomarkers are classified according to the nature of the interaction, level of biological organization and time scale. In this review, we discuss four types of biomarkers: biomarkers of exposure, biomarkers of early biological effects and defence, biomarkers of damage and biomarkers of reproduction (Table 7.1).

Biomarkers of exposure are used to assess an organism's exposure to a particular contaminant or class of contaminants. Usually, measurement of the toxicant or metabolite(s) in target tissues confirms that exposure actually took place. The measurement of adducts to proteins or DNA is another means of assessing exposure. For example, the formation of DNA adducts confirms not only that exposure occurred but also that a biochemical interaction with biological target(s) took place. With respect to physical agents such as ionizing radiation, exposure to this type of agent could be confirmed by the presence of a specific chemical signature (e.g. the formation of pyrimidine dimers in nucleic acids).

Biomarkers of early biological effects and defence provide information on biological disturbance or changes that do not necessarily translate into toxic

Table 7.1 Types of biomarkers used to assess the health of aquatic species.

Type of biomarker	Role	Examples
Exposure	To determine exposure to a class of contaminants and its reactivity to biochemical ligands	Tissue body burdens DNA or protein adducts
Early biological effects	Initial reactions towards the presence of contaminants Defence mechanisms are a special case of early biological effects. They are either adaptive or defensive reactions and do not represent actual toxic damage.	Extrusion of chemicals outside cells by protein pumps Induction of metal-binding proteins or biotransformation enzymes for elimination
Damage	A direct measure of toxic damage at the molecular and cellular levels These effects could compromise the long-term survival of the individual	Lipid peroxidation DNA strand breaks Inflammation Immunosuppression
Reproduction	To assess reproductive status. Alteration of reproduction could have more direct impacts on population status	Decreased vitellogenesis in females or increased vitellogenin levels in males (endocrine disruption)

effects. Defence mechanisms are adaptive biochemical or physiological processes in response to exposure to contaminants. For example, some contaminants are excluded from the intracellular environment by inducing a P-glycoprotein multixenobiotic transporter protein (also called a multixenobiotic resistant transporter) at the cytoplasmic membrane, thus preventing the chemicals from reaching toxic concentrations and altering the normal homeostatic function of cells (Eufemia and Epel, 2000). The expression of these transporters is more of an adaptive response to the presence of xenobiotics than a toxic effect.

Biomarkers of damage are related to the assessment of a specific damage that could compromise normal cell function. For example, the formation of lipid peroxides by pro-oxidants indicates that lipids are damaged by these chemicals either directly (acting as oxidants) or indirectly, i.e. blocking the enzymes involved in the reduction of hydroperoxides in cells (Cossu et al., 2000). Biomarkers of reproduction assess the capacity of the organism to reproduce effectively (Table 7.2). One such biomarker is the capacity of females to produce the egg-yolk protein precursor vitellogenin, the major energy source for the developing embryo.

Biomarkers of early biological effects and damage can be measured at several levels of biological organization, i.e. the molecular, cellular, tissue, systemic and individual levels. Biomarkers should be selected for their usefulness in evaluating the health of an individual and designed to permit preventive measures with respect to population status and maintenance of biodiversity. Indeed, they should

Table 7.2 Overview of biomarkers to assess reproduction status of aquatic organisms.

Biomarker	Sexual differentiation (target: nerve centers)	Gametogenesis (target: gonad)	Fertilization and spawning (target:gonad/gills)
Biochemical parameters	Females: decreased serotonin content and increased estradiol and aromatase activity in nerve centres	Steroidogenesis (progesterone, testosterone and estradiol-17β)	Increased serotonin and cyclooxygenase activity (prostaglandin production) during spawning
	Males: increased serotonin and low oestradiol content in nerve centres. Secretion of a penis forming factor (APGW neuropeptide)	Activation of aspartate transcarbamoylse for pyrimidine synthesis	Fertilization leads to germinal vesicle breakdown and increases free calcium in fertilized eggs
		Increased dopamine levels during gamete development	
		Vitellogenesis in females	
		Spermatogenesis	
		Increased RNA/DNA ration	
Known disruptors	Estradiol-3-benzoate (feminization of oysters)	Estradiol-17β Nonylphenol Tributyltin	Fluoxetine (spawning) Ibuprofen
	Tributyltin (masculinization of clams and gasteropods)		

have some predictive value regarding impaired reproduction, increased susceptibility to diseases and reduced life span (van der Oost et al., 2003). This implies that the effects measured ideally should be characterized in terms of reversibility versus irreversibility. Clearly, an irreversible effect is of greater consequence to the long-term survival of an organism. In addition to these criteria, the effects measured should be predictive or at least provide some information regarding population maintenance. For example, the elevation of metallothioneins in the gills of mussels held in lakes contaminated by mining activities was found to be related to changes at the morphological level, such as reduced growth and weight-to-length ratio (Couillard et al., 1995). These morphological end-points might be influential on population survival.

Bivalves are particularly at risk to contaminants because they are sessile and live for relatively long periods (up to 30 years, depending on the species). The present decline of freshwater mussel populations is believed to be multifactorial. Habitat destruction by siltation, dredging channeling and pollution are among the likely contributors to this decline. Domestic sewage and effluent from paper mills, tanneries, chemical plants, steel mills, acid mine runoff and agriculture all have been implicated in the destruction of mussel populations (Bogan, 1993). The purpose of this review is to highlight the use of biomarkers to assess the health status of bivalves in both freshwater and marine settings. The subject of DNA damage (genotoxicity) will not be discussed here because it is treated in another chapter of this book (Chapter 8). Some consideration will be given to the effects of pollutants on the reproductive status of these organisms. The integration of biomarkers on spatial and temporal surveys and the relevance of biomarkers to the assessment of population maintenance will also be discussed.

7.2 Biomarkers of exposure

Evaluation of exposure usually is performed by measuring the amount of contaminant(s) in the whole organism. This will provide a preliminary indication of the bioavailability of a group of contaminants with respect to the organism. Furthermore, the presence of adducts to proteins or DNA offers the opportunity to assess their respective affinity or reactivity towards biological molecules.

In bivalves, exposure to contaminants usually is performed on the organism as a whole. This approach has some limitations in that no information is provided regarding the location of the major proportion of the contaminant in tissues. Furthermore, the level of contaminants that accumulates in a specific target tissue could be diluted by other organs when assessing the burden carried by the entire body mass. For example, it was shown that mussels exposed for 62 days to a municipal plume accumulated less cadmium and had unchanged silver levels when the metal evaluation was conducted on the whole soft-tissue, but cadmium and silver levels were significantly higher in the gills (Gagné *et al.*, 2002b). Moreover, increased gill metallothionein levels were found in mussels from a site 4 km downstream whereas decreased metallothionein levels were found in the digestive gland. These results suggest that redistribution of contaminants in target tissues could lead to toxic stress, although no apparent changes are observed in the total amount of contaminants in the whole body.

7.3 Biomarkers of early biological effects and defence

Once a chemical enters the vicinity of the cell's environment, it is usually adsorbed to the cell surface or passes through the cell membrane by passive diffusion, pore channels or active transport (e.g. receptors). Although some chemicals could interact with specific targets, others will react with miscellaneous sites depending on their chemical reactivity. Indeed, not all toxicants act through specific targets to produce toxic effects. Some act through hormone receptors (i.e. nonylphenol) or acetylcholinesterase inhibitors (i.e. organophosphorus and carbamate pesticides) and others produce their effects by non-specific interactions to intracellular components. For example, current evidence suggests that binding of cadmium to high-molecular-weight components in the cell leads to cytotoxicity, and protection from cell impairment occurs when the metal is sequestered by metal-binding proteins such as metallothionein.

7.3.1 Defence mechanisms

Cells are equipped with a P-glycoprotein located at the cytoplasmic membrane that extrudes xenobiotics from the cell's intracellular environment. This transporter protein was first identified as a multidrug resistance pump in tumour cells that became resistant to anticancer drugs (van Tellingen, 2001). This pump represents a formidable defence mechanism by which cells could protect themselves, thereby excluding potentially cytotoxic xenobiotics. Environmental contaminants induce this multixenobiotic resistant (MXR) protein in *Mytilus edulis* (Eufemia and Epel, 2000) but other contaminants that could block this MXR transporter were found by Smital and Kurelec (1998) in sediment extracts at polluted sites, rendering the organism highly sensitive to other pollutants. They have shown that exposure of mussels to polluted rivers enhanced the accumulation of rhodamine B dye or a carcinogenic aromatic amine in gills, indicating that contaminants could act as chemosensitizers by enhancing the bioaccumulation of toxic chemicals. This was also shown in another study in which membrane vesicles treated with staurosporine (as MXR transporter blocker) enhanced single-strand DNA broke upon exposure to 2-acetylaminofluorene from the gills of the freshwater clam *Corbicula fluminea* (Waldmann et al., 1995). Organic anion-transporting polypeptides are another class of proteins involved in the extrusion of amphipathic organic solutes, such as steroid conjugates, numerous drugs and organic dyes (Bresler et al., 1999, Hagenbuch and Meier, 2003).

Both thermal- and chemical-induced cellular stress bring about the production of various methionine-rich proteins in molluscs and vertebrates (Veldhuizen-Tsoerkan et al., 1991a; Snyder et al., 2001). These proteins initially were identified as heat-shock proteins (induced by brief, non-lethal exposure to heat) but also were found to be induced by a variety of compounds, including heavy metals such as cadmium and hydrocarbons found in crude oil (Kammenga et al., 2000; Snyder et al., 2001). These proteins display defined molecular weights (23, 60, 70 and 90 kDa) and are supposed to be involved in the maintenance of protein folding, hence their role as molecular chaperones or chaperonins. Protein folding could change upon exposure to heat or chemicals. Indeed, mussels held at contaminated sites for 10 weeks were less likely to produce heat-shock proteins after stress treatments, i.e. aerial exposure and increased temperature (Veldhuizen-Tsoerkan et al., 1991b), indicating that tissues were more susceptible to protein denaturation from heat or chemical exposure.

Heavy metals are well known for inducing the synthesis of a family of metal-binding proteins called metallothioneins (Frazier, 1986). These low-molecular-weight proteins (6000 Da) contain 25–30% cysteine, enabling them to sequester divalent metals such as cadmium, copper, mercury and zinc from the Ib and

IIb groups of the Periodic Table. The sequestration of heavy metals by metallothionein prevents them from binding to high-molecular-weight proteins such as receptors, transporters and enzymes, which is thought to be the major detoxification mechanism. Metallothioneins also are involved in the mobilization (i.e. storage and transport) of essential metals such as zinc and copper in the intracellular environment, hence their role in metal homeostasis in cells. Levels of metallothionein in the bivalve *Corbicula fluminea* paralleled the depuration rates of cadmium and zinc levels, with estimated half-lives of 400 and 52 days, respectively (Baudrimont *et al.*, 2003). This low depuration rate for cadmium was shown in an earlier study with the green-lipped mussel *Perna viridis* (Yang *et al.*, 1995).

In oviparous organisms, metallothioneins were found to be correlated negatively with the synthesis of vitellogenin, the egg-yolk protein precursor that also binds zinc and perhaps other essential metals (Olsson and Kling, 1995; Blaise *et al.*, 2002b). Thus, metallothionein levels could be modulated during the reproductive cycle of organisms (i.e. gametogenesis) and special care should be considered when reporting metallothionein concentrations in temporal field studies. Metallothionein levels were also found to be correlated positively with organism size (Amiard-Triquet *et al.*, 1998; Cosson, 2000).

Metallothioneins are also induced by other factors, such as oxidative stress, heat shock (sometimes considered a heat-shock protein), inflammatory conditions (related to oxidative stress) and glucocorticoid hormones, which limit their specificity towards heavy metal exposure. Thus, tissues should be measured for both metallothioneins and metals in an attempt to relate metallothionein responses to heavy metal exposures. In fact, the identification of metals sequestered by metallothionein might provide additional information. For example, it was found that metallothioneins were saturated with tin in zebra mussels collected from a harbour contaminated by tributyltin (de Lafontaine *et al.*, 2000). In an attempt to define a biomarker of early biological effects more specific to heavy metals, it was proposed that increased labile zinc, an essential metal, could hold promise in this respect when measured in rainbow trout primary hepatocytes (Gagné and Blaise, 1996). The principle is that the entry of toxic metals in cells competes with the binding sites of essential metals such as zinc, thereby increasing the free form (i.e. bound only to low-molecular-weight ligands) of essential metals. Moreover, exposure of rat hepatocytes to the metallothionein-inducing glucocorticoid dexamethasone and the inflammatory mediator interleukin-3 had only a slight impact on the levels of labile zinc (Coyle *et al.*, 1994), suggesting specificity of the response. Microsomal heme oxygenase (another heat-shock protein, HSP 32) was shown to be induced readily in vertebrates (Sunderman, 1987) but this has yet to be demonstrated in bivalves, nor has it been demonstrated that its induction is specific to heavy metals. Heme prosthetic groups are ubiquitous in cells and some of them have been shown to have hydroxindole oxidase activity in mussels (Kampa and Peisach, 1980).

With respect to aromatic hydrocarbons, bivalves, are considered to have low cytochrome P450-dependent biotransformation capacity. Hence, they will be more susceptible to bioaccumulating non-polar organic compounds than eliminating them (Moore and Allen, 2002). Notwithstanding this low biotransformation capacity, benzo[*a*]pyrene (B[*a*]P) hydroxylase activity was found in microsomes of the digestive gland of the blue mussel *Mytilus galloprovincialis* (Akcha *et al.*, 1999). Again, B[*a*]P-like DNA adducts were correlated with whole mussel tissue B[*a*]P concentration. However, exposure of mussels to Arochlor 1254 and hexachlorobiphenyl leads to increased expression of CYP1A-immunopositive protein using hepatic CYP1A of perch, but no increase of CYP1A-like mRNA was observed using cDNA of rainbow trout CYP1A1 (Livingstone *et al.*, 1997). This observation could be explained by the increased enzyme turnover without increasing the total amount of enzyme in tissues and by the increased half-life of CYP1A mRNA.

Soft-tissue homogenate extracts of the digestive gland of *Dreissena polymorpha* were found to have increased 7-ethoxyresorufin *O*-deethylase activity after exposure to a municipal effluent plume for 62 days but no significant change in activity was detected in digestive gland homogenates of *Elliptio complanata* mussels (Gagné *et al.*, 2002b). Furthermore, digestive gland microsomes of *Mytilus galloprovincialis* from a polluted site showed higher levels of CYP1A, CYP2E and CYP4A, as determined by polyclonal antibodies raised against fish species (Peters *et al.*, 1998). Also, they have shown immunoreactivity with at least five cytochrome P450s (CYP1A, CYP2B, CYP2E, CYP3A and CYP4A) in digestive gland microsomes. Phenobarbital is a known inducer of CYP2B cytochromes. Many pharmaceutical compounds, such as carbamazepine and loperamide (discharged in municipal wastewater), are known to be metabolized by CYP3A4. This isoform also possesses testosterone 6β-hydroxylase activity, indicating that these drugs (aromatic amine compounds) could alter the levels of sexual steroids in bivalves. It appears that mussels are equipped with cytochrome P450-dependent monooxygenase enzymes but their induced expression is difficult to demonstrate.

Glutathione *S*-transferase (GST), a noteworthy phase II conjugating enzyme family, is readily found in gill and digestive-gland tissues (Akcha *et al.*, 2000). It catalyzes the conjugation of reduced glutathione to an electrophilic site of the contaminant, thus increasing its hydrophilicity for elimination. Induction of GST is not specific to a particular chemical class but exposure to 100 μg/l furadan for 96 h failed to induce the activity of gill GST in the mussel *Perna perna* and the mangrove oyster *Crassostrea rhizophorae* (Alves *et al.*, 2002). In another study, the activity of GST in gills and hepatopancreas was found to be correlated positively with the levels of chlorinated hydrocarbons but not with chlorinated pesticides in the marine mussel *Perna viridis* (Cheung *et al.*, 2002). Activity of GST in the hepatopancreas appears to be induced readily by high-molecular-weight polyaromatic hydrocarbons (PAHs) (e.g. 5–6 PAHs-ring compounds such as B[*a*]P) but not by low-molecular-weight PAHs (e.g. naphthalene) in

M. edulis mussels (Gowlan *et al.*, 2002). Thus, GST activity correlates with 5–6-ring PAHs rather than with the total burden of PAHs in tissues.

7.4 Biomarkers of damage

This class of biomarkers measures actual toxic effects at the molecular, cellular and physiological levels. It is more closely related to pathological conditions (morbidity) that could lead to death. The oxidative status of cells has been proposed as a universal mechanism leading to cell dysfunction. The normal metabolism (e.g. β-oxidation, pentose shunt pathway, immune function) produces oxidizing precursors such as hydrogen peroxide (H_2O_2), nitric oxide (NO) and peroxynitrite (ONOO) that are rapidly eliminated by non-enzymatic and enzymatic antioxidants to prevent tissue damage.

Hydrogen peroxide is readily eliminated by non-enzymatic antioxidants (such as vitamins A, C and E and reduced glutathione) and by enzymatic pathways. These include catalase, superoxide dismutase, glutathione peroxidase and reductase, which are involved in the elimination of hydroxyl radicals in tissues. This delicate balance could be altered by stress, such as chemical exposure, inflammatory responses and physical agents (heat, radiation). Increased production of these oxidants leads to injury, such as DNA damage (including mutation), lipid peroxidation (LPO), inflammation and tissue necrosis (Canova *et al.*, 1998; Livingstone, 2001; Cheung *et al.*, 2002). Mussels collected from an area contaminated by industrial effluents were shown to have increased lipid peroxidation in tissues (Cossu *et al.*, 2000). They also found that lipid peroxidation was inversely proportional to the activity of antioxidants, such as reduced glutathione, glutathione reductase and peroxidase.

For its part, NO is related to neuro-immune status in bivalves. During phagocytosis, NO is released in hemocytes and combined with superoxide anion from H_2O_2 ($\cdot$OH) to yield peroxynitrite in an attempt to destroy biological agents (e.g. yeasts, viruses and bacteria) (Arumugam *et al.*, 2000). NO is produced by calcium-dependent NO synthase, which is controlled, in part at least, by opioid receptors in microglia (Liu *et al.*, 1996). The general effects of opioids are immuno-inhibitory rather than immuno-stimulatory in hemocytes. Although opioid peptides do not appear to have any influence on NO production, exposure to morphine leads to increased production of NO in hemocytes. Opioid peptides did not stimulate NO release, indicating that the process is controlled by the μ_3-receptor. In the context of the release of pharmaceuticals in the environment by municipal and hospital effluents, the presence of morphine-related compounds is likely to impede these processes. *Mya arenaria* clams, collected at sites close to an urban area, were shown to have increased phagocytosis activity, suggesting that immunostimulating effects could be observed in the field (Blaise *et al.*, 2002b). Of course, it remains to be confirmed whether opioid drugs are a causative factor for the observed effects in these bivalves.

7.5 Biomarkers of reproduction

Significant breakthroughs in knowledge related to bivalve reproduction have been documented in the literature in recent years. Bivalve reproduction consists of many critical steps, beginning in nerve centres and ending in gonad tissues. These steps are sexual development, gametogenesis, fertilization and spawning (Table 7.2). On the whole, sexual differentiation processes are not completely understood in bivalves but some aspects are gaining a better understanding. Serotonin, dopamine and sexual steroids are involved in the sexual differentiation process (Martinez and Rivera, 1994; Croll *et al.*, 1995). In the *Tilapia* brain, serotonin and estradiol are inversely related in females (Tsai *et al.*, 2000). Exposure of nerve tissues to estradiol-17β or to a serotonin synthesis inhibitor was equally effective in increasing the female-to-male ratio in young broods. In the oyster, injection of estradiol-3-benzoate during the undifferentiated stage significantly increased the number of females (Mori *et al.*, 1969). Moreover, the female part of the gonad in the hermaphrodite clam, *Argopecten purpuratus*, contains less serotonin than its male counterpart (Martinez and Rivera, 1994).

Serotonin levels are regulated by monoamine oxidase (MAO), which is the main elimination pathway for monoamines such as dopamine, serotonin, octopamine and noradrenaline. The MAO activity could be induced by a variety of secondary amines in the environment and could likely modulate serotonin levels in nerve tissues and perhaps sex differentiation. For example, MAO activity in the nerve ganglia and gonad was shown to be induced with a concomitant decrease in serotonin and dopamine in mussels exposed for 90 days, 10 km downstream from a primary-treated municipal effluent plume (Gagné and Blaise, 2003). In contrast, scallop MAO activity was shown to be repressed by pharmaceuticals such as the type A and B inhibitors deprenyl, pargyline and clorgyline (Pani and Croll, 1998).

The formation of male sexual organs in gastropods seems to be controlled by the release of the APGW neuropeptide (Ala-Pro-Gly-Trp-NH$_2$) in neurons that innervate gonad tissues. However, this neuropeptide was found also in the central nervous system of the deep sea scallop *P. magellanicus* (Smith *et al.*, 1997). It was present in both juveniles and adults and immunoreactivity was found in ganglia of both sexes. Hence, the exact physiological role of this neuropeptide and other peptides with respect to the gametogenesis of the male reproductive tract in bivalves remains to be fully understood.

Gametogenesis could be followed by biomarkers indicative of gamete activity. During this process, significant amounts of pyrimidine and purine are required to produce sperm and oocytes. Aspartate transcarbamoylase is the rate-limiting enzyme of pyrimidine biosynthesis (Mathieu *et al.*, 1982). The activity of this rate-limiting enzyme was found to be correlated with the gonadosomatic index in the blue mussel *M. edulis*. Moreover, the stimulatory action of cerebral ganglia co-cultivation with mantle tissues is associated with increases of this enzyme activity

(Mathieu, 1987). The RNA/DNA ratio also is readily increased during gamete maturation because the amount of RNA is increased to assist protein synthesis. Dopamine levels also rise during this stage and fall just before spawning, when there is a concomitant rise in serotonin (Osada and Nomura, 1989).

Vitellogenin, the egg-yolk precursor, is regulated by the estrogen receptor in oviparous organisms. It is a high-molecular-weight (in the range 400–600 kDa) glycophospholipoprotein that provides energy for developing embryos. Although vitellogenin is produced in the liver of vertebrates, current evidence indicates that this process occurs in gonad tissues in invertebrates. Under natural conditions, oocytes produce estradiol-17β (E_2) to initiate vitellogenesis, leading to the production of large amounts (ng to mg) of vitellogenin for the developing embryos. During this phase, dopamine levels are steadily increased (Osada and Nomura, 1989). Indeed, a significant increase of dopamine was produced 22 h after the gonad was exposed to E_2, indicating that E_2 is involved in the regulation of this monoamine in the gonad. Other estrogens, such as *p*-nonylphenol, may also activate vitellogenesis and vitellogenin production in males is used as a biomarker to detect the early biological effects of estrogens (Sumpter and Jobling, 1995).

Vitellogenin synthesis in invertebrates (bivalves) seems to occur in the vesicular connective tissue in female gonads, from where it is transported to the oocytes by pinocytosis. Final metabolic transformations are thought to occur in the oocytes (Eckelbarger and Davis, 1996). Thus, vitellogenin production occurs through heterosynthetic (i.e. vesicular connective tissues) and autosynthetic (i.e. oocytes) pathways (Pipe, 1987; Eckelbarger and Davis, 1996). This biosynthetic process was shown to be regulated by estrogens in the oyster *Crassostrea gigas* (Li *et al.*, 1998) and in the clam *Mya arenaria* (Blaise *et al.*, 1999; Gagné *et al.*, 2002a), such that injection of estradiol leads to increased levels of vitellogenin in the gonad. Moreover, the presence of specific and saturable E_2 binding (receptors) in gonad tissues was demonstrated for the first time in *Elliption complanata* mussels (Gagné *et al.*, 2001). Shellfish appear to produce sexual steroids (i.e. estrogens and androgens), which play a major role in the reproductive cycle of these organisms (Delongcamp *et al.*, 1974). Because bivalves are immobile, they may be at risk particularly when exposed continuously to sources of estrogen, such as urban effluents. Indeed, exposure to xenoestrogens throughout their sexual development stages and reproductive cycles could decimate the population by augmenting feminization.

Spawning is usually induced by serotonin or potassium and prepares the oocyte for fertilization. In addition, increased dopamine levels have been measured after injections of E_2 in the sea scallop, but they dropped during the active spawning period (Osada and Nomura, 1989). Moreover, dopamine was shown to inhibit spawning activity in serotonin-treated *D. polymorpha* mussels (Fong *et al.*, 1993), indicating that spawning activity is stimulated by serotonin but negatively controlled by dopamine (i.e. dopamine is linked to gametogenesis rather than spawning and fertilization). In another study, dopamine and noradrenaline levels were low at the spawning stage in bivalve molluscs (Osada *et al.*, 1987). Treatment of gonad tissues with 0.1–1 μM serotonin was found to stimulate egg release in the scallop ovaries (Osada *et al.*, 1992). Moreover, egg release calls for the production of prostaglandins for smooth muscle contraction to assist egg expulsion. Prostaglandins are produced by prostaglandin synthase (or arachidonate cyclooxygenase) and its activity can be used to monitor the spawning process. Interestingly, the cyclooxygenase inhibitor acetylsalicylic acid (aspirin) inhibited the accelerating effect of E_2 on serotonin-induced egg release. In another study with tricyclic antidepressants (i.e. serotonin reuptake inhibitors such as fluoxetine), these compounds were found to reduce spawning activity and fertilization of eggs in zebra mussels (Hardedge *et al.*, 1997).

At the completion of oocyte maturation, oocytes are maintained at the germinal vesicle stage (i.e. prophase I) in the ovary (Fong *et al.*, 1994). Fertilization leads to germinal vesicle breakdown, which involves many mediators, such as serotonin, calcium/calmodulin, cAMP messenger systems, potassium uptake, and the activation of protein kinases for protein phosphorylation and inositol triphosphates. Maturation was induced in *Spisula* oocytes by 5 μM of serotonin (Haneji and Koide, 1988). After 10 and 30 min of incubation, significant incorporation of radioactive phosphates (^{32}P) in proteins from ^{32}P-labelled ATP and GTP was observed on sodium dodecyl sulphate polyacrylamide gel electrophoresis (SDS-PAGE) just before germinal vesicle breakdown. Serotonin also initiates germinal vesicle breakdown and oocytes, to be arrested at metaphase I to await fertilization (Moreau *et al.*, 1996). Serotonin is well known to participate in the maturation of meiose-arrested oocytes. Inhibition of serotonin reuptake by antidepressive drugs (selective serotonin uptake inhibitors) in oocytes could reduce maturation and perhaps fertility (Juneja *et al.*, 1993). Thimerosal, an organic mercury complex used as an antiseptic in many pharmaceutical preparations, was found to have similar serotonergic effects toward germinal vesicle breakdown in oocytes of the clam *Ruditapes philippinarum* (Lippai *et al.*, 1995). Moreau *et al.* (1996) demonstrated that release from metaphase I block is produced by sperm or excess KCl and that fertilization requires functional calcium channels. These studies suggest that environmental contaminants are likely to act at each stage of reproduction.

7.6 Integrating effects

Biomarkers are often used in test batteries to evaluate the effects of exposure to multiple sources of contaminants and to detect responses to various sources of pollution, such as harbours, miscellaneous industrial sites and municipal and hospital wastewaters. Field studies with biomarkers are often plagued by various constraints, such as spatial variation (e.g. change in habitat characteristics), temporal variation (e.g. cycle of reproduction) and availability of organisms that can hamper data acquisition and prevent the use of multivariate methods during statistical analysis. The full potential of biomarker test batteries in monitoring approaches based on effects is often limited by a lack of integrative statistical models over spatial and temporal settings. Only a handful of approaches have been proposed to date, some of which are discussed below.

Biomarker data sets can be plotted on star plots, where the intensity of the responses is directly proportional to the surface area at each site (Figure 7.1). An integrated biomarker response (IBR) can then be calculated simply by adding the surface area for each biomarker response (Beliaeff and Burgeot, 2002). First, this approach calls for estimation of the mean value of the population at each site, where available, or a generic value obtained from a pooled sample of organisms. After standardization of the mean estimates for each site with respect to the mean and standard deviation for all the stations, the standardized mean estimates are scored by adding the minimum value of all the stations. The scored values are then plotted according to their relative area in star plots available in common spreadsheet software, such as Excel (Figure 7.1). Then, the IBR index is obtained as the sum of areas for each biomarker. This approach offers the ability to plot visually a multivariate data set and identify the most affected sites in a spatial study. Moreover, the IBR is based on the sum of the responses observed in organisms and not on mean values. However, this approach still depends on the data distribution because it requires sound mean estimates. The IBR is also practical for observing changes over time by following changes in the surface area of the IBR with respect to sites but natural temporal variation is not discriminated with this approach. This limitation is understandable because no assumption of site quality is factored into the model. In other words, no assumptions are made to identify a reference site or a site supposed to be contaminated only slightly.

A classification scale also based on biochemical markers has been proposed (Narbonne *et al.*, 1999). In this approach, each biomarker response is scaled according to the discriminant factor, defined as the difference between the lowest and highest value of the mean at the site plus the confidence interval divided by the confidence interval of the data set. For example, if the difference between the highest and lowest mean is 15 units for metallothionein and the confidence interval of the data set is 5, then $(15 + 5)/5 = 4$ classes could be generated to scale the metallothionein responses. Then, each biomarker mean is assigned a class value depending on its position with respect to the total number of classes. This approach is applied to each biomarker and the corresponding classes at each site are added, resulting in an integrated biomarker response index. This approach requires parametric data because the classification scale is produced through discriminant factor definition. Thus, this approach is relevant to spatial studies when good estimates of the central value (i.e. good replication is needed) are available at each site for each biomarker.

When the data are non-parametric, even after transformation (e.g. log transformation), a ranked approach can be used to calculate an index (Blaise *et al.*,

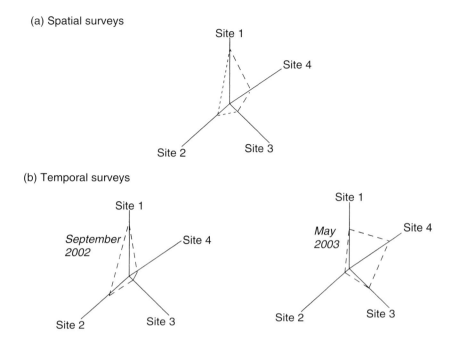

Figure 7.1 Integrative biomarker index as star plots. The integrated biomarker response is obtained from the sum of areas of each biomarker in the test battery. For the spatial survey, it can be observed that sites 1 and 4 are more stressed or affected than the other sites. The general response of biomarkers can be appraised in temporal settings where changes are observed between sites over time.

2002a). The mean value for each site is assigned a rank value with increasing order (i.e. the lower mean corresponds to rank 1 and the highest mean value corresponds to the highest rank value), such that the highest rank number corresponds to the number of sampling sites. Then, each site is tested with a rank-based multicomparison test (e.g. Mann–Whitney U test) to confirm a statistical difference between each site. If the difference is not statistically significant then both sites are assigned the same rank value. Conversely, if the difference is statistically different then rank values of 1 and 2 are assigned accordingly (i.e. rank 2 corresponds to a significantly higher value than rank 1). The process is repeated between rank 2 and rank 3 sites and so forth. Thus, each rank value assigned to a site corresponds to a statistical difference as determined by a non-parametric statistical test. The rank value for each biomarker is then added to yield a biomarker index.

In field studies, data are not always normally distributed or parametric, even with relatively high replication ($N = 10$–16 individuals per site) and the use of data transformation techniques (e.g. logarithmic, reciprocal or arcsine). This could limit multivariate analyses (such as discriminant and principal component analyses) in an attempt to discriminate the most affected sites and the most important or relevant biomarkers. Distribution-free multivariate analysis is possible using rule-based approaches, such as rough set analysis and classification trees (Chèvre et al., 2003b). Rule-based approaches have the advantage of complete independence from data distribution constraints. These approaches are explained in detail in the above reference. With respect to discriminant analysis, rough set analysis and classification trees provided better classifications from a data set having some non-parametric distributions, even after log transformation. Moreover, rough set analysis, like discriminant and principal component analyses, provided classification rules that could identify the most important biomarkers for site identification. Furthermore, rough set analysis offers the possibility of discretization of continuous data into classes. The discretization procedure used by the ROSE2 software (Fayyad and Irani, 1993) is a recursive algorithm that minimizes entropy, i.e. the amount of information necessary to form a class.

Integrating biomarkers in temporal surveys presents additional difficulties compared with spatial surveys, where data are obtained in a given time frame. Indeed, biomarker responses vary throughout the year due to 'natural' variables such as the reproductive state of the organism, sexual differentiation, temperature, nutrient status and habitat modification. The natural temporal variation must be characterized in some way to highlight the effects of pollution, or else confusing results might be obtained. For example, in a molluscan biomarker study performed in littoral regions of the Red, Mediterranean and North Seas, although some biomarkers responded readily at affected sites (e.g. enhanced frequency of DNA lesions, increased expression of multixenobiotic resistance pump) the IBR of Narbonne et al. (1999) failed to show any difference at the various sampling sites (Bresler et al., 1999). The IBR's lack of response may have been attributable to the sampling interval between each site. Indeed, the IBR at the reference site had the lowest IBR value in July 1996 but the highest value in April 1996, indicating that the biomarker responses changed significantly over time. The natural (background) variation of the biomarkers is usually determined by following the various responses over time at pristine sites and comparing them with sites where affects are suspected (Chèvre et al., 2003a).

First, a biomarker-based index is obtained by adding the biomarker responses at each time of sampling at a given site after transformation into classes by a distribution-free approach as explained above (rough set analysis). The biomarker index is then plotted in a control chart format to highlight changes at sites over space and time (Figure 7.2). Because field data are often non-parametric, the median and the 10th, 25th, 75th and 90th centiles were plotted to portray visually the change in the biomarker index outside natural temporal and spatial variations. This approach is useful if the reference site is well identified. In the case of uncertainty or absence of the reference site, the data could be plotted in the same manner, with all sites confounded and the median value of the data set in

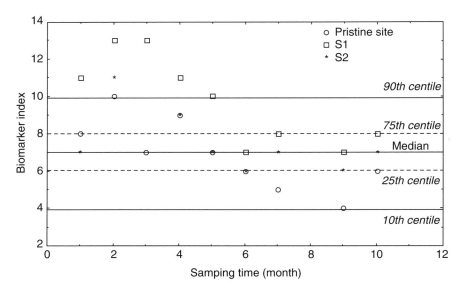

Figure 7.2 The use of a biomarker index in temporal and spatial studies. In this survey, three sites were sampled over ten different time periods. In order to sort out natural variation in time, a control chart approach is used. The median value of the pristine site is plotted with the 10th, 25th, 75th and 90th centiles. Biomarker index values outside the 90th and 10th centiles are most likely to be outside natural variability and could be associated with pollution factors.

Figure 7.2 increased to 8, and with sites outside the first and third quartiles (i.e. 25 and 75%) considered to be affected. In the temporal data, the index was able to identify possible contamination-induced (especially those values outside the 10th and 90th centiles) effects even though the interpretation of temporal variation is complicated by natural variations occurring throughout the year.

7.7 Linking biomarkers at higher levels of biological organization

One of the issues in the use of biomarkers in ecotoxicology is their ecological relevance to population and community structure. Although biomarkers are designed to assess individual health status and reproductive capacity, two issues that differ completely from population maintenance or integrity, they could be used effectively if contamination is the main cause of population loss. Clearly, pollution is not the only cause of population perturbations. Change in habitat characteristics, availability of nutrients, human encroachment (housing and industries) and global environmental changes (e.g. water quantity, global warming, depletion of the ozone layer) are also likely to impede the long-term survival of feral populations. Biomarkers of early biological effects and damage provide information about the susceptibility of organisms to the quality of a given habitat. A change in one or many habitat characteristics could render faunic species more or less susceptible to contaminants without any discernible change in water quality. At this time, bivalve population status is fragile throughout the world and many species are threatened (Bogan, 1993).

Relatively few studies have been performed with bivalves in an attempt to relate effects observed at the individual level with those likely to influence the maintenance of populations. After placing mussels in a lake contaminated by mine tailings for 400 days, gills exhibited increased lipid peroxidation (as determined by thiobarbituric acid reactants) and Ca levels (Couillard et al., 1995). These effects coincided with the appearance of low-molecular-weight Cd (metallothionein) in gills. Mussel growth rate and condition factors were significantly reduced compared with mussels held in a control lake, indicating that the population status could be compromised. In another study, the cytochemical index based on the latency of lysosomal enzymes was found to correlate with growth rates in transplanted blue mussels (Bayne et al., 1979). In a model stream approach for the evaluation of 25(26)-alcohol ethoxylates, species richness and abundance of selected populations such as the mayfly, caddisfly and the Asiatic clam were the most affected (Belanger et al., 2000), indicating that bivalves are good sentinels of benthic community structure. The responses observed with caged mussels in rivers contaminated by the effluent from a bleached kraft pulp and paper mill were more consistent with those observed in the benthic invertebrate survey than with those observed for fish (Martel et al., 2003). They also showed that caged mussels grew more slowly at sites downstream from the plume than they did at upstream sites. In a study with fathead minnows, long-term exposure to ethinylestradiol (the active ingredient in oral contraceptives) was shown to feminize fish (and induce vitellogenin in males) at a threshold concentration of 2 ng/l (Grist et al., 2003). They also found that population growth was reduced at this level of contamination (3.1–3.4 ng/l) and that this effect was more related to reduced fertility than survival rates. Long-term exposure of mussels and barnacles to pollution could significantly reduce genetic diversity, as determined by a randomly amplified polymorphic DNA-polymerase chain reaction (Ma et al., 2000). Moreover, individuals collected at affected sites were more likely to share the same haplotypes than those from clean sites, indicating reduced genetic diversity in bivalves at contaminated sites.

Some trends were observed between biomarker responses and population status in *Mya arenaria* clams from the Saguenay Fjord, Quebec, Canada (Blaise et al., 2003). Clams exhibiting marked responses with metallothionein, lipid peroxidation, vitellogenin-like proteins and gonadosomatic index were found at affected sites where clams generally displayed changes in the development index (i.e. shell length-to-age ratio) and clam density. In contrast, no link could be found between the occurrence of dead animals (empty shells) and site quality. These results suggest that biomarkers of defence, damage and reproduction are statistically associated with changes in population status and that they could be used effectively to monitor evolution of field situations (i.e. restoration or degradation). Correlation analysis of biomarker data and population metrics revealed that the development index was correlated negatively with lipid peroxidation in the digestive gland ($R = -0.51$, $P < 0.05$), metallothionein in the digestive gland ($R = -0.49$, $P < 0.05$), gonadosomatic index ($R = -0.36$, $P < 0.05$) and vitellogenin-like proteins ($R = -0.4$, $P < 0.05$). Moreover, the condition index (total clam weight/shell length) was also correlated negatively with lipid peroxidation in the digestive gland ($R = -0.43$, $P < 0.05$). Some biochemical targets appear to have more relevance to detecting changes at the population level in zebra mussels (Smolders *et al.*, 2004) and *Daphnia magna* (de Coen and Janssen, 2003). Biomarkers related to change in energy reserves (total lipids, carbohydrates and proteins) and cellular respiration rates (electron transport activity) were the most relevant relative to population effects. Indeed, these biomarkers were all correlated with chronic (21 days) threshold effects values (i.e. EC_{10}) based on growth, survival and reproduction of *Daphnia magna (see also Chapter 11)*.

Hence, biomarkers could be used as predictive tools for ecosystem quality in addition to their simple use as indices of exposure to specific pollutants (Vasseur and Cossu-Leguille, 2003). Biomarkers and ecological investigations could be combined together to diagnose ecosystem quality. For example, documented effects with biomarkers together with observed changes in population status would certainly signify a perturbed environment, whereas the absence of these two levels of response would indicate good ecosystem quality. The presence of effects at either level of investigation would indicate a likely fragile and unstable system. Again, the absence of effects at the biomarker level would indicate that pollution is not a contributing factor in the decline of populations or loss of biodiversity.

7.8 Emerging issues

Endocrine disrupters have been studied extensively in fish. A smaller yet increasing number of studies have been conducted recently with bivalves and other benthic invertebrates. Environmental estrogens are by far the most studied endocrine disrupter in aquatic wildlife. However, reproductive success depends not only on estrogens but on other hormonal systems as well, such as gonado-stimulating neurohormones, monoamines (catecholamines and serotonin) and neuropeptides (e.g. the APGW neuropeptides). For example, exogenous opioid peptides were reported to increase dopamine levels in a naloxone-reversible manner in pedal ganglia of the marine mollusc *M. edulis* and this organism was shown to contain specific opiate-binding sites in invertebrates (Kream *et al.*, 1980). Moreover, the presence of μ-type opiate receptors and morphine is expressed in mollusc ganglia and appears to be involved in physiological processes involving adaptation to thermal stress (Cadet *et al.*, 2002). Exposure of mussel populations to a primary-treated municipal effluent plume induces MAO activity with a concomitant reduction in serotonin and dopamine in gonad tissues (Gagné and Blaise, 2003). These studies indicate that environmental pollutants in urban wastewaters are likely to affect other neurohormonal systems in bivalves.

The advent of biotechnology products, such as genetically modified crops, has raised a new issue: biological pollution due to new (micro)organisms and/or their gene or protein products being discharged into aquatic environments. The creation of genetically modified organisms not normally found in nature could constitute a potentially new input of alien species/products with an, as yet, unknown impact on biota. Indeed, genetically modified organisms produce gene expression products not normally found in the wild type, thereby altering the routes of exposure of these products in the environment. For example, corn was modified to express Cry1Ab endotoxin, which is normally found in the soil bacterium *Bacillus thuringiensis* (Bt) in the form of water-insoluble crystals. This endotoxin specifically acts through insects from the *Lepidoptera* and *Diptera* families. The endotoxin expressed in corn is already in its activated more-soluble form in water so that mass cultivation of Bt-corn could release this biological contaminant in soils, sediments and waterways. The increasing use of biotechnology products such as Bt-corn could contribute to an increase of the endotoxin in the environment and could have adverse consequences for non-target organisms. For example, Cry1Ab endotoxin was shown to be toxic to *Chironomus riparius* midge and the monarch butterfly (Kondo *et al.*, 1995; Losey *et al.*, 1999). It is anticipated that cultivation of crops derived from biotechnology is likely to increase steadily in the next 20 years, reaching markets in the billions of US dollars. Thus, the environment is likely to be 'polluted' increasingly by biological contaminants, although possibly less so by chemical pesticides, and these invasive new species will produce miscellaneous by-products, some of which may have unsuspected effects on wildlife.

Recent studies have shown that our surface waters, including drinking water sources, are likely to be contaminated by miscellaneous pharmaceutical, veterinary and personal care products (Halling-Sorensen *et al.*, 1998; Kolpin *et al.*, 2002). These studies have shown that over-the-counter products (e.g. acetylsalicylic acid, ibuprofen, triclosan) and prescription drugs (clofibrates, fluoxetine, carbamazepine) are likely to be found in surface waters near municipal/urban discharges. Moreover, a recent quantitative structure–activity relationship study indicated that these drugs are not likely to have acute effects on fish, daphnids or algae, but long-term effects on reproduction, behaviour, metabolism and bacterial resistance are still uncertain (Sanderson *et al.*, 2003). These drugs may have unsuspected toxicological effects, and biomarkers of defence and effects for these drugs are presently wanting. As discussed above, tricyclic antidepressants have the potential to block the spawning and fertilization of oocytes (Hardedge *et al.*, 1997). Non-steroidal anti-inflammatory drugs such as ibuprofen and aspirin give rise to reduced prostaglandin synthesis by inhibiting cyclooxygenases. However, some of these prostaglandins (PGE_2) assist spawning activity by contracting smooth muscles (Matsutami and Nomura, 1987). Recent advances in proteomic research, which calls for two-dimensional gel electrophoresis, holds promise for identifying novel biomarkers at the protein expression level to detect drug-related effects, because these were designed to have specific biochemical properties. Proteomics also could help to assess the early biological effects of biotechnology products.

References

Akcha, F., Burgeot, T., Narbonne, J.F. and Venier, P. (1999) Relationship between kinetics of benzo(*a*)pyrene bioaccumulation and DNA binding in the mussel *Mytilus galloprovincialis*. *Bull. Environ. Contam. Toxicol.*, **62**, 455–462.

Akcha, F., Izuel, C., Venier, P., Budzinski, H., Burgeot, T. and Narbonne, J. (2000) Enzymatic biomarker measurement and study of DNA adduct formation un benzo(a)pyrene treated mussels. *Aquat. Toxicol.*, **49**, 269–287.

Alves, S.R., Severino, P.C., Ibbotson, D.P., da Silva, A.Z., Lopes, F.R., Saenz, L.A. and Bainy, A.C. (2002) Effects of furadan in the brown mussel *Perna perna* and in the mangrove oyster *Crassostrea rhizophorae*. *Mar. Environ. Res.*, **54**, 241–245.

Amiard-Triquet, C., Rainglet, F., Larroux, C., Regoli, F. and Hummel, H. (1998) Metallothioneins in Arctic bivalves. *Ecotoxicol. Environ. Saf.*, **41**, 96–102.

Arumugam, M., Romestand, B., Torreilles, J. and Roch, P. (2000) *In vitro* production of superoxide and nitric oxide (as nitrite and nitrate) by *Mytilus galloprovincialis* haemocytes upon incubation with PMA or laminarin or during yeast phagocytosis. *Eur. J. Cell Biol.*, **79**, 513–519.

Bayne, B.L., Moore, M.N., Widdows, J., Livingstone, D.R. and Salkeld, P. (1979) Measurement of the responses of individuals to environmental stress and pollution: studies with bivalve molluscs. *Philos. Trans. R. Soc. London. B Biol. Sci.*, **286**, 563–581.

Baudrimont, M., Andres, S., Durrieu, G. and Boudou, A. (2003) The key role of metallothioneins in the bivalve *Corbicula fluminea* during the depuration phase, after *in situ* exposure to Cd and Zn. *Aquat. Toxicol.*, **63**, 89–102.

Belanger, S.E., Guckert, J.B., Bowling, J.W., Begley, W.M., Davidson, D.H., LeBlanc, E.M. and Lee, D.M. (2000) Responses of aquatic communities to 25–6 alcohol ethoxylate in model stream ecosystems. *Aquat. Toxicol.*, **48**, 135–150.

Beliaeff, B. and Burgeot, T. (2002) Integrated biomarker response: a useful tool for ecological risk assessment. *Environ. Toxicol. Chem.*, **21**, 1316–1322.

Blaise, C., Gagné, F., Pellerin, J. and Hansen, P.-D. (1999) Determination of vitellogenin-like properties in *Mya arenaria* hemolymph (Saguenay Fjord, Canada): a potential biomarker for endocrine disruption. *Environ. Toxicol.*, **14**, 455–465.

Blaise, C., Trottier, S., Gagné, F., Lallement, C. and Hansen, P.-D. (2002a) Immunocompetence of bivalve hemocytes evaluated by a miniaturized phagocytosis assay. *Environ. Toxicol.*, **17**, 160–169.

Blaise, C., Gagné, F., Pellerin, J., Hansen, P.-D. and Trottier, S. (2002b) Molluscan shellfish biomarker study of the Quebec, Canada, Saguenay fjord with the softshell clam *Mya arenaria*. *Environ. Toxicol.*, **17**, 170–186.

Blaise, C., Gagné, F. and Pellerin, J. (2003) Bivalve population status and biomarker responses in *Mya arenaria* clams (Saguenay Fjord, Québec, Canada). *Fresenius. Environ. Bull.*, **12**, 956–960.

Bogan, A.M. (1993) Freshwater bivalve extinctions (Mollusca: Unionoida): a search for causes. *Am. Zool.*, **33**, 599–609.

Bresler, V., Bissinger, V., Abelson, A., Dizer, H., Sturm, A., Kratke, R., Fishelson, L., and Hansen, P.-D. (1999) Marine molluscs and fish as biomarkers of pollution stress in littoral regions of the Red Sea, Mediterranean Sea and North Sea. *Helgol Mar. Res.*, **53**, 219–243.

Cadet, P., Zhu, W., Mantione, K.J., Baggerman, G. and Stefano, G.B. (2002) Cold stress alters *Mytilus edulis* pedal ganglia expression of μ-opiate receptor transcripts determined by real-time RT-PCR and morphine levels. *Mol. Brain Res.*, **99**, 26–33.

Canova, S., Degan, P., Peters, L.D., Livingstone, D.R., Voltan, R. and Venier, P. (1998) Tissue dose, DNA adducts, oxidative DNA damage and CYP1A-immunopositive proteins in mussels exposed to waterborne benzo[*a*]pyrene. *Mutat. Res.*, **399**, 17–30.

Cheung, C.C., Zheng, G.J., Lam, P.K. and Richardson, B.J. (2002) Relationships between tissue concentrations of chlorinated hydrocarbons (polychlorinated biphenyls and chlorinated pesticides) and antioxidative responses of marine mussels, *Perna viridis. Mar. Pollut. Bull.*, **45**, 181–191.

Chèvre, N., Gagné, F. and Blaise, C. (2003a) Development of a biomarker-based index for assessing the ecotoxic potential of polluted aquatic sites. *Biomarkers*, 8, 287–298.

Chèvre, N., Gagné, F., Gagnon, P. and Blaise, C. (2003b) Application of rough sets analysis to identify polluted aquatic sites based on a battery of biomarkers: a comparison with classical methods. *Chemosphere*, **51**, 13–23.

Cosson, R.P. (2000) Bivalve metallothionein as a biomarker of aquatic ecosystem pollution by trace metals: limits and perspectives. *Cell. Mol. Biol.*, **46**, 295–309.

Cossu, C., Doyotte, A., Babut, M., Exinger, A. and Vasseur, P. (2000) Antioxidant biomarkers in freshwater bivalves, *Unio tumidus*, in response to different contamination profiles of aquatic sediments. *Ecotoxicol. Environ. Saf.*, **45**, 106–121.

Couillard, Y., Campbell, P.G.C., Pellerin-Massicotte, J. and Auclair, J.C. (1995) Field transplantation of a freshwater bivalve, *Pyganodon grandis*, across a metal contamination gradient. II. Metallothionein response to Cd and Zn exposure, and links to effects at higher levels of biological organisation. *Can. J. Fish. Aquat. Sci.*, **52**, 703–715.

Coyle, P., Zalewski, P.D., Philcox, J.C., Forbes, I.J., Ward, A.D., Lincoln, S.F., Mahadevan, I. and Rofe, A.M. (1994) Measurement of zinc in hepatocytes by using a fluorescent probe, Zinquin: relationship to metallothionein and intracellular zinc. *Biochem. J.*, **303**, 781–786.

Croll, R.P., Too, C.K., Pani, A.K. and Nason, J. (1995) Distribution of serotonin in the sea scallop *Placopecten magellanicus. Invert. Reprod. Dev.*, **28**, 125–135.

de Coen, W. and Janssen, C.R. (2003) The missing biomarker link: relationships between effects on the cellular energy allocation biomarker of toxicant-stressed *Daphnia magna* and corresponding population characteristics. *Environ. Toxicol. Chem.*, 22, 1632–1641.

de Lafontaine, Y., Gagné, F., Blaise, C., Costan, G., Gagnon, P. and Chan, H.M. (2000) Biomarkers in zebra mussels (*Dreissena polymorpha*) for the assessment and monitoring of water quality of the St Lawrence River (Canada). *Aquat. Toxicol.*, **50**, 51–71.

Delongcamp, D., Lubet, P. and Drosdowsky, M. (1974) The *in vitro* biosynthesis of steroids by the gonad of the mussel (*Mytilus edulis*). *Gen. Comp. Endocrinol.*, 22, 116–127.

Eckelbarger, K.J. and Davis, C.V. (1996) Ultrastructure of the gonad and gametogenesis in the eastern oyster, *Crassostrea virginica*. I. Ovary and oogenesis. *Mar. Biol.*, **127**, 79–87.

Eufemia, N.A. and Epel, D. (2000) Induction of the multixenobiotic defence mechanism (MXR), P-glycoprotein, in the mussel *Mytilus californianus* as a general cellular response to environmental stresses. *Aquat. Toxicol.*, **49**, 89–100.

Fayyad, U.M. and Irani, K.B. (1993) Multi-interval discretization of the continuous-value attributes for classification learning. In *Proceedings of the Thirteenth International Joint Conference on the Artificial Intelligence*, 28 August – 3 September 1993, Chambery, France, pp. 1022–1027.

Fong, P.P., Noordhuis, R. and Ram, K.L. (1993) Dopamine reduced intensity of serotonin-induced spawning in the zebra mussel *Dreissena polymorpha* (Pallas). *J. Exp. Zool.*, **266**, 79–83.

Fong, P.P., Kyozuka, K., Abdelghani, H., Hardege, J.D. and Ram, J.L. (1994) *In vivo* and *in vitro* induction of germinal vesicle breakdown in a freshwater bivalve, the zebra mussel *Dreissena polymorpha* (Pallas). *J. Exp. Zool.*, **269**, 467–474.

Frazier, J.M. (1986) Cadmium-binding proteins in the mussel, *Mytilus edulis. Environ. Health Perspectives*, **65**, 39–43.

Gagné, F. and Blaise, C. (1996) Available intracellular Zn as a potential indicator of heavy metal exposure in rainbow trout hepatocytes. *Environ. Toxicol. Water Qual.*, **11**, 319–325.

Gagné, F. and Blaise, C. (2003) Effects of municipal effluents on serotonin and dopamine levels in the freshwater mussel *Elliptio complanata. Comp. Biochem. Physiol.*, **136**C, 117–125.

Gagné, F., Blaise, C., Salazar, M., Salazar, S. and Hansen, P.-D. (2001) Evaluation of estrogenic effects of municipal effluents to the freshwater mussel *Elliptio complanata*. *Comp. Biochem. Physiol.*, **128C**, 213–225.

Gagné, F., Blaise, C., Pellerin, J. and Gauthier-Clerc, S. (2002a) Alteration of the biochemical properties of female gonads and vitellins in the clam *Mya arenaria* at contaminated sites in the Saguenay Fjord. *Mar. Environ. Res.*, **53**, 295–310.

Gagné, F., Blaise, C., Aoyama, I., Luo, R., Gagnon, C., Couillard, Y., Campbell, P. and Salazar, M. (2002b) Biomarker study of a municipal effluent dispersion plume in two species of freshwater mussels. *Environ. Toxicol.*, **17**, 149–159.

Gowlan, B.T., McIntosh, A.D., Davies, I.M., Moffat, C.F. and Webster, L. (2002) Implications from a field study regarding the relationship between polycyclic aromatic hydrocarbons and glutathione *S*-transferase activity in mussels. *Mar. Environ. Res.*, **54**, 231–235.

Grist, E.P.M., Wells, N.C., Whitehouse, P., Brighty, G. and Crane, M. (2003) Estimating the effects of 17β-ethinylestradiol on populations of the fathead minnow *Pimephales promelas*: are conventionnal toxicological endpoints adequate? *Environ. Sci. Technol.*, **37**, 1609–1616.

Hagenbuch, B. and Meier, P.J. (2003) The superfamily of organic anion transporting polypeptides. *Biochem. Biophys. Acta*, **1609**, 1–18.

Halling-Sorensen, B., Nielsen, S.N., Lanzky, P.F., Ingerslev, F., Holten Lutzhoft, H.C. and Jorgensen, S.E. (1998) Occurrence, fate and effects of pharmaceutical substances in the environment. *Chemosphere*; **36**, 357–393.

Haneji, T. and Koide, S.S. (1988) Protein phosphorylation during 5-hydroxytryptamine-induced maturation of *Spisula* oocytes. *Exp. Cell. Res.*, **177**, 227–231.

Hardedge, J.D., Duncan, J. and Ram, J.L. (1997) Tricyclic antidepressants suppress spawning and in the zebra mussel, *Dreissena polymorpha*. *Comp. Biochem. Physiol.*, **118C**, 59–64.

Juneja, R., Ueno, H., Segal, S.J. and Koide, S.S. (1993) Regulation of serotonin-induced calcium uptake in *Spisula* oocytes by tricyclic antidepressants. *Neurosci. Lett.*, **151**, 101–103.

Kammenga, J.E., Dallinger, R., Donker, M.H., Kohler, H.R., Simonsen, V., Triebskorn, R. and Weeks, J.M. (2000) Biomarkers in terrestrial invertebrates for ecotoxicological soil risk assessment. *Rev. Environ. Contam. Toxicol.*, **164**, 93–147.

Kampa, L. and Peisach, J. (1980) Purification and characterization of hydroxyindole oxidase from the gills of *Mytilus edulis*. *J. Biol. Chem.*, **255**, 595–601.

Kolpin, D.W., Furlong, E.T., Meyer, M.T., Thurman, E.M., Zaugg, S.D., Barber, L.B. and Buxton, H.T. (2002) Pharmaceuticals, hormones, and other organic wastewater contaminants in U.S. streams, 1999–2000: a national reconnaissance. *Environ. Sci. Technol.*, **36**, 1202–1211.

Kondo, S., Fujiwara, M., Ohba, M. and Ishii, T. (1995) Comparative larvicidal activities of the four *Bacillus thuringiensis* serovars against a chironomid midge, *Paratanytarsus grimmii* (Diptera: Chironomidae). *Microbiol. Res.*, **150**, 425–428.

Kream, R.M., Zukin, R.S. and Stefano, G.B. (1980) Demonstration of two classes of opiate binding sites in the nervous tissue of the marine molluscs *Mytilus edulis*. *J. Biol. Chem.*, **255**, 9218–9224.

Li, Q., Osada, M., Suzuki, T. and Mori, K. (1998) Changes in vitellin during oogenesis and effect of estradiol on vitellogenesis in the Pacific oyster *Crassostrea gigas*. *Invert. Reprod. Dev.*, **33**, 87–93.

Lippai, M., Gobet, I., Tomkowiak, M., Durocher, Y., Leclerc, C. and Moreau, M. (1995) Thimerosal triggers meiosis reinitiation in oocytes of the Japanese clam *Ruditapes philippinarum* by eliciting an intracellular Ca^{2+} surge. *Int. J. Dev. Biol.*, **39**, 401–407.

Liu, Y., Shenouda, D., Bilfinger, T.V., Stefano, M.L., Magazine, H.I. and Stefano, G.B. (1996) Morphine stimulates nitric oxide release from invertebrate microglia. *Brain Res.*, **722**, 125–131.

Livingstone, D.R. (2001) Contaminant-stimulated reactive oxygen species production and oxidative damage in aquatic organisms. *Mar. Poll. Bull.*, **42**, 656–666.

Livingstone, D.R., Nasci, C., Solé, M., Da Ros, L., O'Hara, S.C.M., Peters, L.D., Fossato, V., Wootton, A.N. and Goldfarb, P.S. (1997) Apparent induction of a cytochrome P450 with immunochemical similarities to CYP1A in digestive gland of the common mussel (*Mytilus galloprovincialis* L.) with exposure to hexachlorobiphenyl and Arochlor 1254. *Aquat. Toxicol.*, **38**, 205–224.

Losey, J.E., Rayor, L.S. and Carter, M.E. (1999) Transgenic pollen harms monarch larvae. *Nature*, **399**, 214.

Ma, X.L., Cowles, D.L. and Carter, R.L. (2000) Effect of pollution on genetic diversity in the bay mussel *Mytilus galloprovincialis* and the acorn barnacle *Balanus glandula*. *Mar. Environ. Res.*, **50**, 559–563.

Martel, P., Kovacs, T., Voss, R. and Megraw, S. (2003) Evaluation of caged freshwater mussels as an alternative method for environmental effects monitoring (EEM) studies. *Environ. Pollut.*, **124**, 471–483.

Martinez, G. and Rivera, A. (1994) Role of monoamines in the reproductive process of *Argopecten purpuratus*. *Invert. Reprod. Dev.*, **25**, 167–174.

Mathieu, M. (1987) Utilization of aspartate transcarbamylase activity in the study of neuroendocrinal control of gametogenesis in *Mytilus edulis*. *J. Exp. Biol.*, **241**, 247–252.

Mathieu, M., Bergeron, J.P. and Danet, A.M.A. (1982) L'aspartate transcarbamylase, indice d'activité gamétogénétique chez la moule *Mytilus edulis* L. *Int. J. Invert. Reprod.*, **5**, 337–343.

Matsutani, T. and Nomura, T. (1987) *In vitro* effects of serotonin and prostaglandins on release of eggs from the ovary of the scallop, *Patinopecten yessoensis*. *Gen. Comp. Endocrinol.*, **67**, 111–118.

Moore, M.N. and Allen, J.I. (2002) A computational model of the digestive gland epithelial cell of marine mussels and its simulated responses to oil-derived aromatic hydrocarbons. *Mar. Environ. Res.*, **54**, 579–584.

Moreau, M., Leclerc, C. and Guerrier, P. (1996) Meiosis reinitiation in *Ruditapes philippinarum* (Mollusca): involvement of L-calcium channels in the release of metaphase I block. *Zygote*, **4**, 151–157.

Mori, K., Muramatsu, T. and Nakamura, Y. (1969) Effect of steroid in oyster-III. Sex reversal from male to female in *Crassostrea gigas* by estradiol-17β. *Bull. Jpn. Soc. Sci. Fish.*, **35**, 1072–1076.

Narbonne, J.F., Daubeze, M., Clerandeau, C. and Garrigues, P. (1999) Scale of classification based on biochemical markers in mussels: application to pollution monitoring in European coasts. *Biomarkers*, **6**, 415–424.

Olsson, P.-E. and Kling, P. (1995) Regulation of hepatic metallothionein in estradiol-treated rainbow trout. *Mar. Environ. Res.*, **39**, 127–129.

Osada, M. and Nomura, T. (1989) Estrogen effect on the seasonal levels of catecholamines in the scallop *Patinopecten yessoensis*. *Comp. Biochem. Physiol.*, **93C**, 349–353.

Osada, M., Matsutami, T. and Nomura, T. (1987) Implication of catecholamines during spawning in marine bivalve molluscs. *Int. J. Invert. Reprod.*, **12**, 241–252.

Osada, M., Mori, K. and Nomura, T. (1992) *In vitro* effects of estrogen and serotonin release of eggs from the ovary of the scallop. *Nippon Suisan Gakkaishi*, **58**, 223–227.

Pani, A.K. and Croll, R.P. (1998) Pharmacological analysis of monoamine synthesis and catabolism in the scallop, *Placopecten magellanicus*. *Gen. Pharmacol.*, **31**, 67–73.

Peters, L.D., Nasci, C. and Livingstone, D.R. (1998) Immunochemical investigations of cytochrome P450 forms/epitopes (CYP1A, 2B, 2E, 3A and 4A) in digestive gland of *Mytilus* spp. *Comp. Biochem. Physiol.*, **121C**, 361–369.

Pipe, R.K. (1987) Oogenesis in the marine mussel *Mytilus edulis*: an ultrastructural study. *Mar. Biol.*, **95**, 405–414.

Sanderson, H., Johnson, D.J., Wilson, C.J., Brain, R.A. and Solomon, K.R. (2003) Probabilistic hazard assessment of environmentally occurring pharmaceuticals toxicity to fish, daphnids and alga by ECOSAR screening. *Toxicol. Lett.*, **144**, 383–395.

Smital, T. and Kurelec, B. (1998) The chemosensitizers of multixenobiotic resistance mechanism in aquatic invertebrates: a new class of pollutants. *Mutat. Res.*, **399**, 43–53.

Smith, S.A., Nason, J. and Croll, R.P. (1997) Detection of APGWamide-like immunoreactivity in the sea scallop, *Placopecten magellanicus*. *Neuropeptides.*, **31**, 155–165.

Smolders, R., Bervoets, L., de Coen, W. and Blust, R. (2004) Cellular energy allocation in zebra mussels exposed along a pollution gradient: linking cellular effects to higher levels of biological organization. *Environ. Pollut.*, **129**, 99–112.

Snyder, M.J., Girvetz, E. and Mulder, E.P. (2001) Induction of marine mollusc stress proteins by chemical or physical stress. *Arch. Environ. Contam. Toxicol.*, **41**, 22–29.

Sumpter, J.P. and Jobling, S. (1995) Vitellogenesis as a biomarker for estrogenic contamination of the aquatic environment. *Environ. Health Perspect.*, **103**, 173–178.

Sunderman, Jr, F.W. (1987) Metal induction of heme oxygenase. *Ann. NY Acad. Sci.*, **514**, 65–80.

Tsai, C-L., Wang, L-H., Chang, C-F., and Kao, C-C. (2000) Effects of gonadal steroids on brain serotonergic and aromatase activity during the critical period of sexual differentiation in Tilapia, *Oreochromis mossambicus*. *J. Neuroendocrind.*, **49**, 894–898.

van der Oost, R., Beyer, J. and Vermeulen, N.P.E. (2003) Fish bioaccumulation and biomarkers in environmental risk assessment: a review. *Environ. Toxicol. Pharmacol.*, **13**, 57–149.

van Tellingen, O. (2001) The importance of drug-transporting P-glycoproteins in toxicology. *Toxicol. Lett.*, **120**, 31–41.

Vasseur, P. and Cossu-Leguile, C. (2003) Biomarkers and community indices as complementary tools for environmental safety. *Environ. Int.*, **28**, 711–717.

Veldhuizen-Tsoerkan, M.B., Holwerda, D.A., van der Mast, C.A. and Zandee, D.I. (1991a) Synthesis of stress proteins under normal and heat shock conditions in gill tissue of sea mussels (*Mytilus edulis*) after chronic exposure to cadmium. *Comp. Biochem. Physiol.*, **100C**, 699–706.

Veldhuizen-Tsoerkan, M.B., Holwerda, D.A., de Bont, A.M., Smaal, A.C. and Zandee, D.I. (1991b) A field study on stress indices in the sea mussel, *Mytilus edulis*: application of the 'stress approach' in biomonitoring. *Arch. Environ. Contam. Toxicol.*, **21**, 497–504.

Waldmann, P., Pivcevic, B., Muller, W.E., Zahn, R.K. and Kurelec, B. (1995) Increased genotoxicity of 2-acetylaminofluorene by modulators of multixenobiotic resistance mechanism: studies with the freshwater clam *Corbicula fluminea*. *Mutat. Res.*, **342**, 113–123.

Yang, M.S., Chiu, S.T. and Wong, M.H. (1995) Uptake, depuration and subcellular distribution of cadmium in various tissues of *Perna viridis*. *Biomed. Environ. Sci.*, **8**, 176–185.

8 Environmental monitoring for genotoxic compounds

Johan Bierkens, Ethel Brits and Luc Verschaeve

8.1 Introduction

The European Union regulation on existing substances aims to map the knowledge that exists with regard to the risks and safety hazards associated with these chemicals, because the knowledge of the tens of thousands of compounds that are commercially available today is by no means comprehensive (Van Wezel, 1999). Therefore, it has been decided to prioritise substances based on their intrinsic properties. Intrinsic properties that have to be taken into account are persistency, toxicity and bioaccumulative potential along with carcinogenicity, mutagenicity and effects on reproduction. (Roex *et al.*, 2001). The latter substances have been implicated as potentially important causal factors in reducing the inter- and intraspecific biodiversity that appears to occur worldwide (World Summit, Rio de Janeiro, 1992). Aside from the obvious human health issues, the concern remains that inheritable mutations lower the reproductive output of affected populations because the exposed individuals may have a decreased viability and fertility. Also, changes in the gene pools induced via direct and indirect effects of genotoxins may affect future generations. As such, the question arises of how the hazards and risks of genotoxic agents can be evaluated.

In this chapter the types of genotoxic effects are discussed and then the techniques currently available for detecting genotoxins are briefly described and evaluated. The emphasis is on screening assays for environmental samples and because the aim of ecotoxicology is to study the effect of pollutants on natural populations and not on individual animals *per se*, the final section discusses the ecological relevance of genotoxic effects and their possible implications for risk assessment.

8.2 Types of genotoxic effect

The genome, defined as the total hereditary material (DNA) contained within the cells of an organism, is made up of individual molecules called nucleotides that contain the bases adenine (A), thymine (T), guanine (G) and cytosine (C). The DNA molecule is a double helix composed of two intertwined nucleotide chains oriented in opposite directions. These chains of nucleotides in DNA are wound up and compacted into the chromosomes that are found in the nucleus of a cell.

Different species may have a different total number of chromosomes. Genes, the functional units of DNA on the chromosomes, encode for the characteristics of a species. In most higher animal and plant species there are two copies of each chromosome (homologous chromosomes) and each gene (allele) is inherited from each parent. Allelic variation, i.e. slight variation between the same genes on homologous chromosomes, causes variation within a species (Russell, 1983). The effects of environmental exposure to genotoxins may be direct alterations in genes and gene expression or indirect effects of pollutants on gene frequencies (Anderson and Wild, 1994; Anderson et al., 1994). Genetic ecotoxicology (also referred to as ecogenotoxicology) studies the interaction of chemicals or physical processes (e.g. radiation) with the genetic material (DNA), the damage response mechanisms (DNA repair) and the subsequent effects, including carcinogenesis, teratogenesis and population effects.

8.2.1 Direct genotoxic effects

Mutagenicity refers to the induction of permanent heritable changes in the amount or structure of the genetic material in individual cells or organisms that is not explicable by recombination of pre-existing genetic variability (Russell, 1983). Mutations may occur in any cell and at any stage in the cell cycle and may involve a single gene or gene segment, a block of genes or whole chromosomes. Three types of mutations can be distinguished.

(1) Gene mutations. A gene that has changed from one allelic form into another has undergone mutation. Spontaneous mutations may occur during the entire life-span of organisms (Drake et al., 1998; Radman, 1999). In man the natural average mutation rate is about one mutation per one hundred thousand to one million cell divisions (Drake et al., 1998). Mutation rates vary greatly among loci (Staton et al., 2001). Gene mutations, when they occur in germ cells, are one of the driving forces of evolution. When they occur in somatic cells they are thought to be the fundamental cause of cancer.

Gene mutations can occur as a result of one of the following four events: a duplication mutation (an exact copy of a DNA sequence is added to the make-up of the gene), a deletion mutation (a DNA sequence is lost from the gene), a substitution mutation (one base is substituted for another in the gene) and an insertion mutation that occurs by the movement of specific sequences and their insertion into a gene. These mutations can occur spontaneously or can be caused by physical, chemical or viral carcinogens (Diffley and Evan, 2000). The efficiency of these mechanisms is known to be sequence dependent, i.e. they are non-random and occur most often at so-called mutation hot spots (Rogozin and Pavlov, 2003).

(2) Chromosome mutations. One speaks of chromosome mutations when the fundamental structure of a chromosome is subject to mutation, which will most

likely (but not only) occur during crossing over at meiosis. There are a number of ways in which the chromosome structure can change, which will result in detrimental changes of the genotype and phenotype of the organism (Russell, 1983). Whereas deletion involves the loss of a portion of a chromosome, a duplication produces an exact extra copy of a specific region of a specific chromosome. Also, inversions that re-order a segment of chromosome backwards, and translocations that occur when a piece of chromosome attaches to another chromosome do occur.

(3) Genome mutations. Genome mutations occur when the total number of chromosomes is altered (aneuploidy) (Russell, 1983). Polyploidy arises when an extra copy of every chromosome is made, and trisomy arises when only one extra copy of a single chromosome is present. If an entire chromosome is absent, the consequent disorder is called a monosomy.

Essential for assessment of the impact of pollutants at population level is the distinction that is made between somatic and germ-line mutations. Mutations in somatic cells may (among other effects) induce neoplasia, but because they do not occur in cells that give rise to gametes the mutation will not be passed on to the next generations.

Cancer is now considered a progressive disease characterized by the accumulation of defects in different genes. Cancer-related genes are classified mainly as either oncogenes or tumour suppressor genes. Mutated proto-oncogenes (i.e. oncogenes) contribute to tumour development by enhancing cell growth. Tumour suppressor genes usually inhibit uncontrolled cell growth and transformation, and their loss of function contributes to tumour development (Migliore and Coppedè, 2002). Also, other types of genes have been found that, if inactivated by mutations, can contribute to carcinogenesis. For example, damaged DNA repair genes affect the carcinogenetic process by destroying the genetic stability of cells, making them more prone to mutational alterations (Devereux *et al.*, 1999). Several chemicals require metabolic activation to exhibit their genotoxic effect; for this reason, differences in metabolic enzymes may account for inter-individual susceptibility to them (Migliore and Coppedè, 2002).

Germ-line mutations that are passed on to future generations may result in gamete loss, embryo mortality, abnormal development (teratogenesis), heritable mutations that affect genetic diversity and heritable mutations that affect gene expression, and consequently Darwinian fitness (Roex *et al.*, 2001). The latter two germ-line mutations may affect future populations and are the most important in studying effects of mutagens on populations (Roex *et al.*, 2001).

8.2.2 DNA repair

A major defence against (environmental) damage to DNA are the DNA repair mechanisms, which are present in all organisms examined, including bacteria, yeast, drosophila, fish, amphibians, rodents and humans. DNA repair is involved in processes that minimize cell killing, mutations, replication errors, persistence of DNA damage and genomic instability. Recently, two papers compiled the data from about 130 human DNA repair genes that were cloned and sequenced (Ronen and Glickman, 2001; Wood *et al.*, 2001). Not all of them, however, have been characterized as yet to their function. Deficiencies in the DNA repair mechanisms have been implicated in tumour induction, aging and human pathologies (von Zglinicki *et al.*, 2001; Digweed, 2003). Damage to DNA is repaired by three distinct systems. The 'mismatch repair' system plays a key role in the correction of errors introduced during DNA replication because of its ability to recognize newly synthesized strands of DNA (Modrich, 1994). By comparison, DNA 'excision repair' removes DNA adducts (Sancar, 1994), and 'transcription-coupled repair' is most active for genes that are undergoing transcription (Hanawalt, 1994). The presence of inducible DNA repair mechanisms may hamper the interpretation of genotoxicity tests because a simple link between exposure to a genotoxin and the incidence of mutations may not exist (Hebert and Murdoch Luiker, 1996).

8.2.3 Indirect genotoxic effects

A broad range of pollutants (not necessarily genotoxic compounds) may affect the genome of future generations and/or the fitness of a population indirectly. A decline in genetic diversity caused by severe fluctuations in population sizes due to contaminant exposure at a site may limit the ability of populations to adapt to a changing environment and/or may lead to increased inbreeding and associated reductions in fertility and offspring viability (Roex *et al.*, 2001). This process of genetic drift (bottlenecks) may increase the chances of extinction for populations.

A second indirect effect of environmental pollutants, genetic adaptation, refers to the advantage of certain genotypes in terms of life-history costs to adapt to environmental changes, which may eventually lead to changes in genotype frequencies (Roex *et al.*, 2001). Although changes in gene frequency are common, the response at single loci often varies among contaminated sites. This suggests that these changes are an indirect consequence of selection rather than providing a single-locus monitor of the effect of contaminant exposure (Hebert and Murdoch Luiker, 1996).

In general, it is assumed that chemical exposure will affect, both directly and indirectly, the genetic variation in natural populations. Populations might respond with increased genetic variation resulting from new mutations directly induced by a mutagen, or with decreased genetic variation resulting from population bottlenecks (genetic drift) or selective sweeps that also affect allele frequencies (Bickham *et al.*, 2000). Such changes in allele frequencies and levels of genetic variability have been described as 'emergent effects' (Bickham and Smolen, 1994). Although the initial effects are at the molecular or cellular level, emergent effects are seen at higher levels of organization but are not predictable solely on knowledge of the mechanism of toxicity (Bickham *et al.*, 2000). Which of the possible outcomes is most likely (i.e. mutation, bottlenecks, selection) is not *a priori* predictable (Belfiore and Anderson, 2001).

8.3 Genotoxicity testing methods

Most of the tests currently in use are developed in the field of human genotoxicology and measure the biochemical and molecular responses discussed above. Genotoxicity tests can be based on both *in vitro* and *in vivo* systems (animal studies). Although *in vitro* tests can be used for the screening of carcinogenic compounds, quantitative calculations on cancer risk can be extrapolated only from *in vivo* experiments because biological repair mechanisms and metabolic activity in *in vitro* tests are not always representative for the *in vivo* situation. Distinction should be made between genotoxicity testing to reveal the intrinsic genotoxic potential of compounds or environmental samples (hazard assessment) and genotoxicity testing required for environmental monitoring of emergent effects (risk assessment). The focus here will be on some fast screening methods available for hazard assessment. However, screening for mutagenic properties of a compound or an environmental sample can be considered a first essential step in genotoxicological risk assessment (Kramers *et al.*, 1992).

8.3.1 Test battery approach

There are a number of well validated tests that must be performed before new compounds can be brought on the market. These tests are also used for other purposes, e.g. human and environmental monitoring, risk assessment, etc. The kind of test(s) that should be performed is greatly dependent on the physicochemical properties of a compound (and consequently its fate in the environment) or on aspects such as the risk for human exposure, but as a general rule a battery of tests is required to allow the detection of different genotoxic events (gene, chromosome, genome mutations). Time and budgetary limitations exclude the use of a large battery of genotoxicity tests for routine screening of environmental samples, therefore a limited number of tests that are technically simple, standardized, inexpensive, fast, ecologically representative and reproducible are necessary (Verschaeve, 2002). Bacterial tests are certainly among the recommended tests because they meet most of the aforementioned requirements. Among them the 'classical' Ames test may be envisaged, but other tests can be applied as well. Such tests are, for example, the umu-C, SOS Chromotest, VITOTOX® test and others. Although DNA is universal and results obtained in bacteria may be more or less predictive for genotoxicity in higher organisms, including man, it may be important also to perform one or several tests on eukaryotic organisms. One of the few tests that may be envisaged for routine screening is the alkaline comet assay. Contrary to cytogenetic methods (investigations of chromosome aberrations, sister chromatid exchange, micronuclei, unscheduled DNA synthesis, etc.), this test can be performed on virtually any cell type, it does not require proliferating cells, large cell populations or labelling techniques, and it allows a cell-by-cell investigation and hence the detection of intercellular differences (Tice, 1995). More details on the bioassays mentioned above and the criteria used to select a proper genotoxicity assay are given in the subsequent sections.

8.3.2 Selection criteria for genotoxicity assays

Several criteria should be considered when selecting genotoxicity tests in product testing or ecotoxicological surveys. The most important criteria can be subdivided into three categories: practicality of the test, acceptability of the test and ecological significance. Some specific criteria for each of these categories are listed in Table 8.1. In order to select appropriate tests, a scoring system should be applied that includes as many of these criteria as possible. According to the purpose of the investigation, i.e. early warning versus establishing causal relationships between the presence of genotoxins and population effects, different weighting factors can be applied to select a test battery of genotoxicity tests. In the latter case more elaborate and thus more expensive tests may be advisable.

8.3.3 Individual fast-screening test systems

As mentioned above several tests can be selected to study the genotoxic potential of substances. As some dozens of assays are currently available, the focus here will be on the most frequently used fast screening assays that can be performed easily in routine biomonitoring programmes (Corbisier *et al.*, 2001). They consti-

Table 8.1 Criteria to consider when selecting genotoxicity tests.

1. *Practicality of the test*
   Feasibility
   Cost-effectiveness
   Rapidity

2. *Acceptability*
   Standardization
   Reproducibility
   Statistical validity
   Good laboratory practice
   Broad chemical responsiveness

3. *Ecological significance*
   Sensitivity
   Type of genotoxic lesion
   Ecological realism
   Biological validity

tute good biomarkers of exposure as the test results correlate well with the exposure concentration. When the focus is on risk assessment, i.e. the effects of genotoxins at species or population level, more advanced techniques in the field of molecular biology (e.g. DNA sequencing, establishing genotype frequencies or DNA fingerprinting) may be necessary.

Short-term genetic bioassays are based on the cellular and subcellular mechanisms underlying mutagenic and carcinogenic processes and are used to ascertain different types of genetic damage. Commonly studied end-points include DNA damage, DNA repair mechanisms, DNA adducts, gene or sister chromatid exchange (exposure markers) and occurrence of micronuclei, chromosomal aberrations, aneuploidy and cell transformation (effect markers). The type of lesion is important because it conveys information about the intrinsic nature of the genetic hazard.

Since about 1970 the field of environmental genotoxicology has expanded quickly, resulting in a wide range of assays often employing bacteria as test organisms. Although a wide range of assays has been developed to evaluate the genotoxicity of pure compounds, a limited number have been utilized successfully for the evaluation of environmental mixtures. A brief, non-exhaustive overview of genotoxicity tests based on markers of exposure and on markers of effect is given below. The distinction between biomarkers of exposure and effect is sometimes arbitrary because it is not always clear what impact the lesion observed in a bioassay will have at a higher level of organization.

8.3.3.1 Genotoxicity tests for monitoring DNA damage and repair (exposure assessment)

Exposure to contaminants can lead to both the modification of nucleotides and the physical disruption of DNA strands. They are not considered to be markers of effect because a clear link cannot always be established with manifest effects at cellular or organism level. Several approaches have been developed to examine the extent of DNA damage by measuring the extent of DNA damage and repair, the extent of breakage in DNA strands, the extent of base-pair modifications (DNA adducts) and by inspection of metaphase chromosomes to examine the variation in rates of strand breakage and reunion.

The bacterial Ames test (Ames et al., 1973; Mortelmans and Zeiger, 2000). The Ames test is the most widely and validated bacterial genotoxicity test. It detects back-mutations in the His$^-$ operon ($\rightarrow$ His$^+$) by growing (mutagen-exposed) *Salmonella typhimurium* bacteria on a histidine-poor medium. Indeed, His$^-$ bacteria cannot grow on a medium that is poor in histidine and will die when histidine is depleted. Only His$^- \rightarrow$ His$^+$ mutants are able to grow on the medium because they can make histidine themselves. The sensitivity of the test is enhanced by the use of particular mutant strains preventing adequate DNA repair or increasing resistance to toxic compounds. Bacteria are grown on a selective medium in the presence of the test compound. After 48 h of incubation at 37°C, mutant colonies are counted and compared with the number of colonies formed in unexposed cultures (spontaneous back-mutations) (Figure 8.1).

Several *Salmonella* strains can be used, each differing in the type of mutation(s) involved, e.g. TA98 and TA100 (base-pair substitution or frame-shift mutations). In the plate incorporation test the sample is brought immediately on the culture (Petri dish) and cultures are initiated. Variants of the test exist, e.g. the pre-incubation test is often preferred because it is more accurate in some instances. A compound is usually considered genotoxic when the mean number of revertants is at least double that found in the solvent control culture and when a dose–effect relationship can be established.

The Ames test is thus typically a bacterial test to assess the genotoxicity of chemicals or environmental compounds but it can be used also to detect the presence of mutagens (and hence reflect a mutagen exposure) in human body fluids, especially in the urine. The Ames test thus can be used as a tool for human biomonitoring studies as well.

A number of bacterial genotoxicity tests discussed below are based on the bacterial SOS response, which helps the cell to save itself in the presence of potentially lethal stresses, such as ultraviolet irradiation, DNA-modifying reagents and inactivation of genes essential to DNA replication (Walker, 1985; Koch and Woodgate, 1998; Janion, 2001). In the SOS system several genes are controlled by a single repressor-operator system. Such a set of unlinked genes, regulated by a common mechanism, is called a regulon. The control elements in the SOS regulon are the products of genes *lexA* and *recA* (Gundas and Pardee, 1975). The RecA protein stimulates DNA strand-pairing during recombination. Remarkably, this small protein has enzymatic activity in addition to the activities involved in

Figure 8.1 His⁻ revertants in the Ames test.

recombination. When bound to single-strand DNA, RecA can stimulate proteolytic cleavage of the proteins encoded by *cI, lexA* and *umuD* (Little *et al.*, 1980; Little and Mount, 1982; Walker *et al.*, 1982). In non-induced cells, the SOS response genes are repressed by the LexA repressor protein. Increasing concentrations of single-stranded DNA, trinucleotides and oligonucleotides cause conversion of the RecA protein into a protease by conformation changes, which splice the LexA repressor protein (Kenyon, 1983). LexA is a repressor that binds to at least 15 different operators scattered about the bacterial genome (Lewis *et al.*, 1994). Each operator controls the transcription of one or more proteins that help the cell to respond after environmental damage that might harm the genetic apparatus. These proteins include the gene products of *uvrA* and *uvrB* involved in nucleotide excision repair, *umuC* and *umuD* involved in error-prone mutagenesis, *sulA* involved in cell division control, *dnaA*, the structural gene for DNA polymerase II, *recA* itself, *lexA* itself and several genes of unknown function, including *dinA*, *dinB* and *dinF* (Koch and Woodgate, 1998; Janion, 2001).

The VITOTOX® test (van der Lelie *et al.*, 1997; Verschaeve *et al.*, 1999). The first test to be discussed, which is based on SOS induction, is the VITOTOX® test. The test is based on bacteria that contain the *lux* operon of *Vibrio fischeri* under transcriptional control of the *recN* gene, which is part of the SOS system. This gene normally is not transcribed (no light production) but will be 'switched on' when the bacteria are exposed to a genotoxic compound (mutagen or 'SOS-inducing' substance). Genotoxicity is thus expressed as light production (Figure 8.2). The signal-to-noise ratio (S/N), i.e. the light production of exposed cells divided by the light production of non-exposed cells, is used to evaluate genotoxicity.

Virtually any bacterial strain can be used but the *Salmonella typhimurium* strains were chosen because they are well known for mutagenicity testing and because the same bacteria can be used for a classical Ames test if required. However, because all *Salmonella* constructs gave very comparable results, only the TA104 construct (called TA104 recN2-4) is used as it was shown sometimes to be a more sensitive than the other hybrid strains. Because it was realized that some compounds act directly on light production (e.g. aldehydes) or enhance the metabolism of the bacteria creating false-positive results, a constitutive light-producing strain with a *lux* operon under control of a strong promoter *pr1* was incorporated (Verschaeve *et al.*, 1999). This is used as an internal control system, which also gives important information on the toxicity of the test compound. Figure 8.3 gives an example of test results. Results are typically obtained within 4 h. As an example of genotoxicity screening on environmental matrices, a test result on the River Musi in Hyderabad (India) is given in Figure 8.4.

Mutatox® test (Johnson, 1991; Microbics, 1995). The Mutatox® test is based on the use of a dark variant of the luminescent bacterium *Vibrio fischeri*. This bacterium is a well known organism used for evaluation of acute toxic effects

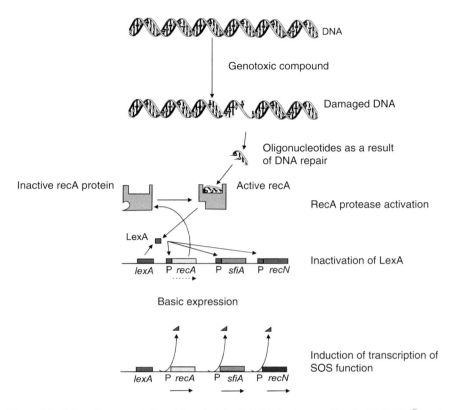

Figure 8.2 Schematic representation of the principle of SOS induction on which the VITOTOX® test is based (insertion of an operon-less *lux* gene next to *recN* results in light production when the bacterial DNA is damaged).

of aquatic environmental samples. The dark variant can be used to detect genotoxic effects in aqueous samples. The presence of genotoxic compounds results in DNA damage and subsequent SOS activation, which leads to the formation of a protease that breaks down a repressor protein of the *lux* pathway, leading to bioluminescence. This restoration of photoluminescence serves as a measure for the genotoxicity of the tested sample. The test can be purchased as a test kit, with light readings being performed using a luminometer. The test can be performed also in microtitre format.

Umu-C test. The umu-c test has been standardized and validated by ISO (ISO 13829, 2000) and is based on the use of the genetically engineered bacteria *Salmonella typhimurium* TA 1535 pSK1002. The test strain was constructed from its precursor, *Salmonella typhimurium* TA98, a *his, rfa, uvrB* and *lacZ* mutant (Ames *et al.*, 1973, 1975; McCann *et al.*, 1975a,b). Genotoxicity effects are

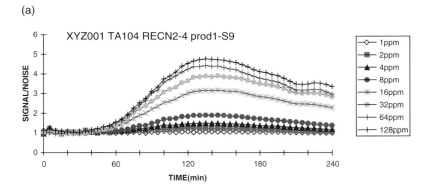

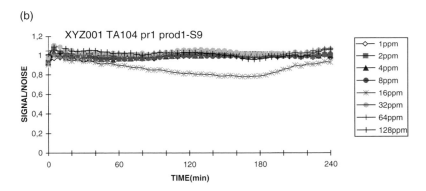

Figure 8.3 Example of results obtained in the VITOTOX® test (with strains TA104recN2-4 and TA104pr1) after exposure to a mutagen (test without addition of a metabolizing rat liver S9 fraction). A dose–response relationship is observed with TA104recN2-4 (a), whereas strain TA104pr1 does not show toxicity or a 'false positive' response (b). A maximum effect with TA104recN2-4 is found after about 140 min.

detected by measuring activation of the SOS response in bacteria and by recording the β-galactosidase activity from an integrated reporter system. With the help of a substrate that is converted into a coloured end-product by β-galactosidase, the amount of DNA damage can be measured. The test is performed in microplates and the coloured end-product is measured using a spectrophotometer.

In Germany, this procedure has often been used for pre-screening and screening of environmental samples and for regulatory purposes (Krumbeck and Hansen, 1995). High sensitivity to genotoxins (individual substances, environmental samples and food) has been described by Oda *et al.* (1985, 1988), Reifferscheid *et al.* (1991) and Ono *et al.* (1992).

SOS Chromotest (Quillardet and Hofnung, 1993). The SOS Chromotest (Quillardet *et al.*, 1982) is based on the use of the intestinal bacterium *Escherichia coli*

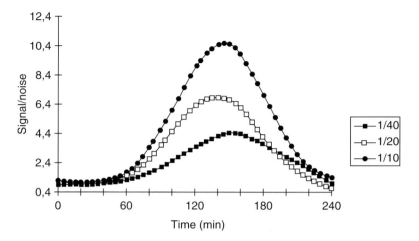

Figure 8.4 VITOTOX® test results expressed as signal-to-noise ratio in *Salmonella typhimurium* TA104r-ecN2-4 strain after exposure to dilutions of water from the Musi river in Hyderabad (India) sampled near Chaitayapuri.

with a reporter gene (*lacZ*) encoding for the enzyme β-galactosidase, coupled to the SOS repair system. With the help of a substrate, a blue chromogen that is converted into a coloured end-product by β-galactosidase, the extent of DNA damage can be measured. The test is available as a test kit with all the necessary materials included. The test detects any primary DNA damage caused by genotoxins and can be used for various kinds of aqueous samples, therefore the test is particularly suitable for testing of environmental samples. The test is performed in microplates and the coloured end-product is measured using a spectrophotometer.

Comet assay (Singh *et al.*, 1988; Tice *et al.*, 1995). The DNA molecules in each cell must undergo continuous maintenance to sustain their integrity. Several of the key mechanisms in this repair process involve the degradation of a short stretch of DNA leading to a transitory break in one DNA strand. A DNA strand breakage can be detected using the alkaline unwinding assay (Shugart, 1988) and agarose gel electrophoresis (Theodorakis *et al.*, 1994).

The DNA alkaline unwinding assay enables the assessment of primary DNA damage in tissue from exposed aquatic test organisms of higher evolutionary order. The test system is based on the fact that DNA of an exposed organism shows a large number of DNA unwinding points in comparison with untreated DNA. As a consequence, the DNA unwinding process is enhanced under alkaline conditions (pH > 12). The hydroxyapatite chromatography technique then is used to separate single- and double-stranded DNA fractions. Basic studies of this procedure have been worked out by Kanter and Schwarz (1978) and Ahnström and Erixon (1981).

The comet assay or 'single-cell gel electrophoresis (SCGE) assay' (Singh *et al.*, 1988) is considered to be a very important alternative for the classical cytogenetic tests and is used worldwide to evaluate the *in vitro* and *in vivo* genotoxicity of chemicals; more recently it has been applied also for environmental biomonitoring. Most frequently, the alkaline version is applied, which detects DNA breakage, alkali-labile sites, open repair sites and cross-links. For this technique cells are mixed with agarose gel, which is spread onto a microscope slide. The cells then are lysed with high salt concentrations and detergents. The remaining nuclear DNA is denatured in an alkali buffer and electrophoresed in the same buffer. The DNA fragments migrate out of the nucleus, towards the positive pole. After electrophoresis, the slides are stained with a fluorochrome such as ethidium bromide. An image analysis system can be used to measure several damage parameters, including tail length and tail DNA content (Figure 8.5).

Ecogenotoxicological surveys using the comet assay have been performed in different organisms, e.g. mussels (Steinert *et al.*, 1998; Wilson *et al.*, 1998), tadpoles (Ralph and Olive, 1997), earthworm coelomocytes (Verschaeve and Gilles, 1993; Salagovic *et al.*, 1996) and cells from plant root and leaves (Koppen, 1996; Koppen & Verschaeve, 1997).

Sister chromatid exchange (Kato, 1974; OECD, 1986). Studies on sister chromatid exchange rely upon the differential labelling of chromatids to permit the recognition of exchanges of DNA among them. Sister chromatid exchange arises as a consequence of breaks in the DNA near the replication fork, followed by an exchange of sister strands rather than the original strands (Figure 8.6). In practice these exchange events are tracked through the incorporation of 5-bromodeoxyuridine (a thymidine analogue) during DNA replication. Because this compound

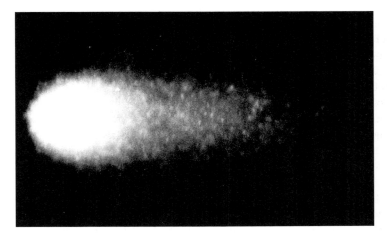

Figure 8.5 Example of a DNA comet.

Figure 8.6 Example of a human metaphase figure showing several sister chromatid exchanges (indicated by arrow heads).

quenches the fluorescence of some DNA-specific stains, its differential display can be used to distinguish sister chromatids. Contaminants that increase the rate of sister chromatid exchange may do so by blocking movement of the replication fork (Tucker *et al.*, 1993).

DNA adducts. Soon after an organism is exposed to toxic chemicals, the presence of an exogenous substance or its interactive product may be detected in the form of a covalently bound DNA adduct. DNA adducts, although directly implicated in chemical carcinogenesis, are relatively early-stage alterations in the progression from genotoxin exposure to disease manifestation. It is only when DNA adducts are mis-repaired or persist through to the DNA replication stage of the cell cycle that manifestations of genotoxicity, including decreased reproductive capacity and the induction of neoplasia, may occur. Quantitative analysis of DNA adducts enables determination of the biologically active levels of exposure to genotoxic chemicals, by taking into account physiological factors such as adsorption, metabolism and detoxification involved in genotoxin adduct formation. Currently, methods of varying sensitivity exist to measure DNA adducts, including ^{32}P-post-labelling, high-performance liquid chromatography (HPLC)/ fluorescence spectrophotometry and immunoassays using adduct-specific polyclonal or monoclonal antibodies.

8.3.3.2 Genotoxicity tests for monitoring cytogenetic effects (effect assessment)

The extent of cytogenetic damage induced by contaminant exposure can be quantified through the analysis of shifts in the incidence of micronuclei or through the study of shifts in genome size distribution or variation in chromosome number or structure (Bickham, 1994).

The in vitro *micronucleus test* (Fenech, 2000). A micronucleus is formed when, during cell division, a chromosome or a chromosome fragment becomes separated from the spindle and therefore is not incorporated into one of the daughter nuclei (i.e. it remains in the cytoplasm and is encapsulated to form a small nucleus) (Di Georgio *et al.*, 1994). This test can detect both clastogenic (chromosome breaking) and aneugenic (e.g. spindle disturbances, genome mutations) events, which can be distinguished using anti-kinetochore antibodies (CREST staining), centromere banding (C-banding) or fluorescent *in situ* hybridization (FISH).

In the *in vitro* micronucleus test usually human peripheral blood lymphocytes are used. To distinguish cells that divided just once in culture, the cultures are treated with cytochalasin-B, a chemical that blocks actin polymerization and, as such, also cytokinesis. After one cell cycle, binucleated cells eventually with one or more micronuclei are obtained (Figure 8.7). It is necessary to distinguish the cells that divided in culture, because cell division is required for the formation of a micronucleus and a very small fraction of lymphocytes may have acquired micronuclei *in vivo*. Furthermore, micronuclei should be detected in cells that divided only once in culture in order to avoid an underestimation of the micronucleus frequency due to cell death.

A number of factors constrain the usefulness of this test, i.e. variability between gender, age and strain, and its limited sensitivity.

Chromosomal aberrations. The study of chromosomal aberrations involves examination of individual chromosomes for deletion, duplication or rearrangement of its normal gene array (Swierenga *et al.*, 1991). Studies are initiated by exposing organisms or cell lines to contaminants, followed by the addition of colchicine to capture cells at metaphase. Chromosome preparations are then stained and the incidence of chromosome aberrations is quantified (Figure 8.8).

Novel methods to detect effects at population level. Ecogenotoxicology has been described as 'an approach that applies the principles and techniques of genetic toxicology to assess the potential of environmental pollution, in the form of genotoxic agents, on the health of the ecosystem' (Shugart and Theodorakis,

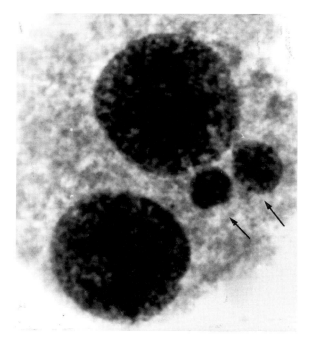

Figure 8.7 A cytochalasin-B-blocked binucleated cell, with two micronuclei.

1994). However, although one might argue that the ultimate goal of ecogenotoxicology is to elucidate whether exposure to mutagens has an impact on biodiversity, experiments showing causal relationships between exposure to genotoxins and loss of biodiversity only become feasible by using novel molecular biological techniques to study genetic variability in natural populations.

The two major categories of methods to assess genetic patterns directly are allozyme electrophoresis and DNA molecular techniques. Allozymes are enzymes (proteins) with varying electrophoretic mobility, encoded by different alleles of single genetic loci. The technique is limited because allozymes are proteins, so alterations at DNA level that do not result in an amino acid substitution are not detected. Despite its limited resolution, allozyme analysis remains the simplest and most rapid technique for surveying genetic diversity in single-copy nuclear genes (Bickham et al., 2000).

Molecular techniques allow the evaluation of molecular-level (DNA and RNA) variation in populations. Most techniques use the polymerase chain reaction (PCR) to amplify short nucleotide sequences from small amounts of tissues. With DNA techniques, one can examine neutral markers (non-coding regions widely dispersed over the entire genome) or patterns at coding loci (regions that are essential for functional differences in important proteins or for gene expression). Also, mitochondrial DNA (mtDNA) (see Lecrenier and Foury, 2000) has

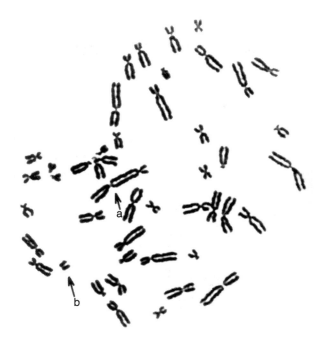

Figure 8.8 Human metaphase figure showing a dicentric chromosome (a) and an acentric fragment (b).

attracted a lot of attention. Mitochondrial DNA is located outside the nucleus, with multiple copies of a single chromosome found in each cell. The mtDNA is maternally inherited and thus does not recombine. However, the rapid turnover rates of mitochondria ensure that the mitochondrial genome undergoes more cycles of replication than the nuclear genome. As such, nucleotide sequence analysis of the genes of mtDNA has become a powerful tool to assess the effect of genotoxins at population level (Belfiore and Anderson, 2001).

Several basic methods of generating data on DNA and mtDNA variation exist. Bickham *et al.* (2000) summarize the technical approaches that can be used to address the effects of chemical contaminants on the genetic diversity in natural populations. They conclude that, because the alternative methods target different segments of the genome, possess differing resolutions and involve varied operating and developmental costs, no single optimal technique exists and the selection of which methodology to use should be guided by the problem under investigation. Essential to this approach is the requirement to discriminate between contaminant-induced genetic change and genetic differences that result from many other, unrelated, variables, i.e. bottleneck effects (reduced survival or fecundity resulting from chronic exposure that lead to an overall reduction in genetic diversity) or selection.

8.4 Exposure assessment of genotoxic compounds

Fast and cheap routine screening methods for the genotoxicity of environmental samples are required in monitoring programmes assessing the 'health status' of the environment. Apart from the assessment of environmental samples, they can be applied also to evaluate the intrinsic properties of new substances or the genotoxic potential of effluents. In the previous sections several of these screening assays for genotoxicity testing have been described. Which one should be used is often a matter of debate and will depend largely on the purpose of the site investigation. In a comparative investigation Verschaeve (2002) demonstrated that because the different tests are based on different molecular induction mechanisms they may, logically, obtain different results. It is therefore not possible to identify the best test to evaluate genotoxicity in environmental compartments such as surface waters, soils and groundwater. Comparisons of different tests are therefore only indicative of which tests could be used reasonably in rapid screening programmes (Verschaeve, 2002). From its comparison the same author concludes that, overall, the Ames test remains a very important tool for environmental monitoring but that tests such as the umu-C and VITOTOX® test may also be considered, especially when no pre-concentration step is required. The comet assay on eukaryotic cell systems may be important as well but it should be realized that DNA comets may result from different biological interaction mechanisms and conditions (including, for example, cell-cycle stages and apoptosis) and is subject to phenomena that should be well understood and controlled (Verschaeve, 2002). In any case, much benefit could be obtained by using a couple of the currently available screening assays in tandem.

Roex *et al.* (2001) compiled some experiments from recent literature that tried to bridge the gap between *in vitro* tests and fitness parameters (e.g. growth, mortality, reproduction) for some ecologically relevant species. From this compilation the authors concluded that because the genotoxicity tests most often were more sensitive compared with the fitness parameters, these tests can act as early warning systems. Moreover, fast screening assays have some advantages over chemical screening methods because they take into account the bioavailability of a compound and the interaction of multiple genotoxins in complex mixtures.

In order to protect our environment, it is required that *in situ* exposure assessment on resident populations for several classes of pollutants, including genotoxins, should be performed on a regular basis. Field investigations in which animal and plant species are collected and investigated for genotoxic lesions (e.g. incidence of tumours or DNA adducts) have the advantage that they can look into chronic exposure integrated over long spans of time. However, because they are time consuming and costly they are applied less often in monitoring programmes compared with fast screening assays. Also, interpretation of the results can be hampered by aspects such as the lack of an adequate reference site or an incomplete match between the species found at different locations.

Numerous studies have demonstrated correlations between elevated levels of carcinogens and the incidence of neoplasia in organisms. Baumann and Harshbarger (1998) reported a long-term trend of liver cancer prevalence in brown bullheads (*Ameirus nebulosus*) inhabiting a river polluted with polycyclic aromatic hydrocarbons (PAHs). Other examples of these are the induction of tumours in oysters (*Crassostrea virginica*) (Gardner *et al.*, 1992) and in planarians (Schaeffer, 1993) following exposure to a mixture of PAHs and polychlorinated biphenyls (PCBs), amines and metals and a range of genotoxins, respectively. Neoplasia in invertebrate species seems to occur rather rarely in comparison with incidences reported for vertebrates (Depledge, 1998). Although these and many more examples point towards a correlation between tumour incidence and exposure to environmental carcinogens, so far no evidence has been found for correlations between high incidences of tumours and a decline in population size in the field (Depledge, 1996).

Apart from studying the incidence of neoplasia, many other markers have been applied in order to assess the exposure to genotoxins in wildlife species, such as DNA adducts, chromosomal aberrations, micronuclei, c-K-*ras* oncogenes, etc. Correlations between contaminant exposure and mutations in oncogenes have been examined in molluscs (Van Beneden, 1994). Pink salmon (*Oncorhynchus gorbuscha*) embryos exposed to crude oil were shown to have high frequencies of mutations in the K-*ras* oncogene (Roy *et al.*, 1999).

Several studies have reported on the suitability of DNA adducts for the determination of chronic exposure to genotoxic compounds in various organisms. Using ^{32}P-post-labelling, Stein *et al.* (1994) demonstrated a dose–response relationship between hepatic DNA adducts in dab (*Limanda limanda*) and flounder (*Platichthys flesus*) and environmental exposure to PAHs. Vanarasi *et al.* (1989) showed a good correlation between the levels of hepatic DNA adducts in the marine flatfish *Parophrys vetulus* (English sole) and the levels of PAHs in sediments at three sites with low, intermediate and high levels of pollution. DNA adducts and micronuclei were shown to be good short-term indicators of effects on fitness in amphibians (Sadinski *et al.*, 1995). Lyons *et al.* (1997) and Harvey *et al.* (1999) demonstated genetic damage, as detected by elevated levels of DNA adducts, in the reproductive organs of the intertidal teleost *Lipophrys pholis* following PAH exposure.

Not all fish species seem equally well suited to study DNA adduct formation. Van der Oost (1997) summarized that about 50% of the species studied under field conditions can be considered as responders by showing significant increases of DNA adduct levels in liver in a polluted environment. In general, in order to obtain meaningful results, the choice of appropriate reference sites is of crucial importance in studies on aquatic organisms. Furthermore, fish from different locations should be matched properly for species, gender, age and sampling time (Kleinjans and van Schooten, 2002). Kirby *et al.* (2000) compared results on hepatic DNA adducts in dab (*Limanda limanda*) and flounder (*Platichthys flesus*) exposed to complex mixtures of PAHs. The results from flounder, being a relative sedentary species, correlated much better with the levels of contamination at the location of capture compared with the results from dab, which is a migratory species. They recommend that when fish are used as sentinels of pollution the use of migratory fish should be avoided.

A comparison of DNA adducts in the blood and liver from different Mugil species collected in a PAH-polluted harbour suggests that DNA adducts in the blood reflect recent exposure as a result of the more rapid turnover of blood cells compared with liver cells (Telli-Krakoc *et al.*, 2001). The level of DNA adducts in the liver has been shown to correlate well with development and the pattern of prevalence of liver tumours in flounders (Vethaak and Wester, 1996).

Other species have been used as environmental sentinels. The frequency of chromosomal aberrations in the gill tissue of *Mytilus edulis* transplanted to field sites contaminated to different extents increased with increasing exposure to contaminants (Al-Sabti and Kurelec, 1985). Pavlica *et al.* (2001) measured a significant increase of the tail length of comets in haemocytes of zebra mussels (*Dreissena polymorpha*) after experimental exposure to polychlorophenol. Reichert *et al.* (1999) showed aromatic DNA adducts in the liver of *Phoca vitulina richardsi* (harbour seals) exposed to petroleum after the Exxon Valdez oil spill.

As for the soil compartment, the comet assay on coelomocytes of earthworms (*Eisenia foetida*) kept in PAH-contaminated soil samples had higher DNA damage than in control samples (Verschaeve, 2002). However no dose–effect relationship was observed. Also, the levels of PAH–DNA adducts in *Lumbricus terrestris*, another earthworm species, kept on industrially contaminated soils increased with exposure time (Van Schooten *et al.*, 1995). Few surveys have been performed on terrestrial plant species, but trifluralin was shown to induce a significant increase in tail length in the comet assay applied on the leaves of *Vicia faba* (Bierkens *et al.*, 1998).

Another means of monitoring the exposure of organisms to genotoxins is to use the induction of biomarkers, i.e. protective enzyme systems and DNA repair mechanisms. An example of this is given by Wirgin and Garte (1994), who have shown a good correlation between levels of expression of CYP1A (cytochrome P450) genes in livers of tomcod and their exposure to hydrocarbon pollution in water and sediments. The levels of these cytochrome P450s are known to be increased by the presence of environmentally significant xenobiotics such as PAHs and PCBs. In a comparative study Kirby *et al.* (2000) found good correlations between the levels of DNA adducts and other biomarkers of PAH exposure, such as bile metabolites and the induction of cytochrome P450 by measuring ethoxyresonifin *O*-deethylase (EROD) activity in dab (*Limanda limanda*) and flounder (*Platichthys flesus*).

8.5 Ecological implications of genotoxic effects

As discussed above, mutations may occur in somatic as well as germ cells. Somatic mutations endanger the survival potential of the individual, but only for animal species with a low reproductive output (e.g. large mammals) may this affect the stability and the size of a population. As such, in order to assess the impact of genotoxins on biodiversity, heritable germ-line mutations are most important because they may change the genetic diversity and fitness of a population. Germ-line mutations are ecologically relevant but are hard to predict from screening on somatic mutations alone. One way to detect possible heritable effects of mutagens is to perform multi-generation toxicity studies in which only the parent generation is exposed and effects on the progeny are determined. An example of this type of experiment is provided by White *et al.* (1999) on fathead minnows *Pimephales promelas* exposed for four months to benzo[*a*]pyrene (B[*a*]P). The hatching frequency, the number of eggs laid by F1 females and the survival of the F2 larvae were shown to be reduced significantly even when no effects were observed in the parental generation. This result points out that heritable effects of pollutants may exist. In two successive studies Vinson *et al.* (1963) and Boyd and Ferguson (1964) showed that mosquitofish (*Gambusia affinis*) exposed to agricultural chemicals were resistant to selected insecticides and when they were transferred to an uncontaminated pond and allowed to breed the offspring showed increased resistance to strobane and chlordane. Also, the offspring of killifish (*Fundulus heteroclitus*) collected at PCB-contaminated areas showed no CYP1A1 responses, in contrast to reference specimens whose parents had not been exposed (Elskus *et al.*, 1999). The authors assume a genetic adaptive response.

As discussed, with the advent of new molecular techniques other types of experiments on the genetic diversity in natural populations became feasible. Methods of detecting genetic variation and several case studies are discussed by Bickham *et al.* (2000) and Belfiore and Anderson (2001). Many of these case studies demonstrate heritable germ-line mutations, reduced genetic variability, shifts in allele frequencies and damaged repair systems in a variety of vertebrate species. For example Wirgin *et al.* (1990) showed genetic diversity at the *c-abl* oncogene and in mitochondrial DNA between populations of Atlantic tomcods (*Microgadus tomcod*) from contaminated and reference sites. Yauk and Quinn (1996) and Yauk *et al.* (2000) demonstrated a high rate of heritable genetic mutations and induced mini-satellite germ-line mutations in herring gulls (*Larus argentatus*) nestling in an industrialized urban site using DNA fingerprinting.

Many heritable mutations are correlated with a reduced fitness, i.e. a reduced survival probability of the offspring of the mutant. The reduction in average fitness with the arrival of new mutations is called mutation load (Crow and Kimura, 1970). However, the more deleterious a mutation is, the sooner it will disappear as a consequence of natural selection (Cronin and Bickham, 1998). It has been shown also that most known lethal mutations in established natural populations are nearly totally recessive (Cronin and Bickham, 1998). This suggests that natural selection removes those mutations that are dominant or partly recessive from natural populations.

It has been hypothesized that heterozygosity is associated with improved survivorship in polluted conditions (Hendrick, 1986; Depledge, 1998). Indeed, some evidence exists to suggest that genetically rich (i.e. heterozygous) species display higher survivorship than genetically poor species after exposure to inorganic and organic pollutants (Nevo et al., 1986; Hawkins et al., 1989; Kopp et al., 1992). Roark and Brown (1996) found a significantly higher proportion of heterozygous individuals of three fish species (*Pimephales notatus*, *Gambusia affinis* and *Fundulus notatus*) at a contaminated site subject to mine-tailing run-off as compared with a reference site. Troncoso et al. (2000) found a positive correlation between multi-loci heterozygosity and survival in young individuals of Chilean scallops (*Argopecten purpuratus*) exposed to copper.

Some caution in interpreting some of these observations should be taken because, as mentioned previously, higher mutation rates caused by chronic exposure to genotoxins may be masked by indirect effects such as selection. Indeed, in contrast with previous examples, Guttman (1994) found a reduced heterozygosity in fish from sites contaminated with heavy metals compared with non-polluted sites. Similar results were obtained from studies on the oligochaete *Limnodrilus hoffmeisteri* from non-polluted and cadmium-polluted sites (Klerks and Levington, 1989). It is assumed that the lower genetic diversity at these heavy-metal-polluted sites reflects selective pressures associated with exposure to heavy metals (Guttman, 1994). This is also illustrated by Patarnello et al. (1991), who found that the allele and genotype frequencies of barnacles (*Balanus amphitrite*) from three locations in the lagoon of Venice did not differ among juvenile populations but differed significantly among adults from the most contaminated location and the two other locations, supporting the hypothesis that post-settlement selection on barnacles occurred in this location.

8.6 Conclusions

Assessment of the genotoxic potential of environmental samples is one of the main tasks of environmental monitoring for the control of pollution. Deposition of genotoxic agents resulting from their continuous accumulation and impact on the environment requires the development of sensitive and rapid assays to monitor their biological relevance. Estimation of genotoxic activity can be carried out by measuring the genetic end-points, which exhibit primary DNA damage. A large number of genotoxicity tests are available for this purpose, several of which have been discussed here. Because no single assay will give full insight into the genotoxic potential of environmental matrices, a test battery approach has been suggested. Moreover, fast screening of the genotoxic potential of environmental samples using the above mentioned bioassays may have some advantages over chemical screening methods because the assays take into account the bioavailability of a compound and the test results reflect the total outcome of the interaction between substances when multiple genotoxins are present in complex mixtures.

Screening for mutagenic properties of compounds is only a first, but essential, step in genotoxicological risk assessment (Kramers et al., 1992). The outcome of these tests only provides information on the potential hazard of a compound or an environmental sample. Therefore, as well as fast-screening monitoring programmes, in situ exposure and effect assessment on resident species and populations is required to evaluate the long-term effects of genotoxins in the environment and to account for both the direct and indirect effects of genotoxins, including genetic drift (bottlenecks) and selection. Numerous studies in the literature have demonstrated a strong correlation between elevated levels of potentially mutagenic substances and tumour incidences in biota. Although neoplasia in fish and invertebrates, as well as genotoxin-induced inheritable mutations, are potentially of great concern, there is often insufficient information to understand causal mechanisms or to quantify environmental risks (Depledge, 1996). Lately, more elaborate experiments studying the genetic diversity of resident populations have been performed. The predictions of bottleneck effects and selection were borne out in several of these studies by using a variety of means of assessing genetic change (Belfiore and Anderson, 2001). In addition to these commonly perceived risks, mutation has been put forward as an important risk to populations and it has been recognized that population-level effects occur at ambient contaminant concentrations (Belfiore and Anderson, 2001). In their review paper Bickham et al., (2000) conclude that 'The potential for an increased mutation rate, especially when combined with population bottleneck and the resulting fixation of deleterious alleles, to contribute to a downward spiral of fitness decline (mutational meltdown) should be of grave concern to ecotoxicologists and conservation biologists'.

References

Ahnström, G. and Erixon, K. (1981) Measurements of strand breaks by alkaline denaturation and hydroxyapatite chromatography. In *DNA Repair: a Laboratory Manual of Research Procedures*, Friedberg, E.C. and Hanawalt, P.C. (eds), vol. 1(B), pp. 403–418. Marcel Dekker, New York.

Al-Sabti, K. and Kurelec, B. (1985) Induction of chromosomal aberrations in the mussel *Mytilus galloprovincialis* watch. *Bull. Environ. Contam. Toxicol*, **35**, 660–665.

Ames, B.N., Durston, W.E.E., Yamasaki, Y. and Lee, F.D. (1973) Carcinogens are mutagens: a simple test system combining liver homogenates for activation and bacteria for detection. *Proc. Natl. Acad. Sci. USA*, **70**, 2281–2285.

Ames, B.N., McCann, J. and Yamasakim, E. (1975) Methods for detecting carcinogens and mutagens with the *Salmonella*/mammalian microsome mutagenecity test. *Mutat. Res.*, **31**, 347–364.

Anderson, S.L. and Wild, G.C. (1994) Linking genotoxic responses and reproductive success in ecotoxicology. *Environ. Health Perspect.*, **102**, 9–12.

Anderson, S.L., Sadinnski, W.J., Shuugart, L.B.P., Depledge, M.H., Ford, T., Stegeman, J., Suk, W., Wirgin, I. and Wogan, G. (1994) Genetic and molecular ecotoxicology: a research framework. *Environ. Health Perspect.*, **102**, 3–8.

Baumann, P.C. and Harshbarger, J.C. (1998) Long-term trends in liver neoplasm epizootics of brown bullhead in the Black river, Ohio. *Environ. Monito. Assess.*, **51**, 213–223.

Belfiore, N.M. and Anderson, S.L. (2001) Effects of contaminants on genetic patterns in aquatic organisms: a review. *Mutat. Res.*, **489**, 97–122.

Bickham, J.W. (1994) Genotoxic responses in blood detected by cytogenetic and cytometric assays. In *Non-destructive Biomarkers in Vertebrates*, Fossi, M.C. and Leonzio, C. (eds), pp. 37–62. Lewis Publishers, Boca Raton, FL.

Bickham, J.W. and Smolen, M.J. (1994) Somatic and heritable effects of environmental genotoxins and the emergence of evolutionary toxicology. *Environ. Health Perspect*, **102** (Suppl. 12), 25–28.

Bickham, J.W., Sandhu, S., Hebert, P.D.N., Chikhi, L. and Athwal, R. (2000) Effects of chemical contaminants on genetic diversity in natural populations: implications for biomonitoring and ecotoxicology. *Mutat. Res.*, **463**, 33–51.

Bierkens, J., Klein, G., Corbisier, P., Van den Heuvel, R., Verschaeve, L., Weltens, R. and Schoeters, G. (1998) Comparative sensitivity of 20 bioassays for soil quality. *Chemosphere*, **37**, 2935–2947.

Boyd, C.E. and Ferguson, D.E. (1964) Spectrum of cross-resistance to insecticides in the mosquito fish, *Gambusia affinis*. *Mosquito News*, **24**, 19–21.

Corbisier, Ph., Hansen, P.-D. and Barcelo, D. (2001) *Proceedings of the BIOSET Technical Workshop on Genotoxicity Biosensing*, VITO Report 2001/MIT/P053. (http://www.vito.be/english/environment/environmentaltox5.htm.). Cronin, M.A. and Bickham, J.W. (1998) A population genetic analysis of the potential for crude oil spill to induce heritable mutations and impact natural populations. *Ecotoxicology.*, **7**, 259–278.

Crow, J.F. and Kimura, M. (1970) *An Introduction to Population Genetics Theory*. Harper and Row, New York.

Depledge, M.H. (1996) Genetic ecotoxicology: an overview. *J. Exp. Mar. Biol. Ecol.*, **200**, 57–66.

Depledge, M.H. (1998) The ecotoxicological significance of genotoxicity in marine invertebrates. *Mutat. Res.*, **399**, 109–122.

Devereux, T.R., Risinger, J.I. and Barret, J.C. (1999) Mutations and altered expression of human cancer genes: what they tell us about causes. In *Carcinogenic Hazard Evaluation*, IARC Scientific Publication No. 146. IARC, Lyon.

Diffley, G.F.X. and Evan, G. (2000) Oncogenes and cell proliferation, cell cycle, genome integrity and cancer: a millennial view. *Curr. Opin. Genet. Dev.*, **10**, 13–16.

Di Georgio, C., De Meo, M.P., Laget, M., Guiraud, H., Botta, A. and Dumenil, G. (1994) The micronucleus assay in human lymphocytes: screening for inter-individual variability and application to biomonitoring. *Carcinogen* **15**, 313–317.

Digweed, M. (2003) Response to environmental carcinogens in DNA-repair-deficient disorders. *Toxicology*, **193**, 111–124.

Drake, J.W., Charlesworth, B., Charlesworth, D. and Crow, J.F. (1998) Rates of spontaneous mutation. *Genetics* **148**, 1667–1686.

Elskus, A.A., Monosson, E., McEloy, A.E., Stegeman, J.J. and Woltering, D.S. (1999) Altered CYP1A1 expression in *Fundulus heteroclitus* adults and larvae: a sign of pollution resistance? *Aquat. Toxicol.*, **45**, 99–113.

Fenech, M. (2000) The *in vitro* micronucleus technique, *Mutat. Res.*, **455**, 81–95.

Gardner, G.R., Pruell, R.J. and Malcolm, A.R. (1992) Chemical induction of tumours in oysters by a mixture of aromatic and chlorinated hydrocarbons, amines and metals. *Mar. Environ. Res.*, **34**, 59–63.

Gundas, L.J. and Pardee, A.B. (1975) Model for regulation of *Escherichia coli* DNA repair functions. *Proc. Natl. Acad. Sci. USA*, **72**, 2330–2334.

Guttman, S.I. (1994) Population genetic structure and ecotoxicology. *Environ. Health Perspect.*, **102** (Suppl. 12), 97–100.

Hanawalt, P.C. (1994) Transcription-coupled repair and human disease. *Science*, **266**, 1957–1958.

Harvey, J.S., Lyons, B.P., Parry, J.M. and Stewart, C. (1999) An assessment of the genotoxic impact of the Sea Empress oil spill by the measurement of DNA adduct levels in selected vertebrate and invertebrate species. *Mutat. Res. (Genet. Toxicol. Environ. Mutagen)*, **441**, 103–114.

Hawkins, A.J.S., Rusin, J., Bayne, B.L. and Day, A.J. (1989) The metabolic/physiological basis of genotype-dependent mortality during copper exposure in *Mytilus edulis*. *Mar. Environ. Res.*, **28**, 253–257.

Hebert, P.D.N. and Murdoch Luiker, M. (1996) Genetic effects of contaminant exposure – towards an assessment of impacts on animal populations. *Sci. Total Environ.*, **191**, 23–58.

Hendrick, P.W. (1986) Genetic polymorphism in heterogeneous environments: a decade later. *Annu. Rev. Ecol. Syst.*, **17**, 535–566.

ISO 13829 (2000) *Water Quality Determination of Genotoxicity of Water and Waste Water Using the Umu Test*. International Organization for Standardization, Geneva.

Janion, C. (2001) Some aspects of the SOS response system – a critical review. *Acta Biochim. Polon.*, **48**, 599–610.

Johnson, B.T. (1991) An evaluation of a genotoxicity assay with liver S9 for activation and luminescent bacteria for detection. *Environ. Toxicol. Chem.*, **11**, 473–480.

Kanter, P.-M. and Schwarz, H.S. (1978) Hydroxylapatite batch assay for quantitation of cellular DNA damage. *Anal. Biochem.*, **97**, 77–84.

Kato, H. (1974) Spontaneous sister chromatid exchanges detected by a BudR labelling method. *Nature*, **251**, 70–72.

Kenyon, C.T. (1983) The bacterial response to DNA damage. *TIBS*, **March**, 84–87.

Kirby, M.F., Lyons, B.P., Waldock, M.J., Woodhead, R.J., Goodsir, F., Law, R.J., Neall, P., Stewart, C., Thain, J.T., Tylor, T. and Feist S.W. (2000) *Biomarkers of Polycyclic Aromatic Hydrocarbon (PAH) Exposure in Fish and their Application in Marine Monitoring* Science Series Technical Report No. 110. Centre for Environment, Fisheries and Aquaculture Science, Lowestoft.

Kleinjans, J.C.S. and van Schooten, F.-J. (2002) Ecogenotoxicology: the evolving field. *Environ. Toxicol. Pharmacol.*, **11**, 173–179.

Klerks, P.L. and Levington, J.S. (1989) Effects of heavy metals in a polluted aquatic ecosystem. In *Ecotoxicology: Problems and Approaches*, Levin, S.A., Harwell, M.A., Kelly, J.R. and Kimball, K.D. (eds), pp. 41–68 Springer Verlag, Berlin.

Koch, W.H. and Woodgate, R. (1998) The SOS response. In *DNA Damage and Repair. Vol. 1, DNA Repair in Prokaryotes and Lower Eukaryotes*, Nickoloff, J.A. and Hoekstra, M.F. (eds), pp. 107–134. Humana Press, Totova.

Kopp, R.L., Guttman, S.I. and Wissing, T.E. (1992) Genetic indicators of environmental stress in central mudminnow (*Umbra limi*) populations exposed to acid deposition in the Adirondack Mountains. *Environ. Toxicol. Chem.*, **11**, 665–676.

Koppen, G. (1996) Protocol of the alkaline comet test on plant cells. *Comet Newsl.*, **4**, 2–4.

Koppen, G. and Verschaeve, L. (1997) The alkaline comet test on plant cells. A new genotoxicity test for DNA strand breaks in *Vicia faba*. *Mutat. Res.*, **360**, 193–200.

Kramers, P.G.N., Knaap, A.G.A.C., van der Heijden, C.A., Taalman, R.D.F.M. and Mohn, G.R. (1992) Role of genotoxicity assays in the regulation of chemicals in the Netherlands: considerations and experiences. *Mutagenesis* **6**, 487–493.

Krumbeck, H. and Hansen, P.-D. (1995) *Auswirkungen der Belastung des Teltowkanals auf die Havelseenkette in Hinblick auf Eutrophierung und Nutzung als Badegewässer: Bericht für Stadtentwickelung und Umweltschutz*, vol. IV. Bundesambt für Naturschutz, Bonn.

Lecrenier, N. and Foury, F. (2000) New features of mitochondrial DNA replication system in yeast and man. *Gene*, **246**, 37–48.

Lewis, L.K., Harlow, G.R., Gregg-Jolly, R.A. and Mount, D.W. (1994) Identification of high affinity binding sites for LexA which define new DNA damage inducible genes in *Escherichia coli*. *J. Mol. Biol.*, **241**, 507–523.

Little, J.W. and Mount, D.W. (1982) The SOS regulatory system in *Escherichia coli*. *Cell* **29**, 11–22.

Little, J.W., Edmiston, S.H., Pacelli, L.Z. and Mount, D.W. (1980) Cleavage of *Escherichia coli* LexA protein by RecA protease. *Proc. Natl. Soc. Sci. USA*, **72**, 3225–3229.

Lyons, B.P., Harvey, J.S. and Parry, J.M. (1997) The initial assessment of the genotoxic impact of the Sea empress oil spill by the measurement of DNA adduct levels in the intertidal teleost *Lipophrys pholis*. *Mutat. Res., (Genet. Toxicol. Environ. Mutagen)*, **390**, 263–268.

McCann, J., Choi, E., Yamasakim, E. and Ames, B.N. (1975a) Detection of carcinogens as mutagens in *Salmonella*/microsome test: Assay of 300 chemicals. *Proc. Natl. Acad. Sci. USA*, **72**(12), 5135–5139.

McCann, J., Spingam, N.E., Kobori, J. and Ames, B.N. (1975b) Detection of carcinogens as mutagens: bacterial tester strain with R factor plasmids. *Proc. Natl. Acad. Sci. USA*, **72**, 979–983.

Micobics (1995) *Microbics Mutatox®, Manual Genotoxicity Test System*. Azur Environmental, UK.

Migliore, L. and Copedè, F. (2002) Genetic and environmental factors in cancer and neurodegenerative diseases. *Mutat. Res.*, **512**, 135–153.

Modrich, P. (1994) Mismatch repair, genetic stability, and cancer. *Science*, **266**, 1959–1960.

Mortelmans, K. and Zeiger, E. (2000) The Ames *Salmonella*/microsome mutagenicity assay. *Mutat. Res.*, **455**, 29–60.

Nevo, E., Noy, R., Lavie, B., Beiles, A. and Muchtar, S. (1986) Genetic diversity and resistance to marine pollution. *Biol. J. Linn. Soc.*, **29**, 139–144.

Oda, Y., Nakamura, S., Oki, I. and Kato, T. (1985) Evaluation of the new system (umu-test) for the detection of environmental mutagens and carcinogens. *Mutat. Res.*, **147**, 219–227.

Oda, Y., Nakamura, S. and Oki, I. (1988) Harman and non-harman induced SOS response and frameshift mutations in bacteria. *Mutat. Res.*, **208**, 39–44.

OECD (1986) *Genetic Toxicology: in vitro Sister Chromatid Exchange Assay in Mammalian Cells*. OECD Guideline for Testing of Chemicals 479. Organization for Economic Cooperation and Development, Paris.

Ono, Y., Somiya, I. and Kawaguchi, T. (1992) Genotoxic evaluation on aromatic organochlorine compounds by using umu-test. *Water Sci. Technol.*, **26**, 61–69.

Patarnello, T., Guinez, R. and Battaglia, B. (1991) Effects of pollution on heterozygosity in the barnacle *Balanus amphitrite* (Cirripedia: Thoracica). *Mar. Ecol. Prog. Ser.*, **70**, 237–243.

Pavlica, M., Klobucar, G.I., Moja, N., Erben, R. and Pape, D. (2001) Detection of DNA damage in haemocytes of zebra mussel using comet assay. *Mutat. Res.*, **490**, 209–214.

Quillardet, P. and Hofnung, M. (1993) The SOS-chromotest: a review. *Mutat. Res.*, **297**, 235–279.

Quillardet, P., Huisman, O., D'Ari, R. and Hofnung (1982) SOS Chromotest, a direct assay of induction of an SOS function in *Escherichia coli* K-12 to measure genotoxicity. *Proceedings of the National Academy of Sciences*, **79**, 5971–5975.

Radman, M. (1999) Enzymes of evolutionary change. *Nature*, **401**, 866–867.

Ralph, E. and Olive, P.L. (1997) The comet assay: alternatives for quantitative analyses, *Mutat. Res.*, **379**, S130–S131.

Reichert, W.L., French, B.L. and Stein, J.E. (1999) Exposure of marine mammals to genotoxic environmental contaminants: application of the ^{32}P-post-labelling assay for measuring DNA-xenobiotic adducts. *Environ. Monit. Assess.*, **56**, 225–239.

Reifferscheid, G., Neil, J., Oda, Y. and Zahn, R.K. (1991) A microplate version of the SOS/umu-test for rapid detection of genotoxins and genotoxic potential of environmental samples. *Mutat. Res.*, **253**, 215–222.

Roark, S. and Brown, K. (1996) Effects of metal contamination from mine tailing on allozyme distributions of populations of great plain fishes. *Environ. Toxicol. Chem.*, **15**, 921–927.

Roex, E.W.M., Traas, T.P. and Slooff, W. (2001) Ecotoxicological hazard assessment of genotoxic substances, Report no. 601503022. RIVM (National Institute of Public Health and the Environment), Bilthoven, The Netherlands.

Rogozin, I.B. and Pavlov, Y.I. (2003) Theoretical analysis of mutation hotspots and their DNA sequence context specificity. *Mutat. Res.*, **544**, 65–85.

Ronen A. and Glickman, B.W. (2001) Human DNA repair genes. *Environ. Mol. Mutagen.*, **37**, 241–283.

Roy, N.K., Stabile, J., Seeb, J.E., Habicht, C. and Wirgin, I. (1999) High frequency of K-ras mutations in pink salmon embryos experimentally exposed to *Exxon Valdez* oil. *Environ. Toxicol. Chem.*, **18**, 1521–1528.

Russell, P.J. (1983) *Genetics*. Little, Brown and Co, Boston.

Sadinski, W.J., Levay, G., Wilson, M.C., Hoffman, J.R., Bodell, W.J. and Anderson, S.L. (1995) Relationship among DNA adducts, micronuclei, and fitness parameters in *Xenopus laevis* exposed to benzo(*a*)pyrene. *Aquat. Toxicol.*, **32**, 333–352.

Salagovic, J., Gilles, J., Verschaeve, L. and Kalina, I. (1996) The comet assay for the detection of genotoxic damage in the earthworms: a promising tool for assessing the biological hazards of polluted sites. *Folia Biol.*, **42**, 17–21.

Sancar, A. (1994) Mechanisms of DNA excision repair. *Science*, **266**, 1954–1956.

Singh, N.P., McCoy, T., Tice, R.R. and Schneider, E.L. (1988) A simple technique for quantification of low levels of DNA damage in individual cells. *Exp. Cell Res.*, **175**, 84–92.

Schaeffer, D.J. (1993) Planarians as a model system for *in vivo* tumourigenesis studies. *Ecotoxicol. Environ. Saf.*, **25**, 1–18.

Shugart, L.R. (1988) Quantitation of chemically induced damage to DNA of aquatic organisms by alkaline unwinding assay. *Aquat. Toxicol.*, **13**, 43–52.

Shugart, L. and Theodorakis, C. (1994) Environmental genotoxicity: probing the underlying mechanisms. *Environ. Health Perspect.*, **102** (Suppl. 12), 13–17.

Staton, J.L., Schizas, N.V., Chandler, G.T., Coull, B.C. and Quattro, J.M. (2001) Ecotoxicology and population genetics: The emergence of 'phylogeographic and evolutionary ecotoxicology'. *Ecotoxicology*, **10**, 217–222.

Stein, J.E., Reichert, W.L. and Varanasi, U. (1994) Molecular epizootiology: assessment of exposure to genotoxic compounds in teleosts. *Environ. Health Perspect.*, **102** (Suppl. 12), 19–23.

Steinert, S.A., Streib Montee, R. and Sastre, M.P. (1998) Influence of sunlight on DNA damage in mussels exposed to polycyclic aromatic hydrocarbons. *Mar. Environ. Res.*, **46**, 355–358.

Swierenga, S.H.H., Heddle, J.A., Sigal, E.A., Gilman, J.P.W., Brillinger, R.L., Douglas, G.R. and Nestmann, E.R. (1991) Recommended protocols based on a survey of current practise in genotoxicity testing laboratories. IV. Chromosome aberration and sister chromatid exchange in Chinese hamster ovary, V79 Chinese hamster lung and human lymphocyte culture. *Mutat. Res.*, **246**, 301–322.

Telli-Krakoc, F., Gaines, A.F., Hewer, A. and Philips, D. (2001) Differences between blood and liver aromatic DNA adduct formation. *Environ. Int.*, **26**, 143–148.

Theodorakis, C.W., D'Surney, S.J. and Shugart, L.R. (1994) Detection of genotoxic insult as DNA strand breaks in fish blood cells by agarose gel electrophoresis. *Environ. Toxicol. Chem.*, **13**, 1023–1031.

Tice, R.R. (1995) The single cell gel/comet assay: a microgel electrophoretic technique for the detection of DNA damage and repair in individual cells. In *Environmental Mutagenesis*, Philips, D.H. and Venitt, S. (eds), pp. 315–339. Bios Scientific Publishers, Oxford.

Troncoso, L., Galleguillos, R. and Larrain, A. (2000) Effects of copper on the fitness of the Chilean scallop *Argopecten purpuratus* (Mollusca: Bivalvia). *Hydrobiologia*, **420**, 185–189.

Tucker, J.D., Auletta, A., Cimino, M.C., Dearfield, K.L., Jackobson-Kram, D., Tice, R.R. and Carrano, A.V. (1993) Sister chromatid exchange: second report of the Gene-Tox program. *Mutat. Res.*, **297**, 101–180.

Vanarasi, U., Reichert, W.L. and Stein, J.E. (1989) ^{32}P-post-labelling analysis of DNA adducts in liver and wild English sole (*Parophrys vetulus*) and winter flounder (*Pseudopleuronectus americanus*). *Cancer Res.*, **49**, 1171–1177.

Van Beneden, R. (1994) Molecular analysis of bivalve tumors: models for environmental/genetic interactions. *Environ. Health Perspect.*, **102**, 81–83.

van der Lelie, D., Regniers, L., Borremans, B., Provoost, A. and Verschaeve, L. (1997) The VITOTOX test, a SOS-bioluminescence *Salmonella typhimurium* test to measure genotoxicity kinetics. *Mutat. Res.*, **389**, 279–290.

van der Oost, R. (1997) Genotoxiciteitsmonitoring in het aquatische milieu (Thesis). Vrije Universiteit Amsterdam, The Netherlands.

Van Schooten, F.J., Maas, L.M., Moonen, E.J., Kleinjans, J.C.S. and van der Oost, R. (1995) DNA dosimetry in biological indicator species living on PAH-contaminated soils and sediments. *Ecotoxicol. Environ. Saf.*, **30**, 171–179.

Van Wezel, A.P. (1999) Overview of international programmes on the assessment of existing chemicals. Report no. 601503015. RIVM (National Institute of Public Health and the Environment), Bilthoven, The Netherlands.

Verschaeve, L. (2002) Genoyoxicity studies in groundwater, surface waters, and contaminated soil. *Sci. World J.*, **2**, 1247–1253.

Verschaeve, L. and Gilles, J. (1995) The single cell gel electrophoresis assay in the earthworm for the detection of genotoxic compounds in soils. *Bull. Environ. Contam. Toxicol.*, **54**, 112–119.

Verschaeve, L., Van Gompel, J., Thilemans, L., Regniers, L., Van Parijs, Ph. and van der Lelie, D. (1999) VITOTOX® genotoxicity and toxicity test for the rapid screening of chemicals. *Environ. Mol. Mutagen.*, **33**, 240–248.

Vethaak, A.D. and Wester, P.W. (1996) Diseases of flounder (*Platichthys flesus*) in Dutch coastal waters with particular reference to environmental stress factors. Part 2. Liver histopathology. *Dis. Aquat. Organ.*, **26**, 99–116.

Vinson, S.B., Boyd, C.E. and Ferguson, D.E. (1963) Resistance to DDT in mosquitofish, *Gambusia affinis*. *Science*, **139**, 217–218.

von Zglinicki T., Bürkle, A. and Kirkwood, T.B.L. (2001) Stress, DNA damage and ageing – an integrative approach. *Exp. Gerontol.*, **36**, 1049–1062.

Walker, G.C. (1985) Inducible DNA repair systems. *Ann. Rev. Biochem.*, **54**, 425–457.

Walker, G.C., Kenyon, C.J., Baggs, A., Elledge, S.J., Parry, K.L. and Shanabruch, W.G. (1982) Regulations and functions of *Escherichia coli* genes induced by DNA damage. In *Molecular and Cellular Mechanisms of Mutagenesis*, Lemont, J.F. and Generoso, W.M. (eds), pp. 43–63. Plenum Publishing, New York.

White, P.A., Robitaille, S. and Rasmussen, J.B. (1999) Heritable reproductive effects of benzo[a]pyrene on the fathead minnow (*Pimephales promelas*). *Environ. Toxicol. Chem.*, **18**, 1843–1847.

Wilson, J. T., Pascoe, P. L., Parry, J. M. and Dixon, D.R. (1998) Evaluation of the comet assay as a method for the detection of DNA damage in the cells of a marine invertebrate, *Mytilus edulis* L. (Mollusca: Pelecypoda). *Mutat. Res.*, **399**, 87–95.

Wirgin, I.I. and Garte, S. (1994) Assessment of environmental degradation by molecular analysis of a sentinel species: Atlantic tomcod. In *Molecular Environmental Biology*, Garte S.J. (ed.), pp.117–132. Lewis Publishers, Boca Raton, FL.

Wirgin, I.I., D'Amore, M., Grunwald, C., Goldman, A. and Garte, S.J. (1990) Genetic diversity an an oncogene locus and in mitochondrial DNA between populations of cancer prone Atlantic tomcod. *Biochem. Genet.*, **28**, 459–475.

Wood, R.D., Mitchell, M., Sgouros, J. and Lindahl, T. (2001) Human DNA repair genes. *Science*, **291**, 1284–1289.

Yauk, C.L. and Quinn, J.S. (1996) Multilocus DNA fingerprinting reveals a high rate of heritable genetic mutations in herring gulls nesting in an industrialised urban site. *Proc. Natl. Acad. Sci. USA*, **93**, 12137–12141.

Yauk, C.L., Fox, G.A., McCarry, B.E. and Quinn, J.S. (2000) Induced minisatellite germline mutations in herring gulls (*Larus argentatus*) living near steel mills. *Mutat. Res.*, **452**, 211–218.

9 Approach to legislation in a global context

A   UK PERSPECTIVE
Jim Wharfe

9A.1   Introduction

The production of chemicals has grown to become one of the largest manufacturing industries in the world and with it, a wide range of chemicals and formulations has become available that spans major business areas concerned with petrochemicals, agricultural chemicals, pharmaceuticals and veterinary medicines, industrial chemicals and associated products. It is perhaps not surprising, therefore, that the regulations concerning the control of chemicals have been developed rather piecemeal and that the accompanying institutional arrangements are complicated.

Environmental law relating to chemicals is relatively new, although laws on environmental issues have been in existence for far longer. Despite the short history, legislation on chemicals is both voluminous and complex, and it increased substantially following the establishment of the European Economic Community in the late 1950s. Figure 9.1 provides an indication of the increase in environmental legislation that has emerged from Europe since the 1970s.

In the opening chapter of this book, reference is made to the hierarchy of some of the relevant international and European legislation and its influence on the development of risk-based approaches to help manage and control hazardous substances. Reference is made also to the central role of ecotoxicity testing in risk assessment.

This part of Chapter 9 sets out a more detailed perspective of how legislation concerning chemicals is implemented in the UK, and it considers how developments have been influenced by traditions. In the context of this chapter, it is not the intention to provide a complete and comprehensive review but rather, by selective illustration, to show how the typical British approach based on informality and regulatory discretion has moved to the more formal and centralised control necessary within the European Union (EU). Where appropriate, reference to other international law is made to indicate where different approaches influence enforcement procedures.

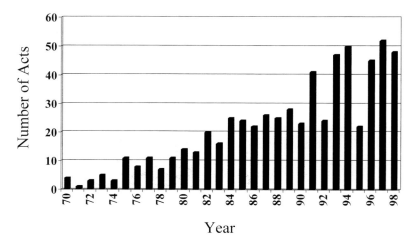

Figure 9.1 Environmental legislation to emerge from Europe since 1970.

9A.2 History and tradition in the UK

In the UK, Central Government is accountable to the monarch of the day and has responsibility for environmental law and policy and, in the first instance, for all primary legislation. This framework allows the transposition of applicable international agreements and European Directives. The current lead for enacting much of the environmental legislation is the Department of the Environment, Food and Rural Affairs (DEFRA). Many delegated bodies, which include government departments, environmental regulatory agencies and local authorities, share the enforcement of such legislation. In the UK the 1995 Environment Act established two new integrated regulatory agencies with a remit covering air, land, water and waste issues: the Environment Agency for England and Wales and the Scottish Environmental Protection Agency. Since the late 1950s following the establishment of the European Economic Community, environmental regulation in the UK, and particularly that associated with chemicals, has been strongly influenced and directed by European agreement. The workings of the community are rule-based with agreement between Member States. This is somewhat in contrast with British tradition, which has been less formal and discretionary. Differences between Member States are recognised and continue to pose challenges to the harmonisation of procedures and the establishment of common baselines. This has undoubtedly hampered progress and by the mid-1990s the European Environment Agency, in its summary document *Environment in the European Union at the Turn of the Century*, questioned the success of environmental policy with factual information indicating little real improvement after 25 years of community policy. Over the last 20 years the UK has complied increasingly with a more centralised and legalistic framework but many issues concerning the control of releases of chemicals into our environment remain; these are discussed further in this chapter.

Early law in the UK concerning pollutants in the environment dates back to medieval history, with restrictions on tannery effluents discharged to rivers and the burning of coal that became a smoke irritant in some local communities. Population levels increased in urban areas during the industrial revolution (Chapter 1), highlighting growing health problems related to emissions and identifying a need to control the siting of installations. The alkali inspectorate was created under the 1863 Alkali Act, primarily to control releases from the chlor-alkali industry, and by the late 19th century, with housing and public health a priority, the 1875 Public Health Act was passed. At the same time a framework for water pollution control was introduced with the 1876 Rivers Pollution Prevention Act, although subsequent enforcement proved extremely difficult.

There were subsequently many similar pieces of national legislation introduced as environmental issues continued to oscillate on and off government agendas and into and out of public concern. Legislation came into force largely to solve particular problems. For example, the issue concerning poor air quality arising from coal burning in heavily populated areas escalated during the industrial revolution and urban smog disrupted daily life and contributed to rising death rates. Further legislation was introduced in the 1920s to reduce smoke from industrial sources but the localised use of coal as a domestic fuel continued to grow. The introduction of the Clean Air Acts of 1956 and 1968 was in response to the great smog of London in 1952. Subsequent legislation introduced since the 1970s targeted releases from vehicle fuel and industrial fuel oils. Traffic congestion, however, continued to grow in urban areas and with it the threat of photochemical smog caused by chemical reactions of pollutant releases with sunlight. These issues contributed to the passage of the 1995 Environment Act that subsequently required the publication in 1997 of the National Air Quality Strategy, setting air quality objectives for local authorities up to 2005.

This example illustrates the often issue-based and *ad hoc* approach to setting national environmental legislation. In recent times, a more integrated assessment has been adopted and the regulatory agencies in the UK have regulatory duties that cover air, land and water to reflect this holistic approach. This model has not been followed by many other European Member States.

The primary environmental legislation in England and Wales that relates to the release and control of waste discharges includes the 1990 Environmental Protection Act, the 1991 Water Resources Act, the 1995 Environment Act and the 1999 Pollution Prevention and Control Act. Some of the more important elements of these acts are discussed below.

The 1990 Environmental Protection Act introduced Integrated Pollution Control (IPC) to releases from prescribed industrial processes. It is divided into Parts A and B, the more technically complex and potentially polluting Part A processes being regulated by the Environment Agency, with some of the less complex processes under Part B regulated locally. The Act requires operators of prescribed processes to apply best available techniques (BAT) to prevent, minimise or render harmless the releases under their control through a system of prior permitting. Additional powers include specific controls on hazardous substances that allow the regulator to obtain relevant information from manufacturers, importers and suppliers about specific chemicals for the purpose of assessing their risk (section 142). Regulations to prohibit or restrict importation into the UK and the use, supply and storage of substances are also possible (section 140). The 1999 Pollution Prevention and Control Act and subsequent 2000 regulations make provision for the implementation of the EU Directive on Integrated Pollution Prevention and Control (IPPC) that replaces and extends the 1990 Act.

The 1991 Water Resources Act relates to the control of discharges to controlled waters. Section 85(1) states 'A person contravenes this section if he causes or knowingly permits any poisonous, noxious or polluting matter or any solid waste matter to enter any controlled waters'. The Act defines controlled waters and provides for the Secretary of State to establish water-quality objectives and the attainment of these objectives. The Act provides powers to the Environment Agency to prosecute a consent-holder if conditions are breached and to serve a works notice on an offender requiring them to remove or alleviate the pollution. Section 190 of the Act also established registers that include details such as: notices of water-quality objectives; applications made for consents; consents and the conditions to which the consents are subject; samples of water or effluent taken by the [Agency] for the purposes of any of the water pollution provisions of this Act; and information produced by analyses of those samples.

The 1995 Environment Act established the Environment Agency for England and Wales and the Scottish Environment Protection Agency, and sets out all provisions for the transfer arrangements of existing legislation.

Although this raft of legislation allows for the transposition of many European Directives, reference to the release of hazardous substances in the primary environmental legislation of the UK is not specific, although the Secretary of State is empowered to make regulations establishing quality objectives and standards. Much of the chemicals regulation in the UK results from European Directives, international conventions and agreements, and is dominated by controls on individual substances. By contrast, some countries make specific reference to the control of hazardous substances in their primary legislation and whole-sample toxicity procedures are embedded in the overall approach. Notable among these are the United States Clean Water Act 1972 and the Canadian Environment Protection Act 1999 (see also Chapter 1, Sections 1.3 and 1.4). For information on the UK Direct Toxicity Assessment (DTA) demonstration programme and the ecotoxicity test methods for effluent and receiving water assessment, the reader is referred to the sources of information given in the note at the end of Section 9A (page 268).

9A.3 Development of chemical regulations in Europe and the UK

The growing international awareness of chemicals in the environment that emerged during the 1970s following such publications as Rachel Carson's *Silent Spring* in 1962 had a profound influence on the development of regulations concerning hazardous substances and subsequently on the development of a hierarchy of international agreements, European Directives and national law. An overview of some of the more important international initiatives and European Directives is provided in Chapter 1.

The management and control of chemicals at all levels is complex and their regulation is made difficult because of:

- The movement of chemicals across national and international political and administrative boundaries.
- Incomplete information on the hazardous properties of many chemicals and on their use, transport and storage.
- The use of single substance-based standards in cases where complex mixtures of many chemical discharges result in compounded toxicity and difficulties in assessing the associated risks.
- The uncertainty surrounding the effects of chemicals on human health and the environment, especially low-level concentrations and long-term exposure.
- The many institutions and organisations involved in different aspects of the management of chemicals.

Important in a national context is the mechanism by which the large amount of legislation concerning the control on chemicals translates into enforcement action in Member States. In the UK this legislation and the associated enforcement are under the control of a large number of regulatory bodies. The more important pieces of national legislation, the associated European Directives and the competent authorities are summarised in Table 9.1. The list is not complete and excludes some additional regulations relating to other issues that concern the transport and storage of chemicals.

Table 9.1 illustrates the complexity of both the legislation and the institutional arrangements. This has resulted in a growing number of priority lists of chemicals of concern, produced largely in response to specific pieces of legislation. The lists often differ in their priorities and lack clear criteria for either selection or de-selection. Monitoring effort associated with the legislative requirements also differs from one Member State to another and there is little or no harmonisation of monitoring programmes.

9A.4 The role of a National Regulatory Agency

With so many pieces of legislation, associated priority lists of substances of concern and different enforcement authorities involved, it is perhaps not surprising that cradle to grave management is far from clear. This is revealed in a closer look at the legislative issues at different points in the life-cycle of chemicals and some of the roles, duties and obligations of a single regulatory agency.

The Environment Agency for England and Wales has specific roles in different regulatory regimes that vary not only for the type of chemical (industrial, biocide, pesticide, veterinary medicines, etc.) but also for the life-cycle stage (production, use, storage, release, waste disposal, etc.). Furthermore, there are different pieces of legislation that apply at the same stage. For instance, the control of chemicals released to the environment through industrial emissions, domestic sewage and waste-tip leachates.

Some of the various roles, duties and powers of the EA are shown in Figure 9.2 and include:

- Enacting part of the UK competent authority, on behalf of DEFRA and together with the Health and Safety Executive (HSE), for the notification of new substances and undertaking environmental risk assessments on existing substances.
- Authorising the release of chemicals to air and water from major industrial sources.
- Consenting discharges to watercourses from sewage treatment plants.
- Assessing the presence and effect of chemicals in, and on, the water environment and exercising powers to meet water-quality objectives.
- Determining operational requirements for landfill sites for hazardous and non-hazardous wastes.
- Administering the consignment system for special waste.
- Acting as the enforcement authority for contaminated land designated as special sites and advising local authorities on other determined sites.
- Reporting on emissions for regulated processes through the Pollution Inventory.
- Monitoring and reporting on the state of the environment for England and Wales.

Ways, other than legislative means, in which control can be exerted are also shown in Figure 9.2. These include codes of good practice and campaigns, fiscal measures such as the landfill tax and the proposed pesticide tax, and the publication of information. The Environment Agency's Pollution Inventory, launched in 1999, provides environmental information to a large audience, particularly the general public. The inventory is a database of emissions of pollutants to the environment from particular industries in the UK.

9A.5 Future developments

Legislation concerned with the assessment and control of chemicals is varied and complex and has resulted in a piecemeal development of the regulatory process.

Table 9.1 Summary of chemicals legislation in England and Wales and associated EU legislation.

Production, Market and Use

Notification of New Substances (NONS) Directive 92/32EEC came into effect following the 6th amendment of the Classification, Packaging and Labelling of Dangerous Substances Directive (CPL 67/548/EC) in 1979 requiring the pre-market notification of chemicals produced after September 1981. Manufacturers and importers are required to submit a notification dossier that includes data on environmental fate and effects to the nominated competent authority of Member States. In England and Wales this duty is shared between the HSE and the Environment Agency acting on behalf of DEFRA. The 7th amendment required risk assessment. During the period awaiting further assessment there is no marketing restriction on the sales of these substances.

Existing Substances Regulation (ESR Council Regulation 793/93) More than 100 000 existing industrial chemicals on the market prior to September 1981 are registered on the European Inventory of Existing Commercial Chemical Substances (EINECS). The ESR establishes priority lists and requires industry to supply all the necessary data for a comprehensive risk assessment. To date there have been four priority lists totalling almost 140 substances of high-volume production for risk assessment. The risk assessment is undertaken by the designated competent authority of the Member State, which in England and Wales is the same for NONS. There is no deadline under the regulation for the risk assessment or for possible trade sanctions where the producer fails to provide the necessary information.

Marketing and Use Directive for Dangerous Substances and Preparations (76/769EEC). The legislative framework for risk management of industrial chemicals enables the European Commission to propose measures to ban or restrict the use of substances. It places a general ban on those substances classified as carcinogens, mutagens or human reproductive toxins in products for use by the general public. Subsequent amendments have introduced daughter directives to place restrictions on specific substances such as polychlorinated biphenyls, mercury, cadmium and lead.

Positive Approvals. Certain product groups are required to meet demanding approval regimes. These include:

- *Plant Protection Products* (Pesticides Directive 91/414/EEC). In England and Wales the Pesticide Safety Directorate (PSD) is responsible for evaluating and processing applications for approvals under the Control of Pesticide Regulations 1986 (COPR). The regimes utilise many of the requirements of other regulations and directives (including CPL, NONS and CHIP – the Chemicals Hazardous Information and Packaging for Supply Regulations) and require information on composition, metabolism, fate and degradation products.
- *Biocidal Products* (Directive 98/8EEC). The HSE is the competent authority for authorisations of non-agricultural pesticides. Under the biocidal products directive, and COPR in England and Wales, the competent authority evaluates risk to human health and the environment on data provided by companies in support of approval applications.
- *Veterinary Medicines* (Directive 81/852/EEC). Veterinary medicines can be approved by the European Medicines Evaluation Agency (EMEA) or at national level. In England and Wales, the Veterinary Medicines Directorate (VMD) is responsible for the licensing of animal medicines under the Marketing Authorisations for Veterinary Medicinal Products Regulations 1994. It also has the power to grant emergency licences and regulate animal testing.
- *Pharmaceuticals* (Directive 65/65/EEC). The EMEA in Europe is responsible for the registration of new pharmaceutical products placed on the market after 1993. In England and Wales the Medicines Control Agency is responsible for licensing

Cont.

Table 9.1 *Continued*

human medicines under the Medicines for Human Use Regulations 1994. The regulations allow trial licences and manufacturer and dealer licenses to be issued.

Worker Protection
The basis for health and safety law in the United Kingdom is provided under *the Health and Safety at Work Act 1974 (HSWA)*, which sets out the general duties that employers have to their employees, and members of the public, and that employees have to themselves and each other. The legislation helps to fulfil wider European requirements for health and safety and places a general duty on manufacturers to ensure that substances are safe when properly used and to carry out such tests as are necessary.

The *Chemicals (Hazardous Information and Packaging for Supply) Regulations (CHIP)* that were enacted in 1993 under the HSWA, together with subsequent amendments, are concerned with the supply of dangerous substances and preparations and sets out how these should be classified, labelled and packaged. There is also a requirement on the supplier to provide safety data sheets.

The *Control of Substances Hazardous to Health Regulations 1994 (COSHH)* are set out under the HSWA and protect workers against risk to health from exposure to substances in the workplace. A risk assessment is required and must be made available to workers, with training given to those who might be exposed to such substances. The HSE is the competent authority for CHIP and COSHH in the United Kingdom.

Accidents
The Control of Major Accident Hazards Regulations 1999 (COMAH) implement the so-called Seveso II Directive 96/82/EEC. The regulations require operators to take steps to prevent the occurrence of accidents and to limit their consequences to people and the environment. Thresholds for storing and using certain dangerous substances are laid down in the Directive. Where higher threshold quantities are stored, a safety report and associated on-site emergency plan are required, together with an off-site plan from the local authority and the communication of safety measures to the public. The HSE and the Environment Agency operate the joint competent authority for England and Wales.

Releases to Air, Land and Water
Extensive legislation exists for controlling releases to the environment. The usual mechanism of control is through either a fixed emission limit and/or environmental quality standards that apply to both point and diffuse sources. Much of the legislation is specific to air, land or water, although integrated pollution control requires releases from major industrial processes to air, land and water to be considered together with the aim of minimising their impact on the environment.

The main legislation is the *Environmental Protection Act (1990)*, the *Water Resources Act (1991)*, the *Environment Act 1995* and the *Pollution Prevention Control Act (1999)* (see main text).

Air
The 1990 legislation introduced *Integrated Pollution Control (IPC)* under which prescribed processes are authorised by the competent authority in such a way as to ensure that operators apply 'best available techniques not entailing excessive cost' (BATNEEC) to prevent, minimise or render harmless releases under their control through a system of prior permitting. In England and Wales the Environment Agency is the competent authority for the larger and more complex Part A installations and Local Authorities for the Part B installations. Registers are kept by the competent authorities, detailing information relating to the authorisation and making this freely

Table 9.1 *Continued*

available to the public. The *Prevention Pollution Control Act 1999* represents a tightening of the 1990 Environmental Protection Act (EPA) and implements The Integrated Pollution Prevention and Control Directive (96/61/EEC). There is more emphasis on improving the environment by requiring industry to use 'best available techniques' (BAT) for pollution prevention – (the words 'not entailing excessive cost' have been dropped, although explanatory guidance states that the intent is unchanged).

The EPA (1990) established a system of Local Authority Air Pollution Control (LAAPC) requiring the competent authorities to ensure that smaller air pollution sources meet specific emission limits. In 1997 Central Government published the National Air Quality Strategy, setting air quality objectives for local authorities up to 2005.

Land
Part IIA of the Environmental Protection Act 1990 – which was inserted into that Act by Section 57 of the Environment Act 1995 – provides a new regulatory regime for the identification and remediation of contaminated land. Under the Contaminated Land (England) Regulations 2000 the local authority decide, in the first instance, whether land within the description of a special site is contaminated land or not. The work of the Environment Agency as enforcing authority starts once that determination is made. However, the statutory guidance on the identification of contaminated land says that, in making that determination, local authorities should consider whether, if land were designated, it would be a special site. If that is the case, the local authority should always seek to make arrangements with the Environment Agency to carry out any inspections of the land that may be needed, on behalf of the local authority.

Operation of the regime is subject to regulations and statutory guidance.
Section 78A(2) defines Contaminated Land for the purposes of Part IIA as:

'any land which appears to the Local Authority in whose area it is situated to be in such a condition, by reason of substances in, on or under the land, that:
(a) Significant harm is being caused or there is a significant possibility of such harm being caused; or (b) Pollution of controlled waters is being, or is likely to be, caused.'

For certain substances, soil quality guidelines have been produced to aid the risk assessment. The regulations require each enforcing authority to keep a public register. The public register is intended to act as a full and permanent record, open for public inspection, of all regulatory action taken by the enforcing authority in respect of the remediation of contaminated land, and will include information about the condition of land.

Water
The Water Resources Act (WRA) 1991 provides for the regulating authority, the Environment Agency in England and Wales, to issue consents that include information on the composition of the discharge, volume, concentration and total amount of substances that can be released. Conditions vary on the use of the receiving watercourse but must ensure no deterioration in water quality. Registers of information are kept and are freely available to the public. The *Water Industry Act 1991* requires sewerage undertakers consenting trade waste to notify the Environment Agency of effluents that are categorised as special if they contain certain hazardous substances. Consent conditions under WRA 1991 are set to achieve environmental quality standards determined as safe levels in the receiving water. The regulations are substance specific and often based on the achievement of a target load or concentration. The regulations enact provisions under a number of European Directives, including: the Dangerous Substances Directive (76/464/EEC) and amendments (1997); the Urban Waste Water Treatment Directive (91/271/EEC); the Nitrate Directive (91/676/EEC); and the Bathing Water Directive 76/160/EEC. Some of the newer

Cont.

Table 9.1 *Continued*

European Directives of particular importance to the water environment move to the achievement of defined environmental outcomes in the form of ecological status (Water Framework Directive 2000/60/EC) and habitat protection (Habitats Directive 92/43/EEC).

The Groundwater Directive (80/68/EEC) refers to the protection of groundwater quality from certain substances. It prohibits the discharge of substances/groups of substances in List 1 and requires pollution to be minimised from other substances in List 2. The *1998 Groundwater Regulations* are enforced by the Environment Agency. List 1 substances are the most toxic and must be prevented from entering groundwater. They include pesticides, sheep dip, solvents, hydrocarbons, mercury, cadmium and cyanide. List 2 substances are less dangerous, but if disposed of in significant amounts they could be harmful to groundwater. They include some heavy metals, ammonia, phosphorus and its compounds. Entry of these substances into groundwater must be restricted to prevent pollution. Control is also required where water is recharged to ground for storage for later drinking-water use. New arrangements will include requirements on monitoring and reporting. A new Groundwater daughter Directive will supplement the provisions of the Water Framework Directive (2000/60/EC).

Waste
The waste management licensing regime is enacted under part 2 of the *Environment Protection Act 1990*, allowing the Environment Agency to grant, modify, suspend or revoke licences for waste management. Chemical waste is covered under the provisions of the *Waste Management Licensing Regulations 1994*, meeting the requirements of the Waste Framework Directive 91/156/EEC. The *Special Waste Regulations 1996*, also made under the Environment Protection Act 1990, implement the requirements of the Hazardous Waste Directive (91/689/EEC) and place additional controls on movements, disposal and treatment of waste containing hazardous material. These regulations will be replaced by new Hazardous Waste Regulations that will introduce specific controls. There are many other Directives associated with chemicals control in waste management (see Chapter 1), including the Waste Incineration Directive (2000/76/EC) and the Landfill Directive (99/31/EC).

Furthermore, the large number of commercially available chemicals with little or no hazard information has impeded risk management and the introduction of targeted environmental monitoring programmes.

Inadequacies in some important elements of the legislative system have been identified at both international and national level. The Commission of the European Communities *White Paper on a Strategy for a Future Chemicals Policy* (2001) and the UK Government strategy *Sustainable Production and Use of Chemicals* (DETR, 1999) both recognise the need for greater transparency and a more timely approach to the regulations with regard to the risk assessment of existing substances. The White Paper proposes a new regulatory system comprising three components: registration, evaluation and authorisation (of chemicals) – REACH. Further information on the proposal is provided in Chapter 1.

The 21st report of the Royal Commission on Environmental Pollution (1998), *Setting Environmental Standards*, called for transparency in the process of setting standards and a programme informed and steered by public values. The most recent (24th) report of the Commission (2003), *Chemicals in Products*, recommends a more fundamental shift in the way that risks from chemicals are managed.

APPROACH TO LEGISLATION IN A GLOBAL CONTEXT 267

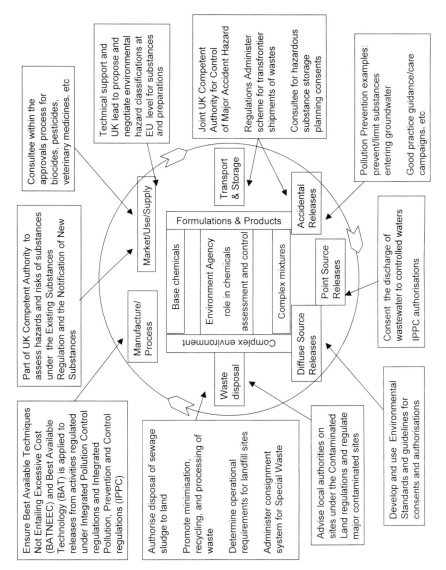

Figure 9.2 Roles, duties and powers of the Environment Agency in chemicals assessment and control.

This report indicates that risk assessment on its own is insufficient, and that monitoring to help understand the fate and effects of chemicals in the environment is not being carried out effectively.

The undoubted benefits of chemicals must be balanced with the serious threat they pose to the environment if left unchecked. Many issues remain and the development of a more strategic approach might consider:

- The institutional arrangements, with a view to reducing unnecessary bureaucracy and improving current understanding.
- The timely application of appropriate regulations for production, marketing and use, and improved understanding of how these sit alongside those used for regulating releases to the environment and waste management.
- The harmonisation of procedures for cradle-to-grave risk assessment of chemicals and control using the most appropriate legislation, and alternative approaches and options other than legislative means of control.
- Criteria for selecting and de-selecting priority substances of concern and the inclusion of monitoring information to aid risk decision-making. In particular, improved understanding of biological responses and their human and ecological significance.
- Exposure to complex mixtures (the usual route of entry to the environment) and the development of techniques and procedures for improved screening and control.
- Accessible information and improved transparency in any approval process. Wider stakeholder involvement should be encouraged without diluting the value of expert judgement.
- Wider stakeholder engagement and socioeconomic implications

Despite these outstanding issues the need for greater international cooperation and harmonisation of approach with respect to legislation concerning the assessment and control of chemicals is recognised by national governments and international organisations. Whole ecosystem management for the benefit of man and wildlife will be necessary to tackle complex issues on biodiversity and habitat protection and these will need the wide involvement of many scientific disciplines to provide the necessary evidence to underpin policy development and decision-making to achieve sustainability.

Notes

Some useful sources of information on the UK Direct Toxicity Assessment (DTA) demonstration programme:

UKWIR (2000) Technical Guidance for the Implementation of Direct Toxicity Assessment (DTA) for Effluent Control: Addressing Water Quality Problems in Catchments where Acute Toxicity is an Issue. Report TX02B 217.

UKWIR (2000) UK Direct Toxicity Assessment (DTA) Demonstration Programme: River Aire Project. Report 00/TX/02/01

UKWIR (2000) UK Direct Toxicity Assessment (DTA) Demonstration Programme: River Esk Project. Report 00/TX/02/02

UKWIR (2000) UK Direct Toxicity Assessment (DTA) Demonstration Programme: Lower Tees Estuary Demonstration Project Parts I and II. Reports 00/TX/02/03 and 00/TX/02/04

UKWIR (2000) UK Direct Toxicity Assessment (DTA) Demonstration Programme: Review of Toxicity Reduction Evaluations at Sewage Treatment Works. Report 00/TX/02/05

UKWIR (2000) UK Direct Toxicity Assessment (DTA) Demonstration Programme: Recommendations from the Steering Group to the Environmental Regulators. Report 00/TX/02/06

UKWIR (2000) UK Direct Toxicity Assessment (DTA) Demonstration Programme: Technical Guidance. Report 00/TX/02/07
(*Note*: UKWIR is UK Water Industry Research Ltd, 1 Queen Anne's Gate, London, SW1H 9BT, UK)
Ecotoxicity Test Methods for Effluent and Receiving Water Assessment: Comprehensive Guidance, October 2001, Environment Agency (Available from Biological Effects Lab, 4 The Meadows, Waterberry Drive, Waterlooville, PO7 7XX, UK)

References

DETR (now DEFRA) (1999) *Sustainable Production and Use of Chemicals*. Department of the Environment, Transport and the Regions, London, UK.
European Commission (2001) *Commission of the European Communities White Paper on a Strategy for a Future Chemicals Policy*, COM(2001) 88 final. ICEC, Brussels.
Royal Commission on Environmental Pollution (1998). *Setting Environmental Standards*. (21st Report). Royal Commission on Environmental Pollution, Westminster, London.
Royal Commission on Environmental Pollution (2003) *Chemicals in Products* (24th Report). Royal Commission on Environmental Pollution, Westminster, London.

B THE NETHERLANDS PERSPECTIVE – SOILS AND SEDIMENTS

Michiel Rutgers and Piet den Besten

9B.1 Developments in soil contamination policy in The Netherlands

In The Netherlands, the policy for the protection of soils and sediments has a history of about 20 years. In the 1980s, the first cases of soil contamination in inhabited areas were discovered and the first remediation plans were put into action. Since then, a vast number of contaminated sites have been discovered. The latest estimates indicate that there are 175 000 seriously contaminated sites (RIVM, 1999).

As a response, a policy framework was developed and introduced in 1986, with an update for the remediation of contaminated sites in 1994, the so-called Soil Protection Act (SPA). The main purpose was to establish the accountability of individuals in contributing to soil and sediment pollution, and to include the question of financial responsibility. In the SPA, the aim to preserve soil quality has a functional basis: the quality of the soil is important due to the functional properties of soil. It is not stated explicitly but it can be assumed that the organisms responsible for soil functions have to be protected. In the SPA, soil and freshwater sediment are considered to be closely linked environmental compartments that require a uniform framework for protection and remediation.

In 1994, intervention values (IVs), target values (TVs) and the methodology to determine the urgency of remediation were formalized in the SPA. The TVs and IVs specify the levels of slightly and seriously contaminated soil, respectively

(Figure 9.3). They are often based on scientifically derived environmental quality criteria. Both human and ecological risk limits are used for deriving these values. The procedure for obtaining these values, and the values themselves, are described in several reports and publications (INS, 1999; Swartjes, 1999; VROM, 2000). When a site is seriously contaminated and when remediation is considered urgent according to the urgency methodology, the remediation objective is to end up with clean soil, i.e. contaminant levels lower than the TV. This approach is multifunctional because below the TV the soil is considered to be clean, allowing all kinds of land-uses and soil functions.

The Ministry of Housing, Spatial Planning and the Environment decided to re-evaluate the clean-up regulations of the SPA in 1996. Cleaning up the soil became too expensive and clean-up operations were not cost-effective enough because too few cases were tackled per unit of time. Furthermore, the size of the soil contamination in relation to the potential resources available for tackling the problem had led to stagnation in spatial planning and economic processes.

In 2002, changes were incorporated in the SPA. The major adjustment was to make remediation cheaper by the introduction of land-use-dependent remediation objectives (LRO; Figure 9.3) (BEVER, 1999). Consequently, the multifunctional approach was abandoned. It is no longer necessary that each remediation project should end in soil with contamination levels below the target values.

9B.2 Environmental quality criteria

The environmental standards for various compounds are derived from toxicity data in the literature. In The Netherlands, quality criteria based on human and

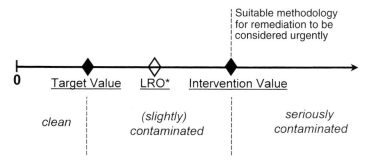

Figure 9.3 Soil, sediment and groundwater quality standards in The Netherlands. LRO = land-use-dependent remediation objective (see text). If the contamination exceeds the IV, remediation is urgent, unless there are no actual risks for dispersion, human health and the ecosystem. In the methodology on remediation urgency (Swartjes, 1999) the exceedance of the HC_{50} values (page 280) is judged against the contaminated area and the land-use. Both parameters influence the priority for remediation. For instance, when the exceedance is smaller than 10 times the local HC_{50} and the contaminated area is smaller than 0.5 km^2 for an industrial site, remediation is considered 'not urgent' with respect to ecological risks.

ecological toxicity are compared and the most stringent values are used in the setting of standards for serious soil contamination. Details about human risk assessments in The Netherlands are described by Swartjes (1999).

In general, ecotoxicological quality criteria in The Netherlands are based on a probabilistic approach via *species sensitivity distributions* (SSDs). Van Straalen and Denneman (1989) have developed this methodology and further advancements are described by Posthuma *et al.* (2002). It was proposed that a cumulative logistic distribution of the logarithm of no-observed-effect concentration (NOEC) data is satisfactory for calculation of the SSD. Historically, the 5th percentile of the distribution has been chosen as the cut-off point, defining the HC_5 (hazardous concentration that is equal or higher than the NOEC for 5% of the exposed species in laboratory tests). This value is called the maximal permissible concentration (MPC), which is subsequently used to derive the target value (TV) in standard setting. The intervention value (IV), the concentration at which the soil is seriously contaminated, is at the 50th percentile of the SSD curve. The TVs and IVs for about 125 compounds and groups of contaminants have been described, including analytical methods for determining these values (VROM, 2000).

9B.3 Towards site-specific approaches

Despite a substantial framework for the derivation of target and intervention values and land-use-dependent remediation objectives, there is a strong recognition of an additional need for site-specific approaches (Ferguson *et al.*, 1998; Nijhof and Koolenbrander, 1998; Rutgers *et al.*, 2000). The vast number of seriously contaminated sites is still increasing and the question arises whether it is necessary to remediate them all. Furthermore, some other problems may arise because of special characteristics of the contaminated site. For instance, many sites in The Netherlands are too big or too complex for clean-up. They also may contain specific cultural, geographical or natural values. For these cases tailor-made solutions should be designed for dealing with the site in a dependable and sensible way, in order to maximize environmental gains and minimize costs for society. The idea was expressed recently in the update of the soil protection policy (VROM, 2001). For instance, in the case of nature and agricultural land-use it is expected that site-specific approaches are the rule rather than the exception. To date, no general Dutch framework for site-specific ecological risk assessment (ERA) in terrestrial ecosystems exists, except for the method to determine the priority for remediation.

9B.4 Risk perceptions and negotiation formats

Many parties are involved in ERA, including local site-owners, local and regional authorities and the public. Consequently, for the entire process of decision-making a transparent procedure should be followed. Each step should be addressed carefully and decided upon by all parties. Essentially, decisions should be made about the land-use, about ecological aspects of interest for the functioning of the ecosystem, about the protection of species or ecotypes and about the instrumentation to be used for ERA (type of investigations and criteria for judgement). In this process different views on risks, nature values, ecological management and sustainable land-use will become manifest. The various ways to cope with this are covered by Power and Adams (1997), the USEPA (1998), Suter et al. (2000), Rutgers et al. (2000) Van de Leemkule (2001), and Den Besten et al. (2003).

9B.5 Many ways to improve site specificity in ERA

The basis for the derivation of environmental quality criteria is a large set of literature toxicity data. Instead of calculating HC_5 and HC_{50} values, these toxicity data can be used to attain effect values for the levels of pollutants at the site. Furthermore, site-specific effect values can be derived while accounting for specific features of the ecosystem, food chain implications, differential bioavailability and summing up the effects of the complete cocktail of contaminants while taking into account natural background concentrations. Discussion about the theoretical background and potency of SSDs for ERA is beyond the scope of this chapter; for this, the reader is referred to a recently published book (Posthuma et al., 2002).

There are many alternative techniques available that are beneficial for site-specific ecological risk assessments, such as the application of bioassays, biomarkers and field ecological monitoring (i.e. biological methods). In contrast to the application of literature toxicity data, the development of frameworks for the application of biological methods in ERA is way behind. In The Netherlands, there are some initiatives for the application of bioassays (reviewed by Den Besten et al., 2003) but not for biomarkers or field monitoring. It is to be expected that these frameworks will be developed soon. The general motivation to use biological methods originates from the idea that, due to so-called conceptual uncertainties, substance-directed approaches sometimes simply fail to notice existing risks (e.g. in the case of an unknown contamination, or complex contaminant mixtures) or sometimes tend to overestimate risks (e.g. in the case of a limited bioavailability). Biological methods integrate the effects of all contaminants present at their actual bioavailability (and detect possible combination or synergistic effects). Conceptual uncertainties also can arise from not taking into account specific ecosystem characteristics, such as sensitive species, or incorrectly addressing the land-use. Ecological risk assessment will always suffer from conceptual uncertainties and the application of biological methods will not solve this (Suter, 1998; USEPA, 1998; Suter et al., 2000). Despite this general limitation, biological methods inevitably will add to the improved understanding of ecological risks (Lancaster, 2000).

9B.6 Weight-of-evidence (WOE) approaches

In order to deal in a pragmatic way with conceptual uncertainties, it was proposed to use weight-of-evidence (WOE) approaches for ERA (Long and Chapman, 1985; Den Besten et al., 1995; Hall and Giddings, 2000; Rutgers et al., 2000; Suter et al., 2000; Burton et al., 2002; Chapman et al., 2002). The WOE rationale is akin to that in the legal process, in that there are many independent ways to arrive at one conclusion. This provides stronger evidence of ecological effects, making ERA less uncertain.

Basically, in the sediment research area the application of WOE started at an early stage and was called the Sediment Quality Triad (Long and Chapman, 1985). For terrestrial ecosystems WOE approaches and the Triad are in a developing stage (USEPA, 1998; Suter et al., 2000; Mesman et al., 2003; Rutgers et al., 2001). The Triad approach is based on the simultaneous and integrated deployment of site-specific chemical, toxicological and ecological information in the risk assessment. The major assumption is that WOE in three independent disciplines will lead to a more precise answer than an approach that is solely based on, for example, the concentrations of pollutants at the site. A multidisciplinary approach will help to minimize the number of false-positive and false-negative conclusions in ERA. It also gives acknowledgement to ecosystems as being too complex for analysing in one-factorial approaches.

In this part of Chapter 9 the current situation and progress towards application of biological based methods in ERA is reviewed, including the development of WOE frameworks. For aquatic ecosystems and sediments, bioassay test systems and WOE frameworks already exist and they are currently being used. For the terrestrial environment the situation is still immature. After briefly presenting the aquatic and sediment frameworks, including key references, the developments for the terrestrial environment are discussed. The preliminary terrestrial framework is demonstrated on the basis of the results of investigations at a contaminated site in The Netherlands.

9B.7 Aquatic ecosystems

Application of bioassays for toxicity testing in aquatic ecosystems is well under way. Recently, a risk framework for a preliminary and a refined effect assessment has been suggested (Straetmans et al., 2003). As part of the toxicity measurements, organic contaminants in aquatic samples are $100\times$ concentrated (using a XAD resin and acetone) and toxicity experiments are performed with dilution series (De Zwart and Sterkenburg, 2002). Major features of this framework are:

- *Preliminary* effect *assessment*. Three acute or chronic bioassays from different taxonomic groups are prescribed. Three cut-off effect levels are discerned, i.e.

a negligible effect level, maximal permissible effect level and a serious effect level. The EC_{50} is expressed in a concentration factor of the sample (i.e. the required sample concentration to obtain 50% response in the respective bioassay) and determines the level of toxicity.
- *Refined* effect *assessment*. The results from four acute or chronic bioassays of different taxonomic groups are expressed in a concentration factor where 50% effect is measured (EC_{50}). These data then are used to construct a sensitivity distribution (De Zwart and Sterkenburg, 2002). The potentially affected fraction (PAF) is determined for the 100% sample (the 'as is' sample). The negligible effect level is at a value for PAF = 5%.

9B.8 Sediments

Ecological risk assessment approaches currently used in The Netherlands for aquatic sediments have been described recently by Den Besten *et al.* (2003). Two main goals for sediment quality assessment in Europe are distinguished:

1. Biological effects-based assessment of *in situ* risks (*in situ* BEBA) at sites where sediment quality and potentially sediment management is to be considered.
2. Biological effects-based assessment of the *ex situ* quality of dredged sediments (*ex situ* BEBA) in order to select sediment management options (e.g. free or confined disposal or treatment options).

In this chapter we will discuss only the assessment of *in situ* risks. For assessment of *ex situ* risks, see Den Besten *et al.* (2003).

Site-specific assessments of the *in situ* risks of sediment pollution in The Netherlands are carried out mainly in freshwater systems. The *in situ* BEBA is then part of a broader evaluation of the risks caused by sediment pollution, aimed at the question of whether the risks make sediment remediation necessary. For this evaluation, a tiered approach is followed:

- *First tier assessment: comparison of levels of priority pollutants with national standards/guidelines*. Chemicals measured routinely are mineral oil, chlorobenzenes, organochlorine pesticides, polychlorinated biphenyls (PCBs: standard group of seven congeners), polycyclic aromatic hydrocarbons (PAHs: 16 from the USEPA) and heavy metals. Contaminant levels are normalized according to the approach described by the CUWVO (1990) in order to compensate for differences in sorption characteristics between sediments. (Standard sediment is defined as having a 25% particle fraction $< 2\,\mu m$ and 10% organic matter on a dry weight basis.) Normalized contaminant levels are then compared with the Dutch sediment quality criteria (developed for first tier assessment of risks for human health and ecosystems). According to the resulting classification, most polluted sediments (class 4 on a scale from 0 to 4) require a risk assessment (second tier).

- *Second tier assessment.* The primary statement for this second tier assessment is as follows: if a priority pollutant exceeds the intervention value (IV), the site needs to be remediated urgently, unless it is shown that there are actually no risks. Thus, there is an assumption of risk until it is disproved. If the data supplied from the second tier show that there is actually no risk at a site where a priority pollutant exceeds the IV, then the need for remediation is no longer considered as urgent. Conversely, if actual risk was confirmed, the next step would review different remediation options that are to be compared for the expected risk reduction. Three main pathways are considered within this tier for achieving a complete risk assessment (Swartjes, 1999), namely human exposure, the risk of transport or dispersion of the contaminants and an *in situ* BEBA approach. The latter is based on a Triad approach. In the Dutch version of the Triad for aquatic sediment, bioaccumulation measurements are also considered, using the results of laboratory tests or, preferably, by measurements using indigenous organisms (Den Besten *et al.*, 1995). Based on the most sensitive parameter, sediments are classified for the categories 'field observations' and 'bioassays' as either '−' (no effect/risk), '±' (moderate effect/risk) or '+' (strong effect/high risk). The goal is to elucidate the relationship between effects on macrozoobenthos and responses of bioassays which, in turn, can be related to levels of chemical pollution. For that purpose, chemical concentrations are converted into 'toxic units' (TU): the ratio between the chemical's normalised concentration and the lowest NOEC reported in the literature among the bioassays included in the battery (Den Besten *et al.*, 1995). High risk is inferred when strong effects are observed in field surveys and/or bioassays that can be related to chemicals present in the sediment.
- *Prioritization.* When the supplied data from the second tier show that there are actually no high risks at a site where a priority pollutant exceeds the IV, the need for remediation is no longer considered urgent. In the case where actual high risks were confirmed, a next step is possible in which different remediation options are considered for the risk reduction that can be achieved. The information from the sediment quality assessment can be used again in setting priorities within the group of locations that urgently need to be remediated. In The Netherlands, some experience exists with the use of multicriteria analysis (MCA; also called the analytic hierarchy process; Saaty, 1980) for this purpose. Multicriteria analysis enables a ranking of sites based on risks for the ecosystem. This method is based on the same classification of results as described above (Den Besten *et al.*, 1995). For *each* criterion (= parameter), standard numerical values (scores) were assigned to the effect/risk classes, from the value 1 for the class representing the strongest effect or highest risk to, for example, 0.5 and 0.25 for the classes representing moderate risk and no risk, respectively. Then the criteria are given a specific place and weight in a hierarchy. The scores are multiplied by the weight of the corresponding criterion and subsequently totalled, resulting in a final score between 0 and 1.

The difference between the final and theoretical score 1 (the score for a site with strong effects/high risk for all parameters) gives an indication of the risks for ecosystem health at each of the sites. For this method all available information from the field surveys can be used, including bioaccumulation data. At a higher level of a decision hierarchy, information from human risk studies, ERA and estimates of contaminant mobility (transport) can be integrated. In the MCA, specific weights can be attributed to the different criteria (= parameters) and, higher in the hierarchy, at branch points. This makes the method useful for decision-makers, who have to deal with all these aspects at the same time and therefore need integrated information. In the near future, estimates of the expected beneficial effects of remedial action will be integrated in the step of prioritization of dredging locations.

The application of bioassays in BEBA provides the risk assessor with more information about the exposure of organisms in contaminated sediment. At the same time, this approach also creates concern with regard to adequate quality assurance of the techniques. Several issues are of great importance when using bioassays for the evaluation of sediment quality. Firstly, bioassays are subject to a number of confounding factors that may have nothing to do with contaminant load (such as grain size, ammonia and countless other issues). Secondly, it is very important to define references and controls that are meaningful for the site under consideration. A third point of concern is the question of whether all relevant modes of action can be covered by a set of bioassays. For instance, if only bioassays are used that measure acute toxicity, sublethal modes of toxicity (effects on fecundity, growth, immunocompetence, etc.) could be overlooked, with important consequences for ecosystem health.

Chemical measurements have developed over the past decade. At present, sophisticated methods are available that can characterise the bioavailability of contaminants (Cornelissen *et al.*, 2001; Burgess *et al.*, 2003). These techniques may prove to be powerful in constructing lines of evidence between contamination and direct effects on organisms living in the sediment.

From the perspective of the EU Water Framework Directive, which focuses primarily on water quality, it is expected that sediment quality assessment will be employed primarily as a diagnostic tool, in case the question is raised of whether sediment pollution should be held responsible for not meeting ecological water quality criteria.

9B.9 Triad in terrestrial ecosystems and selection of biological tests

The weight-of-evidence approach (WOE) is becoming increasingly important for ERA in terrestrial ecosystems. There is a broad range of bioassays, biomarkers and ecological field parameters applicable. Different kinds of implementation levels can be discerned, such as the level of standardization, the ease and costs of analysis, the sensitivity for typical contaminants, the level of biological organisation in the test (sub-organism, individual, population and community-level) and the level of relevance for the site (is the species present at the site?, is the endpoint relevant for the functioning of the ecosystem?, etc.). For many potential possibilities, the reader is referred to other chapters in this book and to the literature (Ferguson et al., 1998; Fairbrother et al., 2003; Markert et al., 2003).

Because there are many selections possible, it is very important to make the goals of the ERA clear and how the experimental set-up is aiming to contribute to these goals (Rutgers et al., 2000; Suter et al., 2000) via a single test or a battery of tests. For instance, when it is important to focus on the cause versus effect relationship (do the 'known' contaminants indeed cause ecological effects?), the selection of tests should focus on this relationship. Suter et al., (2002) formulated a formal system for inference of causality between effects and possible causes based on Koch's postulates. For instance, when a certain test organism responds in a bioassay, it is informative to look for the same species in the field. When certain contaminants are predominantly causing effects via food transfer and bioaccumulation (i.e. some persistent and bioaccumulative pesticides), it makes more sense to look for effects in populations of bioaccumulative organisms (e.g. birds, mammals) than in microbial communities. Therefore, in The Netherlands ERA frameworks for aquatic ecosystems and sediments are designed to focus on the explanation of ecological effects from the presence of contaminants (Den Besten et al., 1995, 2003).

When the aim of ERA is to provide general quality figures for the site, including effects from 'unknown' contaminants or mixtures, a general set of tests should be used that focus on a broad spectrum of contaminant effects. Following this approach, the chance for false negatives is reduced. Also, for general monitoring of activities broadly responding tests should be preferred over specific tests.

Usually, site-specific risk assessment contains one or more tiers: each subsequent tier is used when uncertainty in the risk estimation in the last tier is considered too high (USEPA, 1998; Suter et al., 2000). Usually, in the lower tiers cost-effective and quick biological methods are applied, whereas the higher tiers contain more sensitive and sophisticated bioassays and field monitoring techniques. Ecological risk assessment is finished when the level of uncertainty is considered acceptable for making management decisions about remediation, designing land management, adjusting land-use and the like.

9B.10 Reference data from reference sites, reference samples and literature

A crucial factor in a risk assessment is the quality of the reference data, because the results of the site-specific ecological measurements or calculations are weighed against these data. Reference data can be obtained by including reference sites (preferably more than one) in the sampling scheme, including reference measurements in the experimental set-up or by obtaining reference data from the literature (Bailer et al., 2002; Didden 2003) or expert-based judgement (Chapman et al., 2002). An optimal reference site is in all aspects equal to the contaminated site except for the contamination. In practice, these ideal spots are difficult to find. If there is no or inadequate reference information, effects can be determined only in relative terms by comparison with other sites. This is usually adequate for determining the degree of urgency and further prioritization of clean-up work or the choice of site in the context of spatial planning and the development of areas of natural beauty.

9B.11 Quantification of results from terrestrial tests

Essentially the results from all tests, in particular bioassays and ecological field surveys, should be funnelled into the risk assessment framework. The primary aim is to maximize the utilization of the results of particular tests (as quantitative as possible) and to use results from all tests together in a transparent integrative scheme (e.g. a decision matrix). Burton et al. (2002) reviewed several possibilities for disseminating final WOE findings and concluded that tabular decision matrices are the most quantitative and transparent. Ideally, ERA in The Netherlands for aquatic, sediment and terrestrial systems should follow the same set of conventions for clarity. In practice, however, there are slight differences for the following reasons: (1) there is a wide range of standardization and sensitivity levels in terrestrial methods, making it difficult to define one set of homogeneous 'rules' for interpretation; (2) interpretation of test results in terms of 'effect' or 'no effect' inevitably results in the loss of (sometimes valuable) quantitative information; and (3) preconcentration techniques can be applied to aquatic samples but cannot be applied to soil samples. There is insufficient pore water for concentration.

In order to derive a quantitative decision matrix for easy evaluation and integration of results of different tests in the Triad legs (chemistry, toxicity, ecology), it was proposed to use an effect scale running from 0 (no effect) up to 1 (unlimited effects) in the undiluted samples (100% samples) or in the field. The results from each parameter (bioassay, biomarker or ecological field survey) should be projected on this effect scale according to the best available knowledge or best professional judgements. Different tests obviously will require different approaches. For instance, for a growth test the percentage inhibition can be used directly as the unit for effects. For ecological field monitoring, the results should be scaled relative to the ecological state of the reference site ($= 0$) and a (theoretical) state indicating 100% effects. Projection of test results on this effect scale requires a lot of experience. Fortunately, the WOE approach will help to address and correct mismatches of specific scaling methods due to wrong assumptions (Chapman *et al.*, 2002). Accordingly, lower tiers in the Triad approach should contain tests that are reasonably standardized, whereas at higher tiers the less-standardized tests should be used to improve on the level of site specificity.

Once all results are scaled into a uniform effect value, the overall response of a set of (biological) methods can be calculated. For this, the geometrical mean was used of the 'reversal' effect ($1 -$ effect). Back-transformation of this value gives one integrated effect value per Triad leg. In this way some extra weight is put on results from tests that demonstrate a positive ecological effect. The rationale is that biological methods, especially at the screening level, might be relatively insensitive and produce false-negative results.

In The Netherlands, this Triad approach for terrestrial ecosystem is being tested at a number of contaminated sites. An example of this testing at contaminated sewage fields nearby Tilburg, The Netherlands, is presented in the following sections.

9B.12 Site, sampling, soil characteristics and biological assays

The former sewage field area of about 100 ha is located north of Tilburg, The Netherlands. The site consists of many small fields and a system of inlet and outlet canals. The area is now set up as a forest called the 'Noorderbos'. The field sampling was performed in May 2000. The dimension of the three fields was about 85×25 m, for which a selection was made on the basis of the expected contamination levels (based on former investigations at the site; B. Muijs, pers. comm.). The samples (sandy topsoil 5–20 cm) were transported to the laboratory and soil characteristics were determined according to standard procedures (Table 9.2). Details of this field study can be found in a report (Schouten *et al.*, 2003). The field ecological analyses were carried out partly by Alterra (Wageningen) and the Bedrijfslaboratorium voor Grond en Gewasonderzoek (Wageningen): nematodes, enchytraeids and earthworms. Chemical analysis was carried out by the RIVM and Alterra. Analysis of pollution-induced community tolerance, community-level physiological profiling and the bioassays were carried out by the RIVM.

9B.13 Calculation of toxic pressure from the contamination levels

Species sensitivity distributions (SSD) on the basis of NOEC values in the literature were used to calculate the toxic pressure (TP: an effect value) from the presence of contaminants in the soil or in the pore water (Rutgers *et al.*, 2001; Posthuma *et al.*, 2002). The effect of the contaminants was corrected for the contaminant levels in the local reference samples (field A):

$$TP_{anthropogenic} = 1 - (1 - TP_{sample}) \cdot (1 - TP_{reference})^{-1}$$

Table 9.2 Soil characteristics and contamination levels in soil samples from sewage fields near Tilburg. Sample A can be considered the local reference sample and sample C has the highest contamination levels (Schouton et al., 2003).

Tilburg sewage fields	Sample A	Sample B	Sample C	Local IV[a] (field A)
Soil characteristics				
pH (KCl)	4.5	5.0	4.7	
Organic matter (%)	5.9	5.9	11.4	
Phosphorus (PAl in mg P_2O_5 / 100 g)	55	76.0	146.0	
Clay content (%)	3.0	2.0	3.0	
Moisture (%)	8.0	7.6	19.4	
Chemistry				
Total concentrations (mg/kg dry wt soil)				
Cadmium	0.33	0.48	1.66	8
Chromium	187[b]	923[b]	1713[b]	129
Copper	10	18	60	107
Nickel	5	21	51	78
Lead	27	44	132	340
Zinc	46	88	257	349
Mercury	0.1	0.2	0.9	7
Pore water content (μg/l)				SRC_{eco}[c] (surface water)
Arsenic	40	167	384	890
Cadmium	10	42	47	9.7
Chromium	216	585	971	220
Copper	179	404	1435	19
Nickel	91	337	634	500
Lead	82	83		150
Zinc	325	400	1355	91

[a] Local IV = local intervention value after correction for the soil characteristics based on the HC_{50} (hazardous concentration that is equal to or higher than the NOEC for 50% of the species in laboratory tests).
[b] Exceedance of the HC_{50}.
[c] SRC_{eco} = serious risk concentration, a risk value based on the HC_{50} for surface water, including a correction for the background concentration. Note that the use of surface water toxicity values seems to result in unrealistically high values for calculation of the toxic pressure in pore water in the case of the local reference (sample A). Consequently, calculation of the toxic pressure from pore water concentrations is questionable (see Table 9.4).

Furthermore, the toxic pressure (TP$_{mixture}$) of all contaminants together was estimated from a simple formula:

$$TP_{mixture} = 1 - \Pi_i(1 - TP_i)$$

for i = 1 to n contaminants, following a response addition model for summing up the effects of individual contaminants in mixtures (Posthuma et al., 2002).

9B.14 Determination of toxicity in samples using bioassays

Three bioassays (toxicity tests) were applied to determine the toxicity in the soil samples, i.e. the standard Microtox® toxicity test on soil elutriates, an algal test on soil elutriates using pulse-amplitude-modulated fluorescence (PAM) and a lettuce germination test in soil samples. Details about the test conditions can be found in reports by Rutgers et al. (2001) and Schouten et al. (2003). The results were scaled on an effect scale running from 0 (no toxicity) to 1 (maximal toxicity, i.e. complete inhibition of the response of the test organism). In this case a simple scaling method was used: the fraction of inhibition equals the fraction of effect. The results of the bioassays are summarised in Table 9.3.

9B.15 Ecological field observations

A series of ecological field observations were carried out. Many different types of analyses were performed because the aim of this pilot study was to focus extra attention on this leg of the Triad. The selection of tests was based on the biological indicator for soil quality (BiSQ; Schouten et al., 2000), comprising several microbial parameters (biomass, leucine and thymidine incorporation rate, potential carbon and nitrogen utilisation, microbial community structure), nematode community structure and species abundances, enchytraeid community structure

Table 9.3 Toxicity in soil samples from sewage fields near Tilburg using three bioassays. Sample A is the local reference site and sample C has the highest contamination levels.

Toxicology	Sample A	Sample B	Sample C
Microtox® (EC$_{50}$ in % elutriate) (95% confidence interval)	182[a], (11–2843)	106[a] (34–332)	76 (54–105)
PAM-algal test (% inhibition after 4.5 h exposure)	–4.6	–5.2	–0.6
Lettuce germination test (% after 3 days)	100	95	90

[a] Extrapolated values (EC$_{50}$ was not reached in undiluted sample).

and abundances, earthworm community structure and abundances, and a test on the pollution-induced community tolerance (PICT) of microbial communities (Boivin et al., 2002). In all cases the results were scaled on an effect scale running from 0 (observations at reference site A) to 1 (maximal disturbed situation, based on expert judgement). Details about the analysis can be found in a report (Schouten et al., 2003). The results of the ecological field observations are summarized in Table 9.4.

9B.16 Integration of Triad results and calculation of ecological effects

After conversion of the test results in effect values ranging from 0 to 1, the Triad final assessment (risk) result was derived. A simple weighting algorithm was applied by giving equal weight to each test system. For instance, determination of the taxonomic composition of the nematode community resulted in one effect value for the nematode community. An integrated effect value per Triad leg was calculated from the geometric mean of (1− effect) of all observations (bioassays, ecological field observations):

$$\text{Integrated effect} = 1 - 10^{(1/n \cdot \{\log(1-\text{effect})\text{test1} + \log(1-\text{effect})\text{test2} + \ldots \log(1-\text{effect})\text{testN}\})}$$

In the final stage, the outcome of the Triad was evaluated by the deviation factor. This is the standard deviation of the integrated effects from the Triad legs (chemistry, toxicity, ecology) divided by 0.58 (see Rutgers et al., 2001, for an explanation).

9B.17 Results from the pilot at Tilburg

The focus of the pilot was on the development of a practical framework for the application of biological test methods in a Triad approach. Consequently, the selection of tools for the assessment was based on scientific and pragmatic arguments, for instance by focusing on readily available techniques for determination of the concentration of contaminants in pore water and readily available biological tests such as simple bioassays and the monitoring of soil organisms.

The three fields (A, B and C) clearly demonstrated a trend in contamination levels for most metals, both for total and pore water concentrations (Table 9.2). Chromium concentrations were above the intervention values (IV) for soil. Copper and zinc concentrations were slightly lower than the IV, still indicating significant contamination. Soil pH was similar at all locations, but organic matter content was higher for field C, corresponding to the former inlet of the sewage.

The toxicity in the soil samples of field A, B and C was determined using three bioassays (Table 9.3). For the Microtox® measurements, toxicity was the lowest

Table 9.4 Ecological characteristics in soil samples from sewage fields near Tilburg (sample A is the local reference site and sample C has the highest contamination levels).

Tilburg sewage fields	Sample A	Sample B	Sample C
Micro-organisms			
Bacterial biomass (μg C/g soil)	50	65	91
Thymidine incorporation (pmol/g.h)	39	61	99
Leucine incorporation (pmol/g.h)	333	436	734
Pot. C mineralisation (mg C/kg.wk)	80	57	68
Pot. N mineralisation (mg N/kg.wk)	7.5	8.2	9.7
Number of CFU/g DW soil	7.9×10^7	7.1×10^7	1.7×10^8
Biolog: CFU for 50% substrate conversion	2.1×10^5	6.8×10^4	8.2×10^4
Biolog: Hill slope of substrate conversions	0.52	0.54	0.54
PICT: EC_{50} zinc (mg/l)	81	137	155
PICT: EC_{50} nickel (mg/l)	15	15	46
PICT: EC_{50} copper (mg/l)	3	5	6
PICT: EC_{50} chromium (mg/l)	5	8	7
Nematodes			
Number per 100 g	3930	6560	7300
Number of taxa (genus)	27	24	20
% bacterial-feeders	64	69	67
% fungi-feeders	2.5	1.9	2.5
% plant-feeders	25	20	23
% cp1	9	10	11
% cp2	49	50	79
% cp3+4+5	42	40	11
Maturity index	2.5	2.5	2.2
Enchytraeids			
Number per m^2	1180	1180	235
Number of genera	2	2	1
Earthworms			
Number per m^2	105	97	145
Number of species	1	1	1
Fresh weight (g/m^2)	18.3	27.5	29.7

in sample A and the highest in sample C (EC_{50} is reached after dilution of the sample to 76%). For the PAM-algal test no toxicity was observed in any of the soil elutriates. Slight toxicity was observed in the lettuce germination test; the germination success was decreased by 10% in a sample from field C compared with the reference sample from field A.

A variety of ecological parameters were determined (Table 9.4). Many parameters differed from the reference sample of field A (like the pollution-induced community tolerance (PICT) results) and showed an increase compared with the reference field, but this can be attributed at least partly to increased organic carbon content. It is known that microorganisms (and nematodes) clearly respond positively to the organic matter content of soil (Hassink, 1994; Wardle, 2002). The maturity index (MI) of nematodes was lower in the sample of field C, mainly caused by the decrease in the specialist groups of nematodes (cp3, cp4, cp5).

Scaled results from all the tests were collated in the Triad matrix (Table 9.5). For the reference sample of field A the values were set to zero (local reference), except for the Microtox® result. The reference sample already demonstrated inhibition compared with the blank, and it was decided to use this value. Calculation of effects using pore water ('bioavailable') concentrations of contaminants and toxicity data from aquatic tests did not result in notably lower values for the toxic pressure compared with total concentrations. However, it might be questionable to use aquatic toxicity data for the calculation of effects from pore water concentrations of contaminants.

Table 9.5 Triad decision matrix showing the results from all tests on a quantitative effect scale. The tests are grouped per Triad leg, and the geometric mean of (1−effect) is calculated. The values in bold type are gross effect levels. Back transformation gives the integrated effect value for all tests per Triad leg. In the lower part of the matrix, all integrated results are summarized and the final risk is calculated, including a deviation factor for demonstrating imbalance between results from different Triad legs.

Triad aspect	Parameter		Sample A	Sample B	Sample C
Chemistry	TP$_{\text{mixture, anthropogenic}}$	(Metals – total concentration)	0.00	0.83	0.97
	TP$_{\text{mixture, anthropogenic}}$	(Metals – pore water)	0.00	0.62	0.89
		Integrated effect	**0.00**	**0.75**	**0.94**
Toxicology	Microtox®		0.36	0.48	0.62
	PAM-algae		0.00	0.00	0.00
	Lettuce germination		0.00	0.05	0.10
		Integrated effect	**0.14**	**0.21**	**0.30**
Ecology	Microbial community (CLPP-Biology)		0.00	0.35	0.44
	Community tolerance (PICT)		0.00	0.47	0.71
	Microbial parameters		0.00	0.25	0.42
	Nematods		0.00	0.15	0.32
	Enchytraeids		0.00	0.00	0.68
	Earthworms		0.00	0.15	0.24
		Integrated effect	**0.00**	**0.24**	**0.50**
Summary	Effect assessment chemistry		0.00	0.75	0.94
	Effect assessment toxicology		0.14	0.21	0.30
	Effect assessment ecology		0.00	0.24	0.50
		Final assessment (risk)	**0.05**	**0.47**	**0.73**
		Deviation in triad results	**0.14**	**0.52**	**0.57**

The general trend was that a lot of parameters seem to respond to the contamination levels and, after integration, provide clear indications of effects of contamination. However, the level of risks was lower than on the basis of the chemical-based risk calculations. Nevertheless, the Triad approach for terrestrial ecosystems definitely underpins classical risk assessment based on the presence of contaminants alone, although the deviation between 'classical' risk assessment based on the presence of the contaminants and the results from the biological tests needs to be evaluated and resolved.

9B.18 Issues and recommendations

Successful implementation of a Triad approach is dependent on the availability of a suitable set of instruments: chemical, toxicological (bioassays) and ecological. Without doubt, substance-oriented approaches have received a lot of attention in ecotoxicology and the available techniques are relatively mature (Posthuma *et al.*, 2002). The current focus in this field of research is on improvement of bioavailability considerations, the effect of mixtures and the indirect effects of substances.

Bioassays are becoming increasingly important for ERA and some frameworks are well under way or almost implemented. However, the range of available bioassays is still smaller than the potential range of different matrices for which they have to be used. Consequently, in some cases the application of bioassays is limited. For instance, there is only a narrow set of bioassays available for acidic soils. Organic soils also pose problems with many ordinary bioassays because of the 'natural toxicity' present, which hides the effect of contaminants (Onorati *et al.*, 1998). Nevertheless, the situation will improve rapidly when bioassay testing is adopted as a 'normal' type of end-point in risk assessment because this will direct much attention to the development of new tests. The Water Directive and other initiatives of the European Commission (2002) will probably have a stimulating effect on the application of bioassays because they are assigned as valuable tools in soil, sediment and water-quality assessments.

The level of standardization differs between bioassays. Of course, standardized tools should prevail when there is a choice. In many cases, cheap standardized tests such as many short-term bioassays (e.g. Microtox®) are useful tools, despite the fact that the test species is not present at the site. Test results from short-term bioassays should be regarded as an indication for general toxicity in samples from the site. The ecological relevance might be limited compared with tests on species that are harboured at the site. However, the response of different species within the same taxonomic group is not very different in general, making results from standardized tests with non-native species suitable for ERA (Posthuma *et al.*, 2002). Furthermore, given a limited amount of resources, it is often better to use a cheap standard test because more samples can be analysed and the interpretation of test results is easier because of the available knowledge.

Aforementioned arguments should not be regarded as a barrier to implement new methods in ERA. The reverse is true because there is a lot of improvement possible and this should be stimulated. Before applying a new test some general questions should be answered, namely:

1. Is the end-point ecologically relevant?
2. Is the end-point sensitive for the type of contamination?
3. Is there a plausible cause–effect relationship?

A new discipline in ERA is the use of ecological parameters (Suter, 1998; Ferguson et al., 1998; Suter et al., 2002). There are many potential tools because various biological parameters can be measured easily in the field or in the laboratory on autochthonous organisms or processes. A lot of attention is paid to soil-dwelling organisms such as bacteria, fungi, nematodes, earthworms and arthropods because they are omnipresent, relatively cheap to monitor and live in close contact with the contaminants (Rutgers and Breure, 1999; Schouten et al., 2000). Special attention has to be given to the cause–effect relationship because the majority of ecological parameters are sensitive for contamination and other environmental (natural and anthropogenic) factors (Suter et al., 2002). For these reasons the choice of the reference site or the set of reference data is crucial. When causal relationships are decisive, the demonstration of pollution-induced community tolerance is highly recommended (Rutgers and Breure, 1999; Boivin et al., 2002; Millward and Klerks, 2002).

To be useful for risk assessment, the answers from all tests in a WOE approach should be made comparable, e.g. by a uniform scaling method, preferably without losing quantitative information (Burton et al., 2002; Schmidt et al., 2002). A continuous effect scale running from 0 to 1 (representing a quantitative measure for the fraction of effect on an ecosystem) seems to fulfil quantitative requirements. For this part, many methods are still in development and answers rely heavily on the experts using these methods. In addition, non-objective choices must be made because there is no comprehensive ecosystem theory available. Fortunately, because the Triad approach is based on a WOE procedure, mismatches can be picked up as results that are obviously out of the range, making evaluation and correction conceivable.

Besides the issue of scaling, the issue of weighting different tests should be given attention. Within the conceptual model of the ecosystem, differential weighting may be applied for two reasons: ecological considerations, with the differential weighting on the end-points being defined in the conceptual model that serves as the basis for the ERA (USEPA, 1998); and to account for the uncertainty or variety within the end-points, with tests having a high level of conceptual uncertainty or a high variety in results being given a smaller weight in ERA (Menzie et al., 1996). There is some evidence of the use of different weights in ERA for aquatic systems (Den Besten et al., 1995) following an MCA (see above).

9B.19 Future prospects

Finally, ERA should provide the information necessary for soil and sediment management decisions. The framework should be developed so that the uncertainty is clearly visible. The uncertainty should be reduced after the application of one or more tiers in the Triad. At this moment we have chosen to make the uncertainty visible by calculation of the overall deviation factor of the three Triad legs (Table 9.5). From a number of studies (data not shown) it was concluded that introducing additional information, e.g. in new tiers, generally reduced the deviation factor. It can be expected that the configuration of the biological methods and their integration into the Triad is not yet optimal, considering the limited experiences with biological methods for ERA in the terrestrial environment. Future applications of the Triad will allow for an evaluation of results and also for adjustments. At this time, the proposed Triad system, including the decision matrix, seems to fulfil the current needs, combining different test methods in a WOE approach without losing quantitative information or too much of the transparency.

Acknowledgements

The research reported in this part of Chapter 9 was financed by the Ministry of Public Housing, Spatial Planning and the Environment (RIVM project M/711701) and by the EU (LIBERATION project EVK1-CT-2001-105). Harm van Wijnen and Miranda Mesman are gratefully acknowledged for carefully reading the manuscript.

References

Bailer, A.J., Hughes, R.H., See, K., Noble, R. and Schaefer, R. (2002) A pooled response strategy for combining multiple lines of evidence to quantitatively estimate impact. *Human and Ecological Risk Assessment*, **8**, 1597–1611.

BEVER (1999) *Van trechter naar zeef* (ISBN 9012088437). SDU, Den Haag.

Boivin, M.Y., Breure, A.M., Posthuma, L. and Rutgers, M. (2002) Determination of field effects of contaminants – significance of pollution-induced community tolerance. *Human and Ecological Risk Assessment*, **8**, 1035–1055.

Burgess, R.M., Ahrens, M.J., Hickey, C.W., Den Besten, P.J., Ten Hulscher, T.E.M,. Van Hattum, B., Meador, J.P. and Douben, P.E.T. (2003) An overview of the partitioning and bioavailability of PAHs in sediments and soils. In *PAHs: an Ecological Perspective*, Douben, P.E.T. (ed.), pp. 99–126. John Wiley, Chichester.

Burton, G.A., Chapman, P. and Smith, E.P. (2002) Weight-of-evidence approaches for assessing ecosystem impairment. *Human and Ecological Risk Assessment*, **8**, 1657–1673.

Chapman, P.M., McDonald B.G. and Lawrence, G.S. (2002) Weight-of-evidence issues and frameworks for sediment quality (and other) assessments. *Human and Ecological Risk Assessment*, **8**, 1489–1515.

Cornelissen, G., Rigterink, H., Ten Hulscher, T.E.M., Vrind, B.A. and Van Noort, P.C.M. (2001) A simple Tenax extraction method to determine the availability of sediment-sorbed organic compounds. *Environmental Toxicology and Chemistry*, **20**, 706–711.

CUWVO (1990) Recommendations for the monitoring of compounds on the M-list of the national policy document on water management *Water in the Netherlands: a Time for Action* (in Dutch). Commission for the Implementation of the Act on Pollution of Surface Waters (CUWVO), The Hague.

De Zwart, D. and Sterkenburg, A. (2002) Toxicity-based assessment of water quality. Chapter 18. In *Species Distributions in Ecotoxicology*, Posthuma, L., Suter, G.W. and Traas, T.P. (eds), Chapter 18. CRC Press, Boca Raton, FL.

Den Besten, P.J., Smidt, C.A., Ohm, M., Ruijs, M.M., Van Berghem, J.W. and Van de Guchte, C. (1995) Sediment quality assessment in the delta of the rivers Rhine and Meuse based on field observations, bioassays and food chain implications. *Journal of Aquatic Ecosystem Health*, **4**, 256–270.

Den Besten, P.J., De Deckere, E., Babut, M.P., Power, B., Delvalls, T.A., Zago, C., Oen, A.M.P. and Heise, S. (2003) Biological effects-based sediment quality in ecological risk assessment for European waters. *Journal of Soils and Sediments*, **3**, 144–162.

Didden, W. (2003) Development and potential of a stereotype as a references site in ecological monitoring. *Newsletter on Enchytraeidae*, **8**, 33–40.

European Commission (2002) *Towards a Thematic Strategy for Soil Protection*, COM(2002)-179. Commission of the European Communities, Brussels.

Fairbrother A., Glazebrook, P.W., Tarazona, J.V. and Van Straalen, N.M. (2003) *Test Methods to Determine Hazards of Sparingly Soluble Metal Compounds in Soils*. Society of Environmental Toxicology and Chemistry (SETAC), Pensacola, FL.

Ferguson, C., Darmendrail, D., Freier, K., Jensen, B.K., Jensen, J., Kasamas, H., Urzelai, A. and Vegter, J. (1998) *Risk Assessment for Contaminated Sites in Europe. Volume 1. Scientific Basis*. LQM Press, Nottingham.

Hall, L.W. and Giddings, J.M. (2000) The need for multiple lines of evidence for predicting site-specific ecological effects. *Human Ecological Risk Assessment*, **6**, 679–710.

Hassink, J. (1994) Effect of soil texture on the size of the microbial biomass and on the amount of C and N mineralized per unit of microbial biomass in Dutch grassland soils. *Soil Biology and Biochemistry*, **26**, 1573–1581.

INS (1999) *Integrale Normstelling Stoffen, milieukwaliteitsnormen bodem, water en lucht. Interdepartementale stuurgroep INS*. VROM, Den Haag, The Netherlands.

Lancaster, J. (2000) The ridiculous notion of assessing ecological health and identifying the useful concepts underneath. *Human and Ecological Risk Assessment*, **6**, 213–222.

Long, E.R. and Chapman, P.M. (1985) A sediment quality triad: measures of sediment contamination, toxicity, and infaunal community composition in Puget Sound. *Marine Pollution Bulletin*, **16**, 405–415.

Markert, B.A., Breure, A.M. and Zechmeister, H.G. (2003) *Bioindicators and Biomonitors: Principles, Concepts and Applications*. Elsevier, Oxford.

Menzie, C., Hope Henning, M., Cura, J., Finkelstein, J., Gentile, J., Maughan, J., Mitchell, D., Petron, S., Potocki, B., Svirsky, S. and Tyler, P. (1996) Special report of the Massachusetts weight-of-evidence workgroup: a weight-of-evidence approach for evaluating ecological risks: report of the Massachusetts weight-of-evidence workshop. *Human and Ecological Risk Assessment*, **2**, 277–304.

Mesman, M., Rutgers, M., Peijnenburg, W.J.G.M., Bogte, J.J., Dirven-Van Breemen, M.E., De Zwart, D., Posthuma, L. and Schouten, A.J. (2003) Site-specific ecological risk assessment: the Triad approach in practice. *Conference Proceedings of ConSoil, 8th International FZK/TNO Conference on Contaminated Soil*, pp. 649–656. Forschungszentrum: Karlsrute.

Millward, R.N. and Klerks, P.L. (2002) Contaminant-adaptation and community tolerance in ecological risk assessment: introduction. *Human and Ecological Risk Assessment*, **8**, 921–932.

Nijhof, A.G. and Koolenbrander, J.G.M. (1998) *Assessing Risks from Soil Pollution: Inventory of Bottlenecks and Possible Solutions*, vol. 15. Integrated Soil Research Programme, Wageningen.

Onorati, F., Pellegrini, D. and Ausili, A. (1998) Sediment toxicity assessment with *Photobacterium phosphoreum*: a preliminary evaluation of natural matrix effect. *Fresenius Environmental Bulletin*, **7** (special issue), 596–604.

Posthuma, L., Suter, G.W. and Traas, T.P. (2002) *Species Sensitivity Distributions in Ecotoxicology*. CRC Press, Boca Raton, FL.
Power, M. and Adams, S.M. (1997) Special section – perspectives of the scientific community on the status of ecological risks assessment. *Environmental Management*, **21**, 803–830.
RIVM (1999) *Environmental Balance 1999 – The State of the Dutch Environment*, RIVM Report 251701038. RIVM, Bilthoven.
Rutgers, M. and Breure, A.M. (1999) Risk assessment, microbial communities, and pollution-induced community tolerance. *Human and Ecological Risk Assessment*, **5**, 661–670.
Rutgers, M., Faber, J., Postma, J. and Eijsackers, H. (2000) *Site-specific Ecological Risks: a Basic Approach to the Function-specific Assessment of Soil Pollution*, vol. **28**. Integrated Soil Research Programme, Wageningen.
Rutgers, M., Bogte, J.J., Dirven-Van Breemen, E.M. and Schouten, A.J. (2001) *Site-specific Risk Assessment – Field Investigations with a Quantitative Triad Approach*, Report 711701026. RIVM, Bilthoven.
Saaty, T.L. (1980) *The Analytic Hierarchy Process*. McGraw-Hill, New York.
Schmidt, T.S., Soucek, D.J. and Cherry, D.S. (2002) Modification of an ecotoxicological rating to bioassess small acid mine drainage-impacted watersheds exclusive of benthic macroinvertebrate analysis. *Environmental Toxicology and Chemistry*, **21**, 1091–1097.
Schouten, T., Bloem, J., Didden, W.A.M., Rutgers, M., Siepel, H., Posthuma, L. and Breure, A.M. (2000) Development of a biological indicator for soil quality. *Setac Globe*, **1**, 30–33.
Schouten, A.J., Bogte, J.J., Dirven-Van Breemen, E.M. and Rutgers, M. (2003) *Site-specific Risk Assessment – Field Investigations with a Quantitative Triad Approach*, Report 711701032. RIVM, Bilthoven.
Straetmans, A., Maas, H., Van de Plassche, E. and Vethaak, D. (2003) Toetsingskader voor bioassays in het waterkwaliteitsbeheer. *H2O*, **14/15**, 15–18.
Suter, G.W. (1998) Retrospective assessment, ecoepidemiology and ecological monitoring. *Handbook of Environmental Risk Assessment and Management*, Calow, P. (ed.), pp. 170–217. Blackwell Science, Oxford.
Suter, G.W., Efroymson, R.A., Sample, B.E. and Jones, D.S. (2000) *Ecological Risk Assessment for Contaminated Sites*. CRC Press, Lewis Publishers, Boca Raton, FL.
Suter, G.W., Norton, S.B. and Cormier, S.M. (2002) A methodology for inferring the causes of observed impairments in aquatic ecosystems. *Environmental Toxicology and Chemistry*, **21**, 1101–1111.
Swartjes, F.A. (1999) Risk-based assessment of soil and groundwater quality in the Netherlands: standards and remediation urgency. *Risk Analysis*, **19**, 1235–1249.
USEPA (1998) *Guidelines for Ecological Risk Assessment*, EPA/630/R-95/002F. Risk Assessment Forum, Washington, DC.
Van de Leemkule, M.A. (2001) *Characterizing Land Use Related Soil Ecosystem Health*, Report R15(2001). Technical Committee on Soil Protection, The Hague.
Van Straalen, N.M. and Denneman, C.A.J. (1989) Ecotoxicological evaluation of soil quality criteria. *Ecotoxicology and Environmental Safety*, **18**, 241–251.
VROM (2000) *Target Values and Intervention Values for Soil Sanitation*, Circular DBO/1999226863, 4-2-2000. VROM, The Hague.
VROM (2001) *Kabinetsstandpunt beleidsvernieuwing bodemsanering*. VROM, The Hague.
Wardle, D.A. (2002) *Communities and Ecosystems: Linking the Aboveground and Belowground Components*. Princeton University Press, Princeton, NJ.

C GERMAN PERSPECTIVE

Hans-Jürgen Pluta and Monika Rosenberg

The general approaches to legislation and monitoring practice exhibit obvious differences when comparing water and solid waste/soils. Although the basic concept for wastewater monitoring, including development and standardisation of monitoring methods, has been implemented since 1970, the approaches for monitoring solid waste and soils are still being developed and are under discussion.

9C.1 Wastewater

In Germany, the management of municipal, commercial and industrial wastewater, including the use of bioassays, is based on three main regulations. These date back 25 years. The central basis is provided by the Federal Water Act (Wasserhaushaltsgesetz, WHG) [1] after the fourth amendment in 1976. Pursuant to Article 7a of the WHG, the Wastewater Ordinance (Abwasserverordnung, AbwV) [2] became law. The WHG is supported by the Wastewater Charges Act (Abwasserabgabengesetz, AbwAG) [3], promulgated in 1976.

Taking into account the national, supranational and international requirements and regulations, the WHG, AbwV and AbwAG have been developed and adapted with several amendments. In particular, Directives 76/464/EEC [4] and 91/271/EEC [5], as well as Directive 96/61/EC concerning integrated pollution prevention and control (IPPC Directive) [6], Directive 2000/76/EC on the incineration of waste (Waste Incineration Directive) [7] and Directive 2000/60/EC (Water Framework Directive) [8] have been and are still of great importance with reference to amendments concerning national regulations. A complete and detailed overview of the development of national water law is given by the Federal Ministry for the Environment, Nature Conservation and Nuclear Safety [9].

Based on Article 7a of the WHG and considering the emission-based principle, nationally applicable minimum requirements for the discharges of wastewater into water bodies have been established in terms of general administrative provisions by the Federal Government and in agreement with the Länder (Federal States). These requirements refer to wastewater production, avoidance and treatment using the best available technique (BAT) and limits for monitoring parameters, which allow effective assessment and control of optimisation of avoidance and reduction measures applicable to the discharging companies. The general approach differs from that used at the European Community level and focuses on wastewater types and sectors and not on individual substances or substance groups. By using a sector-specific selection of parameters – key individual parameters and sum parameters in combination with effect parameters (bioassays) – the expenditure (e.g. costs, number of required measurements) for monitoring wastewater is kept within reasonable limits.

Specific requirements for hazardous substances were established in the 5th amendment of the WHG in 1986 in order to optimise avoidance and treatment measures in keeping with the ongoing developments. Hazardous substances have been defined as substances or substance groups that, 'because of concern regarding their toxicity, persistence, ability to accumulate or their carcinogenic, teratogenic or mutagenic effects, must be considered hazardous'. To detect these effects, bioassays have been developed and standardised that can be selected from a predefined set for monitoring purposes, taking into account the wastewater and the state-of-the-art technology to be used in the controlling process.

In accordance with Article 7a of the WHG, the AbwV (Ordinance on Requirements for the Discharge of Wastewater into Waters – Wastewater Ordinance) has been divided into a general framework section, an annex (to Article 4) with analytical methods for parameter determination and sector-specific appendices, including requirements for 53 wastewater sectors [3]. Minimum requirements to be stipulated when granting a permit to discharge wastewater are described, analysis and measurement techniques are specified in the annex and sector-specific requirements in the appendices refer to the analysis and measurement techniques specified in the annex. Article 4 of the AbwV, in conjunction with the annex, lists the analysis and measurement procedures that are to be used to determine the parameters defined in the appendices. Paragraph 2 emphasises that for individual cases other equivalent procedures may be required if compliance with the requirements specified in the ordinance is to be ensured [9]. All analysis and determination procedures are standardised methods (DIN, CEN, ISO).

The general structure of the sector-specific appendices for the AbwV is:

A   Scope of application
B   General requirements
C   Requirements for wastewater at the point of discharge
D   Requirements for wastewater prior to blending
E   Requirements for wastewater at the site of occurrence
F   Requirements for existing discharges

Analysis and measurement procedures (annex) include:

- General procedures (4).
- Anions/elements (12).
- Cations/elements (25).

- Individual substances, sum parameters and group parameters (39), including adsorbable organohalogen (AOX), chemical oxygen demand (COD), total organic carbon (TOC) and total bound nitrogen (TN_b). Total bound nitrogen is defined as the sum of all organically and inorganically bound nitrogen, excluding elemental nitrogen.
- Biological test procedures (12), including

  —fish toxicity (22, 23)
  —*Daphnia* toxicity
  —algae toxicity
  —bacteria luminescence inhibition
  —mutagenic potential
  —fish toxicity (egg)
  —biological oxygen demand (BOD)
  —a range of tests referring to biological degradability

The AbwAG (Act pertaining to charges levied for discharging wastewater into waters – Wastewater Charges Act) is closely linked with the regulatory water law, in particular Article 7a of the WHG, and it complements this law with wastewater charges that act as an economic instrument. The groups subject to such charges include those that directly discharge wastewater into water bodies, i.e. municipal authorities and industrial producers of direct discharges (Article 1 of the Wastewater Charges Act) [9].

Parameters liable to taxation are:

- Sum parameters: COD, AOX
- Ecotoxicological effects: fish toxicity
- Individual parameters: P, N, Hg, Cd, Cr, Ni, Pb, Cu

Analysis and measurement procedures (annex to Article 3) are identical to those listed in the annex to Article 4 of the AbwV.

The charges have increased since 1976 to a current (2004) level of 35.79 euros per damage unit. Noxious substances and groups of noxious substances, including respective threshold values, are given in Table 9.6.

It has to be pointed out that:

- Until 1998, the charge reduction upon compliance with requirements pursuant to Article 7a WHG was set at 75%, which was amended to 50% after 1998.
- Investments in wastewater enhancing measures (e.g. reduction of annual load by optimisation of wastewater treatment) may be credited as soon as parameter-oriented reductions reach 20%.
- Expenditures in the collecting systems for connection to wastewater treatment installations may be credited against charges.
- Parties subject to charges in the New Länder may offset charges for groups of discharges and groups of related investments.

Table 9.6 Noxious substances, groups of noxious substances, respective threshold values and damage units according to AbwAG.

No.	Noxious substances	Damage unit	Threshold concentration and annual load
1	Oxidisable substances expressed as chemical oxygen demand (COD)	50 kg of oxygen	20 mg/l and 250 kg annual load
2	Phosphorus	3 kg	0.1 mg/l and 15 kg annual load
3	Nitrogen	25 kg	5 mg/l and 125 kg annual load
4	Organic halogen compounds as adsorbable organohalogens (AOX)	2 kg of halogen, calculated as organic chlorine	100 µg/l and 10 kg annual load
5	Metals and their compounds		
5.1	mercury	20 g	1 µg/l and 100 g annual load
5.2	cadmium	100 g	5 µg/l and 500 g annual load
5.3	chromium	500 g	50 µg/l and 2.5 kg annual load
5.4	nickel	500 g	50 µg/l and 2.5 kg annual load
5.5	lead	500 g	50 µg/l and 2.5 kg annual load
5.6	copper	1000 g	100 µg/l and 5 kg annual load
6	Toxicity to fish[a]	3000 m^3 of wastewater divided by T_F[b]	$T_F = 2$

[a] The toxic effect is determined in a fish test on the orfe (*Leuciscus idus melanotus*), using various dilutions of wastewater, in conjunction with the method specified in No. 401 of the Appendix of the Wastewater Ordinance. The acute fish test will be substituted by the fish egg test (22, 23) from January 2005.
[b] Dilution factor at which waste water ceases to have a toxic effect on fish.

- Compliance with a declared value may be proved via an authority-authorised measurement programme [9].

The use of charges levied is described in the AbwAG (Wastewater Charges Act), Article 13:

(1) 'The revenue accruing from wastewater charges are intended to be used only for specific purposes in connection with measures for maintaining or improving water quality. The Länder may stipulate that the administrative expenditure associated with the enforcement of this Act and of the Länder's

own supplementary provisions shall be paid out of the revenue accruing from wastewater charges.

(2) The following shall in particular be considered as measures pursuant to para. 1, Art. 13 AbwAG:
1. the construction of wastewater treatment plants
2. the construction of rain water retention basins and facilities for the purification of rain water
3. the construction of circular sewers and retention sewers at and along reservoirs, the shores of lakes and the seashore, and of connecting interceptors permitting the establishment of joint treatment facilities
4. the construction of facilities for the disposal of sewage sludge
5. measures such as low-flow augmentation or oxygenation taken in and at water bodies for observing and improving water quality and measures for maintaining such water bodies
6. research on and development of suitable facilities and techniques for improving water quality
7. basic and further training of operating staff for wastewater treatment plants and other facilities designed to maintain and improve water quality.' [2]

Bioassays are integrated elements used for monitoring and surveillance purposes in the general approach of the WHG, AbwV and AbwAG. Key parameters, sum parameters and effect parameters (bioassays) should be combined and used within the scope of an effective emission control. Sum parameters give general information about the efficiency of wastewater treatment, taking into account our limited knowledge of number and quantity of the components in a wastewater discharge. Even if concentrations of any chemical substance were to be known, no information about possible biological effects of treated effluents on aquatic organisms would be available. Therefore, sum parameters are combined with bioassays, which are valuable instruments for delivering additional information about possible environmental impact and are appropriate tools to monitor the decrease of toxic effects, particularly in complex effluents.

Controlling effluents with biotests according to the Wastewater Charges Act (AbwAG) and the Wastewater Ordinance (AbwV), ecotoxic/biological effects are determined using various dilutions of original wastewater samples following a combined geometric series, based on the numerical values of 2 (i.e. 2, 4, 8, 16 ...) and 3 (i.e. 3, 6, 12, 24 ...) (Table 9.7).

For toxicity testing of wastewater by means of defined dilutions (D), the lowest ineffective dilution (LID) expresses the most concentrated test batch at which no inhibition, or only effects not exceeding the test-specific variability, have been observed. Dilution, D, is expressed as the reciprocal value of the volume fraction of wastewater in the test batch (Table 9.7).

Table 9.7 Defined dilutions (D) and corresponding percentages of wastewater according to DIN EN ISO 5667/16 (Water quality – Sampling – Part 16: Guidance on biotesting of samples).

D	1	2	3	4	6	8	12	16	24	32
% Wastewater	100	50	33.3	25	16.6	12.5	8.3	6.25	4.16	3.12

For example, one-quarter of wastewater (volume fraction of 25%) equals a dilution level of 4 (D = 4). Dilutions are generally fixed, as shown in Table 9.7. Biotests used for the monitoring of wastewater discharges should be

- Standardised
- Reproducible
- Able to show clearly the success of wastewater treatment
- Able to detect clearly a toxic effect to (representative) organisms of the aquatic environment
- Practicable for routine measurements (including equipment)
- Cost-effective
- Rapid

All analysis and measurement procedures cited in the AbwV and AbwAG are standardised methods, either by DIN, ISO or CEN. Development and standardisation of biotests for emission control are carried out by national working groups on biotests (DIN), resulting in DIN standard protocols since 1976. Representatives of federal government, the Länder, industry and universities are usually present in the national working groups.

From the very beginning, experts of national working groups and member bodies have participated in international working groups of ISO/TC 147/SC 5 and CEN/TC 230/WG 2 in order to transfer national standards into DIN EN, DIN ISO or DIN EN ISO standards and to adopt proposals from other member bodies for implementation into national regulations and international standards. Selection of biotests for wastewater monitoring and establishing a set of biotests can be based on the ecotoxic effect, the organisation or the trophic level, as shown in Figure 9.4.

In future, the set of biotests usable for wastewater monitoring will be amended. This can be done by a biotest using a higher water plant (Duckweed growth inhibition test, which is under development by ISO/CEN) or by the fish egg test (22, 23), which will substitute the acute fish test by taking into account the requirements of animal welfare and the Animal Protection Law, respectively.

It has to be noted that the standard protocols for biotesting and chemical analysis originally developed for the use of wastewater and water monitoring are being proposed and used for testing elutriates of soil, waste and sediment samples.

Summarising the experiences and the status of monitoring wastewater using biotests, it can be asserted that:

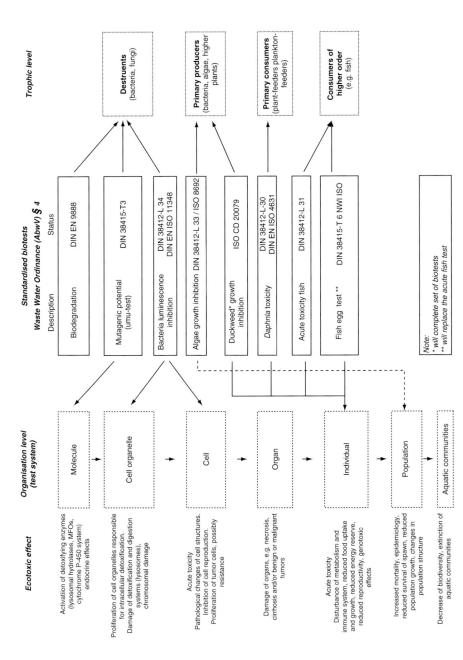

Figure 9.4 Biotests related to ecotoxic effect, organisation level or trophic level.

- Legal regulations, including the use of bioassays in monitoring wastewater discharges, are the Federal Water Act (WHG), the Wastewater Ordinance (AbwV) and the Wastewater Charges Act (AbwAG), where analysis and measurement procedures are listed in the annexes to the AbwV and AbwAG as DIN, DIN EN, and DIN EN ISO standards.
- Over the last 20 years there have been a number of good reports on the use of a combination of sum parameters (COD, AOX) and bioassays – including the measurement of single substances (e.g. P, N and metals) – in emission control. For instance, the percentage of wastewater effluent causing acute fish toxicity has been reduced to approximately 10%. The LID has changed considerably from $D = 30\text{–}60$ to $D = 2\text{–}8$ since 1976.
- The diversity of approaches in applying bioassays for wastewater evaluation in the national, supranational and international regulations, aimed especially to meet the requirements of the European Union, will influence national water law in the future and the general approach in Germany will be discussed within the scope of international harmonisation.
- Apparently, there is a general agreement that the long-term goal should be no measurable (acute, sublethal or chronic) toxicity in wastewater effluents. This might be interpreted as zero toxicity in undiluted effluents.

9C.2 Waste and soil

In contrast to the legal definition of biological tests for the water and wastewater applications, there is no legislation for the use of such established biological procedures either for the characterisation of waste or for the evaluation of soils.

At present, waste and soil samples primarily are analysed and evaluated by means of physical and chemical analysis. This also depends on the fact that it is difficult to draw a general but comprehensive picture of the ecosystem, which may be endangered. Toxic effects of substances, abiotic factors and effects resulting from interactions of various substances and materials present in solid samples have to be considered. A species spectrum has to be selected for effect estimation that is representative of ecological functions and trophic levels, as well as for the route or routes of exposure.

9C.2.1 Waste

In the European Union the aim of legal regulations is to direct waste management towards a sustainable development. Priority is given to prevention before recovery and to recovery before removal. Accordingly, measures have to be planned and accomplished with respect to environmental sustainability and the protection of resources.

The relevant regulations are covered by the:

- Basel Convention [10]
- Council Regulation (EEC) No. 259/93 on the supervision and control of shipments of waste within, into and out of the European Community [11]
- Council Directive 91/689/EEC of 12 December 1991 on hazardous waste [12]
- Council Directive 1999/31/EC of 26 April 1999 on the landfill of waste [13]
- EEC Waste Framework Directive (Council Directive 75/442/EEC of 15 July 1975 on waste) [14]

Reviewing these regulations reveals that there are approaches for legally establishing biological test procedures. For example, the Basel Convention classified waste materials as being hazardous (according to Article 1, Paragraph 1, Point a) if they belong to a group as defined in Annex I, or show characteristics as given in Annex III. One of these so-called H-criteria dealing with hazard characteristics is the H12 criterion of ecotoxicity ('Substances or wastes which if released present or may present immediate or delayed adverse impacts to the environment by means of bioaccumulation and/or toxic effects upon biotic systems'), but it is only listed [10] and not yet specified.

In the EEC Directive of waste movement (259/93) [11], which specifies wastes that are allowed to be translocated within, without and into the European Union, the Organization for Economic Cooperation and Development (OECD) countries or the non-OECD countries, the term 'ecotoxicity' is not mentioned. The term 'environment', however, is specified whereby it predominantly covers environmentally compatible utilisation and duly removal of waste. Only Article 34 deals with the obligation of the waste producer to take all necessary measures in order to maintain environmental quality in the context of the guidelines 75/442/EEC [14] and 91/689/EEC [12].

Implementation of the EEC Directive of waste movement in the German legislation is made by the Abfallverbringungsgesetz [15] (present version from 25 August 1998), which was published in 1994 as the German fulfilment law to the Basel Convention (Deutsches Ausführungsgesetz zum Basler Übereinkommen) [15]. The list of characteristics given in the Basel Convention [10], including criterion H12, was not adopted.

Reviewing the Commission's decision on establishing a list of wastes [16], which was implemented by the 'Abfallverzeichnis-Verordnung' [17] in 2001, it arises that the H14 criterion of ecotoxicity ('Substances and preparations which present or may present immediate or delayed risks for one or more sectors of the environment' is not considered because it is not specified according to the introduction of the Annex, Point 3.1.

In the Council Directive 91/689/EEC of 12 December 1991 on hazardous wastes [12], Article 1, Paragraph 4, hazardous waste is classified in a similar way to that in the Basel Convention [10]. The definition of the H14 criterion of ecotoxicity is not, however, as comprehensive as criterion H12 in the Basel Convention [10].

In further European Union guidelines such as those on landfills, the term 'ecotoxicity' appears in the glossary in Article 2e. However, its relevance is narrowed to, for instance, the 'ecotoxicity of leachates'. A first step to establishing biological test procedures is given in Annex II, Point 3. The suggestion of a three-stage procedure for the characterisation and investigation of waste envisages, among other things, a fundamental characterisation of wastes and agreement on investigations.

An attempt to regulate the evaluation of wastes was presented by the Commission for the Evaluation of Substances Hazardous for Water in March 2003 [18]. It suggests a classification in accordance to the relevant H-criteria of the water-hazard classes for waste. The problem with this approach, however, is that the composition of the waste should be known and focus is given to effects relevant for water-hazard potential. Based on the submitted concept and in accordance with the principle of concern of the Water Management Act, wastes can be treated as not firmly defined substances, such as the substances assigned to the WGK 3 (water-hazard class).

In the current process of establishing biological test procedures on a European level, a standardisation proposal on the 'preparation of water samples for ecotoxicity tests' was submitted to the CEN. Moreover, intensive investigations were carried out that focused on the leaching behaviour of waste, a first step for the application of biological methods to characterise wastes. In addition, methodological requirements for biological testing of waste were established.

Biological test procedures that can be used for solid samples such as waste, soils and sediments have been developed already (see Section 9C.3 below and Chapter 6).

9C.2.2 Soil

In European legislation the comprehensive legal definitions that are available for the classification of waste do not exist for the characterisation of soils, therefore focus is given to German legislation in the following section.

The Bundesbodenschutzgesetz (BBodSchG) [19] and the Bundesbodenschutz-Verordnung (BBodSchV) [20] do not explicitly specify any method for the biological investigation of soils. The indicated methods cover chemical and physical parameters that are listed in Annex I of the BBodSchutzV, Point 3 (investigation procedures).

However, there is some indication for initiatives to establish biotest procedures:

- According to Article 2 of the BBodSchG, soil fulfils natural life support functions and acts as a habitat for humans, animals, plants and soil organisms.
- Article 7 of the BBodSchG obliges the property-owner to take precaution and to prevent soil contamination.

- In Article 8, Abstract 3, clause 1 of the BBodSchG it is stipulated that 'procedures for determination of environmentally harmful substances in soils, in biological material and in other materials' have to be defined. In accordance with Article 22 of the BBodSchG, the federal government can issue decrees to avert harmful soil changes, to remediate soil, landfills and resulting groundwater contamination and to specify measures and appropriate values necessary to establish precaution.
- According to the BBodSchV detailed risk assessment studies are possible, e.g. to investigate the uptake of pollutants by plants and animals.
- Also, Article 9 of the BBodSchV (harmful soil changes in the context of risk assessment), Paragraph 1, Point 2 (considerable accumulation of pollutants that, due to their toxic properties, are particularly relevant for introducing harmful soil changes), may serve as a legal basis for the development of biological test procedures.

9C.3 Biological tests

A strategy for the application of biological tests (partly standardised) and a useful concept for the evaluation of contaminated soils using bioassay data have been established by a DECHEMA working group [21]. Under certain circumstances these tests may be used also to characterise wastes.

For the investigation of soils focus is given to groundwater protection (retention function of soils), e.g. by applying the luminescent bacteria, algae and umu-tests, and evaluation of the habitat function of soils, e.g. by using microorganisms (respiration and nitrification), plants (germination and growth) and invertebrates (earthworms and collembolae), considering also the uptake of pollutants and reproduction success.

Notes

1. Act on the regulation of matters pertaining to water (Federal Water Act – WHG), amended through promulgation of 19 August 2002 (*Federal Law Gazette* I, 3245).
2. Act on charges levied for discharging wastewater into waters (Wastewater Charges Act), amended through promulgation of 3 November 1994 (*Federal Law Gazette*, **I**, 3370), last amended by Article 19 G of 9 September 2001 (*Federal Law Gazette*, **I**, 2331).
3. Ordinance on requirements for the discharge of wastewater into waters (Wastewater Ordinance), amended through promulgation of 15 October 2002 (*Federal Law Gazette*, **I**, 4047), corrected on 16 December 2002 (*Federal Law Gazette*, **I**, 4550).
4. *Council Directive 76/464/EEC* of 4 May 1976 on pollution caused by certain dangerous substances discharged into the aquatic environment of the Community (OJ EC L 129, p. 23).
5. *Council Directive 91/271/EEC* of 21 May 1991 concerning urban wastewater treatment (OJ EC L 135, p. 40).
6. *Council Directive 96/61/EC* of 24 September 1996 concerning integrated pollution prevention and control (OJ EC L 257, p. 26).

7. *Directive 2000/76/EC* of the European Parliament and of the Council of 4 December 2000 on the incineration of waste (OJ EC L 332, p. 91; 2001 L 145, p. 52).
8. *Directive 2000/60/EC* of the European Parliament and of the Council of 23 October 2000 establishing a framework for Community action in the field of water policy (OJ EC L 327, p. 1).
9. Veltwisch, D. (2003) *Wastewater Law; Federal Water Act, Wastewater Charges Act, Wastewater Ordinance*. Text edition with an explanatory introduction. Federal Ministry for the Environment, Nature Conservation and Nuclear Safety, BMU: http://www.bmu.de/files/wastewater.pdf
10. *Basel Convention* on the control of transboundary movements of hazardous wastes and their disposal, adopted by the conference of the plenipotentiaries on 22 March 1989 in the (1994 version), with modifications of 1998.
11. *Council Regulation (EEC) No. 259/93* of 1 February 1993 on the supervision and control of shipments of waste within, into and out of the European Community.
12. *Council Directive 91/689/EEC* of 12 December 1991 on hazardous waste.
13. *Council Directive 1999/31/EC* of 26 April 1999 on the landfill of waste.
14. *Council Directive 75/442/EEC* of 15 July 1975 on waste.
15. Gesetz über die Überwachung und Kontrolle der grenzüberschreitenden Verbringung und Kontrolle der grenzüberschreitenden Verbringung von Abfällen (*Abfallverbringungsgesetz*). Vom 30, September 1994 (als Ausführungsgesetz zum Basler Übereinkommen), zuletzt geändert am 25 August 1998.
16. *2000/532/EC*: Commission Decision of 3 May 2000 replacing Decision 94/3/EC establishing a list of wastes pursuant to Article 1(a) of Council Directive75/442/EEC on waste and Council Decision 94/904/EC establishing a list of hazardous waste pursuant to Article 1(4) of Council Directive 91/689/EEC on hazardous waste.
17. Verordnung zur Umsetzung des Europäischen Abfallverzeichnisses Vom 10.12.2001. *BGBl* I, no. 65 (12 December 2001).
18. Bewertungsmuster der Kommission Wassergefährdende Stoffe (KBwS) zur Stoffeinstufung in Wassergefährdungsklassen im Sinne von *Article 19 G Wasserhaushaltsgesetz*, March 2003.
19. Gesetz zum Schutz des Bodens. Vom 17. March 1998. *BGBi*, **I**, no. 16 (24.03.1998).
20. Bundesbodenschutz- und Altlastenverordnung (BBodSchV). Vom 12. July 1999. *BGBI*, **I**, no. 36 (16 July 1999).
21. Biologische Testverfahren für Boden und Bodenmaterial, hrsg. Von: DECHEMA-Arbeitsgruppe *Validierung biologischer Testmethoden für Böden*, 2001.
22. DIN 38415-6 Suborganismische Testverfahren (Gruppe T) Teil 6: Giftigkeit gegenüber Fischen: Bestimmung der nicht akut giftigen Wirkung von Abwasser auf die Entwicklung von Fischeiern über Verdünnungsstufen (T6). Available as multimedia CD-ROM by Beuth-Verlag GmbH (info@beuth.de), No. 15588, (2002)
22. ISO (2004) ISO CD 15088-1: Water quality – Determination of the non-acute toxicity of waste water to fish eggs. Part 1: Danio rerio.

D USA PERSPECTIVE

Barbara Brown and Margarete Heber

9D.1 History of environmental legislation regulating toxics in water in the USA

Much environmental legislation in the USA is structured by media (air, water, waste) rather than by pollutant or stressor. This part of Chapter 9 focuses primarily on the legislation intended to control pollution in water as an example of how toxic pollution is managed in the USA. In addition, a currently developing methodology to integrate the management of a specific class of toxics, pesticides, is also described.

Early American settlement, both pre- and post-Columbian, was established around water sources. Although the importance of ample water quantity for drinking and other purposes was apparent to these early Americans, an understanding of the impacts of water quality was not well known. In the 19th century, owing to the discoveries of Louis Pasteur, scientists began to gain a greater understanding of the sources and effects of water contaminants (especially those in drinking water) not visible to the naked eye. Starting in the mid-1800s, the impacts of disease-causing pathogens in public water supplies became of public concern. Many states, such as New York, established regulations and commissions to protect the quality of their drinking water. The first US federal regulation of water quality began in 1914, when the US Public Health Service (PHS) set standards for the bacteriological quality of drinking water.

The Water Pollution Control Act (WPCA) of 1948 was the first comprehensive statement of federal interest in clean water programs, as well as the first statute to provide state and local governments with some of the funds needed to solve their water pollution problems. There were no federally required goals, objectives, limits or guidelines. There were no mandatory indicators of whether pollution was indeed occurring. Nonetheless, the US Surgeon General was charged with developing comprehensive programs to eliminate or reduce the pollution of interstate waters. Federal involvement was strictly limited to interstate waters that endangered the health or welfare of a person in a state other than that in which the discharge originated.

Both the US PHS standards and the WPCA were revised and expanded; the last PHS revisions in 1962 covered 28 substances, and the amendments to the water pollution acts increased federal funding assistance to municipal dischargers, as well as moved toward developing water quality standards for interstate waters. However, public concern continued to mount over many environmental and health issues, leading to the passage of key federal environmental and health laws: the Federal Water Pollution Control Act Amendments in 1972 (later named the Clean Water Act with the 1977 amendments) and the Safe Drinking Water Act in 1974. Table 9.8 lists the principal environmental legislation and amendments that now nationally govern water quality, including toxics in water.

9D.2 Current legislative and regulatory framework

9D.2.1 The Clean Water Act

The Clean Water Act (CWA) states that the ultimate objective of the act is '... to restore and maintain the chemical, physical and biological integrity of the

Nation's water.' The CWA consists of two major parts: the provisions that authorize federal financial assistance for municipal sewage treatment plant construction; and the regulatory requirements that apply to industrial and municipal dischargers. Prior to 1987, implementation programs for the CWA were directed primarily at point source pollution, i.e. wastes discharged from discrete sources such as pipes and outfalls. Effluent guidelines are the technology basis for point source controls and exist for many different types of industry (e.g. pulp and paper, petroleum, metal finishing). Permit limits based on effluent are based on either the best available technology or best practicable technology (Heber and Norberg-King, 1996). These permits, however, fit within an overall requirement based on the designated use of the ambient waters into which the industry discharges. All ambient waters have designated uses set by the state, such as suitability for aquatic life use, swimming or fishing. The state water quality standards set for each water body should reflect the designated uses. There are both chemical and biological water quality standards. If the standards for the water body are exceeded, management action must be taken to bring the waters back into compliance, such as by issuing more stringent point source discharge limits on facilities permitted in the water body, or by working to reduce non-point source pollution coming into the water body.

Early emphasis was on controlling discharges of 'conventional pollutants' (e.g. suspended solids or bacteria that are biodegradable and occur naturally in

Table 9.8 US environmental legislation and major amendments for water (Copeland, 2002; Tiemann, 1999).

Year	Act	Public law
1948	Federal Water Pollution Control Act	P.L. 80–845
1956	Water Pollution Control Act 1956	P.L. 84–660
1961	Federal Water Pollution Control Act Amendments	P.L. 87–88
1965	Water Quality Act of 1965	P.L. 89–234
1966	Clean Water Restoration Act	P.L. 89–753
1970	Water Quality Improvement Act	P.L. 91–224, Part 1
1972	Federal Water Pollution Control Act Amendments	P.L. 92–500
1974	Safe Drinking Water Act 1974	P.L. 93–523
1977	Clean Water Act of 1977	P.L. 95–217
1981	Municipal Wastewater Treatment Construction Grant Amendments	P.L. 97–117
1986	Safe Drinking Water Act Amendments	P.L. 99–339
1987	Water Quality Act of 1987	P.L. 100–4
1996	Safe Drinking Water Act Amendments	P.L. 104–182

the aquatic environment). Starting with the 1977 amendments, focus shifted to 'non-conventional' or toxic chemical pollutant discharges. The CWA requires that standards be set based on the risk to humans and the environment. The risk assessments performed to set protective levels evaluate the response of humans and key animals to increasing amounts of a chemical (the 'dose-response relationship'), the likelihood of human exposure to that chemical through various pathways and the mitigating or exacerbating effects of other factors (such as effects from synergistic or antagonistic interactions of multiple toxics, cumulative effects over time, or special susceptibilities of particular subpopulations, such as children or the elderly). Chemical standards are currently set chemical by chemical.

Amendments in 1987 authorized measures to address non-point source pollution (storm water runoff from farm lands, forests, construction sites and urban areas), now estimated by the states to represent the largest remaining water pollution problem in the USA (USEPA, 2000). Because of this, although toxics remain a significant component in the mix of non-point runoff, other major pollutants such as nutrients have risen to the fore of public concern for action. Presently the major thrust in implementation of the CWA is to manage on a watershed basis and integrate all parts of the CWA to operate in an integrated fashion, instead of in isolation.

9D.2.2 The Safe Drinking Water Act

The Safe Drinking Water Act (SDWA) aims to protect public health by regulating the nation's public drinking water supply. Originally, the SDWA focused primarily on treatment as the means of providing safe drinking water at the tap, charging the US Environmental Protection Agency (USEPA) to set national standards that protect consumers from harmful contaminants in drinking water. The 1996 amendments greatly enhanced the law by also recognizing source water protection, operator training, funding for water system improvement, and provision of information to the public as important components in protecting drinking water safety.

For each contaminant requiring federal regulation, the USEPA is charged with establishing a maximum contaminant level goal (MCLG) based, as with the CWA, on the risk to humans. The basic risk assessment process is the same but the protective levels established by the risk management process may incorporate the application of additional uncertainty factors to account for carcinogenicity and mode of action. The USEPA then specifies a maximum contaminant level (MCL), which is the maximum permissible level of a contaminant in drinking water that is delivered to any user of a public water system. These levels are set as close to the goals as feasible, i.e. the level that may be achieved with the use of the best technology, treatment techniques, or other means, taking cost into consideration.

The first interim drinking water standards set in 1975 were for total coliform bacteria and turbidity (as indicators of pathogens), six synthetic organic chemicals (such as pesticides) and ten inorganic chemicals. The 1986 amendments added 83 additional contaminants and established monitoring requirements for unregulated contaminants so that the USEPA could decide whether or not to regulate those contaminants. Particular emphasis has been given successfully in recent years to the management and treatment for lead and copper, as both were used in household plumbing fixtures and pipes for many years, thus increasing the likelihood of water supply contamination.

9D.3 Current implementation: institutional responsibilities (national and state) (USEPA, 2004)

Both the CWA and the SDWA reflect a philosophy of federal–state partnership. Under both laws, the USEPA is required to develop national regulations, guidelines and policies to meet the goals of these acts. Congress realized that protection of water was still primarily a state responsibility, and therefore authorized the USEPA to delegate responsibility for implementation and enforcement of certain programs to those states that met the minimum federal requirements for the stringency of their regulations and the adequacy of their enforcement procedures. The SDWA uses the term 'primacy' to describe this concept, and the CWA uses 'authorization'.

Primacy/authorized state programs operate in lieu of the federal water programs. The SDWA allows states to be granted primacy for two programs: the public water system supervision (PWSS) program and the underground injection control (UIC) program. The CWA allows states to be authorized for two programs: the National Pollutant Discharge Elimination System (NPDES) program, and the Section 404 dredge and fill permit program.

Under the CWA's NPDES permit program, any point source discharge of pollutants to waters must be expressly authorized by a valid NPDES permit. The NPDES program consists of various components, including the base program for municipal and industrial facilities, permitting of federal facilities, pretreatment programs, biosolids programs, and general permitting. Forty-six states have been delegated responsibility for the base program and general permitting. These states have also been delegated the responsibility for various combinations of the other three components. All states but one have primacy for the PWSS program. Only two states have delegated state programs for Section 404 dredge and fill permitting.

Primacy/authorized states implement and enforce state regulations and standards, issue permits, and monitor the activities of the regulated community. All states granted primacy or authorization have adequate enforcement authority to compel facilities to comply with permits, including the authority to apply appropriate regulations and standards to the facilities, sue in court to enjoin violations, enter and inspect facilities, and assess civil or criminal penalties for violations. Referral to the USEPA is used as a last resort when state resources are insufficient to address the issue or when previous state efforts have not been successful. The USEPA may also bring an independent enforcement action in a primacy/authorized state, after appropriate notice, if the state fails to take enforcement action. Citizens also have the right to initiate a court action under the SDWA and the CWA if they believe that the regulations are not being enforced appropriately.

Approval of primacy or authorization is a regulatory action. Where states do not receive primacy/authorization, the USEPA operates the relevant program under federal law. The USEPA also provides oversite to the primacy/authorized state's programs, including taking enforcement action as necessary. The states also must report a variety of information to the USEPA and the public to ensure that their programs are meeting the minimum standards necessary to protect water quality.

To ensure that waters are free from 'toxics in toxic amounts' and to help control discharges of a complex nature, whole effluent toxicity (WET) testing was added to the US NPDES under the CWA (Heber *et al.*, 1996).

State regulatory agencies are required to integrate whole effluent tests, chemical-specific water quality criteria and bioassessment in the receiving environment into NPDES permit writing.

9D.4 Future

9D.4.1 Major issues

As mentioned previously, non-point pollution sources are considered by the states to be the largest remaining water quality problem. Control of these sources will be critical to the development of implementable total maximum daily load (TMDL) requirements, which determine the limits needed to restore impaired watersheds. The current legislative framework, however, allows legal enforcement action against point source dischargers (which can be identified and given consent permits with levels needed to meet the TMDL limits), but provides no mechanism for legal action to curtail non-point source discharges. Restoration therefore depends on setting more stringent controls on point source dischargers. In many cases where non-point sources may constitute the largest pollution sources, agreements to reduce non point-source pollutants will be needed to restore water quality.

Another continuing issue that plagues effective water quality protection is the availability of funding for infrastructure projects needed by public water and waste treatment systems to comply with CWA and SDWA rules. Budgetary constraints on federal aid for state revolving fund programs and the large remaining funding needs suggest that a significant funding gap exists and will continue to grow as the requirements increase and the infrastructure ages.

9D.4.2 New developments: coordinated management of toxics between legislative mandates

This part of Chapter 9 has focused primarily on the management of toxics in one medium – water – as an example of how the US legislative framework regulates toxics. Toxics are also regulated in the USA by other environmental legislation. For example, there is work going on to coordinate and control mercury between the Clean Air Act and the Clean Water Act. Another example is a specific class of toxics – pesticides – that are partially regulated by the Federal Insecticide, Fungicide, and Rodenticide Act (FIFRA), which was first passed in 1947 and extensively amended (primarily in 1972 and most recently 1996) to form the basis of most pesticide regulation today.

The primary focus of FIFRA is to provide federal control of pesticide distribution, sale, and use. The central feature of FIFRA is its pesticide registration program, which requires all pesticides be registered with and approved by the USEPA prior to manufacturing, marketing and distribution to ensure that the pesticide poses no serious threats to human health or the environment when used properly. Registration of new pesticides includes submitting to the USEPA the pesticide's complete formula, a proposed label, and a detailed description of the tests and results upon which the pesticide's manufacturer bases their claims that the pesticide is an effective and safe method of controlling pests.

As with the CWA and the SDWA, the USEPA's decision to register a pesticide is based in part on a risk assessment, using the test results supplied by the manufacturer, of adverse effects on human health and the environment. Registration therefore attempts to ensure that pesticides will be labeled properly and, if used in accordance with specifications, will not cause unreasonable harm to the environment.

The USEPA is developing aquatic life use water quality criteria for atrazine under the CWA. Atrazine is also undergoing re-registration for FIFRA. Both the CWA and FIFRA require a risk assessment of human and ecological effects. Rather than perform separate ecological risk assessments, the USEPA did one risk assessment for both the CWA and FIFRA purposes as part of a pilot to integrate better and improve environmental regulation (Bradbury, 2004 pers. comm.). Risk assessment under the CWA normally follows standard hazard assessment protocols to eight test species spanning multiple phyla. The most recent risk assessment used an end-point of community structure and function based on analysis of literature data from meso- and microcosm experiments, thus developing the risk characterization on a more complex, real-world picture. The meso/microcosm data were used to determine, over a period of many months, that a concentration of 10–20 parts per billion of atrazine would be associated with significant degradation of aquatic community structure and function. However, because of application procedures, atrazine does not usually persist at that concentration in the aquatic environment for that length of time. There may, instead, be a pulse of a higher concentration of atrazine for a much shorter period, depending on the method of application and watershed hydrology. A model was prepared to set time-duration water quality criteria rather than a single number for atrazine, which is a major change in the way water quality criteria are stated. In 2004 and 2005 the registrant will be monitoring a representative sample of watersheds with high potential for atrazine exposure to gather data to evaluate the extent to which any stream systems are experiencing inputs that would exceed the time-duration-based water quality criteria. The overall pilot effort with atrazine, therefore, is intended to develop more realistic water quality criteria under the CWA and greater site specificity for the ecological risk assessment, supporting the re-registration decision under FIFRA.

9D.5 Conclusion

The legislative framework in the USA to protect the environment from toxics is primarily media-specific. The criteria limiting toxic concentrations are developed primarily at the national level by the USEPA based on the risk to human health and the environment, taking into account the control technology available and the cost of the control. Much of the implementation of environmental protection, including toxic control, is delegated to the states. Although criteria development and implementation to date have been based on species toxicity on a media by media basis, pilots are underway to incorporate more real-world conditions. Funding to bring about the goals of the legislative framework remains an ongoing problem.

References

Copeland, C. (2002) *Clean Water Act: A Summary of the Law*, RL 30030. Library of Congress, Congressional Research Service, Washington, DC.

Heber, M.A. and Norbert-King, T.J. (1996) United States Environmental Protection Agency's water-quality based approach to toxics control. In *Toxic Impacts of Wastes on the Aquatic Environment*, Tapp, J.F., Hunt, S.M. and Wharfe, J.R. (eds), pp. 175–187. Royal Society of Chemistry, London.

Heber, M.A., Reed-Judkins, D.K. and Davies, T.T. (1996) USEPA's whole effluent toxicity testing program: a national regulatory perspective. Discussion-Initiation Paper 2.1. In *Whole Effluent Toxicity Testing: an Evaluation of Methods and Prediction of Receiving Environment Impacts*, Grothe, D.R., Dickson, K.L. and Reed-Judkins, D.K. (eds), pp. 9–15. SETAC, Pensacola, FL.

Tiemann, M. (1999) *Safe Drinking Water Act: A Summary of the Law*, RL 30022. Library of Congress, Congressional Research Service, Washington, DC.
USEPA (2000) *National Water Quality Inventory Report to Congress*, EPA-841-R-02-001. Environmental Protection Agency, Office of Water, Washington, DC.
USEPA (2004) *History of Water Regulation in the US*.
http://www.epa.gov/safewater/dwa/electronic/presentations/sdwa/sdwacwa2.pdf

10 Case study: Whole effluent assessment using a combined biodegradation and toxicity approach

Graham F. Whale and Nigel S. Battersby

10.1 Introduction

In many countries, effluent discharges traditionally have been assessed and regulated on the basis of physical and chemical properties such as chemical oxygen demand (COD), biochemical oxygen demand (BOD), pH, suspended solids and concentrations of specific substances of concern. This approach (commonly referred to as the *substance-specific approach*) has led to significant improvements in surface water quality and reduced inputs of environmentally hazardous substances. However, there is increasing recognition by regulators that there are limitations to the substance-specific approach for assessing and controlling effluent discharges. Consequently, a number of regulatory authorities are either using or considering the use of ecotoxicological assessments to provide a more holistic means of assessing the potential impact of effluents in the aquatic environment (Power and Boumphrey, 2004). Further to this move to assess ecotoxicity, there has been a desire to include additional assessments of whole effluents, particularly persistence and bioaccumulation potential. These so-called *whole effluent assessments* (WEA) are being considered in upcoming legislation, such as the EU Water Framework Directive, and within the Oslo and Paris Commissions (OSPAR) as potential mechanisms to control discharges of hazardous substances. For example, the OSPAR Point and Diffuse Sources working group set-up an intersessional expert group (IEG-WEA) in 1999 to examine the value of a WEA approach in helping to achieve the OSPAR objectives for protection of the marine environment. To date, the group has discussed possible ways of using persistence, bioaccumulation and toxicity data for whole effluents, produced reviews of potential methods and conducted a demonstration programme in 2003 to test these methods on a limited number of effluents around Europe (OSPAR, 2004).

The current thinking is that whole effluent toxicity assessment provides a practical, biologically relevant mechanism for providing additional hazard data on the combined effects of all the contaminants present in a complex effluent. There is reasonable confidence that an appropriate set of acute, and to a lesser extent chronic, toxicity assessments exist that have been used in many countries over the past two decades. Consequently, there is guidance on limit conditions for testing and interpreting results. However, this is not the case for methods for assessing the persistence and bioaccumulation hazards of effluent components. One of the problems is that, although there are a number of methods for assessing the ecotoxicity of effluents, the same is not true regarding appropriate methods for measuring persistence and the potential to bioaccumulate. Furthermore, where tests to assess bioaccumulation potential and persistence do exist, they tend to be at an early stage of development and require more practical experience to demonstrate their usefulness and feasibility towards effluents.

Although persistence is an important parameter in assessing the impact of an effluent, it is difficult to ascertain how this can be measured conclusively in such a mixture. It is important to consider how the results of such tests could be used in the context of improving effluent risk assessments or in effluent consent/control schemes. It is difficult to envisage how tests used to determine the persistence of single substances could be used in any meaningful environmental context to assess complex effluents. For example, although criteria for persistence have been proposed by a number of organisations (e.g. OSPAR, UNEP, EC), these are generally based on degradation half-lives for single substances and are of little relevance to effluents (i.e. refinery effluents) that contain mixtures of different components. To compound the problem there is still scientific debate surrounding the justification of how the persistence of a single substance is assessed (in particular, how results from standard tests are interpreted), making it even more difficult to define what 'persistence' means in relation to effluents. This problem is recognised by OSPAR, who state that it is incorrect to refer to the 'persistence of effluents' (OSPAR, 2004).

We recognised these issues several years ago and, in an attempt to resolve this dilemma, approached the problem by trying to improve our understanding of the factors that influence the toxicity of effluents once released into the environment. In order to try to improve the risk assessment of effluents our approach was not to look at persistence *per se*. Rather, the risk assessment of complex petrochemical and refinery effluents was improved by examining the *persistence of toxicity*. This was achieved by undertaking combined biodegradation (respirometer) and toxicity studies. The intention was to provide data that potentially could reduce the imposition of unrealistic safety factors by demonstrating that effluent toxicity was either recalcitrant ('hard' – thus warranting the use of conservative safety factors) or easily biodegradable ('soft' – where longer term environmental risks would be reduced and consequently less stringent safety factors would need to be applied).

In the study described in this chapter we have used this approach to assess the toxicity of three complex effluents:

- An oil refinery process effluent of relatively low toxicity being discharged into a small river.
- A petrochemical effluent of moderate toxicity being discharged to a sewage treatment works.

- An oil refinery waste stream of moderate/high toxicity being ultimately discharged into an estuary.

We describe the methodology adopted for the combined biodegradation and toxicity assessments and discuss how the data can be interpreted and used to improve risk assessment in the context of these three effluents.

10.2 Considerations prior to initiating the study

Prior to undertaking the study it was important to ensure that the methods selected for the biodegradation and toxicity assessment were practical, cost-effective and capable of providing environmentally relevant toxicity and biodegradation data over a reasonably short time period (say <15 days). It was decided that biodegradation of the effluents would be assessed using an electrolytic respirometer – this approach met the above criteria and had several advantages:

1. Biodegradation is followed as biochemical oxygen demand (BOD), which enables descriptors such as BOD_5 to be obtained.
2. The organic matter in the effluent could be measured as chemical oxygen demand (COD) and the extent of biodegradation (BOD) then expressed in terms of COD.
3. In contrast to traditional BOD measurements that use closed bottles, the respirometer replenishes any oxygen (O_2) consumed automatically and low biodegradation due to O_2 depletion does not occur.
4. The instrument is sensitive, enabling the biodegradability of 'weak' effluents to be tested.
5. Oxygen uptake is logged frequently (every 2 h in this study) and automatically – the large number of data points obtained enables changes in the *rate* of biodegradation to be observed, which can often show subtle changes in the course of biodegradation.
6. The respirometer provides sufficient solution to enable chemical analyses and the acute toxicity of the effluents to be monitored during the study.

For the ecotoxicity assessments two types of test were required: to monitor toxicity during biodegradation and to assess the toxicity at the start and end of the biodegradation test. The tests required to *monitor* the toxicity needed to be:

- Able to produce 'quick' results (< 1 day).
- Use relatively small volumes of test media (< 50 ml).
- Be predictive of change of toxicity of ecologically relevant organisms.

For the *initial* and *final* toxicity assessments the tests needed to:

- Use organisms relevant to the environment to be protected.
- Be credible and acceptable to a competent authority (i.e. use internationally recognised tests species and protocols).
- Require less than 1 litre of test media (the combined volume of the duplicate respirometer flasks).

The Microtox® bioluminescence test with the marine bacterium *Vibrio fischeri* (formerly *Photobacterium phosphoreum*) was considered to meet the requirements for the ecotoxicity monitoring method. (Equivalent bioluminescence tests are available from a number of manufacturers.) In this test, the enzyme luciferase (which is intrinsically linked with the metabolism of the bacterium) catalyses a light-emitting reaction, with effects on the bacteria leading to a reduction of light output. Reductions in light emission are quantified photometrically and used to calculate Microtox® EC_{50} values (concentration of test chemical giving a 50% reduction of light output). The Microtox® test requires small sample volumes (1–2 ml) of test solution, is rapid, easy to conduct and results can potentially be correlated to 'more traditional' toxicity studies. For example, Kaiser (1998) found significant correlations between the toxicity of chemicals determined by Microtox® and that determined using other aquatic species such as fish, the water flea *Daphnia* and algae. The advantages of the Microtox® test have been recognised by Rojickova-Padrtova *et al.* (1998) and Repetto *et al.* (2001), who have proposed its inclusion in test batteries to assess environmental hazard.

The Microtox® test also appears to be particularly suitable for assessing hydrocarbon contaminated effluents because Kaiser (1998) has shown that for soluble organic compounds the sensitivity of the Microtox® test is approximately equal to that of *Daphnia*, rainbow trout and fathead minnow bioassays. This has been our experience also in our laboratory, where good correlations ($r^2 = 0.8$) had been found for five sets of Microtox® and oyster embryo tests conducted on one of the refinery effluents assessed in the respirometer study.

For the initial and final ecotoxicity assessments the toxicity test methods selected depended upon the receiving water for the effluent. For effluents A and B, which were ultimately discharged to freshwater courses, *Daphnia magna* was selected to be an appropriate test species. For effluent C, which was discharged to an estuary, the calanoid copepod *Acartia tonsa* (in simplistic terms, a marine equivalent to *Daphnia*) and the oyster *Crassostrea gigas* embryo test were used. These test methods were based on internationally accepted guidelines that had been ring tested and considered suitable for inclusion in the UK Environment Agency's Direct Toxicity Assessment (DTA) programme. Although algal toxicity tests are recommended by the Environment Agency for DTA investigations, these were not used in this study because it was believed that the particulate and often coloured nature of many of the effluents would make such tests difficult to conduct and interpret.

10.3 Materials and methods

10.3.1 Effluents tested

A brief description of the three effluents assessed in this study are given below and their routes into the environment are summarised in Figure 10.1.

10.3.1.1 Effluent A – refinery process effluent

This effluent was derived from ships' ballast water and effluent from an interceptor (mainly site drainage), which provides the feed to a dissolved air flotation (DAF) unit prior to discharge into a freshwater brook. The site drainage consists mainly of what would be regarded as accidentally oil-contaminated water (i.e. spillages onto a concrete apron that are washed into drainage during rainfall events). Because the effluent flow to the DAF is intermittent and subject to shock loads of high salinity when ballast water is discharged, the effluent was sampled on eight occasions with samples stored in the dark at 4°C prior to analysis. The total organic carbon (TOC), inorganic carbon and toxicity to *Vibrio fischeri* (Microtox®) were determined for each sample. These were reasonably consistent, with TOC varying in the range 20–30 mg/l (mean = 25, sd = 2.8), inorganic carbon in the range 23–30 mg/l (mean = 26, SD = 2) and 15-min Microtox® EC_{50} values varying between 10% and 19% (mean = 13, SD = 3) of the effluent. Therefore, a composite of pooled equal volumes of effluent sampled on each occasion was used in this study.

Effluent A was selected for assessment because this was felt to be representative of many effluents that have relatively low levels of oil contamination and

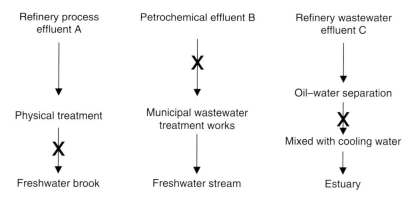

Figure 10.1 Routes of effluents into receiving water.

aquatic toxicity (acute toxicity removed by less than 10-fold dilution). In terms of a local risk assessment, it is assumed that toxicity of this effluent will be diluted rapidly. However, the discharge is one of several industrial discharges into a small watercourse and it would be useful to have some reassurance that the toxicity is rapidly degraded ('soft').

10.3.1.2 Effluent B – petrochemical effluent
This sample was taken from a petrochemical effluent discharged for offsite treatment via a municipal sewage treatment works. This effluent was selected because it was believed to be reasonably representative of a complex petrochemical effluent prior to any biological treatment. The effluent would be considered to be moderately toxic to aquatic organisms (requiring between a 10- and 100-fold dilution to remove acute toxicity). Because this particular effluent was treated offsite it was important to assess whether any toxicity would remain after biological treatment, as this potentially could contribute to any toxicity seen in the effluent from the municipal treatment works. The advantage of the combined study is that it also confirms whether the toxic components of the effluent are susceptible to rapid biodegradation.

10.3.1.3 Effluent C – untreated refinery wastewater
This wastewater was derived from a number of oil–water separation streams, including water separated from an 'oil slops' tank that had been treated only by conventional oil–water separation methods. This effluent would be considered highly toxic to aquatic organisms, requiring a dilution of >100-fold to remove acute toxicity. At the time of the study the effluent was mixed into the refinery cooling water system passing through a final stage of oil–water separation prior to discharge. No toxicity was detected in the mixed cooling water effluent discharge, which received further rapid dilution in the receiving estuary. It should be noted that discharge of this effluent has now ceased.

This effluent was included in this study because it was considered to be representative of the most contaminated refinery wastewater streams. Furthermore, it was considered important to assess that any components responsible for toxicity were not persistent in the receiving environment (i.e. although toxicity had been effectively diluted out by mixing with the cooling water, there could be long-term risks if toxic components were accumulating in the vicinity of the discharge). This latter point was considered important for trying to redress OSPAR concerns, by demonstrating that toxicity is not conservative and does not pose a longer term risk to the marine environment.

10.3.2 Sample collection

Although samples for effluent toxicity assessment ideally should be composite (i.e. pooled hourly spot samples taken over a 24 h period to take into account possible variability), working permits and locations made this difficult logistically.

An autosampler was considered but, owing to concerns that the toxicity of the samples could vary with time due to volatilisation losses and/or possible oxidation while the samples were retained in the device, this approach was not taken in this investigation. Therefore, the adopted strategy was to take a spot sample collected either from a designated sampling point or, for effluent C, using a 10 l stainless-steel bucket lowered into an interceptor pit. Sample bottles (2.5 l brown glass Winchesters) were rinsed with effluent prior to filling. For effluent C, subsamples were taken from the bucket (taking care to avoid taking the surface film). Collection bottles were completely filled and then sealed with 'Parafilm' prior to being transported back to the laboratory in a cold box within a few hours of collection. Storage prior to analysis was in the dark at 4°C.

10.3.3 Assessment of biodegradable and persistent ecotoxicity

10.3.3.1 Biodegradability studies
The biodegradability of each effluent was determined by measuring oxygen uptake (as biochemical oxygen demand, BOD) during incubation under aerobic conditions at 20°C for 14 days. Oxygen uptake was measured using a 20-channel CES electrolytic respirometer (Coordinated Environmental Services, UK). Replicate samples (500 ml) were incubated in stirred flasks and O_2 uptake was measured and replenished automatically by a manometric/electrolysis cell (MEC) attached to each flask. The resolution of each MEC was 0.02 mg of O_2 at standard temperature and pressure (STP) and the maximum O_2 generation rate was 2.1 µg O_2/s.

Prior to incubation, the effluent samples were treated as follows:

1. *Effluent A – refinery process effluent.* The sample was inoculated with 5 ml/l secondary effluent as a mixed population of microorganisms. The inoculum was taken from a sewage works that treats predominantly ($\sim$ 90%) domestic sewage (Chester Sewage Works, Welsh Water).
2. *Effluent B – petrochemical effluent.* The sample was coarse-filtered through medical gauze and split into two. One batch was supplemented with 1 ml/l each of OECD mineral salt medium (OECD, 1992) as a source of inorganic nutrients and inoculated with 5 ml/l secondary effluent from Chester Sewage Works. The other batch was diluted to 25% of its original strength by adding one volume of effluent to three volumes of OECD mineral salt medium. The medium was inoculated with 5 ml/l of secondary effluent. Note that: the final concentrations of inorganic nutrients from the OECD mineral salt solutions were: P, 11.6 mg/l; N, 0.13 mg/l; Na, 8.6 mg/l; K, 12.2 mg/l; Mg, 2.2 mg/l; Ca, 9.9 mg/l; Fe, 0.1 mg/l.
3. *Effluent C – untreated refinery wastewater.* The sample was split into two. One batch was supplemented with 1 ml/l each of OECD mineral salt medium and

the other was diluted to 10% of its original strength by adding one volume to nine volumes of estuarine receiving water (salinity = 20%, COD = 75 mg/l, DOC = 40 mg C/l) and then supplemented with OECD mineral salts as before.

Oxygen uptake (mg O_2/flask) and respiration rate (mg O_2/flask/h) were calculated using the CESAN data analysis program (version 1.1, Coordinated Environmental Services, UK) and plotted using Excel 7.0 for Windows 95.

Initially, and at intervals during incubation, replicate flasks were taken for chemical analysis and acute ecotoxicity tests, as detailed below.

10.3.3.2 Chemical analyses

With the exception of the carbon and oil and grease determinations, analyses were made using commercially available spectrophotometric test methods as summarised in Table 10.1. These had the benefits of being quick, easy to use and relatively inexpensive. Depending on the nature of the effluent and time constraints, not all of the analyses were performed on each sample.

Carbon analysis on the samples was performed using a Dohrmann DC-190 carbon analyser. The TOC concentration of effluent A was determined as *measured total carbon − measured inorganic carbon* on triplicate 200 μl injections of effluent. The analyser was calibrated using a 20 mg carbon/l potassium hydrogen phthalate (KHP) standard. The TOC concentration of effluent B was determined as non-purgable organic carbon (NPOC), where the inorganic carbon is removed

Table 10.1 Chemical analyses performed on the effluents.

Analysis	Method (range)
Chemical oxygen demand (COD)	Dr Lange Cuvette Test kits (5–60, 15–150 or 100–2000 mg COD/l)
$NH_4^+ - N$	Dr Lange Cuvette Test (0.015–2.0 mg/l) or Merck Spectroquant Cell Test (0.2–7.8 mg/l)
$NO_2^- - N$	Dr Lange Cuvette Test or Merck Spectroquant Cell Test (both 0.02–0.6 mg/l)
$NO_3^- - N$	Dr Lange Cuvette Test (1–40 mg/l) or Merck Spectroquant Cell Test (0.1–18 mg/l)
Total N	Dr Lange Cuvette Test (0.015–2.0 mg/l) or Merck Spectroquant Cell Test (0.5–7.8 mg/l)
SO_4^{2-}	Merck Spectroquant Cell Test (100–1000 mg/l)
S^{2-}	Dr Lange Pipette Test (0.1–2.0 mg/l)
Cl^-	Dr Lange Cuvette Test (1–70 mg/l)
F^-	Dr Lange Cuvette Test (0.1–1.5 mg/l)
Phenol	Dr Lange Cuvette Test (0.5–5 mg/l)

by acidification and sparging prior to analysis. Triplicate injections of 200 μl were analysed and the instrument was calibrated against 400 or 50 mg carbon/l KHP standards. The TOC concentration of the refinery wastewater effluent C also was determined as NPOC, using quintuplicate 200 μl injections and a 20 mg carbon/l KHP standard. With all three effluent samples, dissolved organic carbon (DOC) was measured as described above but the samples were filtered through a 0.2 μm inorganic membrane filter prior to analysis.

The oil and grease content of the refinery wastewater effluent C was determined by the partition infrared method (5520 C) according to Standard Methods (1995). A mixture, by volume, of 37.5% *iso*-octane, 37.5% hexadecane and 25.0% benzene was used as the reference oil and infrared absorbance was determined using a Perkin-Elmer 881 infrared spectrophotometer.

10.3.3.3 Aquatic ecotoxicity tests
Vibrio fischeri (trade name Microtox®). The *V. fischeri* used in the Microtox® tests during this study were supplied as freeze-dried batches of reagent by Azur Environmental (SDI), Hook, Hampshire, UK. These were stored at $< -4°C$ prior to use. Microtox® tests were conducted using a Microbics Corporation (SDI) Model 500 Microtox® analyser, following the test protocols for either the 100% screening test or basic test given in the instruction manual supplied with this instrument (Microbics (SDI) 1992). Because the salinity of effluents varied from freshwater to seawater and *V. fischeri* is only viable in saline media, Microtox® osmotic adjustment solution was used to adjust samples of low or zero salinity.

The 5 and 15 min EC_{50} values (those concentrations causing a 50% reduction in light output after 5 and 15 min of exposure, respectively) were calculated using the Microbics (SDI) version 6 software supplied with the instrument.

Daphnia magna (Straus). The culture and toxicity tests with *D. magna* were conducted in general accordance with the OECD 202 (1984) test guideline. The 48 h *D. magna* immobilisation tests were carried out in static, sealed conditions without renewal of the test media. Duplicate groups of 10 *D. magna* (less than 24 h old) taken from a clonal laboratory culture were exposed to 150 ml of each effluent dilution held in sealed 150 ml glass conical flasks. Two flasks containing *D. magna* and reconstituted freshwater only were prepared as controls.

After 24 and 48 hours the numbers of immobilised (i.e. not observed to swim during a 15 second observation period following gentle agitation of the test solution) *D. magna* were recorded.

The reconstituted freshwater used during the *D. magna* culture and tests was prepared following a recipe recommended as being suitable for producing a 'hard' water by the US Environmental Protection Agency (USEPA, 1975). The tests were carried out in a temperature-controlled room set at a nominal $20 \pm 2°C$ with artificial illumination of the test vessels on a 16 h light/8 h dark automatic cycle.

The 24 h and 48 h EC_{50} values (those concentrations causing 50% immobilisation of the *Daphnia* after 24 h and 48 h of exposure, respectively) were calculated with their 95% confidence limits by either Probit analysis (Finney, 1971) or moving average angle analysis (USEPA, 1985).

Marine toxicity assessments. For effluent C, which discharges to a marine environment, the 24 h Pacific oyster (*Crassostrea gigas*) embryo larval test and 48 h immobilisation test with adult copepods (*Acartia tonsa*) were conducted following the procedures described below.

Acartia tonsa (Dana). *A. tonsa* immobilisation tests were conducted following the general guidelines of the ISO TC147/SC5/WG2 protocol, using procedures recommended by the Oslo and Paris Commissions (OSPAR) to provide data for hazard assessment of chemicals intended for use in the North Sea by the offshore oil and gas industry (MAFF, 1994, 1996).

The *A. tonsa* used in this study were obtained from Guernsey Sea Farms (Vale, Guernsey) as age-standardised adults. On arrival they were transferred to a temperature-controlled room ($20 \pm 2°C$) with artificial illumination on a 16 h light/8 h dark automatic cycle. They were maintained under these conditions prior to the start of the toxicity test. The seawater in which the *A. tonsa* were transported was not renewed or aerated during this time. The *A. tonsa* were transferred to a 10 l glass vessel that was topped up with seawater and then they were fed *ad libitum* with a mixed algal diet (*Pavlova lutheri* and *Tetraselmis suecica*) supplied by Guernsey Sea Farms.

The water used for acclimatising and testing the *A. tonsa* was natural seawater collected from the School of Ocean Sciences, Menai Bridge, Anglesey. The quality of the seawater during storage was maintained by circulation through filters and a UV sterilising unit.

Triplicate groups of a nominal 10 *A. tonsa* (18–21 days old at the start of the test) held in 150 ml glass crystallising dishes were exposed to 100 ml of each effluent dilution. Three dishes containing *A. tonsa* and artificial seawater were prepared as seawater controls. Immobilised *A. tonsa* were recorded and removed from the test vessels after 24 h and 48 h of exposure. *A. tonsa* were considered to be immobilised if they failed to respond to touch using the end of a glass pipette. As a quality control procedure a few drops of formalin were added to the test vessels to preserve the remaining *A. tonsa* at the end of the test. This ensures that the remaining *A. tonsa* fall to the bottom of the vessel, enabling them to be counted rapidly to confirm the total number of *A. tonsa* in the test vessel.

The tests were carried out in a temperature-controlled room ($20 \pm 2°C$) with artificial illumination of the test vessels on a 16 h light/8 h dark automatic cycle. The solutions were not renewed or aerated during the tests.

The 24 h and 48 h EC_{50} values for the *A. tonsa* tests were calculated using the methods described previously for the *Daphnia* tests.

Crassostrea gigas (Thunberg). The 24 h oyster (*C. gigas*) embryo/larval development tests were conducted in general accordance with the method described by Thain (1991) for the International Council for the Exploration of the Sea (ICES). The *C. gigas* (conditioned to a pre-spawning state) were supplied by a commercial oyster hatchery (Guernsey Sea Farms, Guernsey, UK).

Gametes were stripped from the sexually mature oysters by rupturing the gonad with a clean glass Pasteur pipette and gently washing seawater over the gonadol tissue into 400 ml of seawater in a fertilisation vessel (600 ml glass beaker). A new pipette and beaker were used for each oyster. Subsamples of gametes from each oyster were examined under a microscope to identify suitable gametes to be used in this test. Gametes were considered to be suitable if the sperm were motile and the eggs well-rounded and free of obvious rupture. The eggs of two to three oysters were pooled and passed through a $100 \pm 10\,\mu m$ mesh to remove any tissue debris. The sperm from two oysters were pooled and passed through a second mesh of the same size.

An estimate of the density of the egg suspension was made using a Sedgewick–Rafter cell and the density of the egg suspension was adjusted to 5×10^3 to 10^4 eggs/ml. Two millilitres of the sperm suspension were added to the 400 ml egg suspension, which was mixed and left for 1–2 h. A sample was examined under the microscope to confirm that early stages of development were occurring and the density of the embryo suspension was adjusted to 3.0×10^3 eggs/ml by dilution with seawater. The density of the embryo suspension was rechecked and 30 ml glass Beatson jars containing 30 ml of the test media were inoculated to give a test density of 50 embryos/ml (nominally 0.5 ml of embryo suspension). A minimum of three replicates for each effluent concentration and 12 controls (containing natural seawater only) were set up for each test. Immediately after inoculation, six of the control vessels were preserved (by the addition of 0.5 ml formalin) and used to estimate the initial embryo density.

The seawater used for the *C. gigas* tests was natural seawater from the laboratory supply, as described previously for the *A. tonsa* tests. Test vessels were incubated for 24 h at $25 \pm 2°C$, after which they were preserved by the addition of 0.5 ml of formalin to each vessel.

To score the test, each preserved sample was mixed and a 2 ml aliquot was pipetted into a flat-bottomed tube. The aliquots were allowed to settle for at least 5 min and then, using an inverted microscope, the number of D-stage veligers (those larvae with fully formed symmetrical shells) was scored.

The results of the oyster embryo test were expressed as the *percentage net risk* (PNR) based on the percentage abnormality at each effluent test concentration. These were calculated as follows:

$$\% \text{ Abnormality} = \frac{I - D}{I} \times 100$$

where: I = mean initial embryo density
D = mean number of D-stage veligers in treatment

Percentage net risk (PNR) was calculated using:

$$\text{PNR} = \frac{\% \text{ Test abnormality} - \% \text{ Control abnormality}}{100 - \% \text{ Control abnormality}} \times 100$$

The PNR was used to calculate a 24 h EC_{50} value and 95% confidence limits using Probit analysis (Finney, 1971) or moving average angle analysis (USEPA, 1985). For these calculations, negative PNR values were set to zero.

10.4 Results

10.4.1 Effluent A – refinery process effluent

The composite sample of refinery effluent A had low levels of COD, carbon and salinity (Table 10.2). The initial sample (day 0) was non-toxic to *Daphnia* and exhibited relatively low levels of toxicity to Microtox® (EC_{50} 20–27% of neat effluent) as summarised in Table 10.3. The biodegradation of the effluent, as measured by the increase in BOD (cumulative oxygen uptake) and change in Microtox® toxicity of the effluent over 14 days, is shown in Figure 10.2. It can be seen that biodegradation of the organic material in the effluent started immediately and was most rapid during the first day of incubation. The rate of biodegradation fell quickly during this period and tailed off to around 0.05 mg $O_2/l/h$ from day 7 onwards. The cumulative O_2 uptake was equivalent to 44% of the initial COD after 5 days of incubation and 67% after 14 days (Table 10.2).

Table 10.3 and Figure 10.2 show that Microtox® toxicity generally decreased with increasing biodegradation and data from the first replicate indicated that the

Table 10.2 Chemical characteristics of refinery effluent A.

Parameter (units)	Initial	After incubation for: 5 days	14 days
COD (mg O_2/l)	63	56	34
Cumulative BOD (mg O_2/l)	–	28	42
Cumulative BOD (% $COD_{Initial}$)	–	**44**	**67**
TOC (mg C/l)	16	15	14
DOC (mg C/l)	16	14	14
Chloride (mg Cl^-/l)	495	–	–
pH	7.1	7.1	8.4

Table 10.3 Results of Microtox® toxicity tests with the refinery process effluent A.

Day	Microtox® 15-min EC_{50} values as percentage of initial sample			
	Replicate 1	Replicate 2	Replicate 3	Composited
0	21[a] (16–27)[b]	27 (21–34)		
1	31 (13–72)	29 (25–34)	>45	40 (16–62)
2	68 (58–80)	17[c] (8.0–35)	64 (59–69)	77 (66–92)
5	>90	51 (48–56)	3.8[c] (1–16)	5.8 (5.7–5.9)
7	>90	>90	>90	>90

[a] Sample prior to aeration and addition of inoculum.
[b] 95% confidence intervals.
[c] Samples believed to be contaminated with copper sulphate from respirometer (EC_{50} values >90% exceed highest effluent concentration assessed in the 100% test).
[d] Composite sample consisted of equal volumes of replicates 1 to 3.
Note: Toxicity increases with decreasing EC_{50} values.

effluent was not toxic after 5 days. However, the data indicate that anomalies in terms of the trend of toxicity were seen on some occasions, particularly on day 5. For example, the toxicity of the third replicate was much higher than the toxicity of the day 0 sample and affected the toxicity of the composite sample. Repeat Microtox® tests confirmed the toxicity of these samples and an investigation was instigated to assess potential causes for these anomalies. The investigation revealed that the procedure followed by one of the operators could have led to the release of copper sulphate from the electrolytic cell of the respirometer, thus contaminating the flask. The procedure for removal of the flasks was subsequently amended to prevent possible recurrence of the problem.

Figure 10.3 compares the loss of COD with cumulative O_2 uptake (cumulative BOD) during incubation. Although the loss of COD followed the BOD, the latter was always greater. Table 10.2 shows that there was only a small decrease in the levels of TOC and DOC during incubation compared with the 45% removal of COD. All the carbon appeared to be in solution, because TOC virtually equals DOC.

The effluent contained approximately 5 mg N/l total nitrogen (inorganic and organically bound N) and 0.4 mg $NH_4^+ - N/l$. This ammonium level fell rapidly during the first 2 days of incubation and was matched by a concomitant rise in $NO_3^- - N$ levels, indicating that nitrification was taking place (Figure 10.4). However, O_2 uptake due to nitrification ($= 4.57 \times$ change in $NO_3^- - N$ levels; OECD, 1992) was only around 2 mg O_2/l [$4.57 \times (0.46 - 0.03)$] and this did not account for the difference between BOD and COD removal shown in Figure 10.3 (ammonium is not oxidised in the COD test). Ammonium levels rose from day 2 onwards as the organic nitrogen in the effluent was mineralised.

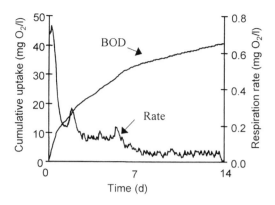

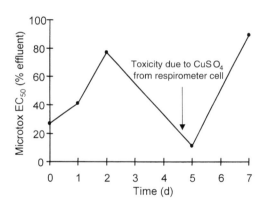

Figure 10.2 Refinery process effluent A: biodegradation and ecotoxicity over 14 days.

10.4.2 Effluent B – petrochemical effluent

In contrast to the refinery process effluent A, the petrochemical effluent B was much higher in COD, carbon and ammonium (Table 10.4). Biodegradation of the undiluted effluent began immediately and continued at a rate of around 3.5–4.0 mgO$_2$/l/h over the 14 day incubation period, with O$_2$ uptake failing to reach a plateau (Figure 10.5).

Analyses therefore concentrated on effluent that had been diluted to 25% of its original strength. It can be seen from Figure 10.5 that the rate of biodegradation of this sample peaked after incubation for 1 day and again over days 2–4. After day 4 the rate of biodegradation tailed off as the cumulative BOD approached a plateau. The 5 and 14 day BOD (BOD$_5$ and BOD$_{14}$) for the diluted effluent represented 47% and

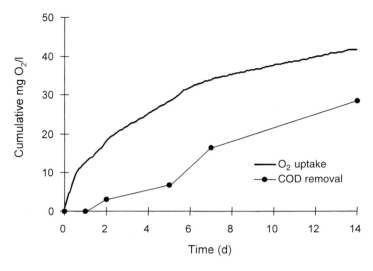

Figure 10.3 Refinery process effluent A: cumulative BOD compared with COD removal during incubation at 20°C (values are the averages for triplicate flasks).

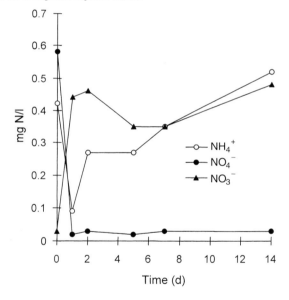

Figure 10.4 Refinery process effluent A: changes in nitrogen levels during incubation at 20°C (values are the averages for triplicate flasks).

67% of the initial COD, respectively (Table 10.4). There was good agreement between the loss of COD and BOD over the incubation period (Figure 10.6), indicating that O_2 uptake was due to biodegradation of the organic matter in the effluent.

Table 10.4 Chemical characteristics of the petrochemical effluent B.

Parameter (units)	Neat effluent	25% Dilution after incubation for:		
		Initial	5 days	14 days
COD (mg O_2/l)	2742	799	320	222
Cumulative BOD (mg O_2/l)	–	–	378	534
Cumulative BOD (% $COD_{Initial}$)	–	–	47	67
TOC (mg C/l)	980	229	123	49
TOC (% Removal)	–	0	46	79
DOC (mg C/l)	726	196	57	47
DOC (% Removal)	–	0	71	76
Phenol (mg/l)	4.2	1.0	–	0.4
Phenol (% Removal)	–	0	–	60
NH_4^+ (mg N/l)	11.3	2.9	0.1	8.3
NO_2^- (mg N/l)	0.1	<0.02	0.1	0.03
NO_3^- (mg N/l)	0.3	0.13	0.2	0.1
Fluoride (mg F^-/l)	<0.5	<0.5	–	–
pH	7.9	7.4	6.1–6.2	7.7–8.1

There was a large decrease in levels of TOC and DOC over the 14 day incubation period. Dissolved material accounted for 74% of the TOC present initially and both fractions were removed by over 70% by day 14 (Table 10.4). The diluted effluent initially contained 4 mg/l phenol, which had been biodegraded by 60% after 14 days of incubation. As seen previously with refinery process effluent A, ammonium levels fell and then rose (Table 10.4) but there was negligible increase in oxidised nitrogen, indicating that little nitrification was occurring during incubation.

As anticipated, petrochemical effluent B was moderately toxic to both *D. magna* and Microtox®, with respective 48 h and 15 min EC_{50} values of 3.2 and 1.4% of the effluent (Table 10.5). Consequently, the effluent would require a 31- to 71-fold dilution to reduce the initial effluent toxicity to the EC_{50} values. There was no significant change in the toxicity for the neat effluent to either *D. magna* or Microtox® over the 14 day respirometer study (Figure 10.5). However, this does not indicate that the toxic components were not biodegradable because, although biodegradation had occurred, this had not reached a plateau. Support that the toxic components were biodegradable comes from the assessment of the effluent that had been diluted to 25% of its original concentration. The day 0 EC_{50} values for the diluted effluent were approximately 25% of those for the neat effluent, indicating that the process for diluting and preparing the effluent sample for the

(a) Biodegradation

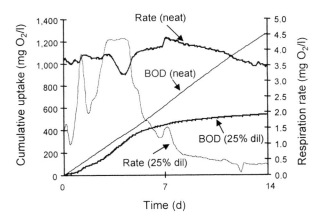

(b) Ecotoxicity

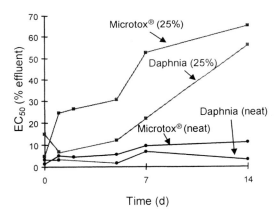

Figure 10.5 Petrochemical effluent B: biodegradation and ecotoxicity over 14 days.

biodegradation study had not influenced its toxicity. Over the 14 day study the toxicity of the effluent decreased to both *D. magna* and Microtox® and was reduced considerably (both tests requiring less than a twofold dilution to reduce the toxicity of the biodegraded effluent below the EC_{50} values) (Figure 10.5).

10.4.3 Effluent C – untreated refinery wastewater

The high salinity of this sample meant that many of the chemical analyses were affected by chloride ion (Cl^-) interference (Table 10.6). This, initially at least,

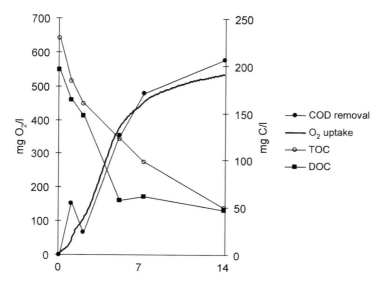

Figure 10.6 Petrochemical effluent B (25% dilution): cumulative BOD, COD removal and levels of total organic carbon (TOC) and dissolved organic carbon (DOC) during incubation at 20°C (values are the averages for duplicate flasks).

could be alleviated by diluting the sample prior to analysis. However, this was not a practical option for the sample that had been diluted already to 10% of its original strength with estuarine water because it reduced the levels of many of the determinants to below their limit of detection. Data are therefore presented for the neat 100% effluent sample only.

Biodegradation of the effluent started immediately and was most rapid after 0.5–1.5 days of incubation (Figure 10.7). Peaks of biodegradation were also observed after 3.5, 4.5 and 5 days, suggesting that preferential biodegradation of different components of the effluent was occurring. Oxygen uptake after 5 and 14 days was 68% and 93% of the initial COD, respectively.

The sample contained 91 mg/l of oil and grease, which was steadily biodegraded by 60% over the 14 day incubation period (Figure 10.8). Levels of TOC and DOC also declined during incubation and final percentage losses were similar to those measured for oil and grease (Table 10.6). If it was assumed that the carbon content of the oil and grease was 86% (i.e. CH_2), then their concentration as carbon was 78 mg C/l in the initial sample. However, the measured level of insoluble carbon in the effluent was 34 mg C/l (insoluble carbon = TOC − DOC). This implies that most of the oil and grease in the effluent was present as micro-emulsions, which could pass through 0.2 μm pores and/or as dissolved material.

The toxicity of effluent C to Microtox® was far greater than the other effluents, with 15 min EC_{50} values for day 0 ranging between 0.19 and 0.37% of the effluent (Table 10.7). A similar level of toxicity was found with the oyster embryo test,

Table 10.5 Results of toxicity tests with the petrochemical effluent B.

Day	EC$_{50}$ values as percentage of initial sample[a]			
	Neat effluent		Effluent diluted 1:4 (25%)	
	Microtox® 15-min EC$_{50}$	D. magna 48-h EC$_{50}$	Microtox® 15-min EC$_{50}$	D. magna 48-h EC$_{50}$
0	1.4 (1.2–1.6)	3.2 (2.6–3.9)	4.6 (3.9–5.4)	15 (12–18)
1	4.7 (4.2–5.4) 5.1 (4.7–5.6)	3.2 (2.6–3.9)	20 (15–26) 29 (23–36)	6.8 (5.5–8.3)
2	5.9 (5.0–7.0) 4.5 (3.6–5.6)	Not tested	23 (19–26) 30 (23–40)	Not tested
5	8.7 (7.6–10) 10 (8.3–12)	1.5 (0.5–2.9)	32 (16–64) 29 (17–48)	12 (11–14)
7	7.8 (6.2–9.8) 11 (11–12)	6.8 (5.5–8.3)	54 (2.7–100+) 61 (10–100)	> 22[b]
14	10 (10–10) 12 (10–14)	3.2 (2.6–3.9)	64 (49–82) >45	56 (50–63)

[a] All Microtox® EC$_{50}$ values > 45% exceed the highest effluent concentration assessed using the Microtox® Basic test. When more than one value is given, this indicates duplicate measurements. Values in parentheses are 95% confidence intervals.
[b] Highest concentration assessed on this occasion.

where the 24 h EC$_{50}$ value for day 0 was 0.41% of the effluent. Consequently, effluent C would require a 240- to 530-fold dilution to reduce the initial effluent toxicity to the EC$_{50}$ values for these tests. However, it was apparent that the effluent was significantly less toxic to *A. tonsa*. (the 48-h EC$_{50}$ value on day 0 was 9.2% of the effluent).

For the oyster and Microtox® tests, the day 0 EC$_{50}$ values for the diluted effluent were not as would have been anticipated on the basis of dilution alone. For example, the oyster embryo EC$_{50}$ value for the diluted effluent was > 22% and therefore was > 50-fold different from the neat effluent, and the Microtox® tests were > 30-fold different. These data indicate that the process for diluting and preparing the effluent

Table 10.6 Chemical characteristics of the refinery wastewater effluent C.

Parameter (units)	Initial	After incubation for: 5 days	After incubation for: 14 days
COD (mg O_2/l)	324	103	Interference[a]
Cumulative BOD (mg O_2/l)	—	221	300
Cumulative BOD (% $COD_{Initial}$)	—	68	93
TOC (mg C/l)	150	73	60
TOC (% Removal)	0	51	60
DOC (mg C/l)	116	56	48
DOC (% Removal)	0	52	59
Oil and grease (mg/l)	91	59	36
Oil and grease (% Removal)	0	35	60
NH_4^+ (mg N/l)	34	5.5	Interference
NO_2^- (mg N/l)	Interference	Interference	Interference
NO_3^- (mg N/l)	2.8	Interference	Interference
Chloride (mg Cl^-/l)	6900	—	—
Sulphide (mg S^{2-}/l)	9.1	Interference	Interference
pH	8.8	7.7–8.2	8.1–8.5

[a] The high chloride concentration of the effluent caused interference with most of the test kits used. Initially, this could be overcome by diluting the samples 20-fold to give a Cl^- concentration of <500 mg/l. However, as the concentration fell during incubation of the sample, this dilution resulted in the concentration of the determinand being less than the limit of detection of the method.

C sample for the biodegradation study had significantly influenced (decreased) its toxicity. In fact, apart from the Microtox® day 0 values, no significant toxicity (i.e. two fold dilution required to reduce toxicity to the EC_{50} value) was recorded in the diluted effluent over the 14 day study (Table 10.7). There was no significant change in the toxicity for the neat effluent to *A. tonsa*. over the 14 day respirometer study, with 48 h EC_{50} values ranging between 9.2% on day 0 to 13% on day 14. However, the toxicity of effluent C to both the oyster and Microtox® decreased significantly from being very toxic ($EC_{50} < 1\%$ effluent) at day 0 to virtually non-toxic to Microtox® (EC_{50} of 91% effluent) and an 24 h EC_{50} value of 10% for the oyster after 14 days (Table 10.7). On examination of the data (Table 10.7) it can be seen that the Microtox® toxicity decreased significantly (by a factor of ~ 37) over the first day of the biodegradation study. This indicates that many of the components responsible for the high levels of initial toxicity were rapidly biodegraded.

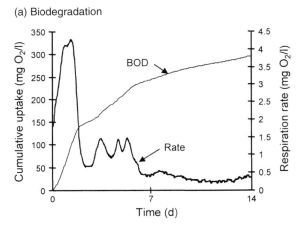

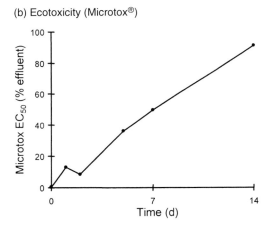

Figure 10.7 Refinery wastewater effluent C: biodegradation and ecotoxicity over 14 days.

10.5 Discussion

The chemical oxygen demand (COD) of an effluent gives a measure of the total amount of oxidisable material present, and will include both biodegradable and persistent organic compounds. In contrast, the five day biochemical oxygen demand (BOD_5) determines how much oxygen (O_2) is consumed by the microorganisms as they metabolise the biodegradable organic matter during incubation at 20°C. The simultaneous measurement of the BOD_5 and COD of an effluent enables an estimate to be made of the fraction of organic matter present that is susceptible to aerobic biodegradation. This fraction, designated F_b, is calculated as (Verstraete and van Vaerenbergh, 1986):

$$F_b = \frac{BOD_5}{(0.65 \times COD)}$$

The F_b values for the three effluents examined in this study ranged from 0.7 to 1.0 (Table 10.8), indicating that a large proportion of the organic matter present was biodegradable. As a comparison, F_b for domestic sewage is 0.9 (Verstraete and van Vaerenbergh, 1986). A knowledge of F_b enables calculation to be made of the

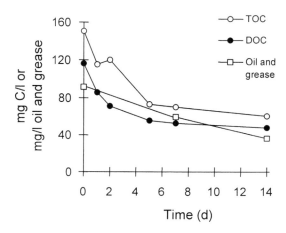

Figure 10.8 Refinery wastewater effluent C: levels of total organic carbon (TOC), dissolved organic carbon (DOC) and oil and grease during incubation at 20°C (values are the averages for duplicate flasks).

amount of organic matter (as COD) that can be removed by biodegradation, either through mineralisation to carbon dioxide (CO_2) or incorporation into new microbial cells (biomass). This 'biodegradable COD' (bCOD) approximates to the BOD that could be achieved after *prolonged* incubation – a value known as the 'ultimate BOD (BOD_∞)' that is calculated as $F_b \times$ initial COD.

A key aspect of this study was to develop a practical approach for categorising effluents in terms of biodegradable or persistent ecotoxicity. Previous studies had used lengthy stabilisation studies of 30–90 days to assess these two types of toxicity (Nyholm, 1996; Gotvajn and Žagorc-Koncan, 1998). In this study, a shorter incubation period of 14 days was used to reduce the effort required. Although biodegradation of the three effluents in terms of O_2 uptake approached but did not reach a plateau by day 14 of incubation (see Figures 10.2, 10.5 and 10.7), the final BOD values were 93–96% of the calculated BOD_∞ (Table 10.8). This indicated that for these types of refinery/petrochemical effluent persistent toxicity could be assessed using a much shorter stabilisation period.

Incorporating combined biodegradation and toxicity studies can improve both hazard and risk assessment of complex 'organic' effluents such as those assessed in this study. For these effluents conventional analysis alone will reveal little about their toxicity, let alone the persistence of organic components. Therefore whole

Table 10.7 Results of marine toxicity tests with the refinery wastewater effluent C.

Day	EC_{50} values as percentage of initial sample			
	Microtox® 15-min EC_{50}		A. tonsa 48-h EC_{50}	C. gigas 24-h EC_{50}
	Neat effluent	Diluted 1:10	Neat effluent	Neat effluent
0	0.19 (0.16–0.22) 0.37 (0.30–0.46)	5.6 (4.3–7.4) 19 (13–27)	9.2 (7.6–10)	0.41 (0.38–0.44)
1	9.0 (6.6–12) 12 (10–14)	>49.5		
2	9.3 (7.8–11) 8.0 (7.2–8.8)	>49.5		
5	32 (26–39) 38 (32–46)	>49.5	10 (10–13)	[a]
7	46 (33–65) 53 (32–87)	>99[b]		
14	91[b] (79–100+) 91[b] (72–100+)	>99[b]	13 (12–14)	10 (9.4–11)

[a] Day-5 oyster test failed.
[b] The EC_{50} values generated using the Microtox® 100% test protocol.

Table 10.8 Biodegradability of the effluents tested.

Effluent	F_b	bCOD ≈ BOD_∞ (mg O_2/l)	BOD_{14} (mg O_2/l)
Refinery process effluent A	0.7	44	42
Petrochemical effluent B	0.7	559	534
Refinery wastewater effluent C	1.0	324	300

effluent assessment (WEA) studies for such effluents can significantly improve the assessment of the potential risk they pose to the receiving environment. In terms of risks posed, most effluent assessment schemes will use an approach based on the

European Chemicals Bureau (ECB) recommendations for determining the risk posed by a chemical or a chemical product. In this approach for the environmental effects assessment, a predicted no effect concentration (PNEC), using the acute or chronic toxicity data and an assessment (or safety) factor, is calculated for species representative of the environmental compartment under investigation (ECB, 2002). Having determined the effluent concentration at which no effects are expected (PNEC), this is compared with the estimated exposure level (referred to as the predicted environmental concentration, PEC). For the environmental protection goals there is considered to be a risk when the PNEC is exceeded (i.e. PEC/PNEC > 1). This can occur in the early stages of effluent assessments because these often follow a tiered approach. Initially, simplistic methods are used to determine the PEC, coupled with conservative assumptions based on a reasonable *worst-case* approach. Consequently, where PEC/PNEC > 1, the first option often is to undertake further testing (either monitoring data to refine exposures or more data on the effects side for further characterisation of the hazard). Owing to the difficulties in determining which components may be responsible for toxicity, most effluent assessments tend to look at refinements of dilution factors and mixing in order to improve derivation of the PEC.

One of the key concerns facing regulators is that any toxicity discharged will be recalcitrant ('hard') and, even if initial toxicity is diluted below any PNEC, effects may occur as 'toxic' material builds up in the environment. Consequently, some regulators are interested in reducing the potential of effluents to cause a problem by reducing the amount of toxicity discharged (a hazard based approach) irrespective of the risks posed.

Whether an effluent control scheme is based on risk or hazard there is merit in using the combined biodegradation toxicity assessment approach. Providing information on the biodegradability of organic components (i.e. obtaining values for the F_b and ultimate BOD) and the persistence of toxicity can be used in risk assessment schemes to ascertain that the toxicity is not conservative and may lead to a reduction of the assessment factors applied. For hazard-based approaches such information could be used to categorise the hazard into 'hard' and 'soft' toxicity (i.e. effluents with 'hard' toxicity being considered the more hazardous).

In looking at the impact of risk assessments it can be seen that the dilution factors required to reduce the effluent to the EC_{50} value decreased considerably for all of the effluents assessed (Table 10.9). All indicated that the potential for longer term effects would be small, because most toxicity was lost within the 14 day period. Some of the most significant reductions were seen with effluent C, where the day 0 data for oyster embryo and Microtox® indicated that a dilution of 240- to 360-fold was required to reduce the effluent to the respective EC_{50} values. This dropped to 10 and < 1 by day 14. One of the advantages of using real-time BOD measurements coupled with regular Microtox® monitoring is that further information on the rate of toxicity reduction can be ascertained. For example,

Table 10.9 Toxicity and dilution factors required to meet EC_{50} values for the effluents before and after 14-day biodegradation study.

Test method	EC_{50} (% effluent)		Dilution factor to achieve EC_{50}	
	Day 0	Day 14	Day 0	Day 14
Refinery process effluent A				
Microtox®	24	>90	4.2	<1
D. magna	>100	>100	<1	<1
Petrochemical effluent B (based on 25% effluent data)				
Microtox®	4.6	>45	22 (88)[a]	<2 (<8)[a]
D. magna	15	56	7 (27)[a]	1.8 (7)[a]
Refinery wastewater effluent C				
Microtox®	0.28	>90	360	<1
A. tonsa	9.2	13	11	8
C. gigas	0.41	10	240	10

[a]For effluent B data in parentheses assume a factor of 4 to adjust to neat effluent.

these data show that a marked reduction in organic components and toxicity occurred within the first two days for both effluents A and C.

The data for effluent B indicate that some care has to be taken when setting up studies. No significant toxicity reduction was seen with the neat effluent. This could have implied that the toxic components were not easily biodegraded. However, because the biodegradation had not reached a plateau, there could have been a number of possible reasons why a decrease in toxicity was not observed. These include the fact that: (a) a decrease in concentration of the toxic components had occurred that was difficult to separate from the inherent variability of the toxicity tests; (b) the partial degradation had produced intermediate degradation products that were still toxic; and (c) the initial phase of the degradation may have favoured the removal of non-toxic components.

If effluent B had not been assessed in both a neat and diluted form it would not have been possible to demonstrate that the toxicity could be removed by biodegradation within 14 days. It is therefore recommended that studies should be designed to incorporate some dilutions in order to maximise the chance of successfully demonstrating whether the toxicity of an effluent can be removed by biodegradation.

10.6 Conclusions

The results of this study demonstrated that there are a number of advantages in using combined toxicity and biodegradation assessments to improve the hazard and risk assessment of refinery/petrochemical effluents. The results indicated that all of the organic components of the effluents were significantly biodegradable and their aerobic biodegradabilities were comparable with that of domestic sewage. The loss of organic components could be linked to toxicity reduction. The toxicity of the two refinery effluents was lost rapidly, with significant reductions within the first two days of the biodegradation study. For the petrochemical effluent, toxicity reduction was only noted for the sample that had been diluted by a factor of four. This was considered to be because the biodegradation of the neat effluent had not been completed by day 14 and consequently still contained significant concentrations of toxic organic components. The data for all the effluents show that the majority of their toxicity was 'soft,' indicating that their potential to cause longer term environmental risks would be significantly lower than that anticipated on the basis of their initial toxicity.

Acknowledgements

The authors would like to thank Joy Worden, Richard Bumpus, Rosemary Eagle, Pascale Deflandre and Sally Wells for their skilled technical assistance.

Disclaimer

The contents of this chapter are based on the personal experiences of the authors and do not necessarily represent the views of the Royal Dutch/Shell Group of companies.

References

ECB (2002) *Existing Chemicals. Step III: Risk Assessment*. European Chemicals Bureau. Institute for Health and Consumer Protection, Ispra, Italy.

Finney, D.J. (1971) *Probit Analysis* (3rd edn), p. 333. Cambridge University Press, Cambridge.

Gotvajn, A.Z. and Žagorc-Koncan, J. (1998) Whole effluent and single substances approach: a tool for hazardous wastewater management. *Water Science and Technology*, **37**, 219–227.

Kaiser, K.L.E. (1998) Correlations of *Vibrio fischeri* bacteria data with bioassay data for other organisms. *Environmental Health Perspectives*, **106**, 583–590.

MAFF (1994) *New Notification Scheme for the Selection of Substances and Preparations to be Used Offshore* (Draft 1 September 1994). MAFF Directorate of Fisheries Research, Burnham-on-Crouch.

MAFF (1996) *Guidelines for the UK Revised Offshore Chemical Notification Scheme in Accordance with the Requirements of the OSPARCOM Harmonised Offshore Chemical Notification Format*. MAFF Directorate of Fisheries Research, Burnham-on-Crouch.

Microbics (SDI) (1992) *Microtox® Manual. A Toxicity Testing Handbook, Volume III, Condensed Protocols*. Microbics Corporation Carlsbad, CAl (now SDI, Newark, DE).

Nyholm, N. (1996) Biodegradability characterization of mixtures of chemical contaminants in wastewater – the utility of biotests. *Water Science and Technology*, **33**, 195–206.

OECD (1984) *Guidelines for Testing Chemicals. Section 2: Effects on Biotic Systems – 202. Daphnia sp. Acute Immobilisation Test and Reproduction Test*. Organization for Economic Cooperation and Development, Paris.

OECD (1992) *OECD Guideline for Testing of Chemicals – 301. Ready Biodegradability*. Organisation for Economic Cooperation and Development, Paris.

OSPAR (2004) *OSPAR Demonstration Programme 2003 on Whole Effluent Assessment (WEA)*, OSPAR Commission SPDS 03/13/7-E, Annex 1. OSPAR Commission, London.

Power, E.A. and Boumphrey, R.S. (2004) International trends in bioassay use for effluent management. *Ecotoxicology*, **13** (5), 377–398.

Repetto, G., Jos, A., Hazen, M.J., Molero, M.L., del Peso, A., Salguero, M., del Castillo, P., Rodriguez-Vicente, M.C. and Repetto, M. (2001) A test battery for the ecotoxicological evaluation of pentachlorophenol. *Toxicology in Vitro*, **15**, 503–509.

Rojickova-Padrtova, R., Marsalek, B. and Holoubek, I. (1998) Evaluation of alternative and standard toxicity assays for screening of environmental samples: selection of an optimal test battery. *Chemosphere*, **37**: 495–507.

Standard Methods (1995) 5520 Oil and grease – 5520 C. Partition-Infrared Method. In *Standard Methods for the Examination of Water and Wastewater* 19th edn, pp. 5–30 – 5–33. American Public Health Association, Washington, DC.

Thain, J.E. (1991) *Techniques in Marine Environmental Sciences No. 11. Biological Effects of Contaminants: Oyster (Crassostrea gigas) Embryo Bioassay*. International Council for the Exploration of the Sea, Copenhagen, Denmark.

USEPA (1975) *Methods for Acute Toxicity Testing with Fish, Macro-invertebrates and Amphibians*, EPA-660/3-75–009. US Environmental Protection Agency, Cincinnati, OH.

USEPA (1985) *Methods for Measuring the Acute Toxicity of Effluents to Freshwater and Marine Organisms* (3rd edn), EPA/600/4-85/013. US Environmental Protection Agency, Cincinnati, OH.

Verstraete, W. and van Vaerenbergh, E. (1986) Aerobic activated sludge. In *Biotechnology. Volume 8. Microbial Degradations*, Schönborn, W. (ed.), pp. 51–54. VCH, Weinheim.

11 Potential future developments in ecotoxicology

Wim De Coen, Geert Huyskens, Roel Smolders, Freddy Dardenne, Johan Robbens, Marleen Maras and Ronny Blust

11.1 Introduction

Since its beginning in the 1960s, the primary goal of ecotoxicology has been to gain knowledge of the fate and effects of toxicants in ecosystems. Originating as a rather descriptive approach revealing as much detail as possible of the behavior and individual (acute) effects of pollutants in the environment, this discipline has evolved into a multidisciplinary field where more accent is put on modeling and predicting the complex interactions of the increasing number of chemicals in various habitats and ecosystems. As a highly multidisciplinary scientific subject, various interests have generated a plethora of tools and technologies, usually driven by regulatory demands or requirements, rather than by scientific advances. This lack of a structured scientific approach was compensated for by the momentum generated by society-driven questions regarding the impacts and threats of certain products or discharges. Ecotoxicology has been growing through interaction between various disciplines, which, as could be expected, has generated different (sometimes conflicting) perspectives and theories on how to deal with the complex interactions at the ecosystem level. This can be visualized easily using a hierarchical view of the different levels of biological organization to describe the various processes and interactions between biotic and abiotic factors in the environment (Figure 11.1). One of the concluding assessments formulated during the past decennium was that this top-down or bottom-up approach creates a paradigm between the mechanistic toxicology and the ecological relevance of environmental toxicology. The classical mammalian toxicology, having profound mechanistic insight and perspective, was not introduced easily into ecotoxicology owing to the substantial variability in species and in their sensitivities, as well as the basic lack of knowledge on the general biochemistry and physiology of most 'standard' ecotoxicity test organisms. On the other hand, from the ecological perspective, mechanistic insights were not required because by nature this research tends to be more descriptive; the overall impact on structures and functions was more important than the underlying mechanism.

Looking towards future developments in ecotoxicology we believe that more quantitative relationships between the various levels of biological organization need to be established in the global implementation of the various

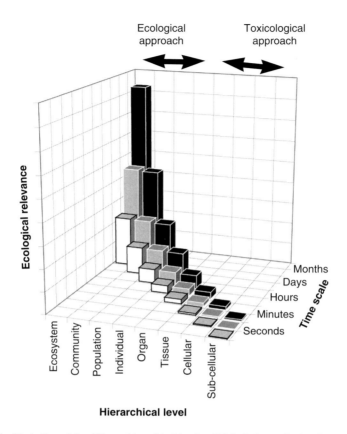

Figure 11.1 Illustration of the different hierarchical levels of biological organization that are studied in ecotoxicology. Each level has characteristic homeostatic (white bar), compensatory (grey bar) and non-compensatory responses (black bar).

ecotoxicological tools and methods. As long as these relationships between the various emerging aspects have not been assessed, the mechanistic insights (e.g. at the sub-organism level) will be nothing more than academically rewarding and of no use for environmental decision-making. If we cannot explain the effects of chemicals on the structure and function of populations, communities and ecosystems, the final decision-making will constitute a 'black-box' approach rather than a scientific management strategy. The necessity for these relationships has been stated strongly by various authors (De Coen and Janssen, 2003; Smolders, 2003) but so far the number of studies addressing this issue at a quantitative level has been sparse. If basic toxicological mechanisms are understood and their relationships with effects at higher level established, one could create a decision support system based on biomarker end-points or even quantitative statistical models predicting emerging scenarios.

Clearly, in order to establish these essential relationships between levels of biological organization, tremendous developments still have to be made. In this chapter we will give an overview of what we believe are crucial developments for future ecotoxicological research, involving the development of mechanistic tools that allow us to understand the modes of action of chemicals and how they deal with the increasing need of realism in ecotoxicological testing. For too long we have neglected the reality of the way chemicals or discharges are tested. Although laboratory validation in the field has been discussed as being essential to comprehend the real-world situation, little has been done to increase the realism of laboratory testing by addressing essential environmental variables (such as food supply, predation, suspended solids) that influence exposure routes, stress conditions or adaptation phenomena. To avoid a mechanistic interpretation of these confounding variables, various extrapolation factors have been used. However, one of the main areas of uncertainty in aquatic toxicology at present is the extrapolation of toxicity data obtained from laboratory testing to effects in real-life ecosystems. Although laboratory testing is essential for eliciting mechanisms of toxicity in a standardized, controlled and understandable fashion, there remains much confusion about the validity of laboratory-to-field extrapolations of toxicity testing (Burton, 1999; Heugens *et al.*, 2001; Selck *et al.*, 2002). Not only does field-based *in situ* biomonitoring provide a measure of the exposure of organisms to a myriad of possible harmful compounds, each of which can elicit a direct toxic effect, but it also incorporates ecological factors such as food availability, trophic dynamics and species interactions that can have an impact on the outcome of toxicity tests as indirect effects (Heugens *et al.*, 2001; Backhaus *et al.*, 2003; Campbell *et al.*, 2003; Fleeger *et al.*, 2003). It is also clear that more mechanistic insights should be generated that can serve as a generic model or tool to extrapolate or model the various biotic and abiotic variables that could affect the outcome of an environmental hazard/risk assessment process.

More and more studies are starting to address the impact of pure chemicals on the different levels of biological organization (Baillieul and Blust, 1999; Bervoets and Blust, 2000; De Smet *et al.*, 2001; Dauwe *et al.*, 2002; De Boeck *et al.*, 2003; Hoff *et al.*, 2003). A major bottleneck in the global testing procedure is that, for reasons of simplification, the impact of new chemicals on the ecosystem is studied as if these chemicals are released into a pristine environment. This process of course adds to a better understanding of the environmental impact of the chemicals as such, but the global procedure is not able to resolve fully the invariable complexity of environmental exposure where a multitude of physicochemical factors act together as a complex mixture. Clearly, more adequate models should be produced that accurately predict the impact of low doses of a mixture of chemicals on the health of the receiving ecosystem.

11.2 Future research needs in ecotoxicology

11.2.1 Biomarkers in ecotoxicology: where do we go from here?

Starting at the lowest level of biological organization, important improvements need to be attained at the biomarker level. Over the last three decades biochemical, physiological and histological measurements have been used increasingly to assess the effects of toxic exposure on biota (Depledge, 1990; McCarthy and Shugart, 1990; Peakall, 1992). Several of these 'clinical' measurements have been proposed as diagnostic tools in ecotoxicological research (Mehrle and Mayer, 1980). Whereas originally much research effort was devoted to the identification of specific chemicals based on biomarker analysis, nowadays more emphasis is being placed on the evaluation of biological effects based on biomarker measurements (McCarty and Munkittrick, 1997). Biomarker techniques offer a number of advantages compared with conventional ecotoxicity tests that generally use lethality as an end-point. In general, biomarkers can be considered as measures of the initial changes caused by toxicological interactions between the chemical and the (biological) receptor site. The effects that are normally studied in conventional toxicity tests (impaired growth, reproduction or survival) can be considered as the final result of the accumulating damage at the sub-organism level.

Initial perturbation caused by toxic exposure will generate responses within the homeostatically regulated mechanism of the unit of study (cell, individual...). These toxicant-induced sub-organism alterations are biologically irrelevant if they do not lead to impaired characteristics at higher levels of biological organization (Versteeg et al., 1983). With a gradual increase of toxic insult, compensatory adjustments are made, leading to deviations from the 'healthy' condition. Within this compensatory phase, the organism's survival and reproduction capacities might start to be affected because the organism has less capacity to withstand additional stress. Eventually, a situation is reached where further repair and/or compensation is impossible and the organism dies. However, the possibility remains that, if the conditions improve and the repair processes are able to restore the compensatory mechanisms, a diseased organism might return to a healthy state. Although the quantitative nature of this relationship is uncertain, this concept underlines the relevance of biomarker analysis because it offers the possibility to determine where the organism is situated in this continuum and can indicate early deviations of the 'normal' functioning of biota (Versteeg et al., 1983; Depledge, 1989). The resulting effects of increased stress-induced disturbance at different levels of biological organization are illustrated in Figure 11.1. Subtle changes occurring at the sub-organism level lead to effects within the homeostatic capacity of the organism, resulting in no adverse effects on growth, survival and reproduction. In an increasing time/concentration continuum, accumulative toxicant-inflicted impact eventually causes loss of species diversity within the communities and finally results in total loss of ecosystem structure and function. Within every increase in hierarchical level, a unit with higher ecological relevance is reached. At every level, however, different types of responses can be distinguished. These range from homeostatic responses (white bar), to compensatory adjustments (grey bar) and eventually non-compensatory responses (black bar). Effects at lower levels can be detected in a much shorter time-span and ultimately will affect higher levels. However, in order to extrapolate effects between different levels of biological organization, a currency has to be identified that links the different levels in a qualitative and quantitative way. The most commonly used currencies to describe this hierarchical transfer are parameters of energy metabolism. As mentioned above, the complementary action of the top-down ecological approach with that of the bottom-up approach of fundamental toxicology should be able to identify which lower level effects might be relevant to predict the direct consequences of biomarker responses for the individual or the population. At present, however, significant uncertainties are involved in the different extrapolations that are performed between the different levels of biological organization (Hastings and Huggins, 1994; Munkittrick and McCarty, 1995).

At the various levels indicated in Figure 11.1, important improvements and new developments are needed to enhance our understanding of the chemical interactions with biota and the environment. In the following section, important examples from each level of biological organization are given where crucial innovations can be expected in the near future. Starting with the lowest level of organization, examples at the gene expression and (post)translational level as well as at the physiological level will be given. All three levels will undoubtedly increase our understanding from the mechanistic point of view. The last two items deal with mixture toxicity and effects of confounding variables, demonstrating what is needed to incorporate more realism into ecotoxicology testing in the coming decennia.

11.2.2 Transgenic systems in ecotoxicology

In environmental risk assessment of chemicals most assays are based on organism end-points such as mortality, growth and reproduction. However, these assays are not adapted to high screening capacities and do not provide information on mechanisms of toxicity. Transgenic systems offer excellent alternatives to circumvent these problems and shortcomings. Such organisms contain a reporter gene fused to a promoter that responds to specific stress caused by a pollutant. Crucial for these transgenic models is the need for a sensitive promoter that is reliable and relevant for a certain stress situation and a reporter that can be detected easily and cost-effectively. It is clear that in the near future the use of these transgenic systems will increase and could become standard practice in environmental toxicity testing. Although the full potential of this transgenic technology has been demonstrated only in part, the specific information obtained with regard to the toxic mechanisms involved and the simplicity of measuring these responses will become key arguments to implement and use this technology in screening and assessing the impact of pollutants on ecosystem health. In the following sections we give an overview of the various sorts of transgenic systems that have been developed recently.

11.2.2.1 Reporter systems
Reporter genes have easily measurable phenotypes that form the basis for sensitive, quantitative and reproducible assays. Unlike the promoter, a reporter can be considered 'universal' in that it functions in different cell systems of both prokaryotic and eukaryotic origin, although it is obvious that some reporters are more suitable for one or other cell type. Reporter systems can be classified on the requirement for a substrate. β-Galactosidase, coded by the *Escherichia coli lacZ* gene, has been used in molecular biology for a long time (Silhavy and Beckwith, 1985). The cytoplasmic location of the protein requires the preparation of cell extract, making the β-galactosidase assay cumbersome. Despite the high background in bacteria, it is still used frequently because of its highly reliable and reproducible nature. The recently developed substrates, such as fluorogenic β-methyl umbelliferyl galactoside (MUG) or fluorescein digalactoside (FDG) (Sussman, 2001), make *lacZ* suitable for application with the new detection tools (Rowland *et al.*, 1999). Bacterial chloramphenicol acetyl transferase (CAT) is another cytoplasmic reporter, used both as a prokaryotic and eukaryotic reporter. Quantification is by radiolabeled substrates such as ^{3}H-labeled acetyl coenzyme A, ^{14}C chloramphenicol or via antibodies and enzyme-linked immunosorbent assay. It can be measured only in extracts of cells and has a narrow dynamic range that limits its use; however its low background and high stability make it ideal for studying cumulative changes in expression (Sussman, 2001).

Alkaline phosphatase from *E. coli* and secreted alkaline phosphatase (SEAP) from mammalian cells are two examples of secreted reporter proteins (Berger, 1988 ; Manoil *et al.*, 1990; Yang *et al.*, 1997). Although these reporters are used less frequently, they have the evident advantage that no cell extracts are required. Different substrates allow detection with different detection tools.

Some 'new' luminescent and fluorescent reporters (some of them even 'non-substrate proteins') are very attractive because of their easy and fast detection, explaining their current frequent use. The bacterial luciferase isolated from the *Vibrio fischeri* lux operon contains luxAB encoding the functional subunits and luxCDE for the synthesis and recycling of the aldehyde substrate (Prosser, 1996). Firefly (*Photinus pyralis*) luciferase, encoded by the *luc* gene catalyses the oxidative carboxylation of beetle luciferin, in which photons are emitted (LaRossa, 1998). Its short half-life and lack of any post-translational modification makes it ideal to look after effects in gene expression (Naylor, 1999). Detection of luminescence using a luminometer or scintillation counter is rapid and evident over a broad linear range. If firefly luciferase is used then luciferin is the substrate, whereas for bacterial luciferase no substrate is needed, even though five proteins must be produced. Bacterial luciferase is slightly more sensitive than firefly luciferase (Kurittu *et al.*, 2000) and evidently has the advantage that no substrate is required. An important disadvantage is the high ATP requirement, and this affects detection owing to the influence of the metabolic state of the cell.

Green fluorescent protein (GFP) is a novel fluorescent reporter system from the jellyfish *Aequorea victoria* (Chalfie *et al.*, 1994, Errampalli *et al.*, 1999; Justus and Thomas, 1999); it is an intrinsically fluorescent protein, meaning that no substrate is required. To increase its sensitivity and usefulness for different cell systems, different mutants have been constructed with a (red) shifted excitation maximum and an enhanced fluorescence. Green fluorescent protein is non-invasive and can be monitored as such in living cells. It is easy to use and does not require any exogenous substrate or cofactor. The recently cloned Dsred from *Discosoma* sp. has excitation and emission peaks at 558 and 583 nm (Matz *et al.*, 1999). The cellular autofluorescence is supposed to be lower at this emission wavelength, which should result in a lower background compared with GFP. Fluorescent proteins have the advantage of being independent of the energy status or physiological condition of a cell.

In a recent study it was reported that luminescence reporters could be detected at lower concentrations and with a faster response time compared with fluorescent reporter proteins. In *E. coli* there is negligible luminescence background but a considerable autofluorescence background, so that larger amounts of protein are required to overcome background levels (Hakkila *et al.*, 2002).

Fluorescent reporters allow detection in a single cell with epifluorescence and confocal laser scanning microscopy (Cormack *et al.*, 1996; Tombolini *et al.*, 1997; Phillips, 2001) or allow quantitative single-cell analysis with flow cytometry (Cormack *et al.*, 1996; Valdivia and Ramakrishnan, 2000), enlarging their potential scope of use.

11.2.2.2 Promoters in (multi)cellular systems

The use of transgenic cells and organisms possessing a specific promoter make it possible to fine tune for the different types of stress caused by a chemical or sample. Unlike reporter genes, promoters are species-specific, requiring promoter isolation from the test organism. As a result, different promoters have been characterized for various test species. Some frequently used promoter systems will be described here.

Prokaryotic promoters. Prokaryotic systems are generally used for rapid screening assays. A well established example is the Ames test that was developed with *Salmonella* in the 1970s (Ames, 1979; McDaniels *et al.*, 1990; Reifferscheid and Heil, 1996) and is still considered as the standard mutagenicity assay.

Since then, different new assays have been developed, mainly in *E. coli*. In *E. coli* the SOS response plays a central role in handling genotoxic stress and RecA is the regulator responsible for the induction of around 20 SOS genes. Therefore, all genotoxic assays are somehow linked to the SOS response, and a long list of assays, all combinations of different SOS promoters with different reporters, have been developed. Thanks to the crucial role of RecA, the recA promoter is used most often (Nunoshiba and Nishioka, 1991; Ptitsyn *et al.*, 1997; Vollmer *et al.*, 1997; Justus and Thomas, 1998, 1999; Min *et al.*, 1999; Kostrzynska, 2002), although the promoters recN (Van der Lelie *et al.*, 1997), umuC (Nakamura *et al.*, 1987; Reifferscheid and Heil, 1996; Schmid *et al.*, 1997; Justus and Thomas, 1998, 1999), sfiA (sulA) (Quillardet *et al.*, 1982; McDaniels *et al.*, 1990; Quillardet and Hofnung, 1993), uvrA (Vollmer, 1997) and alkA (Vollmer *et al.*, 1997) also have been used. The gene *lacZ* is mostly used as a reporter gene. In the Vitotox® assay recN promoter or a mutated (and more sensitive) derivative is used together with the lux operon (Van der Lelie *et al.*, 1997).

Orser developed the Pro-Tox assay to determine the 'fingerprint' of a chemical compound based on 16 different stress promoters, proving the possible broad use of *E. coli* as a transgenic screening system (Orser *et al.*, 1995).

Eukaryotic systems. In eukaryotic cells and organisms, reporter genes similar to those used for prokaryotic cells can be utilized, although for some genes optimized, mutated versions have been developed.

Yeast. Two yeast stress promoters RAD54 and RNR2 were fused transcriptionally to GFP to screen for general stress (Afanassiev *et al.*, 2000). Billinton *et al.* (1998) developed a codon-optimized GFP version for yeast.

Receptor-mediated assay can be developed in yeast, allowing the set-up of very specific assays. *Saccharomyces cerevisiae* was constructed that expressed the aryl hydrocarbon receptor (AhR) and the aryl hydrocarbon nuclear translocator (Arnt) after the Gal promoter. Binding of dioxin to AhR activates it and makes a heterodimer complex that translocates from the cytoplasm to the nucleus, augmenting expression of the genes possessing a xenobiotic response element (XRE). Gene *lacZ* was preceded by a tandem of XRE element, making the yeast a good monitor for dioxin and other arylhydrocarbons (Miller 1997, 1999; Miller *et al.*, 1998; Kawanishi *et al.*, 2003). In the yeast estrogen assay the human estrogen receptor gene was expressed constitutively, whereas on a second plasmid the *lacZ* gene was controlled by the estrogen responsive element (ERE) (Routledge and Sumpter, 1996).

Caenorhabditis elegans. Rather than individual cells, the intact *C. elegans* is used as a multicellular whole-organism biosensor because of its simple structure, short lifespan and its minimal genetic size. Whole-organism assays allow the affected tissue to be determined. Both general and specific transgenes can be developed: transgenic *C. elegans* expressing GFP under control of the CYP35A2 promoter showed a strong and specific induction in the intestine upon β-naphthoflavone addition; and *C. elegans* with the heat shock promoter hsp16 fused to *lacZ* (Candido and Jones, 1996) or *gfp* (Wah and Chow, 2002) can be used to monitor general toxicity in water or soil. The disadvantage of *C. elegans* is its largely impermeable cuticle, so exposure to test substances generally requires ingestion via the pumping action of the pharynx, making the organism rather 'insensitive' to pollutants compared with the conventional 'ecotox' test species.

Fish. Several cell lines from different species have been created to screen for specific stress conditions. Specificity for a stressor lays in the 'responsive element' upstream of the promoter region. The zebrafish cell line ZEM2S containing a plasmid with the responsive elements AHRE, EPRE, MRE and ERE coupled with the luciferase gene gave a dose-dependent induction and specific toxicant response (Carvan et al., 2001). Also, in a rainbow trout gonad cell line an estrogen assay was developed in which an estrogen receptor was expressed with firefly luciferase gene preceded by ERE on a second plasmid (Fent, 2001).

A major drawback of cell lines is that aspects of *in vivo* functioning such as metabolic conversion and breakdown cannot always be incorporated. To circumvent this problem, transgenic fish (e.g. zebrafish) have been 'created'. A transgenic zebrafish was developed in which the XRE (Mattingly, 2001) or AHRE (Carvan et al., 2001) was fused to *gfp*, allowing identification of the affected tissues. Carvan et al. (2001) also created a transgenic zebrafish for metal measurement in which MRE was fused to the luciferase expression gene.

Mammalian cellular systems. The CALUX (chemical activated luciferase gene expression) assay is a dioxin assay with rat hepatoma cells stably transfected with an AhR-controlled luciferase reporter gene containing four dioxin-responsive elements (Murk, 1996). Estrogen-specific assays were developed in human transgenic cell lines MCF-7 or Hela cells in which the ERE was fused to the luciferase reporter gene (Green and Chambon, 1987). A good alternative for the prokaryotic genotoxicity assays was developed in the human cell line HEK293. In human cells p53 is known to be induced upon DNA damage. Translational fusion between the p53 promoter and *gfp*, transfected in the human cell line HEK293, provided a good eukaryotic alternative for the bacterial SOS chromotest because GFP is induced upon DNA damage (Quinones and Rainov, 2001).

Apparently, mammalian cells can also be used for screening different types of stressors, providing a multitude of end-points in a single assay. An excellent example is the Cat-Tox assay, which can be seen as a mammalian alternative for the Pro-Tox assay described above. Here, a series of 14 inducible promoters are fused with the *cat* gene in a human liver cell line (HepG2). This allows determination of a stress response fingerprint of a chemical compound (Todd *et al.*, 1995).

Plants. Plants are rarely used for environmental toxicology screening. Reporter genes that can be visualized easily are GUS, luciferase and GFP. An *Arabidopsis* plant was developed in which an Ni-inducible promoter was fused to β-glucuronidase, allowing for monitoring of metal contamination (Kovalchuk *et al.*, 2001).

11.2.3 Novel markers at the proteome level: glycosylation of proteins

Apart from the effects at transcriptional level, toxic stress can exert effects at the translational and post-translational level. Whereas previously protein analysis was done mainly by focusing on a single gene product, the advent of proteomic technology has opened up almost unlimited toxicological applications. Proteomics is the study of proteins produced by a genome in a cell or tissue at a certain developmental stage and/or physiological state. It is an emerging field that is expected to utilize the rapidly expanding wealth of newly available genomic information for a better understanding of complete biological systems. Genomics has provided a vast amount of information linking gene activity with disease. It is now recognized, however, that there are a number of reasons why gene sequence information and the pattern of gene activity in a cell do not provide a complete and accurate profile of a protein's abundance or its final structure and state of activity. After transcription from DNA to RNA, the gene transcript can be spliced in different ways prior to translation into the protein. Following translation, a fraction of the proteins is changed chemically through post-translational modification, mainly by the addition of carbohydrate and phosphate groups. Such modification plays a vital role in modulating the function of many proteins but this information cannot be deduced from the genetic code of the genes.

Proteomics is a new scientific discipline that directly detects proteins associated with a disease or intoxication by means of their altered levels of expression and/or physical properties between control and diseased/intoxicated states. It enables correlations to be drawn between the range of proteins produced by a cell or tissue and the initiation or progression of a disease/toxic state. Proteomic research furthermore permits the discovery of new protein markers for diagnostic purposes and can be put into force for toxicity assessment. It is a field that has leapt to prominence within recent years and is widely expected to have a major impact on biotechnology. Proteomics encompasses functional proteomics (linking protein expression with gene expression), structural proteomics (determining large numbers of three-dimensional protein structure by spectroscopic or computational techniques) and protein–protein interactions and biochemical pathway studies.

A very new and hardly explored tool to diagnose environmental pollution is the study of glycan patterns on specific blood and/or urine glycoproteins. Over several years, more and more evidence has been gathered indicating that glycan patterns change under the influence of different physiological parameters such as age, development, pregnancy, disease, stress, autoimmunity or inflammation (Mackiewicz and Mackiewicz, 1995; Van den Steen *et al.*, 1998). As quoted by Kornfeld

(1998) 'protein glycosylation is a ubiquitous post-translational modification'. Most proteins in the plasma contain covalently bound carbohydrate units. The number of oligosaccharide units per molecule varies greatly, ranging from only one to hundreds. The structures of the oligosaccharide units present on the proteins are also highly diverse. These oligosaccharides, also called glycans or sugar trees, are sugar polymers constructed by a set of glycosyltransferases and glycosidases that sequentially add or remove sugars one at a time to a growing oligosaccharide unit using either nucleotide-linked sugars or dolichol phosphate-linked sugars as the donors for the transfer reactions. Glycans are N-linked to the amide group of asparagine residues via an *N*-acetylglucosamine residue, or O-linked to hydroxyl groups of serine or threonine residues via an *N*-acetylgalactosamine residue (Van den Steen *et al.*, 1998; Spiro, 2002). Within a population of molecules of a specific glycoprotein, different glycosylated variants are found in different ratios. These variants are called glycoforms. New glycoforms or changed compositions of glycoform populations have been characterized in relation to specific pathological conditions and hence the question is raised whether these glycoforms could be useful biomarkers for acute and/or chronic toxicity, allowing sensitive, specific and reproducible analyses to diagnose toxicity. For understanding the potential of glycosylation status of specific glycoproteins as a toxicity biomarker it is important to understand how glycosylation is critical to the proper functioning of an organism.

It is not long ago that sugar trees were seen merely as decoration without any function for glycoproteins. This idea has been revised with the discovery that glycan structures change under different physiological conditions and that specific oligosaccharide structures are ligands for specific cell-surface lectins (sugar-binding proteins) (McEver, 1997). Recognition by the lectins can be the trigger for diverse signaling pathways. Sugar–lectin interaction is also a cell–cell communication tool or a tool to recruit cells to a site of inflammation (Alper, 2001). Lectins are also involved in the turnover of blood glycoproteins. Owing to the importance of glycans in diverse cellular processes, it was not surprising to find correlations between glycosylation changes and disease, inflammation or stress. It has to be emphasized that only a specific subset of glycoproteins serve as potential glycosylation biomarkers for the physiological conditions mentioned above. Acute-phase proteins such as α_1-acid glycoprotein, haptoglobin, transferrin or α_1-antichymotrypsin are examples of such biomarkers. It is, however, also important to realize that glycosylation changes caused by disease can be substantially different when studying different indicator glycoproteins. This was demonstrated by Turner *et al.* (1995), who analyzed glycan compositions of haptoglobin or antiprotease inhibitor from patients suffering from ovarian cancer. Remarkably different variations in the compositions of glycans within the same disease were found, pointing to the fact that these glycan changes are associated with the disease process itself. Hence, glycoproteins and their glycans have to be studied on a case by case basis.

The reproducibility and specificity of glycan composition changes under specific pathological conditions have been demonstrated by Goodarzi and Turner (1998). Glycan patterns on human haptoglobin were studied using high-performance anion exchange chromatography with pulsed amperometric detection (HPAEC-PAD) (see below) and the profiles of eight healthy persons were obtained and compared with those of groups of patients with Crohn's disease, rheumatoid arthritis, stomach cancer, breast cancer and ovarian cancer. The following observations were made: oligosaccharide profiles were very similar within each group; no glycans specific for any particular disease were detected but profiles were very different and characteristic for some diseases.

Although it is obvious from several studies that acute injury or inflammation events result in specific glycosylation changes, alterations due to chronic disease or chronic exposure have been demonstrated as well. Rydén et al. (2002), for instance, have shown that the α_1-acid glycoprotein fucosylation index provides a highly accurate tool for the diagnosis of liver cirrhosis among patients investigated for liver disease.

In relation to environmental studies, exposures to chemicals resulting in predictable glycosylation changes are scarce. A reduced sialic acid content on human transferrin has been described by De Jong et al. (1995) and Flahaut et al. (2003) as a marker of chronic alcohol consumption. The effects of estrogen exposures were studied in great detail by Brinkman-Van der Linden et al. (1996), who found specific alterations in the glycosylation of α_1-acid glycoprotein due to exposure of women to oral estrogens or male-to-female transsexuals to estrogens in combination with cyproterone acetate. Oral estrogen treatment induced an increase in the degree of branching of sugar trees and a decrease in fucosylation and sialyl Lewis x expression on α_1-acid glycoprotein compared with individuals receiving no estrogens or transdermal estrogen treatment. Tardivel-Lacombe and Degrelle (1991) studied sex steroid-binding protein and also found that glycan compositions changed under the influence of physiological estrogen stimulation. Monnet et al. (1986) demonstrated that rats chronically exposed to phenobarbital not only produce more α_1-acid glycoprotein but this glycoprotein shows alterations in the relative proportions of the sugar moieties and sugar chain compositions. Paul et al. (2001) described glycosylated variants of C-reactive proteins from carp (*Catla catla*) grown in fresh versus polluted aquatic environments. It is clear that exposures to xenobiotics can lead to specific alterations in the glycosylation of specific glycoproteins (Pos et al., 1988; Trabelsi et al., 1997; Davey and Breen, 1998).

Furthermore, it can be demonstrated that, based on known functions of these sugars for the proper functioning of the endocrine system and the immune system, such parameters could serve as reliable biomarkers for chronic exposures. For example, luteinizing hormone, follicle stimulating hormone or placental choriogonadotrophin are glycoproteins in which the glycan structures are crucial to the stability and biological activities of the hormone heterodimers. These hormones consist of α- and β-subunits and assembly to heterodimers is largely determined by the glycan structures on the subunits. As explained by Ulluo-Aguirre *et al.* (2001), estrogens and androgens influence the glycosylation status of these hormones by regulating terminal sialylation and sulfation of the mentioned gonadotrophins. Sialic acid and/or sulfate groups are usually present as terminal charged groups on both O- and N-glycans. Changing the sialic acid and/or sulfate content of the oligosaccharide attachments causes conformational changes of the hormone α- and β-subunits and, in turn, the biological activity of the hormone is changed. Besides affecting the potency of the hormones through conformation changes, glycan compositions are crucial for the survival time in the circulation and for activation of receptor/signal transducers involved in subsequent biological responses. In general, alterations in the glycosylation status of gonadotrophins affect the quality of the gonadotrophin signal delivered to the gonads. Because these glycosylation changes are correlated with estrogens or androgen exposures, the question should be raised whether so-called endocrine disrupting chemicals are also able to influence the glycan patterns synthesized on glycoprotein hormones and, if so, what the consequences might be of these glycan changes to the endocrine system or, more specifically, to the health status of an organism.

A second example to illustrate the importance of glycosylation to the physiology of organisms is given by α_1-acid glycoprotein, also called orosomucoid. This glycoprotein has been the subject of many research projects in humans, linking glycosylation changes to different stages of diseases, inflammation or different physiological stress conditions (Fournier *et al.*, 2000; Van den Heuvel *et al.*, 2000). α_1-Acid glycoprotein is one of the acute phase proteins, the synthesis of which is increased in different organs in response to inflammation. It is predominantly secreted by hepatic cells as an anti-inflammatory and immunomodulatory agent. The hepatic acute phase response is induced by glucocorticoids and cytokines such as interleukin-1 (IL-1) and IL-6. The latter cytokines also induce glycosylation changes of the acute-phase proteins, namely by altering branching and by changing the fucosylation and sialylation degrees of N-glycans. As described by Shiyan and Bovin (1997), different immunomodulatory activities are assigned to different glycoforms of α_1-acid glycoprotein. Proliferation of lymphocytes by particular mitogens such as phytohaemagglutinin is suppressed by α_1-acid glycoprotein and the degree of suppression seems to be correlated to the glycosylation status of α_1-acid glycoprotein. α_1-Acid glycoprotein is also able to inhibit the phagocytic activity of monocytes previously stimulated by a compound such as phorbolmyristyl acetate. Again, the degree of suppression seems to depend on the glycoforms of α_1-acid glycoprotein. α_1-Acid glycoprotein also influences the production of cytokines by monocytes and peripheral blood mononuclear cells and it has been shown that different glycoforms can have opposite effects. It has been suggested that these opposite actions of different glycoforms is a kind of immunological buffering that modulates abrupt changes in immune status. In different conditions of diverse diseases, however, abnormal glycosylation can lead to unfavorable changes in the immune system and contribute to the pathological condition (Rudd *et al.*, 2001). Whether chemical exposures can alter the glycosylation process of diverse acute-phase proteins and, as such, affect the immune system still needs further investigation.

Understanding the variety and functionality of protein glycosylation has improved in recent years, partly due to the fact that methods to characterize glycan structures have been optimized, simplified and substantially refined. A few of these newly developed techniques can be considered very powerful and straightforward. Analysis by HPAEC/PAD, for instance, allows the separation of charged glycans over an anion exchange resin in very basic conditions and has been used to compare glycan patterns of serum proteins in patients suffering from diverse diseases (Goodarzi and Turner, 1998). Mass spectrometry allows extremely sensitive and reproducible characterization of glycan compositions and molecular weights (Dell and Morris, 2001). High-resolution capillary gel electrophoresis has been developed as a simple tool to separate fluorescently labeled reducing oligosaccharides according to their molecular weight, charge and dynamic volume (Guttman *et al.*, 1996). In order to perform this kind of analysis, oligosaccharides are released from their carrier proteins by using specific glycanases or through hydrazinolysis. Unraveling of the carbohydrate structures is possible through treatments with glycosidases of different specificities, followed by analysis of the trimmed products. Fluorophore-assisted carbohydrate electrophoresis (FACE) technology has been introduced as a simple tool for molecular biologists who want to study the glycan compositions and structures by separating them over polyacrylamide gels (Raju, 2000). This technique has been refined by the use of DNA-sequencing equipment to obtain ultrasensitive profiling and sequencing of oligosaccharides (Callewaert *et al.*, 2001).

In conclusion, glycosylation of specific serum glycoproteins has been shown to change under different pathological conditions. Because chronic exposures to xenobiotics often lead to disease, it can be expected to find that the glycan compositions of selected glycoproteins serve as sensitive, reliable and easy to analyze biomarkers of toxic exposures.

11.2.4 Organism-level effects: mechanisms of reproductive toxicology

Since the 1990s, concerns about a possible decline in sperm counts and increased incidence of male genital abnormalities in humans (Sharpe and Skakkebak, 1993), the occurrence of alligators with abnormal male genitalia in pesticide polluted lakes (Guillette *et al.*, 1994) and the occurrence of male fish showing female characteristics in rivers receiving effluents from water treatment plants (Purdom *et al.*, 1994) have emphasized the fact that many of these chemicals can be harmful at levels well below those that cause mortality.

The effects of endocrine disruptors have clearly indicated that exposure at low concentrations during 'critical windows' of exposure can have a severe impact on the reproduction of species. Looking at the complexity of the endocrine system, it is clear that measuring single end-points (e.g. such as vitellogenin in fish to screen for estrogen-like compounds) is largely insufficient to cover the various sorts of perturbations that can occur in the endocrine system.

In future, ecotoxicological research methods should be developed that can help to elucidate mechanisms of reproductive toxicity. One such potential reproductive aspect deals with gamete physiology. For example, the effects of aquatic pollution on the reproductive endocrine system of fish are well documented (Kime, 1995, 1998) but there have been very few studies of its effects on gamete quality. A sufficient level of gamete quality is evidently vital for successful reproduction. It is clear that reproductive dysfunction at sublethal concentrations can be caused either by modulation of the endocrine system so that gamete development takes place in an abnormal hormonal environment or directly by action on the gametes themselves. It is therefore probably not necessary to sharply differentiate between the two possible targets of toxicants – endocrine glands or gametes – when reproductive dysfunction is discussed (Kime and Nash, 1999). Whether the chemical's effect is localized at the hypothalamus, pituitary, gonad, liver or directly at the gametes, the net result may be a change in gamete quantity or quality. These are the only end-points that really matter because production of sufficient numbers of viable, good quality gametes is the ultimate goal of the whole reproductive system. In extreme situations there may be a complete arrest in gametogenesis, whereas at lower toxicant levels poor quality gametes may be produced, resulting in low fertilization success or decreased survival of the offspring. Examination of gamete viability thus may provide a useful end-point of reproductive dysfunction at any of these sites and show effects at concentrations below those that completely inhibit gamete development (Kime and Nash, 1999).

Within an ecotoxicological perspective, mammalian methods cannot be extrapolated automatically to other species. Sperm of teleosts, for example, differs in a few very important aspects from that of mammals. Unlike mammalian sperm it is not motile on ejaculation and attains motility only on contact with water. After activation it only moves for a few minutes (for freshwater fish typically around 1 min). Furthermore it enters the egg via the micropyle rather than through an acrosomal reaction. The first minute after the start of motility is therefore crucial to its success in fertilizing an egg, and even when it is deposited on the egg's surface it still has to move fast enough in the right direction to reach the micropyle. Clearly the fertilizing ability of fish sperm is very dependent on its motility, and any pollutant that decreases this movement may be expected to affect fertilization.

Successful reproduction is clearly essential for the propagation of every species, prerequisites for which are high quality and properly functioning gametes. Spermatozoa, however, are known to be very sensitive to environmental pollution, as shown in studies by Billard and Roubaud (1985), Khan and Weis (1987a,b), Anderson *et al.* (1991), Rurangwa (1999), and Van Look (2001). The motility of fish sperm can be affected by pollutants in several ways: (1) by alteration of the internal hormonal environment during its development, which may be due to hormonal activity of the pollutant itself or by its interference with any possible site in the reproductive endocrine system; (2) by alteration in the function of the Sertoli cells, responsible for nurturing the developing germ cells; (3) by changes in the seminal fluid or sperm maturation; or (4) by direct effects on the sperm itself, which can result in either cytological damage or changes in the efficiency of the mitochondrial energy production. Sperm can come into direct contact with pollutants accumulated in the gonads or with chemicals present in the surrounding water during spawning. Measurement of sperm motility therefore provides an assessment of the final impact of any pollutant on the male reproductive system as a whole (Kime, 1998).

Undoubtedly much of the past research has been centred on the impact of environmental estrogens, especially on the male, because effects on sperm cells, which are produced continually in large quantities, are much more amenable to quantification than effects on eggs. Although egg quality can be measured only by laborious assessment of the fertilization rate and larval survival, sperm quality, as in mammals, is amenable to easy measurement. Sperm motility nowadays can be measured easily, rapidly, quantitatively and objectively by means of computer-assisted sperm analysis (CASA). Owing to the advantages of using sperm rather than ova, research is increasingly focusing on sperm quality as an indicator of reproductive dysfunction caused by environmental pollution. Until recently, assessment of the motility of sperm relied on subjective estimates, often using laborious methods. The parameters commonly applied to describe sperm motility are the percentage of motile spermatozoa, the swimming vigour and the duration of motility. These parameters were estimated and then given a motility score corresponding to an arbitrary scale of criteria (Boyers *et al.*, 1989).

In the 1990s CASA systems were developed and used initially for monitoring male fertility in clinical andrology laboratories. In recent years, however, CASA has been used increasingly to examine the sperm motility of other mammals and, more recently, of fish (e.g. Kime *et al.*, 1996; Rurangwa *et al.*, 1998). The CASA systems have numerous advantages over earlier methods for sperm motility analysis, providing an objective, precise, accurate and efficient measurement of sperm motility. The short duration of fish sperm motility and very rapid change in velocity and numbers of motile sperm with time necessitate analysis of these parameters over a very short period. This was very difficult with the earlier, subjective methods but is feasible with CASA systems. The CASA system used in our studies is the Hobson Sperm Tracker (HST) (Hobson Vision Ltd, Baslow, Derbyshire, UK). The major advantage of this tracking system over other available CASA systems is that it can perform continuous tracking in real time. It can simultaneously follow 200 cells and generate 16 different motility parameters. It can also track sperm for a minimum tracking interval of 5 seconds.

Because sperm may be held immotile in an 'extender' for up to 24 h without significant loss of quality (Kime *et al.*, 1996; Rurangwa *et al.*, 1998), it is possible to test the effects of exposure for this period by using an extender to which a pollutant has been added. Exposing the sperm to a pollutant in an extender and immediately analyzing the motility is analogous to direct aquatic pollution, where sperm comes into contact with polluted waters during spawning, and to pollutant in the seminal plasma or testes (Kime *et al.*, 1996). Under natural conditions, however, sperm may be exposed to a pollutant *in vivo* for a much longer time from spermatogenesis to spermiation.

In the past, most of the research on effects of pollutants on sperm motility has been limited to this system of *in vitro* exposure of sperm. Using this technique, mercury was shown to decrease significantly the sperm motility at only 1 μg/l (Rurangwa *et al.*, 1998), which is the permitted level in drinking water. Because fish bioaccumulate mercury and tissue levels of over 0.5 μg/g are measured (Lockhart *et al.*, 1972), these data suggest that environmentally realistic levels of mercury may be having an impact on the reproduction of fish in the wild. The same was shown for fish sperm exposed *in vitro* to environmentally realistic concentrations of tributyl tin (TBT) (Figure 11.2) (details available from the author upon request). After 1 min of exposure goldfish sperm was the most sensitive to TBT. Complete inhibition of motility was observed at 100 μg/l, whereas trout sperm was immotile only at 1000 μg/l. Carp sperm, however, was still motile at 1000 μg/l, the highest concentration tested. The straight-line velocity of the sperm of all three species was significantly lower after immediate

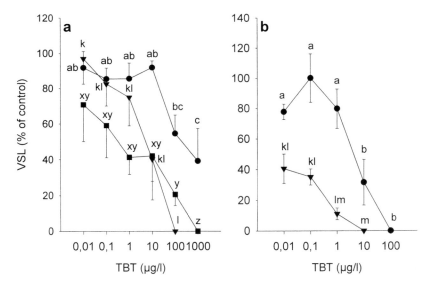

Figure 11.2 Straight-line velocity (VSL) of (●) carp, (▲) goldfish and (■) trout sperm exposed for 1 min (a) and 24 h (b) to different concentrations of TBT (as TBTCl) via an extender. Values are means ±SEM (n = 6). Different letters (a, b, c for carp; k, l, m for goldfish; x, y, z for trout data) indicate significant differences between different treatments.

exposure to 100 μg/l. After 24 h of exposure goldfish sperm was still more sensitive compared with carp. Complete immobility occurred at 10 μg/l whereas the same could be observed only at 100 μg/l for carp sperm. The straight-line velocity of goldfish spermatozoa was negatively affected at 1 μg/l, whereas for carp sperm it decreased at 10 μg/l.

Tributyltin is found in coastal waters and marinas in concentrations of 0.1–11.7 μg/l (Maguire *et al.*, 1986; Tolosa *et al.*, 1996; Axiak *et al.*, 2000). In temperate waters, high TBT levels have been found during the boating season, which lasts approximately from May until August (Holm *et al.*, 1991). This period coincides with reproduction and recruitment of many organisms such as fish and snails. Hence, it is possible that the spawning and offspring of a number of organisms may be affected by TBT.

It is evident that species reproduction is not only affected by direct effects on sperm cells and thus the prediction of long-term effects of pollutants based on *in vitro* tests is not feasible at the present time. For example, pollutants may be broken down in the organism to less toxic, and sometimes even more toxic, compounds or the bioavailability of a pollutant can be altered by binding to proteins. The possible sites at which pollutants may affect species reproduction *in vivo* have been mentioned above. Recently, a few researchers have included sperm motility measurement using CASA successfully in their examination of potential reproductive effects of pollutants on zebrafish exposed in the laboratory (details available from the author upon request). The same approach was used on fish captured from polluted rivers to determine whether the sperm produced had decreased motility (Jobling et al., 2002).

Although studies performed by us and others clearly show an effect of pollutants on sperm velocity or duration of movement, it is necessary to demonstrate that this decreased motility also causes a decreased fertilizing ability if it is to have an application in monitoring the reproductive health of fish in polluted rivers. Correlation of sperm motility with fertilization rates during natural spawning in fish such as the zebrafish may provide valuable information in this respect, as has been proven in preliminary studies (details available from the author upon request).

Future research on effects of pollution on species reproduction, such as fish, in our opinion has to focus on an integrated approach linking sperm motility measurements with as many other end-points as possible. Relevant measurements in exposed populations are, in addition to sperm quality, steroid and vitellogenin levels, reproductive behavior, egg production, fertilization and hatching rates and gene expression analysis in endocrine glands to elucidate mechanisms of action of tested chemicals. Only such integrated toxicological studies will enable us to assess fully the impact of pollutants on the reproductive output of species. Furthermore, we believe that modeling and understanding the effects of pollutants on reproductive parameters of organisms will provide quantitative links to estimate impacts at higher levels of biological organization (e.g. populations).

11.2.5 Realistic effect assessments: predicting effects of mixtures rather than single chemicals

Mixture toxicity studies are relatively new in ecotoxicology and hence our know-how on how to deal with the vast amount of chemicals and their combined effects is limited. Most of today's environmental legislation is based on risk assessment of single compounds or on the concentration addition model, which has only limited validity, as will be explained further. When an organism is exposed to different chemicals, many different scenarios are possible. The interaction between different compounds either in a direct chemical way or through interference with cellular systems such as transport, receptor binding, etc. can cause mixtures to react through different scenarios compared with their individual compounds. Different toxicants brought together in one mixture can react in a variety of ways and can show additive, synergistic, potentiating, masking or other combinatorial effects. It is generally accepted that two models – concentration addition and independent action (or response addition) – have a broader application and are suited to predict mixture toxicity from the toxicity of the single toxicants. Concentration addition is applied when the constituent chemicals act through the same cellular mechanism on the same target, i.e. have the same mode of action, and was first described in 1926 (Loewe and Muischnek, 1926). In this case the effect of the mixture can be predicted from the known toxic units (i.e. the concentration of a compound divided by its effective concentration EC_x at a given level x) of all compounds in the mixture. This also means that, given two compounds with the same toxicity, one can replace the other in the mixture without any effect on the overall toxicity, i.e. the sum of all toxic units in a mixture is equal to unity (see Formula 1).

Formula 1: Bases of the concentration addition model

$$\sum_{i=1}^{n} \frac{c_i}{EC_{x_i}} = 1$$

Where: c = concentration of EC_x of mixture
EC_x = effect concentration at x%

The alternative model of independent action (Bliss, 1939) is valid for mixtures of toxicants with different mode of action and different targets, assuming no overlap or influence between one another. In this case the overall effect of the mixture can be calculated from the effects of the individual toxicants at their respective concentrations (Formula 2).

Formula 2: Basis of the independent action model

$$E(c_{mix}) = 1 - (\prod_{i=1}^{n}(1 - E(c_i)))$$

Where: $E(c_{mix})$ = total effect of the mixture
$E(c_i)$ = effect of compound i

Both models assume that the mixture under study is fully described in its chemical composition and that the dose–response curves of all compounds in the mixture are known. Both models are extensively covered in the literature both theoretically (Teuschler *et al.*, 2002; Backhaus *et al.*, 2003; Vighi *et al.*, 2003) and when applied in case studies (Nirmalakhandan *et al.*, 1997; Altenburger *et al.*, 2000; Richardson *et al.*, 2001; Backhaus *et al.*, 2003). Using and interpreting either of the models, the reader has to be aware that both are mere simplifications of a very complex reality. A lot of cases can be found in the literature illustrating mixture effects not accounted for by the aforementioned models (Preston *et al.*, 2000; Chu and Chow, 2002) and validity should be assessed case by case. A critical paper on how to deal with different mixture risk definitions has been written by Hertzberg and MacDonell (2002).

Assessing the effect of mixture toxicity basically starts from the need of increased fundamental know-how or from the need to measure the impact of anthropogenic or natural pollution on an ecosystem. As mentioned above, substantial advantage can be obtained from measurements at a lower level of biological organization. In an above subsection on transgenic organisms, the existence of Pro-Tox was described from a technological viewpoint. Here we will show that bacterial gene profiling assays can generate new insights into the way mixtures of chemicals interact at the molecular level. Because gene profiling assays give a quantitative and qualitative view on the actual mode of action of a compound or mixture, they are a powerful tool to assess the toxicity of both known and unknown mixtures. Moreover, the quantitative aspects of the test allow a molecular evaluation of the mixture toxicity in view of different models such as concentration and effect addition, masking, potentiation, etc.

Figure 11.3 illustrates the resolving power of gene profiling for interpreting compound interactions in mixtures. The MerR signal, indicative of heavy metal stress, is clearly potentiated in the simultaneous presence of lindane and cadmium as illustrated by the fold induction above 500 in the mixture, but is less than 40 and 5 for cadmium and lindane respectively. The underlying mechanism is not yet understood and is given here to demonstrate the need for mechanistic understanding of mixture toxicity. We believe that in the future such mechanistic studies will be essential to evaluate the power of the different theoretical models (concentration addition, independent action and others). These molecular mechanisms will show us for which classes of chemicals and for which combinations of these pollutants the theoretical models are able to convey and predict the effects occurring in the real-world environment.

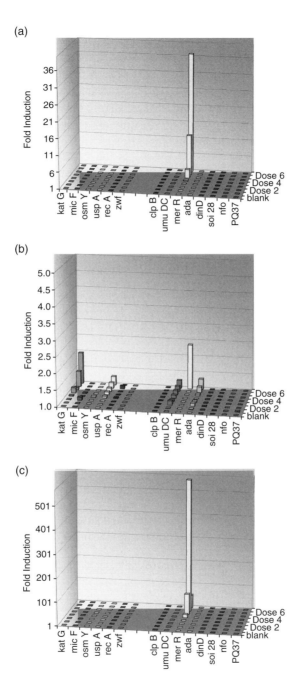

Figure 11.3 Bacterial gene profiling assay of cadmium (a), lindane (b) and the mixture of both compounds (c). The x-axis denotes the stress gene promoters indicative for the different end-points.

11.2.6 Ecological complexity in toxicity testing: interactions between pollutant stress and food availability

Energy budgets and bioenergetics have been identified as very useful currencies to link different levels of biological organization both under laboratory and field conditions because they can provide a causal relationship between different levels of biological organization, potentially relating rapid and sensitive responses at a cellular level to ecologically relevant end-points such as growth, condition and reproduction (Calow, 1991; De Coen and Janssen, 1997, 2003; Kooijman 2000; Smolders *et al.*, 2003a,b). The central concept, known as the metabolic cost hypothesis, states that exposure to a stressor is energetically demanding for organisms, so when in contact with a stressor an increased amount of energy is relocated to basal maintenance, leaving less energy available for growth and reproduction (Calow and Sibley, 1990; De Coen and Janssen, 2003; Smolders *et al.*, 2003a,b). Under laboratory conditions, the link between stress and reduced growth, declining condition or lower reproduction has been established for a wide range of organisms under both freshwater and marine environments, ranging from plants (Yang *et al.*, 2002; Cleuvers, 2003), invertebrates (De Coen and Janssen, 2003; Verslycke *et al.*, 2003), bivalves (Toro *et al.*, 2003; Smolders *et al.*, 2004) to fish (Rowe, 2003; Smolders *et al.*, 2003b), illustrating the usefulness of energy budgets as an extrapolation currency.

Although this concept is very well defined under laboratory conditions, the usefulness under field conditions is less straightforward. Direct effects of toxicant exposure can be elicited under laboratory conditions using strictly defined test protocols, but the ecological relevance of these tests may be questionable at times. Indirect effects of toxicant exposure, i.e. effects through trophic dynamics and food availability, are often ignored in laboratory tests but can have a significant impact on the outcome of toxicity tests (Preston, 2002; Campbell *et al.*, 2003; Fleeger *et al.*, 2003). In internationally accepted test protocols for the chronic exposure of aquatic organisms to stressors under laboratory conditions, organisms are traditionally fed a fixed amount of food (OECD, 1993; USEPA, 1994). However, because food availability, quality and trophic dynamics are not controlled under field conditions, these test protocols do not necessarily reflect the ecological relevance of field exposure.

The fact that food availability matters can be illustrated by Figure 11.4. These data are taken from Smolders (2003) and reflect the effect of effluent exposure for 28 days on the condition and lipid content of carp (*Cyprinus carpio*). Figure 11.4a represents the evolution of Fulton's condition factor (FCF = weight $\times$ length^{-3}) and lipid budget in a constant flow-through on-line monitoring system receiving 0%, 50%, 75% and 100% effluent; see Smolders *et al.* (2002) for a full description of the on-line monitoring system. The test was performed following the OECD Guideline 204 for chronic fish toxicity tests (OECD, 1993) and a fixed amount of food was added daily to all aquaria. As observed previously with zebrafish

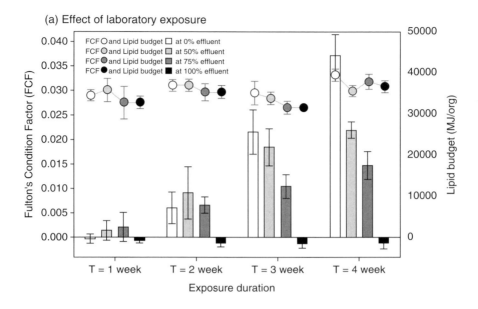

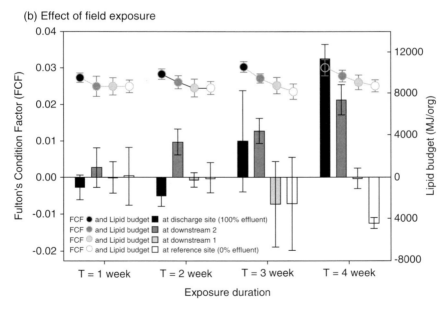

Figure 11.4 Effect of laboratory (a) and field (b) exposure on Fulton's condition factor (FCF) and lipid budgets in carp. The FCF is indicated in the line plot and lipid budgets in the vertical bars. Shading indicates effluent percentage from 0% (white) to 100% (black). Grey bars are intermediate effluent percentages. Data are average values ± standard deviations (n=5).

(Smolders *et al.*, 2002, 2003b), exposure to increasing effluent concentrations caused a significant decrease in the condition and lipid budget of carp.

At the same time, fish were exposed in the field at the discharge site and further downstream during an *in situ* biomonitoring campaign. Contrary to the on-line monitoring exposure, carp at the effluent discharge point (receiving 100% effluent) had a significantly higher FCF and lipid budget than fish at sites further downstream and at the (clean) reference site (Figure 11.4b). These data illustrate that under laboratory conditions the effluent exposure reduced the FCF and lipid budget of fish, whereas under field exposure conditions the opposite effect was observed.

This contradictory response in carp condition and lipid budget can be explained by an increased food availability at the effluent discharge point compared with the other field exposure sites. Analysis of the benthic macroinvertebrate community indicated that at the effluent discharge point there was a high dominance in both numbers and percentages of the freshwater isopod *Asellus aquaticus*, which is a preferred food source for carp (Khan, 2003). At the other locations there was a much lower dominance of one particular macroinvertebrate species and much lower food availability, although the biological water quality was much higher. The direct effect of effluent exposure observed in the on-line monitoring was completely reversed by the indirect effect of the effluent on the macroinvertebrate community.

Interaction between the adverse effect of toxicants (under laboratory conditions, when food availability is kept constant) and variable food availability (quality and quantity) is essential for understanding the ecological complexity of field-based *in situ* biomonitoring and underlines the necessity for integrated research, where toxicological, ecological and physicochemical data are integrated within one framework (Heugens *et al.*, 2001; Fleeger *et al.*, 2003). A better understanding of how indirect effects of toxicant exposure, such as trophic interactions, food availability and hence the physiological condition of organisms, can influence the outcome of ecotoxicological testing under real-life, ecologically relevant exposure conditions should be one of the main topics of future research if we want to improve our ability to extrapolate toxicity data from laboratory testing to effects on real-life ecosystems. Within this need, energy budgets are useful for linking different levels of biological organization, thus providing an ecologically relevant, integrative and holistic framework to detect both direct and indirect effects of pollutant exposure (Kooijman and Bedaux, 1997; Campbell *et al.*, 2003; Smolders *et al.*, 2003a).

However, experimental observations show that there is no fixed pattern in which food availability is affected by toxicant exposure. The data presented in Figure 11.4 indicate that food availability can have a beneficial effect on exposed organisms, as reported by several authors. In particular, downstream of bleached kraft mill effluents, increases in energy budgets, growth and/or condition of fish were reported for white sucker (*Catostomus commersoni*) (Hodson *et al.*, 1992), spoonhead sculpin (*Cottus ricei*) (Gibbons *et al.*, 1998) and redbreast sunfish (*Lepomis auritus*) (Adams *et al.*, 1992). In a comparison of 51 mills in Quebec, reports indicated that if significant differences in the condition of fish were observed downstream of pulp and paper mill effluents, in 90% of the cases the fish exposed to these effluent discharges showed a significantly better condition than fish from the control site. Food availability might have been one of the confounding factors, because effluent discharge areas also had a significantly higher abundance of macroinvertebrates as a food source compared with the reference sites (Langlois and Dubuc, 1999). This toxicity reduction in the presence of high food levels has been observed also for chironomids (Sibley et al., 1997; Stuijfzand et al., 2000; de Haas et al., 2004), where the suitability of Chironomus riparius as a test organism for whole-sediment bioassays was questioned because of its food level dependency (De Haas et al., 2004). Also for daphnids, the impact of food levels on an organism's physiological state was identified as an essential factor in determining the toxicity (Enserink et al., 1995; Heugens et al., 2001; Herbrandson et al., 2003).

In contrast, a reduction in fitness of organisms with increasing food availability has been observed for yellow perch (*Perca flavescens*) (Campbell et al., 2003), daphnids (Kluttgen and Ratte, 1994; Taylor et al., 1998; Antunes et al., 2004) and bivalves (Björk and Gilek, 1997). Remarkably, these authors all attributed the increased toxicity at high food levels to the sorption of pollutants to the food sources, thus altering the route of exposure of the toxicants from water-borne to particle adsorbed. This indicates that the main route of exposure, i.e. through the water, sediment or food, may determine the interaction between food availability and toxicant responses. It is obvious that the effect of exposure route on the body burden and the toxicity of pollutants, the effect of food availability and interactions between these can have serious repercussions on current risk assessment procedures. Clarifying these interactions provides one of the main future challenges for increasing the ecological relevance of aquatic toxicology.

Finally, this interaction between food availability and pollutant stress is a clear indication of how the ecological complexity of ecosystems may interfere with toxicological end-points. Chapman (2000) recently argued that, often, toxicological studies are conducted independent of ecological considerations, neglecting the functioning of populations and communities and the processes that affect all these parameters. Combined *in situ* and laboratory exposure, in association with physicochemical and ecological information to account for ecosystem complexity, will provide considerable benefit in order to understand and predict the effects of chemicals on natural communities under realistic exposure conditions and to improve the quality of laboratory-to-field extrapolations (Chapman, 2000; Morris and Keough, 2002; Anderson et al., 2003).

11.3 Conclusions

In this chapter we have demonstrated various themes that will provide generic yet fundamental insights into the complex web of interactions between chemicals, biota and their surrounding environments. These are of course general examples where we think major breakthroughs can be expected in the coming years. Furthermore, at this very moment, large advantages can be expected from the various genome projects that are undertaken worldwide. At present, the advent of genomics and bioinformatics in the science of environmental toxicology remains largely under-exploited. These emerging technologies are being used in the pharmaceutical industry but the applications in environmental sciences are rather new. It is clear that the complexity of the large numbers of chemicals with varying structures and activities represents another dimension compared with studying a set of carefully selected pharmaceuticals.

A major challenge for the future is to optimize the knowledge related to the increased volume of toxicological data to construct molecular (and genetic) pathways that will explain the mechanisms of toxic events. Owing to increased synthesizing activities in the chemical and pharmaceutical industry, large volumes of chemicals and wastes are discharged into the environment. A major problem for new as well as existing chemicals is the poor knowledge of the impact and effects of chemicals on health and the environment. This is true not only for individual chemicals but also for chemical mixtures, irrespective of their use in production or in products or of their distribution in waste or in wasted material. As a result of the intensive use of chemicals in our daily life, humans and biota are exposed during a full life-cycle to a mixture of pollutants, usually at low concentrations. It is therefore critical that we rapidly increase our understanding of the consequences of such exposures.

In principle, toxicogenomics and proteomics could become a tool to guide legislation and guidelines to regulate pollutant levels in the environment. This will ensure that, in the future, new and prior undiscovered adverse effects of chemical exposure will become detected *a posteriori*. It can be expected that in the future toxicogenomics/proteomics will have an impact on the risk and exposure assessment of chemicals. Toxicogenomics/proteomics has the potential to improve our understanding of complex exposure scenarios. This should uncover dangerous conditions and should improve preventive strategies for chemical handling. Another major application of gene expression profiling is to understand the genetic variability and susceptibility to (toxic) stress within ecosystems. Impact assessments of ecosystems should help us to define the most vulnerable species within the different trophic levels.

However, before toxicogenomics/proteomics can meet all of the above challenges there are many problems yet to be solved. The primary problem at present is the lack of substantial genomic information for those species that are widely used in ecotoxicology. Most of the standard OECD species (except for zebrafish) have no characterized genome and are not expected to become the next decoded species in future genome projects. Compared with humans or yeast, for example, there are no commercially available DNA arrays that could be used in ecotoxicological applications. Substantial sequencing efforts therefore should be undertaken to fill in these crucial gaps in the knowledge for species such as algae, daphnids and other benchmark ecotoxicity species. Furthermore, it will require a tremendous effort to evaluate the 'ecological relevance' of toxicogenomics/proteomics data. In other words, the link between changes at the transcriptional level and at the population, community or ecosystem level will have to be established. These quantitative relationships between sub-organism events and long-term impact on the reproduction, growth and survival of species will be crucial for the successful application of these methodologies in environmental risk assessment procedures. Within the next century it can be expected that this discipline will mature into a complementary tool to unravel the emerging environmental toxicological problems.

References

Adams, M.S., Crumby, W.D., Greeley, M.S., Shugart, L.R. and Saylor, C. (1992) Responses of fish populations and communities to pulp mill effluents: a holistic assessment. *Ecotoxicology and Environmental Safety*, **24**, 347–360.

Afanassiev, V., Sefton, M., Anantachaiyong, T., Barker, G., Walmsley, R. and Wolfl, S. (2000) Application of yeast cells transformed with GFP expression constructs containing the RAD54 or RNR2 promoter as a test for the genotoxic potential of chemical substances. *Mutation Research* **464**, 297–308.

Alper, J. (2001) Searching for medicine's sweet spot. *Science* **291**, 2338–2343.

Altenburger, R., Backhaus, T., Boedeker, W., Faust, M., Scholze, M. and Grimme, L.H. (2000) Predictability of the toxicity of multiple chemical mixtures to *Vibrio fischeri*: mixtures composed of similarly acting chemicals. *Environmental Toxicology and Chemistry*, **19**, 2341–2347.

Ames, B.N. (1979) Identifying environmental chemicals causing mutations and cancer. *Science*, **204**, 587–593.

Anderson, B.S., Middaugh, D.P., Hunt, J.W. and Turpen, S.L. (1991) Copper toxicity to sperm, embryos and larvae of topsmelt *Atherinops affinis*, with notes on induced spawning. *Marine Environmental Research*, **31**, 17–35.

Anderson, B.S., Hunt, S.W., Phillips, B.M., Nicely, P.A., de Vlaming, V., Connor, V., Richard, N. and Tjeerdema, R.S. (2003) Integrated assessment of the impacts of agricultural drainwater in the Salinas River (California, USA). *Environmental Pollution*, **124**, 523–532.

Antunes, S.C., Castro, B.B. and Gonçalves, F. (2004) Effect of food level on the acute and chronic responses of daphnids to lindane. *Environmental Pollution*, **127**, 367–375.

Axiak, V., Vella, A.J., Aius, D., Bonnici, P., Cassar, G., Cassone, R., Chircop, P., Micallef, D., Mintoff, B. and Sammut, M. (2000) Evaluation of environmental levels and biological impact of TBT in Malta (central Mediterranean). *Science of the Total Environment*, **258**, 89–97.

Backhaus, T., Altenburger, R., Arrhenius, A., Blanck, H., Faust, M., Finizio, A., Gramatica, P., Grote, M., Junghans, M., Meyer, W., Pavan, M., Porsbring, T., Scholze, M., Todeschini, R., Vighi, M., Walter, H. and Grimme, L.H. (2003) The BEAM-project: prediction and assessment of mixture toxicities in the aquatic environment. *Continental Shelf Research*, **23**, 1757–1769.

Baillieul, M. and Blust, R. (1999) Analysis of the swimming velocity of cadmium-stressed *Daphnia magna*. *Aquatic Toxicology*, **44**, 245–254.

Berger, J., Hauber, J., Hauber, R., Geiger, R. and Cullen, B.R. (1988) Secreted placental alkaline phosphatase: a powerful new quantitative indicator of gene expression in eukaryotic cells. *Gene*, **66**, 1–10.

Bervoets, L. and Blust, R. (2000) Effects of pH on cadmium and zinc uptake by the midge larvae *Chironomus riparius*. *Aquatic Toxicology*, **49**, 145–157.

Billard, R. and Roubaud, P. (1985) The effects of metals and cyanide on fertilization in rainbow trout (*Salmo gairdneri*). *Water Research*, **19**, 209–214.

Billinton, N., Barker, M.G., Michel, C.E., Knight, A.W., Heyer, W.D., Goddard, N.J., Fielden, P.R. and Walmsley, R.M. (1998) Development of a green fluorescent protein reporter for a yeast genotoxicity biosensor. *Biosensors and Bioelectronics* **13**, 831–838.

Björk, M. and Gilek, M. (1997) Bioaccumulation kinetics of PCB 31, 49 and 153 in the blue mussel, *Mytilus edulis* L. as a function of algal food concentration. *Aquatic Toxicology*, **38**, 101–123.

Bliss, C.I. (1939) The toxicity off poisons applied jointly. *Annuals of Applied Biology*, **26**, 585–615.

Boyers, S.P., Davis, R.O. and Katz, D.F. (1989) Automated sperm analysis. *Current Problems in Obstetrics, Gynecology and Fertility*, **12**, 172–200.

Brinkman-Van der Linden, C.M., Havenaar, E.C., Van Ommen, C.R., Van Kamp, G.J., Gooren, L.J. and Van Dijk, W. (1996) Oral estrogen treatment induces a decrease in expression of sialyl Lewis x on alpha 1-acid glycoprotein in females and male-to-female transsexuals. *Glycobiology*, **6**, 407–412.

Burton, G.A. (1999) Realistic assessments of ecotoxicity using traditional and novel approaches. *Aquatic Ecosystem Health Management*, **2**, 1–8.

Callewaert, N., Geysens, S., Molemans, F. and Contreras, R. (2001) Ultrasensitive profiling and sequencing of N-linked oligosaccharides using standard DNA-sequencing equipment. *Glycobiology*, **11**, 275–281.

Calow, P. (1991). Physiological costs of combating chemical toxicants: ecological implications. *Comparative Biochemistry and Physiology*, **100c**, 3–6.

Calow, P. and Sibley, R.M. (1990) A physiological basis of population processes: ecotoxicological implications. *Functional Ecology*, **4**, 283–288.

Campbell, P.G.C., Hontela, A., Rasmussen, J.B., Giguère, A., Gravel, A., Kraemer, L., Kovesces, J., Lacroix, A., Levesque, H. and Sherwood, G. (2003) Differentiating between direct (physiological) and food-chain mediated (bioenergetic) effects on fish in metal impacted lakes. *Human and Ecological Risk Assessment*, **9**, 847–866.

Candido E.P. and Jones, D. (1996) Transgenic *Caenorhabditis elegans* strains as biosensors. *Trends in Biotechnology*, **14**, 125–129.

Carvan, M.J., Sonntag, D.M., Cmar, C.B., Cook, R.S., Curran, M.A. and Miller, G.L. (2001) Oxidative stress in zebrafish cells: potential utility of transgenic zebrafish as a deployable sentinel for site hazard ranking. *Science of the Total Environment*, **274**, 183–196.

Chalfie, M., Tu, Y., Euskirchen, G., Ward, W.W. and Prasher, D.C. (1994) Green fluorescent protein as a marker for gene expression. *Science*, **263**, 802–805.

Chapman, P.M. (2002) Integrating toxicology and ecology: putting the 'eco' into ecotoxicology. *Marine Pollution Bulletin*, **44**, 7–15.

Chapman, P. M. (2000) Whole effluent toxicity testing – usefulness, level of protection, and risk assessment. *Environmental Toxicology and Chemistry*, **19**, 3–13.

Chu, K.W. and Chow, K.L. (2002) Synergistic toxicity of multiple heavy metals is revealed by a biological assay using a nematode and its transgenic derivate. *Aquatic Toxicology*, **61**, 53–64.

Cleuvers, M. (2003) Aquatic ecotoxicity of pharmaceuticals including the assessment of combination effects. *Toxicology Letters*, **142**, 185–194.

Cormack, B.P., Valdivia, R.H. and Falkow, S. (1996) FACS-optimized mutants of the green fluorescent protein (GFP). *Gene*, **173**, 33–38.

Dauwe, T., Bervoets, L., Blust, R. and Eens, M. (2002) Tissue levels of lead in experimentally exposed zebra finches (*Taeniopygia guttata*) with particular attention on the use of feathers as biomonitors. *Archives of Environmental Contamination and Toxicology*, **42**, 88–92.

Davey, F.D. and Breen, K.C. (1998) Stimulation of sialyltransferase by sub-chronic low-level lead exposure in the developing nervous system. *Toxicology and Applied Pharmacology*, **151**, 16–21.

De Boeck, G., Ngo, T.T.H., Van Campenhout, K. and Blust, R. (2003) Differential metallothionein induction patterns in three freshwater fish during sub-lethal copper exposure. *Aquatic Toxicology*, **65**, 413–424.

De Coen, W.M. and Janssen, C.R. (1997) The use of biomarkers in *Daphnia magna* toxicity testing IV. Cellular energy Allocations: A new biomarker to assess the energy budget of toxicant-stressed *Daphnia* populations. *Journal of Aquatic Stress and Recovery*, **6**, 43–55.

De Coen, W.M. and Janssen, C.R. (2003) The missing biomarker link: relationships between effects on the cellular energy allocation biomarker of toxicant-stressed *Daphnia magna* and corresponding population characteristics. *Environmental Toxicology and Chemistry*, **22**, 1632–1641.

de Haas, E.M., Paumen, M.L., Koelmans, A.A. and Kraak, M.H.S. (2004) Combined effects of copper and food on the midge *Chironomus riparius* in whole-sediment bioassays. *Environmental Pollution*, **127**, 99–107.

De Jong, G., Feelders, R., Van Noort, W.L. and Van Eijk, H.G. (1995) Transferrin microheterogeneity as a probe in normal and disease states. *Glycoconjugate Journal*, **12**, 219–226.

De Smet, H., Blust, R. and Moens, L. (2001) Cadmium-binding to transferrin in the plasma of the common carp *Cyprinus carpio*. *Comparative Biochemistry and Physiology C*, **128**, 45–53.

Dell, A. and Morris, H.R. (2001) Glycoprotein structure determination by mass spectrometry. *Science*, **291**, 2351–2356.

Depledge, M.H. (1989) The rational basis for detection of the early indicators of marine pollutants using physiological indicators. *Ambio*, **18**, 301–302.

Depledge, M.H. (1990) New approaches in ecotoxicology: can inter-individual physiological variability be used as a tool to investigate pollution effects? *Ambio*, **19**, 251–252.

Enserink, E.L., Kerkhofs, M.J.J., Baltus, C.A.L. and Koeman, J.H. (1995) Influence of food quantity and lead exposure on maturation in *Daphnia magna* – evidence for a trade-off mechanism. *Functional Ecology*, **9**, 175–185.

Errampalli, D., Leung, K., Cassidy, M.B., Kostrzynska, M., Blears, M., Lee, H. and Trevors, J.T. (1999) Applications of the green fluorescent protein as a molecular marker in environmental microorganisms. *Journal of Microbiology Methods*, **35**, 187–199.

Fent, K. (2001) Fish cell lines as versatile tools in ecotoxicology: assessment of cytotoxicity, cytochrome P4501A induction potential and estrogenic activity of chemicals and environmental samples. *Toxicology In Vitro*, **15**, 477–488.

Flahaut, C., Michalski, J.C., Danel, T., Humbert, M.H. and Klein, A. (2003) The effects on the glycosylation of human transferrin. *Glycobiology* **13**, 191–198.

Fleeger, J.W., Carman, K.R. and Nisbet, R.M. (2003) Indirect effects of contaminants in aquatic ecosystems. *Science of the Total Environment*, **317**, 207–233.

Fournier, T., Medjoubi-N.N. and Porquet, D. (2000) Alpha-1-acid glycoprotein. *Biochimica Biophysica Acta*, **1482**, 157–171.

Gibbons, W.N., Munkittrick, K.R. and Taylor, W.D. (1998) Monitoring aquatic environments receiving industrial effluents using small fish species 1: response of spoonhead sculpin (*Cottus ricei*) downstream of a bleached-kraft pulp mill. *Environmental Toxicology and Chemistry*, **17**, 2227–2237.

Goodarzi, M.T. and Turner, G.A. (1998) Reproducible and sensitive determination of charged oligosaccharides from haptoglobin by PNGase F digestion and HPAEC/PAD analysis: glycan composition varies with disease. *Glycoconjugate Journal*, **15**, 469–475.

Green, S. and Chambon, P. (1987) Estradiol induction of a glucocorticoid-responsive gene by a chimeric receptor. *Nature*, **325**, 75–78.

Guillette, L.J.J., Gross, T.S., Masson, G.R., Matter, J.M., Percival, H.F. and Woodward, A.R. (1994) Developmental abnormalities of the gonad and abnormal sex hormone concentrations in juvenile alligators from contaminated and control lakes in Florida. *Environmental Health Perspectives*, **102**, 680–688.

Guttman, A., Chen, F.A., Evangelista, R.A. and Cooke, N. (1996) High-resolution capillary gel electrophoresis of reducing oligosaccharides labeled with 1-aminopyrene-3,6,8-trisulfonate. *Analytical Biochemistry*, **233**, 234–242.

Hakkila, K., Maksimow, M., Karp, M. and Virta, M. (2002) Reporter genes lucFF, luxCDABE, gfp, and dsred have different characteristics in whole-cell bacterial sensors. *Analytical Biochemistry*, **301**, 235–242.

Hastings, A. and Huggins, K. (1994) Persistence of transients in spatially structured ecological models. *Science*, **263**, 1133–1136.

Herbrandson, C., Bradbury, S.P. and Swackhamer, D.L. (2003) Influence of suspended solids on acute toxicity of carbofuran to *Daphnia magna*: II. An evaluation of potential interactive mechanisms. *Aquatic Toxicology*, **63**, 343–355.

Hertzberg, R.C. and MacDonell, M.M. (2002) Synergy and other ineffective mixture risk definitions. *Science of the Total Environment*, **288**, 31–42.

Heugens, E.H.W., Hendriks, A.J., Dekker, T., van Straalen, N.M. and Admiraal, W. (2001) A review of the effects of multiple stressors on aquatic organisms and analysis of uncertainty factors for use in risk assessment. *Critical Reviews in Toxicology*, **31**, 247–284.

Hodson, P.V., McWhirter, M., Ralph, K., Thievierge, D., Carey, J.H., Vanderkraak, G., Whittle, D.M. and Levesque, M.C. (1992) Effects of bleached kraft mill effluents on fish in the St-Maurice river, Quebec. *Environmental Toxicology and Chemistry*, **11**, 1635–1651.

Hoff, P.T., Van Dongen, W., Esmans, E.L., Blust, R. and De Coen, W.M. (2003) Evaluation of the toxicological effects of perfluorooctane sulfonic acid in the common carp (*Cyprinus carpio*). *Aquatic Toxicology*, **62**, 349–359.

Holm, G., Norrgren, L. and Linden, O. (1991) Reproductive and histopathological effects of long-term experimental exposure to bis(tributyltin)oxide (TBTO) on the three-spoiled stickleback, *Gastrosteus aculeatus* Linnaeus. *Journal of Fish Biology*, **38**, 373–386.

Jobling, S., Coey, S., Whitmore, J.G., Kime, D.E., Van Look, K.J.W., McAllister, B.G., Beresford, N., Henshaw, A.C., Brighty, G., Tyler, C.R. and Sumpter, J.P. (2002) Wild intersex roach (*Rutilus rutilus*) have reduced fertility. *Biology of Reproduction*, **67**, 515–524.

Justus, T., and Thomas, S.M. (1989) Evaluation of transcriptional fusion with green fluorescent protein versus luciferase as reporters in bacterial mutagenicity tests. *Mutagenesis*, **14**, 351–356.

Justus, T., and Thomas, S.M. (1998) Construction of a umuC'–luxAB plasmid for the detection of mutagenic DNA repair via luminescence. *Mutation Research*, **398**, 131–141.

Kawanishi, M., Sakamoto, M., Ito, A., Kishi, K. and Yagi, T. (2003) Construction of reporter yeasts for mouse aryl hydrocarbon receptor ligand activity. *Mutation Research*, **540**, 99–105.

Khan, A.T. (2003) Dietary studies on exotic carp (*Cyprinus carpio* L.) from two lakes of western Victoria, Australia. *Aquatic Sciences*, **65**, 272–286.

Khan, A.T. and Weis, J.S. (1987a) Effects of methylmercury on sperm and egg viability of two populations of killifish (*Fundulus heteroclitus*). *Archives of Environmental Contamination and Toxicology*, **16**, 499–505.

Khan, A.T. and Weis, J.S. (1987b) Toxic effects of mercuric chloride on sperm and egg viability of two populations of mummichog, *Fundulus heteroclitus*. *Environmental Pollution*, **48**, 263–273.

Kime, D.E. (1995) The effects of pollution on reproduction in fish. *Reviews in Fish Biology and Fisheries*, **5**, 52–95.

Kime, D.E. (1998) *Endocrine Disruption in Fish*. Kluwer Academic Publishers, Boston.

Kime, D.E. and Nash, J.P. (1999) Gamete viability as an indicator of reproductive endocrine disruption in fish. *Science of the Total Environment*, **233**, 123–129.

Kime, D.E., Ebrahimi, M., Nysten, K., Roelants, I., Rurangwa, E., Moore, H.D.M. and Ollevier, F. (1996) Use of computer assisted sperm analysis (CASA) for monitoring the effects of pollution on sperm quality of fish; application to the effects of heavy metals. *Aquatic Toxicology*, **36**, 223–237.

Kluttgen, B. and Ratte, H.T. (1994) Effects of different food doses on cadmium toxicity to *Daphnia magna*. *Environmental Toxicology and Chemistry*, **13**, 1619–1627

Kooijman, S.A.L.M. (2000) *Dynamic Energy and Mass Budgets in Biological Systems*. Cambridge University Press, Cambridge.

Kooijman, S.A.L.M. and Bedaux, J.J.M. (1997) Analysis of toxicity tests on fish growth. *Water Research*, **30**, 1633–1644.

Kornfeld, S. (1998) Diseases of abnormal protein glycosylation: an emerging area. *Journal of Clinical Investigation*, **101**, 1293–1295.

Kostrzynska, M., Leung, K.T., Lee, H. and Trevors, J.T. (2002) Green fluorescent protein-based biosensor for detecting SOS-inducing activity of genotoxic compounds. *Journal of Microbiological Methods*, **48**, 43–51.

Kovalchuk, I., Kovalchuk, O. and Hohn, B. (2001) Biomonitoring the genotoxicity of environmental factors with transgenic plants. *Trends in Plant Science*, **6**, 306–310.

Kurittu, J., Karp, M., Korpela, M. (2000) Detection of tetracyclines with luminescent bacterial strains. *Luminescence*, **15**, 291–297.

Langlois, C. and Dubuc, N. (1999) *Pulp and Paper Environmental Effects Monitoring (EEM). Results Synthesis for the 47 Cycle 1 Studies Conducted in Quebec.* Environmental Canada, Montreal, Canada.

LaRossa, E.A. (1998) *Bioluminescence Methods and Protocols.* Humana Press, Totowa, NJ.

Lockhart, W.L., Uthe, J.F., Kenny, A.R. and Mehrle, P.M. (1972) Methyl mercury in northern pike (*Esox lucius*): distribution, elimination, and some biochemical characteristics of contaminated fish. *Journal of the Fisheries Research Board of Canada*, **29**, 1519–1523.

Loewe, S. and Muischnek, H. (1926). Über combinationsirkungen. 1. Mitteilung: hilfsmittel der fragestellung. *Archives of Experimental and Pathological Pharmacology*, **114**, 313–326.

Mackiewicz, A. and Mackiewicz, K. (1995) Glycoforms of serum a1-acid glycoprotein as markers of inflammation and cancer. *Glycoconjugate Journal*, **12**, 241–247.

Maguire, R.J., Tkacz, R.J., Chau, Y.K., Bengert G.A. and Wong, P.T.S. (1986) Occurrence of organotin compounds in water and sediment in Canada. *Chemosphere*, **15**, 253–274.

Manoil, C., Mekalanos, J.J. and Beckwith, J. (1990) Alkaline phosphatase fusions: sensors of subcellular location. *Journal of Bacteriology*, **172**, 515–518.

Mattingly, C.J., McLachlan, JA. and Toscano Jr, W.A. (2001) Green fluorescent protein (GFP) as a marker of aryl hydrocarbon receptor (AhR) function in developing zebrafish (*Danio rerio*). *Environmental Health Perspectives*, **109**, 845–849.

Matz, M.V., Fradkov, A.F., Labas, Y.A., Savitsky, A.P., Zaraisky, A.G., Markelov, M.L. and Lukyanov, S.A. (1999) Fluorescent proteins from nonbioluminescent Anthozoa species. *Nature Biotechnology*, **17**, 1227.

McCarthy, J.F. and Shugart, L.R. (1990) *Biomarkers of Environmental Contamination.* Lewis Publishers, Boca Raton, FL.

McCarty, L.S. and Munkittrich, K.R. (1997) Environmental biomarkers in aquatic toxicology: fiction, fantasy or functional? *Human and Ecological Risk Assessment*, **2**, 268–274.

McDaniels, A.E., Reyes, A.L., Wymer, L.J., Rankin, C.C. and Stelma, G.N. (1990) Comparison of the *Salmonella* (Ames) test, umu tests, and the SOS Chromotests for detecting genotoxins. *Environmental and Molecular Mutagenesis*, **16**, 204–215.

McEver, R.P. (1997) Selectin–carbohydrate interactions during inflammation and metastasis. *Glycoconjugate Journal*, **14**, 585–591.

Mehrle, M. and Mayer, F.L. (1980) Clinical tests in aquatic toxicology: state of the art. *Environmental Health Perspectives*, **34**, 139–143.

Miller, C.A. (1997) Expression of the human aryl hydrocarbon receptor complex in yeast. Activation of transcription by indole compounds. *Journal of Biological Chemistry*, **26**, 32824–32829.

Miller, C.A. (1999) A human aryl hydrocarbon receptor signaling pathway constructed in yeast displays additive responses to ligand mixtures. *Toxicology and Applied Pharmacology*, **160**, 297–303.

Miller, C.A., Martinat, M.A. and Hyman L.E. (1998) Assessment of aryl hydrocarbon receptor complex interactions using pBEVY plasmids: expression vectors with bi-directional promoters for use in *Saccharomyces cerevisiae*. *Nucleic Acids Research*, **26**, 3577–3583.

Min, J., Kim, E.J., LaRossa, R.A. and Gu, M.B. (1999) Distinct responses of a recA::luxCDABE *Escherichia coli* strain to direct and indirect DNA damaging agents. *Mutation Research*, **442**, 61–68.

Monnet, D., Feger, J., Biou, D., Durand, G., Cardon, P., Leroy, Y. and Fournet, B. (1986) Effect of phenobarbital on the oligosaccahrin structure of rat alpha-acid glycoprotein. *Biochimica et Biophysica Acta*, **881**, 10–14.

Morris, L. and Keough, M.J. (2002) Organic pollution and its effects: a short-term transplant experiment to assess the ability of biological endpoints to detect change in a soft sediment environment. *Marine Ecoloology Progress Series*, **225**, 109–121.

Munkittrick, K.R. and McCarty, L.S. (1995) An integrated approach to aquatic ecosystem health: top-down, bottom-up or middle out? *Journal of Aquatic Ecosystem Health*, **4**, 77–90.

Murk, A.J., Legler, J., Denison, M.S., Giesy, J.P., van de Guchte, C. and Brouwer, A. (1996) Chemical-activated luciferase expression (CALUX): a novel *in vitro* bioassay for Ah receptor active compounds in sediments and pore water. *Fundamental and Applied Toxicology*, **33**, 149–160.

Nakamura, S., Oda, Y., Shimida, T., Oki, I. and Sugimoto, K. (1987) SOS-inducing activity of chemical carcinogens and mutagens in *Salmonella typhimurium* TA 1535/pSK1002: examination with 151 chemicals. *Mutation Research*, **192**, 239–246.

Naylor, L.H. (1999) Reporter gene technology: The future looks bright. *Biochemical Pharmacology*, **58**, 749–757.

Nirmalakhandan, N., Xu, S., Trevizo, C., Brennan, R., and Peace, J. (1997) Additivity in microbial toxicity of nonuniform mixtures of organic chemicals. *Ecotoxicology and Environmental Safety*, **37**, 97–102.

Nunoshiba, T. and Nishioka, H. (1991) 'Rec–lac test' for detecting SOS-inducing activity of environmental genotoxic substance. *Mutatation Research*, **254**, 71–77.

OECD (1993) *Fish, Prolonged Toxicity Test: 14-day Study. Guideline 204. OECD Guidelines for the Testing of Chemicals*, Vol. 1. OECD Paris.

Orser, C.S., Foong, F.C.F., Cpaldi, S.R., Nalezny, W., Mackay, W., Benjamin, M. and Farr, S.B. (1995) Use of prokaryotic stress promotors as indicators of the mechanisms of chemical toxicity. *In Vitro Toxicology*, **8**, 71–85.

Paul, I., Mandal, C., Allen, A.K. and Mandal, C. (2001) Glycosylated molecular variants of C-reactive proteins from the major carp *Catla catla* in fresh and polluted aquatic environments. *Glycoconjugate Journal*, **18**, 547–556.

Peakall, D. (1992) In *Animal Biomarkers as Pollution Indicators*, p. 291. Chapman and Hall, New York.

Phillips, G.J. (2001) Green fluorescent protein – a bright idea for the study of bacterial protein localization. *FEMS Microbiology Letters*, **204**, 9–18.

Pos, O., Van Dijk, W., Ladiges, N., Linthorst, C., Sala, M., van Tiel, D. and Boers, W. (1988) Glycosylation of four acute-phase glycoproteins secreted by rat liver cells *in vivo* and *in vitro*. Effects of inflammation and dexamethasone. *European Journal of Cell Biology*, **46**, 121–128.

Preston, B.L. (2002) Indirect effects in aquatic ecotoxicology: implications for ecological risk assessment. *Environmental Management*, **29**, 311–323.

Preston, S., Coad, N., Townend, J., Killham, K., and Paton, G.I. (2000) Biosensing the acute toxicity of metal interactions: are they additive, synergistic, or antagonistic? *Environmental Toxicology and Chemistry*, **19**, 775–780.

Prosser, J.I., Killham, K., Glover, L.A. and Rattray, E.A.S. (1996) Luminescence-based systems for detection of bacteria in the environment. *Critical Reviews in Biotechnology*, **16**, 157–183.

Ptitsyn, L.R., Horneck, G., Komova, O., Kozubek, S., Krasavin, E.A., Bonev, M., and Rettberg, P. (1997) A biosensor for environmental genotoxin screening based on an SOS lux assay in recombinant *Escherichia coli* cells. *Applied Environmental Microbiology*, **63**, 4377–4384.

Purdom, C.E., Hardiman, P.A., Bye, V.J., Eno, N.C., Tyler, C.R. and Sumpter, J.P. (1994) Estrogenic effects of effluents from sewage treatment works. *Chemistry and Ecology*, **8**, 275–285.

Quillardet, P. and Hofnung, M. (1993) The SOS chromotest: a review. *Mutation Research*, **297**, 235–279.

Quillardet, P., Huisman, O., Ari, R.D. and Hofnung, M. (1982) SOS chromotest, a direct assay of induction of an SOS function in *Escherichia coli* K12 to measure genotoxicity. *Proceedings of the National Academy of Sciences USA*, **7**, 5971–5975.

Quinones, A. and Rainov, N.G. (2001) Identification of genotoxic stress in human cells by fluorescent monitoring of p53 expression. *Mutation Research*, **494**, 73–85.

Raju, T.S. (2000) Electrophoretic methods for the analysis of N-linked oligosaccharides. *Analytical Biochemistry*, **283**, 125–132.

Reifferscheid, G. and Heil, J. (1996) Validation of the SOS/umu test using test results of 486 chemicals and comparison with the Ames test and carcinogenicity data. *Mutation Research*, **369**, 129–145.

Richardson, J.R., Chambers, H.W. and Chambers, J.E. (2001) Analysis of the additivity of in vitro inhibition of cholinesterase by mixtures of chlorpyrifos-oxon and azinphos-methyl-oxon. *Toxicology and Applied Pharmacology*, **172**, 128–139.

Routledge, E.J. and Sumpter, J.P. (1996) Estrogenic acitivity of surfactants and some of their degradation products assessed using a recombinant yeast assay. *Environmental Toxicological Chemistry*, **15**, 241–248.

Rowe, C.L. (2003) Growth responses of an estuarine fish exposed to mixed trace elements in sediments over a full life cycle. *Ecotoxicological and Environmental Safety*, **54**, 229–239.

Rowland, B., Purkayastha, A., Monserrat, C., Casart, Y., Takiff, H. and McDonough, K.A. (1999) Fluorescence-based detection of lacZ reporter gene expression in intact and viable bacteria including *Mycobacterium* species. *FEMS Microbiology Letters*, **179**, 317–325.

Rudd, P.M., Elliott, T., Cresswell, P., Wilson, I.A. and Dwek, R.A. (2001) Glycosylation and the immune system. *Science*, **291**, 2370–2375.

Rurangwa, E. (1999) Heavy metal pollutant disruption of growth and reproduction in the African catfish, *Clarias gariepinus*. PhD Thesis, Katholieke Universiteit Leuven, Leuven, Belgium.

Rurangwa, E., Roelants, I., Huyskens, G., Ebrahimi, M., Kime, D.E. and Ollevier, F. (1998) The minimum effective spermatozoa to egg ratio for artificial insemination and the effects of mercury on sperm motility and fertilization ability in *Clarias gariepinus*. *Journal of Fish Biology*, **53**, 402–413.

Rydén, I., Pahlsson, P. and Lindgren, S. (2002) Diagnostic accuracy of a1-acid glycoprotein fucosylation for liver cirrhosis in patients undergoing hepatic biopsy. *Clinical Chemistry*, **48**, 2195–2201.

Schmid, C., Reifferscheid, G., Zahn, R.K. and Bachmann, M. (1997) Increase of sensitivity and validity of the SOS/umu-test after replacement of the beta-galactosidase reporter gene with luciferase. *Mutation Research*, **394**, 9–16.

Selck, H, Riemann, B, Christoffersen, K, Forbes, V.E., Gustavson, K., Hansen, B.W., Jacobsen, J.A., Kusk, O.K. and Petersen, S. (2002) Comparing sensitivity of ecotoxicological effect endpoints between laboratory and field. *Ecotoxicological and Environmental Safety*, **52**, 97–112.

Sharpe, R.M. and Skakkebaek, N.E. (1993) Are estrogens involved in falling sperm counts and disorders of the male reproductive tract? *Lancet*, **B41**, 1392–1395.

Shiyan, S.D. and Bovin, N.V. (1997) Carbohydrate composition and immunomodulatory activity of different glycoforms of a1-acid glycoprotein. *Glycoconjugate Journal*, **14**, 631–638.

Sibley, P.K., Benoit, D.A. and Ankley, G.T. (1997) The significance of growth in *Chironomus tentans* sediment toxicity tests: relationship to reproduction and demographic endpoints. *Environmental Toxicological Chemistry*, **16**, 336–345.

Silhavy, T.J. and Beckwith, J.R. (1985) Use of lac fusions for the study of biological problems. *Microbiological Review*, **49**, 398–418.

Smolders, R. (2003) The impact of an industrial effluent on the receiving aquatic ecosystem: a physico-chemical, ecological and toxicological analysis. PhD thesis, University of Antwerp, Antwerp, Belgium.

Smolders, R., Bervoets, L., De Boeck, G. and Blust, R. (2002) Integrated condition indices as a measure of whole effluent toxicity in zebrafish (*Danio rerio*). *Environmental Toxicology and Chemistry*, **21**, 87–93.

Smolders, R., Bervoets, L., Wepener, V. and Blust, R. (2003a) A conceptual framework for using mussels as biomonitors in whole effluent toxicity. *Human Ecological Risk Assessment*, **9**, 741–760.

Smolders, R., De Boeck, G. and Blust, R. (2003b) Changes in cellular energy budget as a measure of whole effluent toxicity in zebrafish (*Danio rerio*). *Environmental Toxicology and Chemistry*, **22**, 890–899.

Smolders, R., Bervoets, L., De Coen, W. and Blust, R. (2004) Cellular energy allocation in zebra mussels exposed along a pollution gradient: linking cellular effects to higher levels of biological organization. *Environmental Pollution*, **129**, 99–112.

Spiro, R.G. (2002) Protein glycosylation: nature, distribution, enzymatic formation, and disease implications of glycopeptide bonds. *Glycobiology*, **12**, 43–56.

Stuijfzand, S.C., Helms, M., Kraak, M.H.S. and Admiraal, W. (2000) Interacting effects of toxicants and organic matter on the midge *Chironomus riparius* in polluted river water. *Ecotoxicology and Environmental Safety*, **46**, 351–356.

Sussman, H.E. (2001) Choosing the best reporter assay. *Scientist*, **15**, 25.

Tardivel-Lacombe, J. and Degrelle, H. (1991) Hormone-associated variation of the glycan microheterogeneity pattern of human sex steroid-binding protein (hSBP). *Journal of Steroid Biochemistry and Molecular Biology*, **39**, 449–453.

Taylor, G., Baird, D.J. and Soares, A.M.V.M. (1998) Surface binding of contaminants by algae: consequences for lethal toxicity and feeding to *Daphnia magna* Straus. *Environmental Toxicology and Chemistry*, **17**, 412–419.

Teuschler, L., Klaunig, J., Carney, E., Chambers, J., Conolly, R., Gennings, C., Giesy, J., Hertzberg, R., Klaassen, C., Kodell, R., Paustenbach, D. and Yang, R. (2002) Support of science-based decisions concerning the evaluation of the toxicology of mixtures: a new beginning. *Regulatory Toxicology and Pharmacology*, **36**, 34–39.

Todd, M.D., Lee, M.J., Williams, J.L., Nalezny, J.M., Gee, P., Benjamin, M.B., and Farr, S.B. (1995) The Cat-Tox (L) assay – a sensitive and specific measure of stress-induced transcription in transformed human liver-cells. *Fundamental And Applied Toxicology*, **28**, 118–128.

Tolosa, I., Readman, J.W., Blaeovet, A., Ghilini, S., Bartocci, J. and Horvat, M. (1996) Contamination of Mediterranean coastal waters by organotins and Trgarol 1051 used in antifouling paints. *Marine Pollution Bulletin*, **32**, 523–535.

Tombolini, R., Unge, A., Davey, M.E. and De Bruijn, F.J. (1997) Flow cytometry and microscopic analysis of GFP-tagged *Pseudomonas fluorescens* bacteria. *FEMS Microbiology Ecology*, **22**, 17–28.

Toro, B., Navarro, J.M. and Palma-Fleming, H. (2003) Relationship between bioenergetics responses and organic pollutants in the giant mussel, *Choromytilus chorus* (Mollusca: Mytilidae). *Aquatic Toxicology*, **63**, 257–269.

Trabelsi, N., Greffard, A., Pairon, J., Bignon, J., Zanetti, G., Fubini, B. and Pilatte, Y. (1997) Alterations in protein glycosylation in PMA-differentiated U-937 cells exposed to mineral particles. *Environmental Health Perspectives*, **105** (Suppl. 5), 1153–1158.

Turner, G.A., Goodarzi, M.T. and Thompson, S. (1995) Glycosylation of alpha-1-proteinase inhibitor and haptoglobin in ovarian cancer: evidence for two different mechanisms. *Glycoconjugate Journal*, **12**, 211–218.

Ulluo-Aguirre, A., Maldonado, A., Damian-Matsumura, P. and Timossi, C. (2001) Endocrine regulation of gonadotropin glycosylation. *Archives of Medical Research*, **32**, 520–532.

USEPA (1994) *Short-term Methods for Estimating the Chronic Toxicity of Effluents and Receiving water to Freshwater Organisms* (3rd edn), EPA/600/4-91/002. USEPA, Cincinnati, OH.

Valdivia, R.H. and Ramakrishnan, L. (2000) Applications of gene fusions to green fluorescent protein and flow cytometry to the study of bacterial gene expression in host cells. *Methods in Enzymology*, **326**, 47–73.

Van den Heuvel, M.M., Poland, D.C.W., De Graaff, C.S., Hoefsmit, E.C.M., Postmus, P.E., Beelen, R.H.J. and Van Dijk, W. (2000) The degree of branching of the glycans on a1-acid glycoproteins in asthma. A correlation with lung function and inflammatory parameters. *American Journal of Respiratory and Critical Care Medicine*, **161**, 1972–1978.

Van den Steen, P., Rudd, P.M., Dwek, R.A., Van Damme, J. and Opdenakker, G. (1998) Cytokine and protease glycosylation as a regulatory mechanism in inflammation and autoimmunity. *Advances in Experimental Medicine and Biology*, **435**, 133–143.

Van der Lelie, D., Regniers, L., Borremans, B., Provoost, A. and Verschaeve, L. (1997) The VITOTOX test, an SOS bioluminescence *Salmonella typhimurium* test to measure genotoxicity kinetics. *Mutation Research*, **389**, 279–290.

Van Look, K.J.W. (2001) The development of sperm motility and morphology techniques for the assessment of the effects of heavy metals on fish reproduction. PhD Thesis, University of Sheffield, Sheffield.

Verslycke, T., Vercauteren, J., Devos, C., Moens, L., Sandra, P., and Janssen, C.R. (2003) Cellular energy allocation in the estuarine mysid shrimp *Neomysis integer* (Crustacea: Mysidacea) following tributyltin exposure. *Journal of Experimental Marine Biology and Ecology*, **288**, 167–179.

Versteeg, D.L., Graney, R.L. and Giesy, J.P. (1983) Field utilization of clinical measures of xenobiotic stress in aquatic organisms. In *Aquatic Toxicology and Hazard Assessment*, Vol. 10, ASTM STP 971, Adams, W.J., Chapman, G.A. and Landis, W.G. (eds), pp. 289–306. American Society for Testing and Materials, Philadelphia, PA.

Vighi, M., Altenburger, R., Arrhenius, A., Backhaus, T., Bodeker, W., Blanck, H., Consolaro, F., Faust, M., Finizio, A., Froehner, K., Gramatica, P., Grimme, L. H., Gronvall, F., Hamer, V., Scholze, M., and Walter, H. (2003) Water quality objectives for mixtures of toxic chemicals: problems and perspectives. *Ecotoxicology and Environmental Safety*, **54**, 139–150.

Vollmer A.C., Belkin S., Smulski D.R., Van Dyk, T.K. and LaRossa R.A. (1997) Detection of DNA damage by use of *Escherichia coli* carrying recA'::lux, uvrA'::lux, or alkA'::lux reporter plasmids. *Applied and Environmental Microbiology*, **63**, 2566–2571.

Wah, C.H. and Chow, K.L. (2002) Synergistic toxicity of multiple heavy metals is revealed by a biological assay using a rematode and its transgenic derivative. *Aquatic Toxicology*, **61**, 53–64.

Yang, S., Wu, R.S.S. and Kong, R.Y.C. (2002) Physiological and cytological responses of the marine diatom *Skeletonema costatum* to 2,4-dichlorophenol. *Aquatic Toxicology*, **60**, 33–41

Yang, T.T., Sinai, P., Kitts, P.A. and Kain, S.R. (1997). Quantification of gene expression with a secreted alkaline phosphatase reporter system. *Biotechniques*, **23**, 1110–1114.

Glossary

ACR	acute to chronic ratio
AF	application factors
AFNOR	Association Française de Normalisation (French standardisation organisation)
AHH	aryl hydrocarbon hydroxylase
ALA – D	δ-amino levulinic acid dehydratase
ALA – S	δ-amino levulinic acid synthetase
ANOVA	analysis of variance
AP-PCR	arbitrarily primed PCR
ARDRA	amplified ribosomal DNA restriction analysis
ASTM	American Society for Testing Materials
ATP	adenosine triphosphate
AVS	acid volatile sulfide
BAT	best available technology
BATEA	best available treatment economically achievable
BATNEEC	best available technique not entailing excessive cost
BEBA	biological effects-based assessment
BOD	biological oxygen demand
BPT	best practical technology
CAT	chloramphenicol acetyl transferase
CASA	computer assisted sperm analysis
CDNA	complementary DNA
CEN	European Committee for Standardisation
ChE	cholinesterase
CHIP	Chemicals (Hazardous Information and Packaging for Supply Regulations)
CLPP	community level physiological profiling
COD	chemical oxygen demand
COMAH	Control of Major Accident Hazards
COPEC	chemicals of potential ecological concern
COSHH	Control of Substances Hazardous to Health
CP	chlorinated paraffins
CSTEE	Scientific Committee for Toxicity, Ecotoxicity and the Environment
CWA	Clean Water Act (1972)
DDT	1,1,1-trichloro-2,2-bis(p-chlorophenyl)ethane
DEFRA	Department of the Environment, Food and Rural Affairs
DIN	Deutsche Institut für Normung (German standardisation organisation)

DGGE	denaturing gradient gel electrophoresis
DNA	deoxyribonucleic acid
DOC	dissolved organic carbon
DTA	Direct Toxicity Assessment
EC	effective concentration
ECB	European Chemicals Bureau
ECOD	ethoxycoumarin-o-dealkylase
EDTA	ethylenediaminetetraacetic acid
EINECS	European Inventory of Existing Commercial Chemical Substances
ERA	ecological risk assessment
EROD	ethoxyresorufin-O-deethylase
FAO	Food and Agriculture Organization
FCF	Fulton's condition factor
FDG	fluorescein digalactoside
FIFRA	Federal Insecticide, Fungicide and Rodenticide Act
GABA	gamma amino butyric acid
GLP	Good Laboratory Practice
GR	glutathione reductase
GST	glutathione S-transferase
HBB	hexabromobenzene
HCH	hexachlorocyclohexane
HPAEC-PAD	high performance anion exchange chromatography with pulsed amperometric detection
HSE	Health and Safety Executive
HSP	heat shock proteins
HSWA	Health and Safety at Work Act
HQ	hazard quotient
IBR	integrated biomarker response
ICES	International Council for the Exploration of the Sea
IFCS	Intergovernmental Forum on Chemical Safety
IPC	Integrated Pollution Control
IPPC	Integrated Pollution Prevention and Control
ISO	International Organization for Standardization
IV	intervention value
LABRAP	Laboratory for Biological Research in Aquatic Pollution
LAAPC	Local Authority Air Pollution Control
LC	lethal concentration
LID	lowest ineffective dilution
LTU	lethal toxic unit
MANOVA	multivariate analysis of variance
MATC	maximum acceptable toxicant concentration
MCL	maximum contaminant level
MCLG	maximum contaminant level goal
MFO	mixed function oxidase system
MRL	maximum residue levels
mRNA	messenger RNA
MT	metallothioneins

mtDNA	mitochondrial DNA
MUG	β-methyl umbelliferyl galactoside
MXR	multixenobiotic resistant
NADPH	nicotinamide adenine dinucleotide phosphate
NMMP	National Marine Monitoring Programme
NMR	nuclear magnetic resonance
NOEC	no-observed-effect concentration
NONS	Notification of New Substances Regulations
NPDES	National Pollutant Discharge Elimination System
NPOC	non-purgeable organic carbon
NTAC	National Technical Advisory Committee
OECD	Organization for Economic Cooperation and Development
OSPAR	OSPAR Convention (formerly the Oslo and Paris Commissions)
PAF	potentially affected fraction
PAH	polycyclic aromatic hydrocarbons
PAM	pulse–amplitude modulated (fluorescence)
PBB	polybrominated biphenyls
PBDE	polybrominated diphenyl ethers
PBP	pentabromophenol
PBFR	polybrominated flame retardants
PC	phytochelatins
PCB	polychlorinated biphenyls
PCDD	polychlorinated dibenzodioxins
PCDF	polychlorinated dibenzofurans
PCP	pentachlorophenol
PCR	polymerase chain reaction
PEC	predicted environmental concentration
PEEP	potential ecotoxic effects probe
PentROD	pentotyresorufin-O-deethylase
PICT	pollution induced community tolerance
PLFA	phospholipid fatty acid
PNEC	predicted no effect concentration
PNR	percentage net risk
POPs	persistent organic pollutants
POTW	publicly operated treatment works
PPP	Plant Protection Products Directive
PTD	polyethylene tube dialysis
Px	peroxidase
QA	quality assurance
QC	quality control
QSAR	quantitative structure activity relationships
QSPR	quantitative structure–property relationship
RAPD	randomly amplified polymorphic DNA
REACH	Registration, Evaluation and Authorisation of Chemicals
RNA	ribonucleic acid
RT-PCR	reverse transcriptase polymerase chain reaction
SAR	structure–activity relationships

SDSD	Safety Data Sheet Directive
SDS-PAGE	sodium dodecyl sulphate polyacrylamide gel electrophoresis
SDWA	Safe Drinking Water Act
SEAP	secreted alkaline phosphatase
SIR	substance induced respiration
SOD	superoxide dismutase
SPA	Soil Protection Act
SPE	solid phase extraction
SPMD	semipermeable membrane devices
SSCP	single strand confirmation polymorphism
SSD	species sensitivity distributions
SWR	Special Waste Regulations
TBBPA	tetrabromobisphenol-A
TBP	2,4,6-tribromophenol
TCDD	2,3,7,8-tetrachlorodibenzo-*p*-dioxin
TEC	toxicologically effective concentration
TEF	toxic emission factor
TEPP	tetraethyl pyrophosphate
TEQ	toxic equivalent
TER	toxicity emission rate
TGGE	temperature gradient gel electrophoresis
TIE	toxicity identification evaluation
TN_b	total bound nitrogen
TNT	2,4,6-trinitrotoluene
TOC	total organic carbon
TOEC	threshold-observed-effect concentration
TMDL	total maximum daily load
TRE	toxicity reduction evaluation
T-RFLP	terminal-restriction fragment length polymorphism
TSD	Technical Support Document (for Water Quality based Toxics Control)
TTR	transthyretin
TV	target value
UKWIR	United Kingdom Water Industry Research
UN	United Nations
UNECE	United Nations Economic Commission for Europe
UNEP	United Nations Environment Programme
UNIDO	United Nations Industrial Development Organization
UNITAR	United Nations Institute for Training and Research
USEPA	United States Environmental Protection Agency
WEA	whole effluent assessment
WET	whole effluent toxicity testing
WHO	World Health Organization
WOE	weight of evidence
WPCA	Water Pollution Control Act (1948)
WWTP	wastewater treatment plants

Index

Note: **bold** signifies an entry in a table, *italic* signifies a figure.

abalone **135**
Acartia tonsa **49**, 313, 319
Accipiter nisus 81
2-acetylaminofluorene 210
acetylsalicylic acid 216
acid-volatile sulphide 150, 151, 155, 156
acute to chronic ratio (ACR) 103
adenine 229
adsorbable organohalogen (AOX) 292, **293**, 297
Aequorea victoria 343
affinity factor 76
AFNOR standard 54
agrochemical 5
Aiolopus thalassinus 180
Alces alces 68
aldrin 4, **7**, 79, 81, 140
algae 12, **14**, **15**, 71, 96, 121, 134, **135**, 223, **284**, 292, *296*, 300, 313, 363
algal test **108**, 165, **281**, 283
Algaltoxkit 110
Alkali Act 259
alkyl phthalate 4
allelopatic substance 61
aluminium 68, 90
Ambersorb resin 150
ambient
    conditions 35
    contaminant concentrations 251
    media 155
    sample 37, 39, 42, 155
    testing 147
    water 134, 303
Ameirus nebulosus 247
Americamysis bahia **135**, 136, 138, **145**
American Society for Testing and Materials (ASTM) 11, **12**, 43, 144, **145**
aminopeptidase 177

ammonia 136, 141, 142, 146, 155, **266**, 276
Ampelisca abdita **135**, 136, 138, **145**
amphipod **135**, **145**, 151
anabasine 4
analysis
    accuracy 46
    carbon 317
    census 189–91
    chemical 36, 46, 95, *98*, 100, 119, 122, 154, 163, 192, 279, 295, 297, 317
    data 34, 38, 119, 192
    nucleic acid 181, 182
    PICT 170
    statistical 217
    water 94
Analysis of Variance (ANOVA) 40
Animal Protection Law **293**, 295
anthropogenic 67, 68, 83–5, 167, **284**, **286**, 356
    compound 155
    substance 72
anthracene 171
antibiotics 87, 88
anticonvulsant 87
    carbamazepine 87, 212, 223
anti-inflammatory 223, 349
Aporrectodea caliginosa 167, 176, 178
application factors 102, 103
Aqua Survey 136
aquatic
    bioassay 46
    ecosystem 68, 88, 89, 107, 273, 277
    ecotoxicology 185
    environment 131 *et seq.*
    media 110
    monitoring 34
    toxicity 9, 11, 48, 49, 115, 132–40, 166, 284, 315

Arbacia punctulata **135**, 136
Arenicola brasiliensis 148
Arenicola marina 148
Argopecten purpuratus 214, 250
Arochlor 73, 212
aromatase 72, **208**
aryl hydrocarbon hydrolase (AHH) 76, 176
arsenic 2, 3, **6**, 174, 178, **280**
Asellus aquaticus 360
aspartate transcarbamoylase **208**, 214
aspirin 216, 223
assay
   bacterial gene profiling *357*
   bacterial luminescence inhibition 110
   bioluminescence 48
   Cat-Tox 345
   chemical activated luciferase gene
      (CALUX) 345
   comet 175, 240, 241, 246, 248, 250
   dioxin 345
   fluorogenic 5′ nuclease 185
   β-galactosidase 342
   immobilisation 110
   immunosorbent 342
   lysosomal membrane stability 193
   mutagenicity 343
   neutral red retention time 174
   oestrogen 345
   Pro-Tox 344
   receptor mediated 344
   single-cell gel electrophoresis 175, 241
   VITOTOX 344
   yeast oestrogen 344
Atherinops affinis **135**
Atlantic tomcod 248, 249
atrazine **7**, 148, 307, 308

Bacillus thuringiensis (Bt) 223
bacterial chemoreceptor 96
bacterial luminescence inhibition assay 110
Balanus amphitrite 250
Baltic Marine Environment Protection
    Commission 20
Baltic seal 74
barium 142, 150
barnacle 221, 250
basal maintenance 358
Bathing Water Directive **19**, **265**
battery of tests 95, 107, 233, 277
BBSK project 190
benomyl 172

benzene, 141, 175, 176, 188, 318
   1, 4-bis[2(3, 5-dichloropyridoxyl)] 176
best available technology (BAT) 131, 303
best available treatment (BAT) 98
best available treatment economically
    achievable (BATEA) 98
best practical technology (BPT) 131, 303
benzo[a]pyrene 85, *86*, 175, 212, 249
bezafibrate 87
bicarbonate 89
bioaccumulation 63, 64, 73, 83, *98*, 133, **145**,
    148, 149, 155, 210, 275–7, 298, 310, 311
   survey 10
   testing 144, 146, 147
bioassay 9, 13, **14**, **15**, 34–9, 42, 46–8, 55, 95,
    97, 100, 105, 107, **108**, 110–13, 123,
    165, 166, 169, 173, 182, 192, 193, 235,
    251, 272, 273, 275–90, 297–300, 313
   aquatic 9, 46
   biomarker 65, 163, 164
   data variability 43–7, 52, 55
   investigation 94
   PEEP index 106
   performance 100
   quality 42
   reliability of test data 47
   sediment 36
   water column 36
   whole effluent 361
bioavailability
   assessment 143, 146–8
   estimating 148–50, 154
   factors controlling 143, 150, 152
biocenosis 99
biochemical oxygen demand (BOD) 39, *98*,
    116, 292, 310, 312, 316, 321–3, *324*,
    325, *327*, **329**, 331, 333
Biocidal Products Directive **17**, **263**
bioconcentration factor (BCF) 68, 69
biodegradation
   assessment 312, 334
   data 312, 321
   process 94
   rate 321, 323
   study 310, 311, 326, 329, 334, 335
bioindicator 10, 163, 191
bioinformatics 362
biological
   detector 132
   diversity 143
   effects-based assessment (BEBA) 274–6

biological (*cont'd*)
  indicator 37, 99, 281
  indices **98**, 99
  relevance 37, 250
  survey 99, 119, 121, 122
  testing 34, 46, 96, 163, 164, 276, 282, 292, 297–300
BIOLOG plates 170
biomagnification 10, 63, 73, 81, 83
biomarker 206 *et seq*.
  biological effect 207, 211, 215
  data 217, 222
  definition 164
  end-point 339
  index **218**, 219, 220
  reproduction 214
  research 10
biosensor
  method 192
  technology 11
  transgenic organism 344
biotic survey 99
bipyridylium herbicide 4, **7**
bivalve 137, **145**, 206, 208, 209, 211–16, 221–3, 358, 361
β-blocker 87
Blue Book 131
blue mussel 149, 212, 214, 221
Brachionus calyciflorus 54, 110
Brassica rapa 183
bromine 72
brown bullhead 247
bulk sediment 144, 146, 147
Burkholderia spp 188

Caenorhabditis elegans 180, 344
calanoid copepod, **49**, 313, 319
calcium 66, 68, 69, 142, 151, **208**, 216
California sea mussel **135**
Canadian Environmental Protection Act (CEPA) 116
Cancer pagarus 68
carbamate 4, **7**, 174, 178, 209
  carbaryl **7**
  isolan **7**
  methiocarb 174
  pyrolan **7**
carbamazepine 87, 212, 223
carbon tetrachloride 65
carcinogenicity 13, 168, 229, 304

carcinogens, mutagens or reprotoxic substances (CMRs) 21
Carson, Rachel 4, **7**, 15, 261
catalase (CAT) 172, 177, 213
Catostomus commersoni 360
Ceriodaphnia dubia 106, **109**, 134, **135**, 138, **145**
Champia parvula **135**, 136
channel catfish **145**
chaperonins 210
characterisation studies 43
chemical
  analysis 36, 46, *98*, 100, 119, 163, 192, 279, 295, 297, 312, 317, 326
  dependency 1, 2
  development 3, 261
  environmental impact 9, 36
  exposure 21, 169, 179, 182, 232, 350
  oxygen demand (COD) *98*, 292, **293**, 310, 312, 317, 329
  pest control 223
  regulation 1, **7**, **8**, 15, **17**, 19, 22
  tests 98, 140, 313, 354
Chemicals (Hazardous Information and Packaging for Supply Regulations) (CHIP) **17**, **263**, **264**
Chemicals in Products Report 266
chemiluminescent 185
chemistry of toxicants 61 *et seq*.
Chilean scallop 250
Chimiotox 113, 114
Chironomus riparius 223, 361
Chironomus tentans **135**, **145**, 149
cholinesterase (ChE) 39, 178, 193, 209
  activity 39
  inhibition 178, 193
chlordane 4, **7**, 79, 81, 249
chlorine 4, 5, 69, 72–4, 76, 142, 151, 155, 274, **293**
chlorpyrifos 178
chromosome 229–31, 233, 235, 243, *245*
  aberration 233, 243
  mutation 230, 231
chrysanthemum 2, 3
clam 178, 210, 214–16, 221, 222
Clean Air Act 259, 307
Clean Water Act (CWA) **14**, 131, 132, 260, 302–308
Clean Water Restoration Act **303**
Clophen 73

Clostridium perfringens 107
cobalt 142, 150, 153
Cognottia sphagnorum 168
Conseil Européen pour la Normalisation (CEN) 291, 295, 299
Comet assay 175, 233, 240, *241*, 246, 248
community level physiological profiling (CLPP) 170, **284**, 279
computer assisted sperm analysis (CASA) 352, 354
confidence limits 39, 123, 319, 321
congeners *73*, 74, *75*, 76, 77, *78*, 274
contaminated land 191, 192, 262, 265
Control of Major Accident Hazards (COMAH) 264
Control of Pesticides Regulations **17**
Control of Substances Hazardous to Health (COSHH) **264**
copper arsenite **6**
copper sulphate **6, 322**
Corbicula fluminea 210, 211
Corophium volutator **135**
Cottus ricei 360
crab 68, 183
Crassostrea gigas **135**, 215, 313, 319, 330
Crassostrea virginica 247
crustacean test **108**
cyclodiene 81
cyclooxygenase **208**, 216, 224
cyclophosphamide 87
Cyprinodon variegates **135**, 136, **145**
cytochrome 64, 76, 85, 176, 184, 212, 248, *296*
cytosine 229
cytostatics 87

dab (*Limanda limanda*) 247, 248
Dangerous Preparations Directive **17**
Dangerous Substances Directive **18**, **263**, **265**
Daphnia magna **48**, **49**, 50, 107, 110, 134, **135**, 138, **145**, 222, 313, 318
Daphnia pulex 134, **135**, 138, **145**
Daphtoxkit 110
Dendraster excentricus **135**
Department of the Environment, Food and Rural Affairs (DEFRA) 258, 262, **263**
derris plant 3
Detergents Directive **19**
Developments (Future) 337 *et seq.*
diazinon **7**, 172, 178

3, 4-dichloroaniline 48, **51**, 52, **54**
dichlorodiphenyltrichloroethane (DDT) 4, **6**, 66, 79, *80*
2, 4-dichlorophenoxyacetic acid (2, 4-D) 4, **7**, 175
dieldrin 4, **7**, 79, 81, 151
diflubenzuron 170, 171
dilution water 138
DIN (Deutsche Institut für Normung) 291, **295**, *296*, 297
dioxin **18**, *75*, 76, 77, 82, 175, 344, 345
Direct Toxicity Assessment (DTA) 35, 52, 260, 313
disease 2, 5, 81, 189, 208, 231, 242, 302, 340, 346–50
dissolved organic carbon (DOC) 151, 154, 317, 318, **321**, 322, **325**, *327*, **329**, *331*
DNA
    adducts 175, 183, 206, **207**, 212, 231, 232, 235, 242, 246–8
    alterations 175, 176
    analytical techniques 185, 235, 244, 350, 362
    damage 175, 183, 209, 213, 219, 230, 232, 235, 236, *238*, 239, 240, 248, 345
    repair 230, 232, 235, *238*, 241, 248
    restriction analysis 181, 182
    synthesis 175, 234, 236
dopamine **208**, 214, 215, 222, 223
Dreissena polymorpha 212, 248
dwarf surf clam **135**
dyes 5, 96, 174, 185, 210

Earth Summit, Rio de Janiero **7**, 16, 229
earthworm, 165–8, 175–8, 180, 183, 186–7, 190, 193, 248, 279, 282, **283–4**, 286, 300
ecoassessment 97, *98*, 99, 119, 121, 122, *125*
ecogenotoxicology 230, 243, 244
ecological
    field survey 38, 39, 98, 278, 281, 282
    indicator 77, 94, 95, 163, 164, 169, 192, 358
    parameters 276, 283, 286
    relevance 22, 36, 37, 42, 169, 174, 187, 188, 220, **234**, 246, 249, 285, 286, 312, 337, 338, 341, 358, 361
    risk assessment (ERA) 10, 22, 97, 117, 119, 125, 166, 192, 270–72, 274, 277, 307, 308

ecotoxicity 1, 9, 20, 22, 24, 36, 46, 47, 52, 53, 66, 164, 257, 260, 298, 299, 310–13, 316–18, **323**, **330**, 340, 363
  data sets 11
  testing 1, 9, 20, 22, 24, 46, 47, 52, 53, 66, 164, 257, 260, 317, 318, 340
ecotoxicology 1, 11, 164, 167, 169, 184, 185, 187, 189, 206, 220, 229, 230, 285, 337, *338*, 340, 341, 355, 362
effective concentration (EC) 39, 53, 102, 106, **109**, 111, 120, 355
Eisenia foetida 248
electrophoresis
  allozyme 244
  fluorophore-assisted carbohydrate (FACE) 350
  gel 175, 182, 186, 187, 216, 224, 240, 241, 350
Elliptio complanata 212
elk 68
elutriate testing 146
Enchytraeus albidus 168
Enchytraeus buchholzi 184
Enchytraeus crypticus 176
endocrine disruption 5, 77, 85, 88, **207**
endosulphan 81, 176
end-point
  biological 22, 144, 208, 250, 339
  selection 117, 139
English sole *(Parophrys vetulus)* 247
Environment Act 258–60, 264, **265**
Environment Agency 52, 258, 260, 262, **263–66**, 313
Environment Canada 11, **49**, 101, 106, 116
Environment in the European Union at the Turn of the Century 258
environmental
  assessment 97, 136, 262, 341, 363
  effects 9, 99, 100, 230, 333
  factors 50, 174, 177, 179, 180, 286
  impact 3, 42, 107, 119, 294, 339
  legislation 1, 15, 20, 22, 23, 257, *258*, 259, 260, 301, 302, **303**, 307, 355
  management 13, 16, 35, 41, 97, 116
  monitoring 13, **14**, 34, 35, 38, 40–42, 55, 87, 233, 240, 246, 266
  pathway 5
  safe levels 13
  samples 33, 42, 43, 95, 105, 168–70, 229, 233, 238–40, 246, 250, 251
  toxicant 61, 62, 83, 84, 87

toxicology 34, 63, 66, 83, 85, 97, *98*, 186, 312, 337, 341, 346, 362
Environmental Protection Act 35, 116, 259, **264**, **265**
Environmental Protection Agency (USEPA) 11, 37, **45**, 95, 96, 100, 101, 104, 115, 117, 118, 131, 132, 134, 137–40, 142, 144, **145**, 146, 149, 150, 154, 155, 258, 272–4, 277, 286, 304–308, 318, 319, 321, 358
envirotoxicant 63–7, 72, 74, 76, 78, 81–4
enzyme 65, 66, 96, 136, 137, 144, 169, 173, 177, 184, **207**, 211, 212, 221, 231, 244, 248, *296*, 313
  activity 71, 97, 165, 172, 176, 177, 214, 342
  antioxidant 177
  function 33, 171, 240
  inhibition 178, 214
Eohaustorius estuarius **145**
Escherichia coli 188, 239, 342–4
E-screen test 96
estrogen, *see* oestrogen
ethoxycoumarin *O*-dealkylase (ECOD) 176
ethoxyresorufin *O*-deethylase (EROD) 76, 85, 176, 248
ethinylestradiol 88, 226
ethylene diamine tetraacetic acid (EDTA) 141, 151
Eudrilus eugeniae 167
European Chemicals Bureau (ECB) 333
European Commission **8**, **17**, 21, 263, 285
European Inventory of Existing Commercial Chemical Substances (EINECS) 8, **17**, 21, **263**
European Union Water Framework Directive **8**, **18**, 22, **266**, 276, 290, 310
Existing Substance Regulation **17**
extraction methods 148, 156

Falco peregrinus 81
fathead minnow 134, **135**, 138, **145**, 188, 221, 249, 313
Federal Insecticide, Fungicide and Rodenticide Act (FIFRA) 307, 308
Federal Ministry for the Environment, Nature Conservation and Nuclear Safety 290
Federal Water Act 290, 297
Federal Water Pollution Control Act 302, **303**
fertilisation 2, 6, **208**, 214–16, 223, 320, 351, 352, 354

fertiliser 2, **6**, 68, 87
firefly 342, 343, 345
fish acute toxicity measures 10
Fisheries Act **14**
flounder 247, 248
fluorescein digalactoside (FDG) 342
fluorine 72
fluoxetine 87, **208**, 216, 223
Folsomia candida 167
Folsomia fimetaria 167
freshwater clam 210
Freshwater Fish Directive **19**
freshwater isopod 360
fucosylation 348, 349
Fulton's condition factor (FCF) 358, *359*, 360
fulvic acid 151
fumigant 2–4, **6**
Fundulus heteroclitus 249
fungicide 3, 4, **7**, 69, 75, 82, 165, 167, 307
   anabasine 4
   benomyl 172
   pentachlorophenol 75

β-galactosidase 189, 239, 240, 342
Gambusia affinis 249–50
gametogenesis **208**, 211, 214, 215, 351
gamma amino butyric acid (GABA) 81, 82
Gammarus pulex 39, **135**
Gemfibrozil 87
gene 169, 179, 183–6, 223, 229, 230, 232, 233, 237, 342, 344–6, 356, *357*
   arrays 11, 243
   expression 76, 180, 184, 185, 223, 230, 231, 244, 341, 342, 345, 346, 354, 362
   lux 188, *238*
   mutation 168, 230
   reporter 240, 344, 345
genetic drift 232, 251
genome mutation 183, 231, 235, 243
genotoxic
   agent 64, 95, 107, 175, 229, 243, 250
   assay 94, 234, 235, 344, 345
   compound 86, 229, 232, 236, 237, **238**, 242, 247
   effect 85, 94, 95, 229–32, 238, 249, 296
   exposure assessment 233, 246, 251
   testing methods 106, 107, 232, 233, **234**, 235–7, 243, 246
genotoxicity 229 *et seq.*
geochemistry 133, 143

giant kelp **135**
glucocorticoid dexamethasone 211
glucosephosphate isomerase 177
glutathione 64, 65, 212, 213
   peroxidase 177, 213
   reductase 177, 213
   S-transferase (GST) 176, 177, 184, 212, 213
glycoprotein 180, 207, 210, 346–50
glycosidase 347, 350
glycosylation 346–50
glycosyltransferase 347
Gold Book 131
Good Laboratory Practice (GLP) 44, **234**
grasshopper 4, 178, 180
grass shrimp **145**
green algae **49**, 71, 106, 110, 134, **135**
Green Book 131
green-lipped mussel 211
Groundwater Directive **18**, **266**
guanine 229

Habitats Directive 19, 22, **266**
haemoglobin synthesis 71, 74
   δ-aminolevulinic acid dehydratase (ALA-D) 71
   δ-aminolevulinic acid synthetase (ALA-S) 71
Haliotus rufescens **135**
harbour seal 248
hazard quotient (HQ) 120
hazardous substances 1, **7**, 8, 13, 19, 257, 260, 261, **265**, 291, 310
Hazardous Waste Directive **266**
H-criteria 298, 299
Health and Safety at Work Act (HSWA) **264**
Health and Safety Executive (HSE) 262, **263**, **264**
heat shock proteins (HSP) 178, 179, 184, 186, 188, 189, 210, 211
Helicoverpa zea 177
Helix pomatia 180
heptachlor 81
herbicide 3, 6, 7, 177
   bipyridilium 4, 7
   chlorinated phenoxy 75
   2, 4-dichlorophenoxyacetic acid (2, 4-D) 4, **7**, 175
   paraquat 177
   sodium trichloroacetate 177

herbicide (cont'd)
  triazine 7
  2, 4, 5-trichlorophenoxyacetic acid (2, 4, 5-T) 75
herring gull 249
heterogeneity 189
  environmental 50
  genetic 50
hexabromobenzene (HBB) 76
hexachlorobiphenyl 212
hexachlorocyclohexane (HCH) **6**, 79, 81
Hexagenia limbata 145
high performance anion exchange chromatography with pulsed amperometric detection (HPAEC-PAD) 348
Homarus gammarus 68
Homeland Security 136
homeostasis 66, 70, 89, 90, 179, 180, 188, 211, 371
hormesis effect 96
hormone 5, 64, 74, 77, 184, 209, 211, 222, 348, 349
Hyalella azteca **135**, **145**
hydrocarbons 66, 72, 83, **84**, 133, 142, 165, 170, 172, 210, 212, 247, **266**, 274, 344
  halogenated 72, 83, 212
hydrogen cyanide 6
hydrogen peroxide 72, 213
hydroxindole oxidase 211
hypersaline brine 139

ibuprofen **208**, 223
Ictalurus punctatus **145**
immune system 174, 348, 350
immunosuppressant 74
indium 142
information cycle 35
inland silverside 135, **145**
insecticide 2–4, 6, 39, 78, 80, 249, 307
  aldrin 79, 81
  alkyl phthalate 4
  chlordane 4, **7**, 79, 81, 249
  cyclodiene 81
  diazinon **7**, 172, 178
  dieldrin 4, **7**, 79, 81, 151
  endosulphan 81
  endrin 81, 176
  heptachlor 81
  hexachlorocyclohexane (HCH) **6**, 79, 81
  isolan **7**

lindane 79
malathion 4, **7**, 178
pyrethroid 4, **7**, 155, 171
pyrethrum 3, 6
pyrolan **7**
rotenone 3, 6
tetraethyl pyrophosphate (TEPP) 4, **6**
thiodan **7**
1, 1, 1-trichloro-2, 2-bis(*p*-chlorophenyl) ethane (DDT) 4, **6**, 66, 79, *80*
integrated biomarker response (IBR) 217, *218*, 219
Integrated Pollution Control (IPC) 259, 264
Integrated Pollution Prevention and Control (IPPC) 18, 35, 260, **265**, 290
Intergovernmental Forum on Chemical Safety (IFCS) 16
interleukin-3 211
International Organization for Standardization (ISO) 11, 164
intertidal teleost 247
isoelectric focusing 186
isopod 172, 176, 177, 179, 180, 190, 360
isozymes 64, 74, 177

jellyfish 343

kerosene **6**
killifish 249
Kruskall–Wallis test 40

Laboratory for Biological Research in Aquatic Pollution (LABRAP) 110
lacewing 177, 178
landfill 18, 87, 95, 166, 172, 262, **266**, **298**, 300
Landfill Directive **18**, **266**, **298**
Larus argentatus 249
lead
  arsenate 6
  tetraethyl 70
legislation 257 *et seq.*
Lepomis auritus 360
Lepomis macrochirus **145**
Leptocheirus plumulosus **145**
lethal concentration (LC) 39, 102, 104, 112, 351
Leuciscus idus melanotus **293**
Limanda limanda 247, 248
Limnodrilus hoffmeisteri 250
lime sulphur 3, **6**

lindane 79
lipid peroxidation (LPO) **207**, 213, 221, 222
Lipophrys pholis 247
litterbags 169–71
lobster 68
Local Authority Air Pollution Control (LAAPC) 265
loperamide 212
lowest ineffective dilution (LID) 294
luciferase 313, 342, 343, 345, 346
lugworm, 148
Lumbricus rubellus 167
Lumbricus terrestris 167, 248
Lumbriculus variegates 151
lysosomal membrane stability 169, 174, 193

Macoma edulis **145**
Macoma nasuta **145**
Macrocystis pyrifera **135**, 136
malathion 4, **7**, 178
manganese 142, 150, 151, 188
mangrove oyster 212
Mann-Whitney Test 40, 218
marine copepod 178
Marketing and Use Directive for Dangerous Substances and Preparations 263
mass spectrometry 186, 187, 350
maturity index 191, 283, 284
maximum acceptable toxicant concentration (MATC) 102
maximum contaminant level (MCL) 304
  goal (MCLG) 304
maximum residue levels (MRL) **17**
membrane analog method 149
Mendel, laws of heredity 2, **6**
Menidia beryllina **135**, 136, **145**
mercuric chloride 6
mercury
  medical 69
  methyl 69, 148
  poisoning 5, 70
metabonomics 187
metal-binding proteins 70, 137, 179, 180, 207, 210
Methods for Aquatic Toxicity Identification Evaluation: Phase I Toxicity Characterization Procedures 141
methylation 67, 71, 175
methyl bromide 4, **6**
β-methyl umbelliferyl galactoside (MUG) 342

metoprolol 87
Microgadus tomcod, 248–9
Micromus tasmaniae 177
micronucleus test 243
Microtox® 50, 51, 109, 110, 136, 188, 193, 285, 325–7, 329, 333, **334**
  assay, 166, 187
  methodology 48
  test 48, 49, *51*, *54*, 108, 136, 169, **281**, 282, **284**, 313, 314, 318, 321, **322**, **328**, 332
midge **135**, **145**, 223
millipede 173, 179
minicontainers 170, 171
Mining Effluent Regulations **14**
Ministry of Housing, Spatial Planning and the Environment 270
mixed function oxidase system (MFO) 64, 176
molecular-level activity 11
monoamine oxidase (MAO) 214
Montreal Protocol **7**, 19
mosquito fish 249
mud shrimp **135**
Mulinia lateralis **135**
Multivariate Analysis of Variance (MANOVA) 40
multixenobiotic resistant (MXR) 210
mussel **135**, **145**, 149, 208–16, 221–23, 241, 248
mutagenic 64, 85, 233, 235, 251, 291, 292, **296**
mutagenicity 13–15, 96, 168, 229, 230, 237, 343
Mutatox® 109, 110, 168, 237
Mya arenaria 213, 215, 221
mysid shrimp **135**, **145**
Mysidopsis bahia 136
Mytilus californianus **135**
Mytilus edulis **145**, 210, 248
Mytilus galloprovincialis **135**, 212

naphthalene 85, 188, 213
naphthoflavone 186, 345
National Air Quality Strategy 259, **265**
National Marine Monitoring Programme (NMMP) 36
National Pollution Discharge Elimination System (NPDES) **14**, 132, 305, 306
nematode 165, 173, 179, 180, 183, 188, 191, 279, 281, 282, **283**, **284**, 286
Nenathes arenaceodentata **145**

nephrotoxic 69
Neries diversicolor 152
Neris virens **145**
neurological response **33**, 77
neurotoxic 71, 80
nicotine 3, 6
Nitrate Directive 19, 265
nitric acid 89
nitric oxide 213
nitrification 165, 171, 172, 300, 322, 325
nitrogen 90, 152, 165, 171, 191, 193, 281, 292, **293**, 322, *324*, 325
    fixation 172
    oxides (NOx) 89
non-purgable organic carbon (NPOC) 317
nonylphenol **208**, 209, 215
no observed effect concentration (NOEC) 102, 271
noradrenaline 214, 215
Notification of New Substances Regulations (NONS) **17**, **263**
nuclear magnetic resonance spectroscopy (NMR) 187
nucleotides 229, 235, 237, *238*
nutrient cycling 36

obligatory notification system 8
octopamine 214
oestradiol **208**
oestrogen 5, 66, 74, 80, 95, 184, 185, 215
oil spill 5
    Amoco Cadiz 5
    detergent use 5
    Exxon Valdez 248
    Sea Empress 5
    Torrey Canyon 5, **7**
oligochaete **135**, **145**, 151, 173, 250
Oncorhynchus gorbuscha 247
Oncorhynchus mykiss 49, 134, **135**, **145**, 211, 212, 313, 345
Oniscus asellus 176, 179
opioid peptide 213, 222
orfe **293**
organic carbon 148, 150–52, 154, 156, 284, 292, 314, 317, 318, *327*, *331*
organic compounds 3, 4, 37, 70, 82, 96, 115, 149, 154, 155, 165, 171, 175, 181, 182, 212, 313, 329
Organization for Economic Cooperation and Development (OECD) 11, 164, 298
organochlorine 4, 5, 151, 155, 274

organophosphate 4, 209
organophosphorus 4, 209
oribatid mite 173, 190
OSPAR (Oslo–Paris) **7**, 19, 310, 311, 319
β-oxidation 213
ozone **7**, 19, 206, 220

paint 4, 5, 68, 70–73, 83
Palaemonetes spp **145**
paraffin 83
parathion 4, **7**
Paris green **6**
Parophrys vetulus 247
partition coefficient 62, 154
Pavlova lutheri 319
pentabromophenol (PBP) 76
pentachlorophenyl (PCP) 75, 82
pentotyresorufin-*O*-deethylase (PentROD) 176
Perca flavescens 361
Perna viridis 211, 212
peroxidase 172, 177, 213
peroxynitrite 213
persistent organic pollutants (POPs) **7**, 19
pesticide
    chlordane 4, **7**, 79, 81, 249
    chlorpyrifos 178
    parathion 4, **7**
    toxaphene 4, **7**, 79
petrochemical 5, 257
    effluent 311, 314–16, *323*, **325**, *326–8*, **332**, **334**, 335
Phaeodactylum tricornutum **49**
pharmaceutical 5, 85, 87, 212–14, 216, 223, 257, **263**, 362
phenobarbital 212, 358
Pheretima posthuma 176
Phoca vitulina richardsi 248
phosphoglucomutase 177
phospholipid fatty acid (PLFA) 181, 193
Photinus pyralis 342
Photobacterium phosphoreum 106, 110, 313
phytochelatins 179, 180
phytoplankton 152
pickle-jar tests 9
Pimephales promelas 134, **135**, 138, **145**, 188, 221, 249, 313
Plant Protection Products Directive (PPP) **17**, 263
plastics 5, 68, 72, 73, 83, 169
Platichthys flesus 247, 248

Plectus acuminatus 179
pollution-induced community tolerance (PICT) 170, 282, **283**, **284**
Pollution Prevention and Control Act 18, 35, 259, 260, 264, 265
polybrominated biphenyl (PBB) 76
polybrominated diphenyl ether (PBDE) 76
polybrominated flame retardant (PBFR) 76, 77
polychaete **145**, 148, 152
polychlorinated biphenyl (PCB) 66, 73, *74*, 75–8, 82–4, *98*, 133, 148, 150, 155, 165, 167, 171, 175, 177, 247, 249, 274
polychlorinated dibenzodioxins (PCDD) 74, 75
polychlorinated dibenzofurans (PCDF) 75
polycyclicaromatic hydrocarbons (PAH) 84, 85, *86*, *98*, 148, 149, 151, 166, 247, 248
polyethylene tube dialysis system (PTD) 149
polymerase chain reaction (PCR) 181–5, 221, 244
polymeric organic carbon 152
Porcellio scaber 176, 177
pore water assessment 146, 147, 149, 150, 156
potential ecotoxic effects probe (PEEP) 105–107, **108**, 113
potentially affected fraction (PAF) 274
predicted environmental concentration (PEC) 333
predicted non effect concentration (PNEC) 333
propranolol 87
prostaglandin **208**, 216, 224
protein expression 11, 185, 187, 224, 346
proteinurea 69
proteomic 23, 186, 187, 193, 224, 346, 362
protoxkit 110
Psammechinus miliaris **135**
Pseudokirchneriella subcapitata **135**
Pseudomonas fluorescens 148, 188
Pteronarcys californica 180
Public Health Act 259
publicly operated treatment works (POTW) 132
Pulp and Paper Regulations 14
pulse-amplitude-modulated-fluorescence (PAM) **281**, 283, **284**
purple urchin **135**
pyrene 85, 176, 188

pyrethroid 4, **7**, 155, 171
pyrethrum 3, **6**

radiation 5, 187, 206, 213, 230, 236
rainbow trout 49, 134, **135**, **145**, 211, 212, 313, 345
Raphidocelis subcapitata 110, **135**
ratio-to-reference values 123
reagent salt formula 139
Red Book 131
redbreast sunfish 360
red macroalgae **135**
refinery 311, 314
  effluent 311, 313, *314*, 316, **321–4**, 325, 332, 334, 335
  wastewater 312, *314*, 315, 316, 318, 326, **329**, *330–32*, 334
Registration, Evaluation and Authorisation of Chemicals (REACH) **8**, **17**, 21
regulatory authorities 4, 310
reproductive impairment 13
respirometer 311–13, 316, *322*, 325, 329
Rhepoxynius abronius **145**
Rhizobium leguminosarum 172
ring-testing 50, 52, 53
Rivers Pollution Prevention Act 259
rodenticide **6**, 307
rotenone 3, **6**
Rotoxkit 110
Rotterdam Convention **7**, 16
Ruditapes philippinarum 216

Safe Drinking Water Act (SDWA) 302, **303**, 304
Safety Data Sheet Directive (SDSD) 17
salinity 43, **45**, 139, 314, 317, 318, 321, 326
salmon (pink) 247
Salmonella typhimurium 168, 235–8, 240, 343
sample
  collection 42, 94, 107, 315
  environmental 33, 43, 168–70, 229, 233, 238, 239, 246, 250, 251
  frequency 38, 39, 42
  handling 164, 184, 317
  pretreatment 42, 43
sampling 41, 42
sand dollar *(Dendraster excentricus)* **135**
Saccharomyces cerevisiae 344

Scientific Committee for Toxicity,
    Ecotoxicity and the Environment
    (CSTEE) 36
screening tests 95, 96, 110
seasonal Kendall test 41
sea urchin 135, **145**
secreted alkaline phosphatase (SEAP) 342
Sediment Quality Triad 273
Selenastrum capricornutum **49**, 106, 110,
    134, **135**
selenium **6**, 191
semi permeable membrane devices
    (SPMD) 149
serotonin **208**, 214–16, 222, 223
Setting Environmental Standards 266
sex steroid 72, 87, 348
sheepshead minnow 135, **145**
Shellfish Water Directive **19**
Shewart Control Chart 46
shore urchin **135**
sialylation 349
Silent Spring 4, **7**, 15, 261
sister chromatid exchange 235, 241, 242
Skeletonema costatum **49**
snail 173, 180, 183, 354
sodium dodecyl sulphate polyacrylamide
    gel electrophoresis (SDS-PAGE) 186,
    216
sodium thiosulphate 141
soil 163 *et seq.*
    enzyme activity 165, 172
    functional assessment 164
Soil Protection Act (SPA) 269, 270
SOILPACS study 190
solid phase extraction (SPE) 141
SOS chromotest 106, 107, 233, 239, 345
sparrow hawk 81
Spearman's rank correlation test 107
Special Waste Regulation (SWR) 35, **266**
species sensitivity distributions (SSD) 271,
    272, 279
sperm 214, 216, 320, 350–54
    maturation 352
    motility 320, 351–4
Sphagnum spp 89
Spirillum volutans **109**
spoonhead sculpin (*Cottus ricei*) 360
staphylinid beetle 173
staurosporine 210
steroidogenesis **208**
Stockholm Convention 19

stonefly 180
Strategy for a Future Chemicals Policy 21,
    266
Strongylocentrotus purpuratus **135**, 136
strontium 142
structure-activity relationships (SAR)
    114–16
    quantitative (QSAR) 115
sublethal
    end-point 10, 103, 167
    level 81, 82, 99
    toxicity 102, 103, 141, 276, 297
sulphur 2, 3, **6**, 90
sulphur dioxide 69, 89
sulphuric acid 64, 89
superoxide dismutase (SOD) 177, 213
Surface Water Abstraction Directive **19**
*Sustainable Production and Use of
    Chemicals* 266

*Technical Support Document for Water
    Quality based Toxics Control*
    (TSD) 132
Tenax TA method 149
teratogenic effects 64, 69
test
    algal toxicity 165, 313
    ambient 147
    Ames 168, 233, 235, *236*, 237, 246, 343
    aquatic invertebrate 166
    aquatic plant 54, 165
    bait lamina 169–71, 193
    battery approach 233, 251
    bulk sediment 144, 147
    chamber 134, 138
    Dr Lange cuvette **317**
    earthworm reproduction 167, 168, 193
    Enchytraeid reproduction 168
    fast screening 233, 234, 246, 247, 251
    growth inhibition 166
    guidelines 44, 53
    Lemna 166
    life-cycle 182, 183
    Merck Spectroquant cell 317
    method development 10, 44, 53, 183
    micronucleus 243
    Microtox® 48, *51*, *54*, 313, 318, 322, 328
    nitrogen mineralisation (N-MIN) 165,
        193
    replicates 48
    ring 50, *51*, 52–4, 173, 313

springtail reproduction 167, 168
standardisation 42, 44, 55
substance induce respiration (SIR) 165
terrestrial invertebrate 166, 173, 174
terrestrial plant 165, 166, 192
types 144, **145**
water column 36, 143, 144, 146
testosterone 72, 85, **208**, 212
tetrabromobisphenol A (TBBPA) 76
2, 3, 7, 8-tetrachlorodibenzo-*p*-dioxin (TCDD) 75, 76, 82
tetraethyl lead 70
tetraethyl pyrophosphate (TEPP) 4, **6**
Tetrahymena thermophila 110
Tetraselmis suecica 319
textiles 5
Thamnocephalus platyurus 111
thamnotoxkit 111
thiobarbituric acid 221
thiodan **7**
threshold observed effect concentration (TOEC) 103, 106
thymine 229
thymus 74, 76
thyroxine 74, 77
Tigriopus brevicornis 178
timber preservative 4
time-series 41, 42
Tisbe battagliai **45, 49**
tobacco leaves 3
topsmelt **135**
total bound nitrogen (TN_b) 292
total maximum daily load (TMDL) 306
total organic carbon (TOC) 292, 314, *327*, 331
toxaphene 4, **7**, 79
toxic
  emissions 116
  metal 67, 68, 142, 184, 211
  response 105–107, 111, 137, 138, 185
  units 104–106, 112, 120, 275, 355
toxicants, chemistry of 61 *et seq.*
toxicity equivalent (TEQ) 76, 84
toxicity identification evaluation (TIE) 95, 100, 140
toxicological data 10, 68, 97, 100, 116, 362
transcriptomics 185
trans-stilbene oxide 176
transthyretin (TTR) *74,* 77
Treaty of Rome **7**
triad 97, *98,* 99, 123, *125,* 273, 275, 276, 278, 279, 281, 282, 284–7

triazine, 4, **7**,
2, 4, 6-tribromophenol (TBP) 76
tributyl tin 5, 353
1, 1, 1-trichloro-2, 2-bis(*p*-chlorophenyl) ethane (DDT) 4, **6**, 66, 79, *80*
trichloroethylene 186, 191
2, 4, 5-trichlorophenoxyacetic acid (2, 4, 5-T) 75
Triclosan 223
trifluralin 248
triiodothyronine *78*
trophic-level testing 10, **14, 15**

Umu-C test 238
United Nations Economic Commission for Europe (UNECE) 19
United Nations Environmental Program (UNEP) 16, 311
United Nations Food and Agriculture Organization (FAO) 16
United Nations Industrial Development Organization (UNIDO) 16
United Nations Institute for Training and Research (UNITAR) 16
Urban Waste Water Treatment Directive **265**
US Environmental Protection Agency (USEPA) 11, **45**, 115, 117, 132, 140, **145**, 154, 274, 304–308, 318
US Public Health Service (PHS) 302

validation
  field 100, 101
  laboratory 339
  program 102
values
  intervention (IV) 269, 271, 275, **280**, 282
  target (TV) 269–71
veterinary medicines 85, 88, 89, 223, 257, 262, **263**
Veterinary Medicines Directive **263**
Vibrio fischeri 48, 106, 110, 136, 169, 187, 237, 313, 314, 318
Vicia faba 248
vitellogenin 85, 184, 185, **207**, 211, 215, 221, 222, 351, 354
VITOTOX test 233, 237, *238, 240,* 246, 344

warfarin **7**
Waste Incineration Directive **18, 266**, 290

wastewater
- refinery 312, *314*, 315, 316, 318, 326, **329**, *330–32*, 334
- toxicity 131, 132, 138–41, 150, 152, **293**, 294, 297
- treatment plant (WWTP) 88, 131, 294

Wastewater Charges Act **14**, 290, 292–4, 297

Wastewater Ordinance **14**, 290, 291, **293**, 294, 297

water column testing 36, 143, 144, 146

water flea (*Daphnia* spp) **48**, **49**, 50, 107, 110, 134, **135**, 138, **145**, 222, 313, 318

Water Framework Directive **8**, **18**, 22, **266**, 276, 290, 310

Water Pollution Control Act (WPCA) 302, **303**

Water Quality Act **303**

water quality criteria 131, 143, 276, 306–308

Water Quality Improvement Act **303**

Water Resources Act 35, 259, 260, 264, **265**

weight of evidence (WOE) 47, 118–20, 122, 123, 140, 146, 156, 273, 276

White Paper on a Strategy for a Future Chemicals Policy 266

white sucker 360

whole-effluent
- assessment (WEA) 310 *et seq.*
- toxicity testing (WET) **14**, 132, 306

woodlice 173, 179

World Health Organization (WHO) 16

XAD resin 149, 273

xenobiotics 61–5, 84, 206, 207, 210, 219, 248, 344, 348, 350

xenoendocrine 66, 77

xenoestrogen 80, 184, 185, 215

xylene 141

yeast 186, 213, 231, 344, 362

yellow perch 361

zebra mussel 211, 216, 222, 248

zinc
- mining 68
- sulphate 52